Plasmas are charged particle gases found throughout the solar system in environments such as the solar atmosphere, the solar wind, and the planetary magnetospheres and ionospheres. *Physics of Solar System Plasmas* provides a comprehensive introduction to these ionized gases of the solar–terrestrial environment. The author places emphasis on the solar wind and magnetospheres – regions of space around the Earth and other planets that are controlled by the planetary magnetic field and are shielded from the solar wind.

The text includes a broad introduction to plasma physics, incorporating discussions of kinetic theory and magnetohydrodynamics (MHD). This leads into a thorough description of the Sun and the solar wind, building on the plasma physics and MHD introduced earlier in the book. Finally, the author discusses magnetospheric physics. As an introductory text, the explanations of space plasma phenomena use fluid theory rather than plasma kinetic theory and emphasize physical processes rather than observational details.

Problem sets at the end of each chapter make this a useful text for advanced undergraduate students in astrophysics, geophysics, or atmospheric sciences. Graduate students and researchers will also find it a valuable source of information.

PHYSICS OF SOLAR SYSTEM PLASMAS

THOMAS E. CRAVENS
University of Kansas

CAMBRIDGE
UNIVERSITY PRESS

PUBLISHED BY THE PRESS SYNDICATE OF THE UNIVERSITY OF CAMBRIDGE
The Pitt Building, Trumpington Street, Cambridge, United Kingdom

CAMBRIDGE UNIVERSITY PRESS
The Edinburgh Building, Cambridge CB2 2RU, UK
40 West 20th Street, New York NY 10011–4211, USA
477 Williamstown Road, Port Melbourne, VIC 3207, Australia
Ruiz de Alarcón 13, 28014 Madrid, Spain
Dock House, The Waterfront, Cape Town 8001, South Africa

http://www.cambridge.org

First published 1997
First paperback edition 2004

Typeset in Times Roman

A catalogue record for this book is available from the British Library

Library of Congress Cataloguing-in-Publication Data
Cravens, Thomas E., 1948–
Physics of solar system plasmas / Thomas E. Cravens.
p.   cm. – (Cambridge atmospheric and space science series)
ISBN 0 521 35280 0 (hardback)
1. Solar wind.   2. Magnetosphere.   3. Astrophysics   I. Title.
II. Series.
QB529.C7   1997
523.5′8 – dc21   96-48929
                                                         CIP

ISBN 0 521 35280 0 hardback
ISBN 0 521 61194 6 paperback

PHYSICS OF SOLAR SYSTEM PLASMAS

Cambridge atmospheric and space science series

Editors

Alexander J. Dessler
John T. Houghton
Michael J. Rycroft

Titles in print in this series

To my wife JD and my parents

Errata

1.   p. 17, equation 2.20, top part, insert on the right-hand side of the equation the factor:
$$[m_s/(2\pi k_B T)]^{3/2}$$

2.   p. 64, line 8 in text material, "less" should be "greater".

3.   p. 117, last line, the following words should be switched: "oil" and "water".

4.   p. 421, Figure 8.42, the arrows on the magnetic field lines in the right-hand magnetic islands in the top two panels are in the wrong direction.

5.   p. 466, line 2 ($z < 0$) of equation (A.24), the minus sign should be a plus sign.

Contents

Preface

This book is an introduction to the physics of the solar wind and magnetosphere. These regions of space are filled with charged particle gases called plasmas. The study of solar system plasmas is commonly called space physics. This book started as lecture notes for courses that I have taught at the University of Michigan and the University of Kansas. The book is an introductory textbook aimed at advanced undergraduate and graduate students who possess an undergraduate physics background but have not taken any plasma physics courses. An introduction to plasma physics, including the topic of magnetohydrodynamics, is included in order to make the book self-contained. Undergraduate-level electromagnetic theory and mechanics are extensively used, and the Appendix provides a very brief review of the first topic.

The book can be divided into three parts. The first part, consisting of Chapters 1 through 4, provides an introduction to plasma physics. In particular, Chapter 1 gives a brief introduction to space plasma physics, kinetic theory is discussed in Chapter 2, and Chapter 3 is concerned with single particle motion in electric and magnetic fields. Chapter 3 also contains material on energetic particle motion in the radiation belts. Magnetohydrodynamics (MHD) is introduced in Chapter 4. Examples dealing with phenomena in the solar wind and magnetosphere are provided. Students who have already taken a standard plasma physics course can skip over much of the first part of the book.

The second part of the book contains two chapters focused on the Sun and the solar wind. Chapter 5 presents a short review of solar physics. The solar wind is the topic of Chapter 6.

Magnetospheric physics is the subject of the third part of the book. Chapter 7 describes how the solar wind interacts with the terrestrial magnetosphere and also other planets and bodies in our solar system. An introduction to ionospheric physics is also included in this chapter. Chapter 8 is devoted to the internal dynamics of the terrestrial and Jovian magnetospheres.

Because this book is an introductory text, emphasis is placed on explanations of space plasma phenomena using fluid theory (i.e., MHD) rather than plasma kinetic theory, which tends to be more mathematical. Physical processes rather than observational details are emphasized, although data are shown where appropriate. In addition, each chapter contains a list of books and articles, as well as a selection of problems. The books and articles at the back of each chapter are not meant to be exhaustive but were mainly included to permit ambitious students and readers to further pursue the topics presented.

I am deeply grateful to the many colleagues and friends who have made very helpful suggestions and who have provided much needed encouragement. In particular, I appreciate the support of my good friends and colleagues Andy Nagy and Tom Armstrong. Janet Luhmann, Fran Bagenal, Tamas Gombosi, Andy Nagy, and Tom Armstrong provided detailed and valuable feedback on the manuscript, most of which was gratefully adopted and which made this a better book. I thank James Van Allen for his encouragement and I also thank Lou Frank, Jack Brandt, Freeman Miller, Gary Linford, and H. Zirin for photographs that have enhanced this book. Support from the National Science Foundation for portions of the book involving the solar wind interaction with nonmagnetic planets is also acknowledged.

I would also like to acknowledge Alex Dessler, editor of the Series, for getting me started on this book, as well as Catherine Flack of Cambridge University Press for her advice on the manuscript and figures. Thanks also go to Tizby Hunt-Ward for patiently working with me on the manuscript and figures and to the many students who have put up with various versions of this book over the years.

The support and encouragement of my wife and children over the many years this project took are deeply appreciated.

Lawrence, Kansas   *T. E. Cravens*

Physical constants

Speed of light	c	2.9979×10^{8}	m/s
Planck constant	h	6.6262×10^{-34}	$J \cdot s$
Elementary charge	e	1.6022×10^{-19}	C
Permittivity of free space	ε_0	8.85×10^{-12}	$C^2 \, N^{-1} \, m^{-2}$
Permeability of free space	μ_0	$4\pi \times 10^{-7}$	$N \cdot s^2 \, C^{-2}$
Rest mass of proton	m_p	1.6726×10^{-27}	kg
Rest mass of electron	m_e	9.1095×10^{-31}	kg
Boltzmann constant	k_B	1.3806×10^{-23}	$J \cdot K^{-1}$
Gravitational constant	G	6.672×10^{-11}	$N \cdot m^2 \, kg^{-2}$
Radius of the Earth	R_E	6.37×10^{3}	km
Radius of the Sun	$R_\odot$	6.96×10^{5}	km
Astronomical Unit (radius of Earth orbit)	AU	1.496×10^{8}	km

Vector calculus identities

$$\mathbf{A} \cdot (\mathbf{B} \times \mathbf{C}) = \mathbf{B} \cdot (\mathbf{C} \times \mathbf{A}) = \mathbf{C} \cdot (\mathbf{A} \times \mathbf{B})$$
$$\mathbf{A} \times (\mathbf{B} \times \mathbf{C}) = \mathbf{B} \cdot (\mathbf{A} \cdot \mathbf{C}) - \mathbf{C}(\mathbf{A} \cdot \mathbf{B})$$

$$\nabla \cdot (\nabla \times \mathbf{A}) = 0$$
$$\nabla \times (\nabla f) = 0$$
$$\nabla \times (\nabla \times \mathbf{A}) = \nabla(\nabla \cdot \mathbf{A}) - \nabla^2 \mathbf{A}$$

$$\nabla(fg) = f(\nabla g) + g(\nabla f)$$
$$\nabla(\mathbf{A} \cdot \mathbf{B}) = \mathbf{A} \times (\nabla \times \mathbf{B}) + \mathbf{B} \times (\nabla \times \mathbf{A}) + (\mathbf{A} \cdot \nabla)\mathbf{B} + (\mathbf{B} \cdot \nabla)\mathbf{A}$$
$$\nabla \cdot (f\mathbf{A}) = f(\nabla \cdot \mathbf{A}) + \mathbf{A} \cdot (\nabla f)$$
$$\nabla \cdot (\mathbf{A} \times \mathbf{B}) = \mathbf{B} \cdot (\nabla \times \mathbf{A}) - \mathbf{A} \cdot (\nabla \times \mathbf{B})$$
$$\nabla \times (f\mathbf{A}) = f(\nabla \times \mathbf{A}) - \mathbf{A} \times (\nabla f)$$
$$\nabla \times (\mathbf{A} \times \mathbf{B}) = (\mathbf{B} \cdot \nabla)\mathbf{A} - (\mathbf{A} \cdot \nabla)\mathbf{B} + \mathbf{A}(\nabla \cdot \mathbf{B}) - \mathbf{B}(\nabla \cdot \mathbf{A})$$

1

Space physics

Most of the visible matter in the universe exists as a fluid composed of electrically charged particles rather than as a gas made of neutral atoms or molecules. Gas mixtures of electrically charged particles, such as electrons and ions, are called *plasmas*. Plasmas are found in the following solar system environments: the solar atmosphere, the interplanetary medium, planetary magnetospheres, and planetary ionospheres. Most of the interstellar medium is also plasma, as are most other regions of our galaxy.

Most of the plasma found in our own solar system is accessible to in situ measurements made by instruments onboard spacecraft. Since the advent of the space age in the late 1950s, space probes have visited Mercury, Venus, Mars, Jupiter, Saturn, Uranus, Neptune, and comets Giacobini–Zinner, Halley, and Grigg–Skjellerup. The space environment surrounding the Earth has also been extensively studied by experiments onboard rockets and satellites. The Sun and astrophysical plasma environments outside our own solar system are not subject to direct measurements but must be observed remotely with sophisticated instruments located either at ground-based observatories or on orbiting observatories. An exception to this are the very energetic particles called *cosmic rays*, which can be observed using Earth-based or balloon-borne experiments. *Solar cosmic rays* have energies up to about 100 million electron volts (100 MeV) and originate in the solar corona. (Note that $1\,\text{MeV} = 10^6\,\text{eV}$ and $1\,\text{eV} = 1.602 \times 10^{-19}\,\text{J}$.) *Galactic cosmic rays* permeate our galaxy and have energies that can greatly exceed 1 GeV ($1\,\text{GeV} = 10^9\,\text{eV}$). The plasma in the terrestrial *ionosphere* can also be studied remotely using either radio wave reflection or radar backscatter techniques. In fact, the ionosphere was first discovered by means of radio reflection early in this century. The study of solar system plasmas is commonly given the name *space physics* or *space plasma physics*, and this is the subject matter of this textbook.

1.1 Plasmas

Plasmas behave differently than neutral gases mainly as a consequence of the *Lorentz force*. You will recall from your basic physics education that in the presence of electric and magnetic fields a particle possessing charge q is subject to the Lorentz force

$$\mathbf{F} = q(\mathbf{E} + \mathbf{v} \times \mathbf{B}).\tag{1.1}$$

SI units are employed in this book (see the Appendix). The force $\mathbf{F}$ has units of Newtons [N] and the charge q is in units of coulombs [C]. $\mathbf{E}$ is the *electric field intensity* [units of V/m] and $\mathbf{B}$ is the *magnetic flux density* [units of Wb/m^2 or tesla]. Typically, $\mathbf{E}$ and $\mathbf{B}$ will simply be referred to as the electric and magnetic fields, respectively. (See the appendix for a very brief review of electromagnetic theory.) The particle velocity is denoted $\mathbf{v}$.

The Lorentz force according to Equation (1.1) states that a charge moving in a magnetic field experiences a force at right angles to both $\mathbf{v}$ and $\mathbf{B}$ if $\mathbf{E} = 0$. Let us consider a simple example. Suppose that a uniformly magnetized region of space is located adjacent to an unmagnetized region ($\mathbf{B} = 0$), as illustrated in Figure 1.1. A proton moving from the unmagnetized region into the magnetized one is deflected as shown and leaves the magnetized region.

This simple example shows that charged particle motion is "inhibited" in the presence of magnetic fields. Suppose that a large number of protons (a plasma) are present in the unmagnetized region. Then, all the protons have trajectories similar to the one shown in Figure 1.1, and the plasma on the left-hand side "avoids" the magnetized region. Similarly, if we start off many protons (or electrons for that matter) deep in the magnetized region then they will move in circular trajectories and thus remain in the same region. In this sense, the magnetic field and plasma are "frozen" together.

The magnetic field configuration pictured in Figure 1.1 must be created by a system of electrical currents in such a way that *Ampère's law* is obeyed (the appendix contains a brief review of all of *Maxwell's equations* including Ampère's law). Contributions by currents, both internal and external to the plasma system, must in general be considered. The currents inside natural plasmas are due to the motions of the charged particles that make up that plasma. An example of a magnetic field created by external currents is the geomagnetic field generated by electrical currents

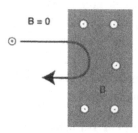

Figure 1.1. Schematic of a proton trajectory moving from a region with no magnetic field to a region with a uniform magnetic field B that is directed out of the page.

in the Earth's interior. For our simple example, a layer of current must exist at the interface between the two regions, and a current system external to both regions that creates a uniform magnetic field must also be present.

We can see from this simple example that the physics required to understand plasmas is quite basic: *classical electromagnetic theory* and *Newton's laws*. We can envision solving the equation of motion for each and every charged particle in a system, including not only the Lorentz force but other forces such as gravity. We could find the electric and magnetic fields as functions of spatial position and time by solving Maxwell's equations in which the source terms (i.e., the *electric charge density* and the *current density*) are specified using the position and velocity vectors of *all* the particles in the system. As you might expect, there is a slight problem with this procedure. The number of particles in any natural plasma environment, such as the terrestrial magnetosphere, greatly exceeds 10^{30}, presenting us with a formidable accounting problem. Clearly, there is too much information to keep track of. This accounting problem is handled using "statistical" techniques in the field of *plasma kinetic theory*. A brief review of this field is given in Chapter 2 and a more complete treatment is provided by Gombosi (1994) or Bittencourt (1986).

Even kinetic theory often provides more information than we really want about a plasma, and a *fluid approach* is called for, in which only macroscopic variables such as density and temperature are retained. The combination of *fluid theory* and *classical electrodynamics* yields the field of *magnetohydrodynamics* (or *MHD* for short). MHD theory will be the main tool employed in this text to explain solar system plasma environments such as the solar wind and magnetosphere.

1.2 Brief overview of the solar wind and magnetosphere

The outermost portion of the solar atmosphere is called the *solar corona* and contains plasma with a temperature of approximately 10^6 K (as will be discussed in Chapter 5). The corona is ordinarily observed in the visible part of the spectrum only during solar eclipses or with special instruments called coronagraphs because the light from the solar *photosphere* (the visible Sun) overwhelms the coronal visible light. However, the corona is much brighter than the photosphere at extreme ultraviolet (EUV) and X-ray wavelengths and can be observed with Earth-orbiting satellites such as the *Yohkoh* X-ray observatory (see Figure 1.2). The coronal gas expands outward into the solar system and fills interplanetary space, where it is called the *solar wind*.

The existence of the solar wind was recognized only rather recently. The first hints of the propagation of something other than light from the Sun to the Earth came in 1859 when the British astronomer Richard Carrington observed a solar flare that was followed a half day later by a disturbance in the geomagnetic field. An auroral display was also seen at the same time. Small disturbances in the Earth's magnetic field were first detected in the eighteenth century and were

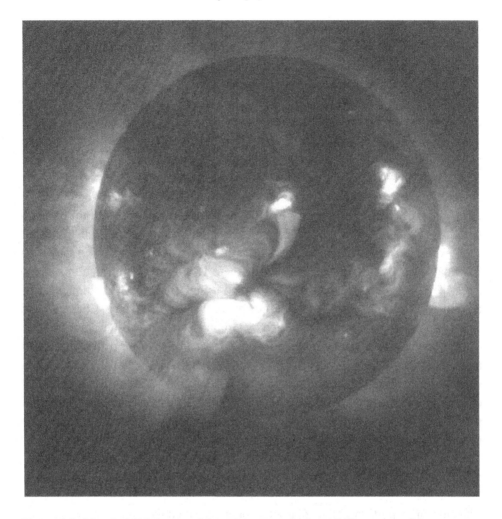

Figure 1.2. May 8, 1992, image of the solar corona in the soft X-ray part of the spectrum from the *Yohkoh* satellite, launched from Japan in1991. The bright regions are hotter (i.e., temperatures of a few million degrees Kelvin) than the darker regions (temperatures of only about 10^6 K). The solar X-ray images are from the *Yohkoh* mission of the Institute for Space and Astronautical Sciences (ISAS) of Japan. The X-ray telescope was prepared by the Lockheed Palo Alto Research Laboratory, the National Astronautical Observatory of Japan, and the University of Tokyo with the support of NASA and ISAS.

called *magnetic storms*. Following Carrington's discovery, the Norwegian scientist Kristian Birkeland proposed that beams of electrons were responsible for the occurrence of magnetic storms and auroral displays following flares. However, the existence of a more or less continuous wind of neutral plasma (equal number densities of both negative and positive particles) was not established until the early 1950s when the solar wind's existence was deduced from the observations of cometary ion tails by L. Biermann. The presence of an *interplanetary magnetic field* (or IMF) was also deduced, and it is the draping of the IMF field lines around the head of a comet that creates cometary ion tails. The modern concept of the solar wind as the outflowing extension of the corona was presented by Eugene Parker of the University of Chicago in 1958. Particle detectors onboard *Explorer 10* made the first actual measurements of solar wind plasma in 1961, and measurements made by detectors onboard *Mariner 2* clearly established that the solar wind was a continuous medium (Synder et al., 1963).

The solar wind is quite variable spatially and temporally, but on average at a distance from the Sun of 1 Astronomical Unit (AU) its number density is $10 \, \text{cm}^{-3}$, its flow speed is 400 km/s, and its temperature is about 30,000 K. An Astronomical Unit is equal to the radius of Earth's orbit. The solar wind is primarily made up of equal numbers of protons and electrons. The sound speed in the solar wind is only about 50 km/s, and hence the solar wind flow is quite supersonic. The interplanetary magnetic field strength at 1 AU is typically about 5 nT, where $1 \, \text{nT} = 10^{-9} \, \text{T}$. The origin and properties of the solar wind will be the subject of Chapter 6.

The solar wind inevitably encounters, and interacts with, the various planets and objects in the solar system. A cavity is carved out of the solar wind by the Earth's magnetic field. This cavity has been given the name *magnetosphere*. The first suggestion of such a cavity (sometimes called the *Chapman–Ferraro cavity*) was made by the British scientist Sidney Chapman and his colleague Vincent Ferraro (Chapman and Ferraro, 1930), although their suggestion was originally applied to "clouds" of plasma in interplanetary space since it was not then appreciated that the solar wind was continuous.

A simple schematic of the magnetosphere and the solar wind being diverted around it is shown in Figure 1.3. The region inside the magnetosphere contains

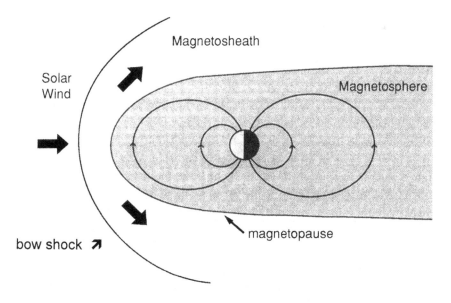

Figure 1.3. Schematic of the Earth's magnetosphere and the solar wind flow around it (not to scale).

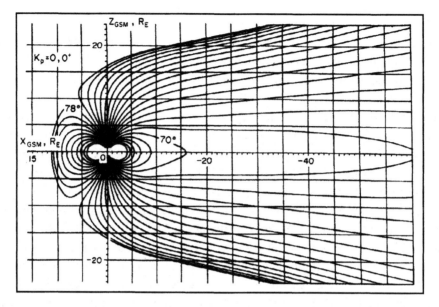

Figure 1.4. Magnetic field lines from the empirical model (i.e., based on extensive sets of data from magnetometer experiments on board Earth-orbiting satellites) of Tsyganenko (1987). The figure shows the noon–midnight cross-sectional plane and is appropriate for magnetically quiet time periods. (From Stern and Alexeev, 1988.)

the Earth's magnetic field, which is close to being dipolar, although the magnetic field lines in the region downstream of the Earth (a region called the *magnetotail*) are stretched out of dipole shape by electrical currents flowing across the tail. A more realistic depiction of magnetic field lines in the magnetosphere is shown in Figure 1.4. A simple explanation of why the Earth's magnetic field acts as an obstacle to the solar wind was suggested by Figure 1.1. The charged particles that make up the solar wind, both electrons and protons, are repelled by the Earth's magnetic field and are not able to penetrate it. As a consequence, the solar wind exerts a force on the magnetosphere at its outer boundary, which is called the *magnetopause*. A layer of current is present at the magnetopause. The force balance at the magnetopause will be examined in Chapter 7 using MHD concepts developed in Chapter 4.

A *bow shock* forms upstream of the magnetopause for the same reason that a shock wave precedes the passage of a supersonic jet. The solar wind flow is supersonic, and when it encounters an obstacle such as the magnetosphere a standing shock wave forms, downstream of which the plasma slows down and becomes hotter. The downstream plasma is subsonic and is able to flow around the obstacle. The region of space located between the magnetopause and the bow shock is called the *magnetosheath*.

The magnetosphere is not a vacuum, although the plasma density inside it is generally lower than the density of the solar wind. The internal structure and dynamics of the magnetosphere are complex. The structure of the outer magnetosphere, including the magnetotail, is the subject of Chapter 8. The structure of the inner magnetosphere is mainly discussed in Chapter 3. *Trapped particle radiation belts* (or *van Allen belts*) circle the Earth in the inner magnetosphere at radial distances between 1.2 and 4 Earth radii and contain energetic electrons, protons, and heavy ions (such as oxygen and helium) with energies ranging from about a kilovolt (keV) up to 100 MeV. These particles are trapped in the *magnetic bottle* associated with Earth's dipole magnetic field. The radiation belts were discovered in 1958 by James van Allen of the University of Iowa using cosmic ray detectors onboard the *Explorer 1* and *2* satellites. These were the first two U.S. satellites to orbit the Earth. They were preceded in orbit by the Soviet *Sputnik 1*, which was launched on October 4, 1957, but did not make measurements of the radiation belts. The fascinating history of these discoveries can be found in Van Allen's book *Origin of Magnetospheric Physics* (see the bibliography).

Plasma is also found in the *ionospheres* of the Earth and other planets. Ionospheric plasmas are cold in comparison with the plasma found in the solar corona or the magnetosphere, with temperatures of a few thousand Kelvins. Ionospheres contain *partially ionized plasmas* (i.e., mixtures of neutral gas and charged particle gases), whereas the hot magnetospheric or solar coronal plasmas have temperatures of millions of degrees or greater and are *fully ionized*. The terrestrial ionosphere is located at an altitude above about 100 km, and the maximum electron density of about 10^6 cm^{-3} in the daytime is typically found near an altitude of about 300 km. Ionospheric plasma is created by the ionization of the neutral atoms and molecules present in the tenuous upper atmosphere of the Earth or other planets. Ionization is mainly due to photoionization of the atmospheric neutrals by extreme ultraviolet photons, which are produced in the hot coronal and chromospheric plasmas of the solar atmosphere. However, in the polar regions energetic particle bombardment of the upper atmosphere can also ionize neutral atoms or molecules, as well as electronically excite these neutrals, resulting in the emission of radiation. For example, the *aurora borealis* in the northern hemisphere and the *aurora australis* in the southern hemisphere are caused by the bombardment, or precipitation, into the polar atmosphere of electrons with energies of several thousand electron volts (that is, several keV). These energetic electrons are accelerated at higher altitudes and originate in the magnetic tail region of the magnetosphere. They are guided by geomagnetic field lines into the *auroral zones*, which are located near 70 degrees latitude in both hemispheres.

Those readers with an interest in history might want to read the book *Cosmic Rays* by Friedlander (1989), the book *Origins of Magnetospheric Physics* by Van Allen (1983), the article "A brief history of magnetospheric physics before the spaceflight era" by Stern (1989), and the article "A brief history of magnetospheric physics during the space age" by Stern (1996).

A good background in plasma physics, beyond what is found in Chapters 2–4 of this text, can be found in the books by Chen (1974), Bittencourt (1986), Melrose (1986), or Krall and Trivelpiece (1973). I also recommend Rees (1989) on the physics and chemistry of the upper atmosphere and Zirin (1988) on solar physics. Other relatively recent books on different aspects of space plasma physics include Hargreaves (1992), Kivelson and Russell (1995), Walt (1994), Volland (1984), Kelley (1989), and Parks (1991). Further references will be provided throughout this text.

Bibliography

Bittencourt, J. A., *Fundamentals of Plasma Physics*, Pergamon Press, Oxford, 1986.

Chapman, S. and V. C. A. Ferraro, A new theory of magnetic storms, *Nature*, **126**, 129, 1930.

Chen, J., *Introduction to Plasma Physics*, Plenum Press, New York, 1974.

Freidlander, M. W., *Cosmic Rays*, Harvard Univ. Press, Cambridge, 1989.

Gombosi, T. I., *Gaskinetic Theory*, Cambridge Univ. Press, Cambridge, UK, 1994.

Hargreaves, J. K., *The Solar-terrestrial Environment*, Cambridge Univ. Press, Cambridge, UK, 1992.

Kelley, M. C., *The Earth's Ionosphere – Plasma Physics and Electrodynamics*, Academic Press, San Diego, 1989.

Kivelson, M. G. and C. T. Russell, eds., *Introduction to Space Physics*, Cambridge Univ. Press, Cambridge, UK, 1995.

Krall, N. A. and A. W. Trivelpiece, *Principles of Plasma Physics*, McGraw-Hill, New York, 1973.

Melrose, D. B., *Instabilities in Space and Laboratory Plasmas*, Cambridge Univ. Press, Cambridge, UK, 1986.

Parks, G. K., *Physics of Space Plasmas – An Introduction*, Addison-Wesley, Redwood City, CA, 1991.

Rees, M. H., *Physics and Chemistry of the Upper Atmosphere*, Cambridge Univ. Press, Cambridge, UK, 1989.

Snyder, C. S., M. Neugebauer, and V. R. Rao, The solar wind velocity and its correlation with cosmic ray variations and with solar geomagnetic activity, *J. Geophys. Res.*, **68**, 6361, 1963.

Stern, D. P., A brief history of magnetospheric physics before the spaceflight era, *Rev. Geophys.*, **27**, 103, 1989.

Stern, D. P., A brief history of magnetospheric physics during the space age, *Rev. Geophys.*, **34**, 1, 1996.

Stern, D. P. and I. I. Alexeev, Where do field lines go in the quiet magnetosphere?, *Rev. Geophys.*, **26**, 782, 1988.

Tsyganenko, N. A., Global quantitative models of geomagnetic field in the cislunar magnetosphere for disturbance levels, *Planet. Space Sci.*, **35**, 1347, 1987.

Van Allen, J. A., *Origins of Magnetospheric Physics*, Smithsonian Institution Press, Washington, DC, 1983.

Volland, H., *Atmospheric Electrodynamics*, Springer-Verlag, Berlin 1984.

Walt, M., *Introduction to Geomagnetically Trapped Radiation*, Cambridge Univ. Press, Cambridge, UK, 1994.

Zirin, H., *Astrophysics of the Sun*, Cambridge Univ. Press, Cambridge, UK, 1988.

2

Introduction to kinetic theory

A gas consisting of charged particles is called a *plasma*, although the use of the term is often restricted to charged particle gases in which collective phenomena, such as plasma oscillations, are more important than collisional phenomena. Collisions generally involve the short-range interactions of discrete particles, whereas collective phenomena involve large numbers of particles working in unison. The charged particle species in most plasmas are positive ions and negative electrons, although negative ions are also present in the D-region of the terrestrial ionosphere. *Fully ionized plasmas* contain only charged particles, whereas *partially ionized plasmas* also contain neutral gas. The solar wind plasma – that is, the interplanetary medium – is a fully ionized plasma; the ionosphere is a partially ionized plasma. A variety of methods have been developed to describe plasmas. *Kinetic theory* uses particle distribution functions to describe plasmas, whereas *fluid theory* (which includes *magnetohydrodynamics* or MHD) only uses a few macroscopic quantities derived from the full particle distribution functions. Because the subject of kinetic theory is largely outside the scope of an introductory book on space physics, this book will primarily use fluid theory to explain plasma phenomena in the solar system. However, a short introduction to kinetic theory and the derivation from kinetic theory of the fluid equations is provided in this chapter. More detailed treatments of kinetic theory can be found in the references listed at the end of the chapter.

2.1 The Boltzmann and Vlasov equations

Classically, a gas can be described by specifying the position and velocity vectors of each particle in that gas (i.e., all the electrons and ions). Let us denote a particular particle by the index $\alpha = 1, \ldots, N$, where N is the total number of particles in the system (or gas). Typically, $N/2$ of the particles are electrons and $N/2$ are ions. The position vector of particle α is $\mathbf{x}_\alpha = (x_\alpha, y_\alpha, z_\alpha)$ and the velocity vector is $\mathbf{v}_\alpha = (v_{x\alpha}, v_{y\alpha}, v_{z\alpha})$, where both $\mathbf{x}_\alpha(t)$ and $\mathbf{v}_\alpha(t)$ are functions of time. Each particle α obeys Newton's second law for a classical (nonquantum) system:

$$m_\alpha \frac{d\mathbf{v}_\alpha}{dt} = \mathbf{F}_\alpha \quad \text{and} \quad \frac{d\mathbf{x}_\alpha}{dt} = \mathbf{v}_\alpha, \tag{2.1}$$

where m_α is the mass of particle α and $\mathbf{F}_\alpha$ is the total force on particle α at the location of the particle at time t. The force $\mathbf{F}_\alpha$ includes contributions from gravity as well as electric and magnetic forces. In a plasma, the most important force is the *Lorentz force*,

$$\mathbf{F}_\alpha = q_\alpha(\mathbf{E} + \mathbf{v}_\alpha \times \mathbf{B}), \tag{2.2}$$

where q_α is the charge of particle α, and where the electric field intensity, $\mathbf{E}$, and the magnetic flux density, $\mathbf{B}$, must be evaluated at the time and location of the particle.

Vectors $\mathbf{E}$ and $\mathbf{B}$ in general include both contributions from sources external to the plasma and from sources associated with *all* other particles within the plasma. Let us consider a plasma in which only the internal electrostatic force is important and the force on the particle, α, is just the Coulomb force, $\mathbf{F}_\alpha = q_\alpha \mathbf{E}$. The electric field is evaluated at $\mathbf{x}_\alpha(t)$ and is given by

$$\mathbf{E}(\mathbf{x}_\alpha(t)) = \frac{1}{4\pi\varepsilon_0} \sum_{\substack{\beta=1 \\ \beta \neq \alpha}}^{N} \frac{q_\beta(\mathbf{x}_\alpha - \mathbf{x}_\beta)}{|\mathbf{x}_\alpha - \mathbf{x}_\beta|^3}. \tag{2.3}$$

The sum in this equation is over all particles other than α itself.

The full particle description requires far too much information ($6N$ numbers at each time t) to be practical for realistic-sized systems where $N > 10^{19}$, even for a small-sized plasma. The chief goal of the field of kinetic theory is to reduce this information to manageable proportions using statistical methods, while preserving the essential information associated with macroscopically observed quantities. A key product of kinetic theory is the Boltzmann equation, which will be discussed below, but which will not be derived here – this being beyond the scope of this book.

2.1.1 Single-particle distribution function

Most of the essential information about a plasma is contained in the *single-particle distribution function*, $f_s(\mathbf{x}, \mathbf{v}, t)$, where the position vector $\mathbf{x}$, the velocity vector $\mathbf{v}$, and time t are all independent variables. A separate distribution function is required for each species of plasma particle, s, such as electrons or ions of a particular mass. Phase space encompasses both ordinary space ($\mathbf{x}$) and velocity space ($\mathbf{v}$) and has six dimensions. Thus, to specify a point in phase space, $(\mathbf{x}, \mathbf{v})$, one requires six quantities (x, y, z, v_x, v_y, v_z). The single-particle distribution function (or just distribution function), $f_s(\mathbf{x}, \mathbf{v}, t)$, is defined as the number of particles per unit volume (of phase space) that are present, at time t, in an infinitesimally small

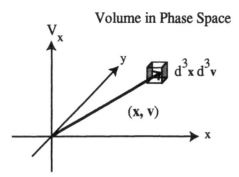

Figure 2.1. Schematic showing part of phase space and a small volume centered at the point $(\mathbf{x}, \mathbf{v}) = (x, y, z, v_x, v_y, v_z)$.

volume of phase space ($\Delta V = \Delta x \Delta y \Delta z \Delta v_x \Delta v_y \Delta v_z$), centered at the point in phase space, $(\mathbf{x}, \mathbf{v})$:

$$f_s(\mathbf{x}, \mathbf{v}, t) = \lim_{\Delta V \to 0} \frac{\# \text{ particles of type } s \text{ in } \Delta V}{\Delta V}. \tag{2.4}$$

In other words, f_s is the number of type-s particles situated between x and $x + \Delta x$, and also between y and $y + \Delta y$, and z and $z + \Delta z$, and between v_x and $v_x + \Delta v_x$, etc. We are dealing with a statistical distribution here – it does not matter which particles of type s (i.e., which electrons) are in this volume because all these particles are identical. The schematic in Figure 2.1 illustrates the concept of this phase space density. The units of f_s are s^3/m^6.

The integral of $f_s(\mathbf{x}, \mathbf{v}, t)$ over all velocity space yields the number density of particles of type s. The density n_s is a function of position and time:

$$n_s(\mathbf{x}, t) = \int f_s(\mathbf{x}, \mathbf{v}, t) \, d^3\mathbf{v}$$

$$= \int_{-\infty}^{\infty} \int_{-\infty}^{\infty} \int_{-\infty}^{\infty} f_s(x, y, z, v_x, v_y, v_z, t) \, dv_x \, dv_y \, dv_z. \tag{2.5}$$

The single particle distribution function can be found by solving *the Boltzmann equation*, which can itself be derived from the individual particle picture using the techniques of kinetic theory.

2.1.2 The Boltzmann equation

The Boltzmann equation can be written

$$\frac{\partial f_s}{\partial t} + \mathbf{v} \cdot \nabla f_s + \mathbf{a} \cdot \nabla_\mathbf{v} f_s = \left(\frac{\delta f_s}{\delta t} \right)_{\text{collision}}, \tag{2.6}$$

where $\mathbf{a}$ is the acceleration of a particle of type s, located at position $\mathbf{x}$ and possessing velocity $\mathbf{v}$. Note that $\mathbf{a}$ includes the effects of all noncollisional forces on the particles including gravity. For a plasma, the most important force is the Lorentz force; the acceleration of a particle of species s with charge q_s is given by

$$\mathbf{a}(\mathbf{x}, \mathbf{v}, t) = (q_s/m_s)[\mathbf{E}(\mathbf{x}, t) + \mathbf{v} \times \mathbf{B}(\mathbf{x}, t)]. \tag{2.7}$$

The electric and magnetic fields in expression (2.7) are *not* the complete fields, such as those given by Equation (2.3), but are *macroscopic*, or average, fields that do not include the microscopic fields associated with discrete particle collisions. These macroscopic fields include long-range average contributions from the plasma particles in a statistical sense. $\mathbf{E}$ and $\mathbf{B}$ can be found using Maxwell's equations; the charge density and current density in these equations are also macroscopic quantities, which can be specified in terms of the single-particle distribution function f_s, as will be discussed later. $\mathbf{E}$ and $\mathbf{B}$ can also include contributions from external sources.

Collisional effects (and the microscopic details of the electromagnetic field) are included in the collision term on the right-hand side of the Boltzmann equation. Coulomb collisions between charged particles involve the long-range electrostatic force, which varies as the inverse square of the particle separation (Equation (2.3)), and this type of interaction requires careful treatment. Formally, the collision term includes the two-particle correlation function, which is proportional to the probability of two particles strongly interacting with each other (Nicholson, 1983; Krall and Trivelpiece, 1973). We will only use simplified versions of the collision term in this book; the reader is referred to the references listed at the end of the chapter for a more complete treatment of the Boltzmann equation and the collision term. An especially simple form of the collision term is the *Krook collision term* (Bhatnagar, Gross, and Krook, 1954), or BGK collision term:

$$\left(\frac{\delta f_s}{\delta t}\right)_{\text{collision}} = -\frac{f_s - f_{s\text{M}}}{\tau_{\text{coll}}}. \tag{2.8}$$

Here $f_{s\text{M}}$ is the *Maxwell–Boltzmann distribution function* (or *Maxwellian*) for particle species s. τ_{coll} is a collision time – that is, the average time between collisions for a given particle. The BGK collision term is inaccurate but does have the desirable property that a distribution function that is non-Maxwellian at some initial time evolves into a Maxwellian in a period of time of the order of τ_{coll}. The Maxwellian distribution will be described mathematically in a later section. Collisions do indeed have this effect on the distribution function, although the details of this evolution are more complex than is suggested by expression (2.8).

Now let us consider a few aspects of mathematical notation. $\nabla = \hat{\mathbf{x}}\,\partial/\partial x + \hat{\mathbf{y}}\,\partial/\partial y + \hat{\mathbf{z}}\,\partial/\partial z$ is the gradient operator, and $\nabla_{\mathbf{v}} = \hat{\mathbf{x}}\,\partial/\partial v_x + \hat{\mathbf{y}}\,\partial/\partial v_y + \hat{\mathbf{z}}\,\partial/\partial v_z$ is the gradient operator in velocity space. The product $\mathbf{v} \cdot \nabla$ can be also be expressed in component notation as

$$\mathbf{v} \cdot \nabla = v_x \frac{\partial}{\partial x} + v_y \frac{\partial}{\partial y} + v_z \frac{\partial}{\partial z} = v_j \frac{\partial}{\partial x_j} = \sum_{j=1}^{3} v_j \frac{\partial}{\partial x_j}. \tag{2.9}$$

All the expressions in Equation (2.9) are equivalent. Note that the index j runs from 1 to 3. For example, $v_1 = v_x$, $v_2 = v_y$, and $v_3 = v_z$. The second to last form uses the summation convention in which the presence of repeated indices (e.g., j) is interpreted as a summation over that index from 1 to 3.

The left-hand side of Equation (2.6) is just equal to the total derivative of f_s in phase space:

$$\frac{D_{ps} f_s}{D_{ps} t} = \left(\frac{\delta f_s}{\delta t} \right)_{\text{collision}} \tag{2.10}$$

The Boltzmann equation simply states that the total time derivative of the single-particle distribution function equals the time rate of change of the distribution function due to collisions. If the plasma, or gas, is collisionless, then the right-hand side of Equation (2.10) is zero and the total derivative in phase space is zero. In this case, the phase space density of particles of type s in a small volume of phase space remains constant as this volume moves through phase space on a trajectory specified by the 6-dimensional "vector in phase space" $(\mathbf{v}, \mathbf{a})$. The Boltzmann equation with the collisional term equal to zero is called the *collisionless Boltzmann equation*, or also the *Vlasov equation*:

$$\frac{\partial f_s}{\partial t} + \mathbf{v} \cdot \nabla f_s + \frac{q_s}{m_s} (\mathbf{E} + \mathbf{v} \times \mathbf{B}) \cdot \nabla_\mathbf{v} f_s = 0. \tag{2.11}$$

The Vlasov equation provides the starting point for much of plasma physics. It is a deceptively simple-appearing equation that proves surprisingly difficult to solve for most situations.

2.1.3 The convective derivative

Some insight into the total derivative in phase space can be obtained by studying the total derivative (or *convective derivative*) in ordinary space. Consider a fluid (such as air, water, or a plasma) for which the flow velocity is given by $\mathbf{u}(\mathbf{x})$. $\mathbf{u}(\mathbf{x})$ is the bulk, or average, velocity of a small volume of the fluid (i.e., the "wind" velocity) located at position $\mathbf{x}$. The total derivative of some quantity Q (e.g., density, temperature, etc.) is the time rate of change of Q in a *frame of reference moving with the fluid* (with velocity $\mathbf{u}$). The convective derivative is equal to

$$\frac{DQ}{Dt} = \frac{\partial Q}{\partial t} + \mathbf{u} \cdot \nabla Q, \tag{2.12}$$

where $Q = Q(\mathbf{x}, t)$ is in general a function of both $\mathbf{x}$ and t.

In one dimension (e.g., $\mathbf{u} = u_x \hat{\mathbf{x}}$) and for a steady-state situation (in which $\partial Q/\partial t = 0$), the total derivative becomes $DQ/Dt = u_x \partial Q/\partial x$. The total time derivative of Q now equals the rate of change of x of a fluid parcel moving with the fluid ($u_x = $ "dx/dt"), multiplied by the derivative of Q with respect to x. For example, for a person on a raft floating down a river with speed u_x, the rate of

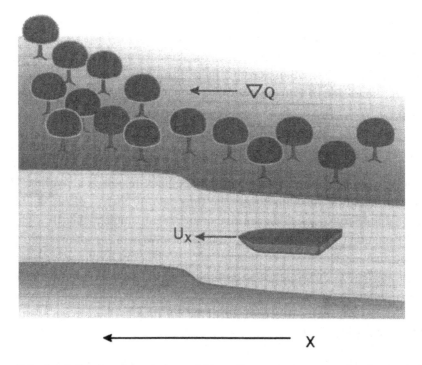

Figure 2.2. A raft floating down a river with speed u_x passes trees on the riverbank. The tree density Q is greater downstream than it is upstream. The raft is carried by the river from a region of low tree density to one of high tree density; that is, the tree density increases with time in the raft's frame of reference.

change – for the rafter – of the density of trees, Q, on the river bank adjacent to the raft is just equal to u_x multiplied by $\partial Q/\partial x$. (See Figure 2.2.) Note that $\partial Q/\partial t = 0$, since trees are immobile in the frame of reference of the Earth (although not in the raft's frame of reference).

2.2 The Maxwell–Boltzmann distribution function

2.2.1 Examples of distribution functions

An infinite number of different distribution functions satisfy the Boltzmann equation. For example, a uniform gas of electrons all moving in the x direction with speed u_{x0} is described by the *beamlike* distribution function,

$$f_e(\mathbf{x}, \mathbf{v}, t) = n_0\delta(v_x - u_{x0})\delta(v_y)\delta(v_z), \tag{2.13}$$

where $\delta(x)$ is the delta function. The constant n_0 is the electron number density as can be seen by substituting the distribution (2.13) into Equation (2.5) (see Problem 2.1).

The *shell* distribution is another example of a distribution function. For example, consider a uniform gas of ions that move in all directions with equal probability (i.e., an *istotropic* distribution). All ions in a shell distribution have the same speed $v = |\mathbf{v}| = v_0$, which can thus be written as

$$f_i(\mathbf{x}, \mathbf{v}, t) = (n_0/4\pi v_0^2)\delta(v - v_0), \tag{2.14}$$

where n_0 is the ion number density. Although beam and shell distributions are solutions of the Boltzmann equation at a given time, they do not remain beam and shell distributions at later times but eventually evolve into Maxwellian distributions in a time period roughly equal to the collision time.

2.2.2 Maxwellian distribution functions

The Maxwellian distribution for particles of type s is given by the expression

$$f_{s\,\text{M}}(\mathbf{x}, \mathbf{v}, t) = n_s(\mathbf{x}, t)\left(\frac{m_s}{2\pi k_{\text{B}} T_s}\right)^{3/2} \exp\left[-\frac{\frac{1}{2}m_s v^2}{k_{\text{B}} T_s}\right], \tag{2.15}$$

where $v^2 = v_x^2 + v_y^2 + v_z^2$. The density n_s is a function only of position and time and can be found by integrating $f_{s\text{M}}$ over all velocity space (see Problem 2.2). The temperature of species s is also a function of position and time, $T_s = T_s(\mathbf{x}, t)$. The probability of finding a particle of type s decreases exponentially with increasing v^2 (or, equivalently with increasing kinetic energy) for a Maxwellian. The function $f_{s\text{M}}$ falls off more rapidly for a cold gas (low temperature) than for a hot gas (high temperature) according to Equation (2.15). Note that the Maxwellian distribution is isotropic – that is, $f_{s\text{M}}$ depends only on the magnitude of the velocity vector $\mathbf{v}$ and not on its direction.

A distribution that is closely related to the Maxwellian distribution is the *drifting Maxwellian* distribution for which the gas as a whole moves in some direction with a constant velocity. In this case, the distribution function still looks like the Maxwellian specified by Equation (2.15) but in a frame of reference moving with the uniform velocity $\mathbf{v}_0$. The distribution function for a drifting Maxwellian is given by Equation (2.15), but with v^2 replaced by $|\mathbf{v} - \mathbf{v}_0|^2$.

A reduced, one-dimensional Maxwellian distribution can be obtained by integrating $f_{s\text{M}}$ over all y and z velocity components:

$$\begin{aligned}
g_{s\text{M}}(\mathbf{x}, v_x, t) &= \int_{-\infty}^{\infty}\int_{-\infty}^{\infty} f_{s\,\text{M}}(\mathbf{x}, \mathbf{v}, t)\, dv_y dv_z \\
&= n_s(\mathbf{x}, t)\left(\frac{m_s}{2\pi k_{\text{B}} T_s}\right)^{1/2} \exp\left[-\frac{m_s v_x^2}{2k_{\text{B}} T_s}\right].
\end{aligned} \tag{2.16}$$

The function $g_{s\text{M}}$ represents the probability of finding a particle of type s with x component of the velocity lying between v_x and $v_x + dv_x$. Like the full Maxwellian,

Reduced Maxwellian Distribution

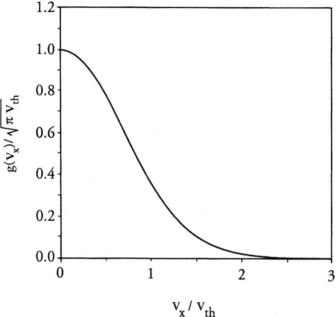

Figure 2.3. Reduced Maxwellian distribution function g_{sM} divided by $n_s[\pi v_{th}]^{1/2}$ plotted versus v_x divided by the thermal speed v_{th}.

g_{sM} is an exponential function (see Figure 2.3). The maximum of g_{sM} is located at $v_x = 0$ and g_{sM} decreases one e-folding as v_x increases from 0 to $v_x = v_{th}$. The thermal speed is defined by

$$v_{th} = [2k_B T_s/m_s]^{1/2}. \qquad (2.17)$$

The thermal speed is inversely proportional to the square root of the mass; thus, electrons have a much larger thermal speed than do ions with the same temperature.

2.2.3 Equilibrium Maxwell–Boltzmann distribution

The Maxwellian distribution function given by Equation (2.15), in general, is not an equilibrium distribution function and is a function of time via the time dependence of the density and temperature. In contrast, the Maxwell–Boltzmann distribution for a gas in thermodynamic equilibrium is independent of time and can be expressed in terms of the total energy E of a particle as

$$f_{sM}(\mathbf{x}, \mathbf{v}) = f_{sM}(0, 0) \exp[-E/k_B T], \qquad (2.18)$$

where $f_{sM}(0, 0)$ is a reference distribution function. The particle energy, E, for a static electric field is given by

$$E = 1/2 m_s v^2 + q_s V(\mathbf{x}), \qquad (2.19)$$

where $V(\mathbf{x})$ is the electrostatic potential at position $\mathbf{x}$. Equation (2.18) can be re-written as

$$f_{sM}(\mathbf{x}, \mathbf{v}) = \left[n_0 e^{-q_s V(x)/K_B T} \right] e^{-m_s v^2/2 k_B T}$$

$$n_s(\mathbf{x}) = n_0 e^{-q_s V(\mathbf{x})/k_B T}, \qquad (2.20)$$

where n_0 is the density at $\mathbf{x} = 0$. The density is proportional to n_0, as well as to an exponential called the *Boltzmann factor*. We will use this thermal equilibrium distribution near the end of the chapter to derive a shielding length scale in a plasma.

2.3 Macroscopic variables

For many applications we do not need to know the details of the distribution function. It is sufficient to work with a limited number of macroscopic variables. *Velocity moments* of the distribution function have traditionally been used (cf. Burgers, 1969; Gombosi, 1994) to define these basic *macroscopic* (or *fluid*) *variables*. These macroscopic variables are functions only of position (and not velocity) and can be considered to be "measurable" quantities. The nth moment of the single particle distribution function $f_s(\mathbf{x}, \mathbf{v}, t)$ is defined by

$$M^n f_s = \int \mathbf{v}^n f_s(\mathbf{x}, \mathbf{v}, t) \, d^3 \mathbf{v}. \qquad (2.21)$$

The zeroeth moment ($n = 0$) is just the number density, $n_s(\mathbf{x}, t)$, as given earlier by Equation (2.5). The first moment ($n = 1$) is just the (net) *particle flux* of species s:

$$\Gamma_s(\mathbf{x}, t) = \int \mathbf{v} f_s(\mathbf{x}, \mathbf{v}, t) \, d^3 \mathbf{v}. \qquad (2.22)$$

The particle flux has units of $\mathrm{m}^{-2}\,\mathrm{s}^{-1}$ and is a vector quantity. The integrals in Equations (2.21) and (2.22) are over all of velocity space. In component notation ($j = 1$–3), Equation (2.22) becomes

$$\Gamma_{sj} = \int v_j f_s(\mathbf{x}, \mathbf{v}, t) \, d^3 \mathbf{v}$$

$$= \int_{-\infty}^{\infty} \int_{-\infty}^{\infty} \int_{-\infty}^{\infty} v_j f_s(x, y, z, v_x, v_y, v_z, t) \, dv_x \, dv_y \, dv_z. \qquad (2.23)$$

The *bulk flow velocity* of species s is defined as the average velocity and is expressed in terms of the flux as

$$\mathbf{u}(\mathbf{x}, t) = \langle \mathbf{v} \rangle = \Gamma_s(\mathbf{x}, t)/n_s(\mathbf{x}, t). \qquad (2.24)$$

The velocity $\mathbf{u}$ will often carry the species subscript s ($\mathbf{u}_s$) if the context is not clear. Note that the average of a quantity Q is defined as

$$\langle Q \rangle = \frac{1}{n_s} \int Q(\mathbf{x}, \mathbf{v}, t) f_s(\mathbf{x}, \mathbf{v}, t) \, d^3\mathbf{v}. \tag{2.25}$$

The second moment of the distribution function is related to the *pressure tensor* of species s. The pressure tensor is best found by using the *peculiar – or random – velocity*, which is defined in terms of the total individual particle velocity and the bulk flow speed:

$$\mathbf{c} = \mathbf{v} - \mathbf{u}. \tag{2.26}$$

The peculiar velocity and the total velocity are the same if the gas as a whole is stationary. The pressure tensor is proportional to the second moment of f_s and can be calculated using $\mathbf{c}$ rather than $\mathbf{v}$:

$$\tilde{\mathbf{P}}_s(\mathbf{x}, t) = m_s n_s \langle \mathbf{c}\,\mathbf{c} \rangle = m_s \int \mathbf{c}\,\mathbf{c} f_s(\mathbf{x}, \mathbf{v}, t) \, d^3\mathbf{v}. \tag{2.27}$$

P_{sij} constitutes a 3×3 matrix. In component notation, the (i, j) element of the pressure tensor $\tilde{\mathbf{P}}_s(\mathbf{x}, t)$ is given as

$$P_{sij}(\mathbf{x}, t) = m_s \int c_i c_j f_s(\mathbf{x}, \mathbf{v}, t) \, d^3\mathbf{v}. \tag{2.28}$$

In Equation (2.28) the substitution $\mathbf{v} = \mathbf{c} + \mathbf{u}$ can be made, and the integral can be evaluated using the peculiar velocity – $d^3\mathbf{c}$ – instead of $d^3\mathbf{v}$. A scalar pressure p_s is defined as one third of the trace of P_{sij}:

$$p_s = \frac{1}{3} \operatorname{Tr} \tilde{\mathbf{P}}_s = \frac{1}{3} \sum_{j=1}^{3} P_{sjj} = \frac{1}{3} P_{sjj}. \tag{2.29}$$

The pressure tensor for an isotropic distribution contains only one independent quantity (i.e., the scalar pressure) and can be expressed in terms of the following matrix:

$$\tilde{\mathbf{P}}_s = \begin{pmatrix} p_s & 0 & 0 \\ 0 & p_s & 0 \\ 0 & 0 & p_s \end{pmatrix}, \tag{2.30}$$

where the scalar pressure $p_{xx} = p_{yy} = p_{zz} = p_s$ is given by

$$p_s(\mathbf{x}, t) = \frac{1}{3} m_s \int_0^\infty c^2 f_s(\mathbf{x}, c, t) 4\pi c^2 dc. \tag{2.31}$$

The distribution function in Equation (2.31) is isotropic and is solely a function of the magnitude of the peculiar velocity. A physical interpretation of pressure will be given later. The pressure p_s has units of N/m^2, which is the same as J/m^3. Equations

(2.30) and (2.31) can be applied to the shell distribution given by Equation (2.14) as well as to the Maxwellian distribution. Using expression (2.31), the scalar pressure for the Maxwellian distribution can be shown (Problem 2.3) to be equal to $p_s = n_s k_B T_s$; this is the *equation of state* of an ideal gas, which relates the pressure of species s to the density n_s and the *temperature* T_s.

The heat flux vector for species s is closely related to the third moment of the distribution function and is given by

$$\tilde{\mathbf{Q}}_s(\mathbf{x}, t) = \frac{1}{2} m_s \int \mathbf{c}\, c^2 f_s(\mathbf{x}, \mathbf{v}, t)\, d^3\mathbf{v}. \tag{2.32}$$

The SI unit for the heat flux is W/m^2. Most fluid theories in practical use do not go beyond five moments. Rather than carry out the integral in Equation (2.32), it is usually easier to represent the heat flux with a simple phenomenological expression, as will be done later in the chapter.

2.4 The fluid conservation equations

Now that macroscopic variables have been defined, we need some prescription for determining them, without having to undertake the very difficult task of solving the Boltzmann equation. Equations that are easier to solve than the Boltzmann equation can be found by taking moments of the Boltzmann equation. The nth moment of the Boltzmann equation can be represented by

$$\int \mathbf{v}^n \left[\frac{\partial f_s}{\partial t} + \mathbf{v} \cdot \nabla f_s + \mathbf{a} \cdot \nabla_{\mathbf{v}} f_s = \left(\frac{\delta f_s}{\delta t} \right)_{\text{collision}} \right] d^3\mathbf{v}. \tag{2.33}$$

We will only consider the zeroeth, first, and second moments here. Furthermore, little attention will be devoted to the collision terms.

2.4.1 Continuity equation

Now we evaluate the zeroeth moment of each term of the Boltzmann equation. The first term of Equation (2.33) with $n = 0$ becomes

$$\int \frac{\partial f_s}{\partial t} d^3\mathbf{v} = \frac{\partial}{\partial t} \int f_s\, d^3\mathbf{v} = \frac{\partial n_s}{\partial t}. \tag{2.34}$$

Note that the order of the time derivative and the integral over velocity has been reversed – this is allowed because time and velocity are independent variables. The definition of density given by Equation (2.5) was also used.

The zeroeth moment of the second term of Equation (2.33) gives

$$\int \mathbf{v} \cdot \nabla f_s\, d^3\mathbf{v} = \int \frac{\partial}{\partial x_j} (v_j f_s)\, d^3\mathbf{v}$$

$$= \frac{\partial}{\partial x_j} \int v_j f_s\, d^3\mathbf{v}. \tag{2.35}$$

Component notation and the summation convention were used in the first line of Equation (2.35), and in the second line the order of the derivative and integral were switched. The integral on the second line is just the definition of the net particle flux, as given by Equations (2.22) and (2.23). The particle flux is equal to $n_s \mathbf{u}_s$, where $\mathbf{u}_s$ is the average (or bulk) flow velocity of species s, whose jth component is u_{sj}. Equation (2.35) then becomes

$$= \frac{\partial}{\partial x_j} [n_s \langle v_j \rangle] = \frac{\partial}{\partial x_j} [n_s u_{sj}]$$
$$= \nabla \cdot (n_s \mathbf{u}_s). \tag{2.36}$$

Vector notation is again used in the second line of Equation (2.36). This equation indicates that the second term of the zeroeth moment of Boltzmann's equation is equal to the divergence of the net particle flux of species s.

We can write the acceleration term in Equation (2.33) as

$$\int \mathbf{a} \cdot \nabla_{\mathbf{v}} f_s \, d^3 \mathbf{v} = \int a_j \frac{\partial f_s}{\partial v_j} \, d^3 \mathbf{v}$$
$$= \int_{-\infty}^{\infty} \int_{-\infty}^{\infty} \int_{-\infty}^{\infty} \left(a_x \frac{\partial f_s}{\partial v_x} + a_y \frac{\partial f_s}{\partial v_y} + a_z \frac{\partial f_s}{\partial v_z} \right) dv_x dv_y dv_z. \tag{2.37}$$

It is sufficient for us to evaluate the first term of Equation (2.37) (the other two terms give identical results):

$$\int_{-\infty}^{\infty} \int_{-\infty}^{\infty} \left[\int_{-\infty}^{\infty} a_x \frac{\partial f_s}{\partial v_x} dv_x \right] dv_y \, dv_z. \tag{2.38}$$

The integral over v_x (i.e., the term in brackets in this expression) is easy to evaluate if the x component of the acceleration, a_x, does not depend on v_x, as is indeed true for the Lorentz force. Then

$$\left[\int_{-\infty}^{\infty} a_x \frac{\partial f_s}{\partial v_x} dv_x \right] = \left[a_x \int_{-\infty}^{\infty} \frac{\partial f_s}{\partial v_x} dv_x \right] = a_x [f_s]_{-\infty}^{\infty} = 0, \tag{2.39}$$

where we have used the fact that the distribution function is zero at $v_x = \pm\infty$. Thus, the acceleration term of the zeroeth moment of the Boltzmann equation equals zero.

We are left with the task of determining the zeroeth moment of the collision term, which we call S_s. It will be demonstrated below that S_s is the *net source* per unit volume of particles of type s (regardless of velocity) due to collisions. Putting together Equations (2.34) through (2.39), we obtain the familiar *continuity equation*,

$$\frac{\partial n_s}{\partial t} + \nabla \cdot (n_s \mathbf{u}_s) = S_s. \tag{2.40}$$

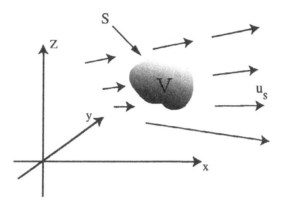

Figure 2.4. Volume of space V in a fluid of species s, with flow speed $\mathbf{u}_s$. The surface of the volume is designated S. The total number of particles of species s in the volume is N_s

The integral form of the continuity equation is useful for physical interpretation. We integrate both sides of Equation (2.40) over some fixed volume V for a fluid moving with bulk flow speed $\mathbf{u}_s$ (Figure 2.4). The volume integral of the time derivative of the number density, $\partial n_s(\mathbf{x}, t)/\partial t$, is just the time derivative of the total number of particles of species s in the volume, N_s. The volume integral of the divergence term can be converted to a closed surface integral using Gauss's integral theorem, which gives

$$\int_V \nabla \cdot (n_s \mathbf{u}_s)\, d^3\mathbf{x} = \oint_S n_s \mathbf{u}_s \cdot d\mathbf{S}, \tag{2.41}$$

where S designates the closed surface and the vector $d\mathbf{S}$ is a differential surface element oriented normal to the surface. Recall that $n_s \mathbf{u}_s$ is the particle flux Γ_s, and thus Equation (2.41) gives the total net flux of particles (units of s^{-1}) through the closed surface S – this is positive for a net flux of particles of type s *out* of the volume V and is negative for a net flux into the volume V.

Now we consider the right-hand side of the continuity equation. The volume integral of the net production rate per unit volume, S_s, gives the total net production rate *inside* the volume:

$$\tilde{S}_s = \int_V S_s\, d^3\mathbf{x}. \tag{2.42}$$

The integral form of the continuity equation becomes

$$\frac{\partial N_s}{\partial t} = \tilde{S}_s - \int_S \Gamma_s \cdot d\mathbf{S}. \tag{2.43}$$

Equation (2.43) states that the time rate of change of the total number of particles (of type s) in the volume is equal to the change due to the flux of particles across the surface plus the total net production of particles within the volume ($\tilde{S}_s$). If the fluid is stationary and $\mathbf{u}_s = 0$, then N_s changes only if there is a net collisional creation or destruction of particles of species s locally within the volume. For example, in a partially ionized plasma such as that found in the terrestrial ionosphere, ions and electrons can be created by the photoionization of neutrals by solar extreme ultraviolet radiation (i.e., collisions with photons). And ions of one species (e.g., O^+) can be converted to ions of another species (e.g., NO^+) by chemical reactions (e.g., ion–neutral collisions) such as

$$O^+ + N_2 \longrightarrow NO^+ + N. \tag{2.44}$$

Electrons in the ionosphere can be removed by recombination with ions. All these "collisional/chemical" production and loss terms enter the continuity equation via the net production terms S_s. The net production rate can be expressed as the difference between the production rate, P_s, and the loss rate, L_s; that is, $S_s = P_s - L_s$.

The density is constant in time and space (i.e., $n_s = constant$) for the important special case of an *incompressible* fluid. Furthermore, if there is no production or loss of particles ($S_s = 0$), the continuity equation simply becomes

$$\nabla \cdot \mathbf{u}_s = 0. \tag{2.45}$$

This simple incompressible continuity equation is generally applicable to air in the lower atmosphere or to liquids, but most plasmas are *compressible* and the full continuity equation (2.40) must be used.

2.4.2 Momentum equation

Now let us determine the first moment of the Boltzmann equation and obtain an equation for the flow speed $\mathbf{u}_s$. This is the momentum equation, which is a vector equation with three components. The jth component of the momentum equation is given by

$$\int v_j \{\text{Boltzmann equation}\} \, d^3\mathbf{v}, \tag{2.46}$$

where j runs from 1 to 3 (for the x, y, and z components, respectively).

The first moment of the time-derivative term in Equation (2.33) (or Equation (2.46)) is

$$\boxed{1} = \int v_j \frac{\partial f_s}{\partial t} \, d^3\mathbf{v} = \frac{\partial}{\partial t} \int v_j f_s \, d^3\mathbf{v} = \frac{\partial}{\partial t}(n_s u_{sj}), \tag{2.47}$$

where the order of the time derivative and the integral over velocity were switched. The definition of the flow velocity, Equation (2.24), was employed for the last step. The vector version of (2.47) is given by

$$\frac{\partial}{\partial t}(n_s \mathbf{u}_s). \tag{2.48}$$

The jth component of the second term (the advection term) is

$$\boxed{2} = \int v_j \mathbf{v} \cdot \nabla f_s \, d^3\mathbf{v} = \int v_j v_k \frac{\partial f_s}{\partial x_k} \, d^3\mathbf{v} = \frac{\partial}{\partial x_k} \int v_j v_k f_s \, d^3\mathbf{v}, \qquad (2.49)$$

where the removal of the derivative to outside the integral in the last step in Equation (2.49) was made possible because x_k, v_k, and v_j are all independent variables.

Next, we use the definition of peculiar velocity given in Equation (2.26) to write the jth component of the velocity as $v_j = c_j + u_j$. Putting this into Equation (2.49), we find that

$$\boxed{2} = \frac{\partial}{\partial x_k} \int v_j v_k f_s \, d^3\mathbf{v}$$

$$= \frac{\partial}{\partial x_k} \left\{ \underbrace{\int c_j c_k f_s \, d^3\mathbf{v}}_{(i)} + \underbrace{\int (c_j u_k + c_k u_j) f_s \, d^3\mathbf{v}}_{(ii)} + \underbrace{\int u_j u_k f_s \, d^3\mathbf{v}}_{(iii)} \right\}.$$

$$(2.50)$$

The first integral, (i), can be evaluated by using the definition of the pressure tensor, as given by Equation (2.28):

$$(i) = \int c_j c_k f_s \, d^3\mathbf{v} = n_s \langle c_j c_k \rangle = (1/m_s) P_{s,jk}, \qquad (2.51)$$

where $P_{s,jk}$ is the j, k component of the pressure tensor for species s. Integral (ii) is zero because u_k (or u_j) is independent of $\mathbf{v}$, and the average of c_j or c_k is zero, due to the definition of peculiar velocity. The product $u_j u_k$ can be removed outside the integral (iii), in which case we find

$$(iii) = \int u_j u_k f_s \, d^3\mathbf{v} = u_j u_k \int f_s \, d^3\mathbf{v} = n_s u_j u_k. \qquad (2.52)$$

When we put all the parts of Equation (2.50) back together, we find

$$\boxed{2} = \frac{\partial}{\partial x_k} \left(\frac{1}{m_s} P_{s,jk} + n_s u_j u_k \right). \qquad (2.53)$$

Expression (2.53) can be rearranged in several ways, including

$$\boxed{2} = \frac{1}{m_s} \frac{\partial P_{s,jk}}{\partial x_k} + u_j \frac{\partial}{\partial x_k} (n_s u_k) + n_s u_k \frac{\partial u_j}{\partial x_k}. \qquad (2.54)$$

Let us convert Equation (2.54) to vector form. The first two derivatives in this equation turn into divergence operators and the last derivative becomes the gradient operator:

$$\boxed{2} = \frac{1}{m_s} \nabla \cdot \tilde{\mathbf{P}}_s + \mathbf{u}_s \nabla \cdot (n_s \mathbf{u}_s) + n_s (\mathbf{u}_s \cdot \nabla) \mathbf{u}_s. \qquad (2.55)$$

We now evaluate the first velocity moment of the third term on the left-hand side of the Boltzmann equation – the acceleration term. The jth component of this moment is

$$\boxed{3} = \int v_j a_k \frac{\partial f_s}{\partial v_k} d^3\mathbf{v}$$

$$= \int v_j a_x \frac{\partial f_s}{\partial v_x} d^3\mathbf{v} + \int v_j a_y \frac{\partial f_s}{\partial v_y} d^3\mathbf{v} + \int v_j a_z \frac{\partial f_s}{\partial v_z} d^3\mathbf{v}, \qquad (2.56)$$

where the summation over k was carried out explicitly. Two of the three integrals in expression (2.56) have indices $j \neq k$, and these two integrals are zero (see Problem 2.5). The x component ($j = 1$) of the other integral can be integrated by parts over v_x and is equal to

$$\int v_x a_x \frac{\partial f_s}{\partial v_x} d^3\mathbf{v} = -n_s \langle a_x \rangle, \qquad (2.57)$$

where $\langle a_x \rangle$ is the average of the x component of the acceleration of a particle of species s. The y and z components can similarly be determined. Equation (2.56) can now be written

$$\boxed{3} = -n_s \langle a_j \rangle. \qquad (2.58)$$

We have finished taking the first moment of the left-hand side of the Boltzmann equation. We should now evaluate the velocity moment of the right-hand side of the Boltzmann equation (i.e., the collision term). This represents the change in momentum per unit volume of the fluid species s due to collisions. However, we skip over this very difficult task and simply designate the jth component of the collisional change in momentum as $\delta M_{s,j}/\delta t$. A simple heuristic expression for this collision term will be given later.

The velocity moment equation (i.e., the momentum equation) can now be written in vector form by combining $\boxed{1} + \boxed{2} + \boxed{3}$:

$$\frac{\partial}{\partial t}(\rho_s \mathbf{u}_s) + \nabla \cdot (\rho_s \mathbf{u}_s \mathbf{u}_s) + \nabla \cdot \tilde{\mathbf{P}}_s - \rho_s \langle \mathbf{a}_s \rangle = \left(\frac{\delta \mathbf{M}_s}{\delta t} \right)_{\text{Collision}}, \qquad (2.59)$$

where the mass density is related to the number density by $\rho_s = n_s m_s$. The average acceleration of a plasma fluid parcel is related to the average force by $\langle \mathbf{a}_s \rangle = \langle \mathbf{F}_s \rangle / m_s$. This average force should include the Lorentz force associated with internal electric and magnetic fields and should include any external forces on the fluid:

$$\langle \mathbf{a}_s \rangle = \frac{q_s}{m_s}[\mathbf{E}(\mathbf{x}, t) + \mathbf{u}_s \times \mathbf{B}(\mathbf{x}, t)] + \mathbf{a}_{\text{external}}. \qquad (2.60)$$

The electric and magnetic fields are understood to be average, or "macroscopic," fields that can be found from Maxwell's equations using macroscopically defined sources (see Section 2.3); all microscopic fields are incorporated into the collision term. An example of an external acceleration is gravity, $\mathbf{a}_{\text{ext}} = \mathbf{g}$. The acceleration due to gravity near the surface of the Earth is $\mathbf{g} = -9.88\,\hat{\mathbf{z}}\,\text{m/s}^2$.

The momentum equation can be written in several forms, including the following form, which can be derived using the continuity equation plus Equation (2.59) (Problem 2.6):

$$\rho_s \left[\frac{\partial \mathbf{u}_s}{\partial t} + \mathbf{u}_s \cdot \nabla \mathbf{u}_s\right] + \nabla \cdot \tilde{\mathbf{P}}_s = \rho_s \langle \mathbf{a}_s \rangle + \left(\frac{\delta \mathbf{M}_s}{\delta t}\right)_{\text{Collision}} - m_s \mathbf{u}_s S_s. \quad (2.61)$$

Recall that S_s is the net source of particles of species s that appears on the right-hand side of the continuity equation (2.40). The last term in equation (2.61) represents the effect of adding mass to the fluid (this is the "mass-loading" term). The term in brackets is the convective derivative of the flow velocity, which is just the acceleration of a parcel of fluid, $D\mathbf{u}_s/Dt$. Rearranging Equation (2.61) so that only $D\mathbf{u}_s/Dt$ appears on the left-hand side, the right-hand side then specifies this acceleration.

2.4.3 The pressure gradient force

The fluid acceleration not only includes $\langle \mathbf{a}_s \rangle$ and collisional effects but also a contribution due to any nonuniform pressure distribution in the fluid – this is, the *pressure gradient force*. The pressure gradient force per unit mass on a fluid parcel is $-(1/\rho_s)\nabla \cdot \tilde{\mathbf{P}}_s$. For isotropic pressure, this becomes $-(1/\rho_s)\nabla p_s$, where p_s is the scalar pressure. The physical meaning of the pressure gradient force can be illustrated by considering the pressure force on a finite volume of fluid (Figure 2.5). We find the change in the momentum in the volume by taking the volume integral of the divergence of the pressure tensor. This volume integral can be transformed by means of the divergence theorem (or Gauss's theorem) into a surface integral over the surface of the volume:

$$\left(\frac{D(\text{momentum in } V)}{Dt}\right)_{Pressure} = -\int_v \nabla \cdot \tilde{\mathbf{P}}_s d^3\mathbf{x} = -\int_s \tilde{\mathbf{P}}_s \cdot d\mathbf{S}. \quad (2.62)$$

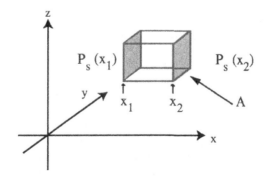

Figure 2.5. A cube-shaped volume (i.e., fluid parcel) is shown, each surface of which has area A. A pressure gradient across this volume in the x direction results in a force on the parcel.

Equation (2.62) represents the force on the fluid parcel due to the pressure gradient force. Consider a simple scalar pressure that is only a function of x: $p_s = p_s(x)$. The only nonzero elements of the pressure tensor are the diagonal terms, which are all equal to p_s. The force on the fluid volume is then solely in the x direction and is equal to $-[p_s(x_2) - p_s(x_1)]A$, where x_1 and x_2 define the x extent of the volume and A is the surface area. Simply put, a net force on a volume results when the pressure is greater on one side of the volume than on the other side. Pressure is due to the random motion of the particles; hence, the pressure gradient force is associated with an excess of collisions on one surface of a volume in comparison with the collisions on the opposite surface.

2.4.4 The collision term

The first moment of the Boltzmann collision term can formally be expressed as

$$\left(\frac{\delta \mathbf{M}_s}{\delta t}\right)_{\text{collision}} = \int \mathbf{v}\, d^3\mathbf{v}\left(\frac{\delta f_s}{\delta t}\right)_{\text{collision}} \tag{2.63}$$

In general, $(\delta f_s/\delta t)_{\text{collision}}$ includes contributions from creation and loss of particles of type s, Coulomb collisions with other charged particles (including species s), and collisions with neutral atoms and molecules.

The term $(\delta f_s/\delta t)_{\text{collision}}$ has the form of a *Fokker–Planck equation* for the case of a fully ionized plasma, because for the long-range Coulomb force, significant changes in f_s can result from a very large number of small-angle "collisions" (see references at the end of this chapter including Spitzer, 1962, and Bittencourt, 1986). However, "collisions" still implies "discrete" interactions of individual particles even if these interactions are long range. For most fully ionized space plasmas, such as the solar wind plasma and plasma in the solar corona, the collision term in the momentum equation can be entirely neglected with little loss of accuracy. However, both Coulomb and ion–neutral collisions are important for the colder and denser plasmas found in planetary ionospheres.

The change in momentum due to collisions can be written (without derivation) as

$$\left(\frac{\delta \mathbf{M}_s}{\delta t}\right)_{\text{collision}} = -\sum_{t \neq s} \nu_{st}\rho_s(\mathbf{u}_s - \mathbf{u}_t) + \rho_s\eta_s\nabla^2\mathbf{u}_s + P_s m_n \mathbf{u}_n - L_s m_s \mathbf{u}_s. \tag{2.64}$$

The first term on the right-hand side of Equation (2.64) represents the change in fluid momentum due to collisions of species s with all other species (hence, the summation over index t) including charged particle species and neutral species; this is the friction term. The second term is the viscosity term (only an approximate version has been included here), which can usually be neglected for space plasmas; this term handles the change in momentum due to velocity shears in the presence of collisions. The third term accounts for the creation of fluid momentum of species

s via ionization of a neutral species n, which has flow velocity $\mathbf{u}_n$ and mass m_n. The last term in Equation (2.64) accounts for the loss of momentum associated with the chemical loss of species s. Note that the last two terms of Equation (2.64), when combined with the last term of the momentum equation (2.61), yield the following "mass-addition" or "mass-loading" term appropriate for Equation (2.61): $(m_n\mathbf{u}_n - m_s\mathbf{u}_s)P_s$.

A cross section represents the probability of some type of collisional process taking place. For example, suppose that a broad "beam" of projectiles (e.g., pellets fired from a gun) are fired at a target (e.g., a basketball). Then the probability that a pellet will hit the basketball is proportional to the cross-sectional area (or *cross section*) of the basketball, as well as being proportional to the flux of projectiles. The *collision frequency* for some process is the rate at which this process takes place (with units of s^{-1}) and is proportional to the product of the cross section, the relative speed of the projectile and target, and the number density of the targets. For Equation (2.64), the relevant process is the transfer of momentum between colliding species. For the continuity equation, (2.40), the relevant process is that of creation or destruction of species s.

The term ν_{st} is the *momentum transfer collision frequency* between species s and species t and has units of inverse seconds. It is thus apparent that Equation (2.64) must have units of $[N/m^3]$. The momentum transfer collision frequency can be found by means of a suitable averaging of the momentum transfer cross section over the distribution functions of the two colliding species:

$$\nu_{st} = \frac{m_t}{m_t + m_s} \langle g\sigma_{\mathrm{m}}(g) \rangle_{\mathrm{cm}} n_t. \tag{2.65}$$

Here n_t is the number density of the target species; m_s and m_t are the masses of the "projectile" and "target" species, respectively; $g = |\mathbf{v} - \mathbf{v}'|$ is the relative velocity between particles of type s and t; and σ_{m} is the momentum transfer cross section as a function of g. $\langle\ \rangle_{\mathrm{cm}}$ is an average in the center of mass reference frame:

$$\langle g\sigma_{\mathrm{m}}(g) \rangle_{\mathrm{cm}} \equiv \frac{1}{n_s n_t} \int d^3\mathbf{v} \int d^3\mathbf{v}' |\mathbf{v} - \mathbf{v}'| \sigma_{\mathrm{m}}(g) f_s(\mathbf{v}) f_t(\mathbf{v}'). \tag{2.66}$$

The cross section depends on the nature of the interaction between the two species of particles. And the distribution functions in Equation (2.66) are generally assumed to be Maxwellians. For electron–neutral collisions, one can write the collision frequency as $\nu_{en} = k_{en} n_n$, where n_n is the neutral density and where the collision coefficient k_{en} depends on the particular neutral species and on the electron energy. Typically, $k_{en} \approx 10^{-8}\ \mathrm{cm^3\ s^{-1}}$ for electrons with energies of a few electron volts or less. Similarly, for ion–neutral collisions, the collision frequency can be written as $\nu_{in} = k_{in} n_n$ with $k_{in} \approx 10^{-9}\ \mathrm{cm^3\ s^{-1}}$ for most neutral species and for ion energies of the order of several eV or less. Charge transfer is one of the most important types of ion–neutral interactions for space plasmas and can be represented by the reaction

$$A^+ + B \longrightarrow A + B^+, \tag{2.67}$$

where A and B represent different atomic or molecular species. For example, A might be atomic hydrogen (H) and B might be atomic oxygen (O), in which case reaction (2.67) is just the accidentally resonant charge transfer reaction, $H^+ + O \rightarrow O^+ + H$. For this process, ion A^+ is neutralized and replaced by ion B^+. However, the momentum is largely retained by the neutral particle A. If the neutral particle B is slow, then the final ion B^+ is also slow regardless of the speed of the ion A^+. If the species A and B are the same, then the net result of the reaction (2.67) is a large loss of momentum for ion species A^+. Charge transfer cross sections are typically about 10^{-15} cm^2 for most species. For a more detailed discussion of collisional processes and their consequences you can consult a couple of the references cited at the end of the chapter (Banks and Kockarts, 1973; Rees, 1989).

Now we very briefly consider the collision frequency for interactions between two different charged species; again, for a detailed treatment of this topic consult one of the references listed at the end of the chapter. The momentum transfer cross section for Coulomb collisions varies as the inverse power of g^4, and the collision frequency is (in SI units)

$$\nu_{st} = \frac{e^4 n_t \ln \Lambda}{4\pi \varepsilon_0^2 m_s \mu_{st} \langle g_{st}^3 \rangle}, \tag{2.68}$$

where $\mu_{st} = m_s m_t / (m_s + m_t)$ is the reduced mass for species s and t, and $\ln \Lambda$ is the *Coulomb logarithm*, which has a value of roughly $\ln \Lambda \approx 20$ for space plasmas. Note that ν_{st} is proportional to the average of the product of the speed g and a cross section that varies as g^{-4}. Hence, as the particle speed increases, the Coulomb collision frequency decreases. The average $\langle \rangle$ was defined above and can be evaluated for a Maxwellian gas, in which case Equation (2.68) gives the following collision frequency formula as a function of temperature:

$$\nu_{st} = \frac{n_t e^4 \ln \Lambda \sqrt{\mu_{st}}}{\sqrt{2\pi}\, 32 \varepsilon_0^2 m_s (k_B T_s)^{3/2}} \quad \text{[SI units]}. \tag{2.69}$$

Using Equation (2.69), the following useful expressions for electron–electron (ν_{ee}), electron–ion (ν_{ei}), and ion–ion (ν_{ii}) momentum transfer collision frequencies can be obtained (Banks and Kockarts, 1973):

$$\begin{aligned}
\nu_{ee} &= 54\, n_e / T_e^{3/2} \\
\nu_{ei} &= \nu_{ee} \\
\nu_{ie} &= (m_e / m_i)\nu_{ei} \\
\nu_{ii} &= (m_e / m_i)^{1/2}\nu_{ee}.
\end{aligned} \tag{2.70}$$

The units of density in expression (2.70) are cm^{-3}. The electron–ion and electron–electron collision frequency formulae are essentially the same (very small differences have been neglected), but the ion–electron and ion–ion collision frequencies are much smaller than the electron–electron or electron–ion collision frequencies.

2.4.5 Energy equation

The continuity and momentum equations can be used to determine the density ρ_s and flow speed $\mathbf{u}_s$ for species s, respectively. The momentum equation, (2.61), contains the pressure p_s, which cannot be simply expressed in terms of density or flow speed from what we have done up to now. Another equation – the *energy equation* – is required to describe how the pressure (or thermal energy density) of a fluid should behave. Pressure has units of force per unit area $[N/m^2]$, but the units $[N/m^2]$ are the same as the units of energy density $[J/m^3]$. In fact, pressure is essentially thermal energy density.

The energy equation for species s is obtained by taking the second moment of the Boltzmann equation. The energy equation takes many forms, several of which will be shown in this section. We will not derive the energy equation; the method is essentially the same as was used to derive the continuity and momentum equations. To give an idea of how this might work, the second moment of the first term of the Boltzmann equation yields

$$\left\langle v_i v_j \frac{\partial f_s}{\partial t} \right\rangle = \frac{1}{m_s} \frac{\partial \tilde{\mathbf{P}}_{s,ij}}{\partial t} + \frac{\partial}{\partial t}(n_s u_i u_j). \tag{2.71}$$

Note that both the pressure tensor and bulk flow velocity appear in this expression.

For an isotropic Maxwellian distribution, the pressure of species s can be related to the density and temperature of species s by means of the *equation of state* for an ideal gas:

$$p_s = n_s k_B T_s, \tag{2.72}$$

where T_s is the temperature of species s, which appears in the Boltzmann factor discussed earlier. For a non-Maxwellian gas, expression (2.72) can be used to define an "effective" temperature using the pressure and the density of the gas. Equation (2.72) can also be written as $p_s = \rho_s R T_s$, where R is the gas constant for species s and $\rho_s = m_s n_s$ is the mass density of species s.

The *conservative* form of the energy equation has the appearance of a continuity equation for the total energy density of the fluid $[J/m^3]$. The total energy density includes: (1) *internal energy* density $(\rho_s U_s)$, where U_s is *specific internal energy* (units of J/kg), (2) bulk kinetic energy density ($\rho_s u_s^2/2$), and (3) potential/field energy density ($\rho_s U_{\text{pot}}$). The energy equation is

$$
\frac{\partial}{\partial t}\left[\rho_s\left(U_s + \frac{1}{2}u_s^2 + U_{\text{pot}}\right)\right]
$$
$$
+ \nabla\cdot\left[\rho_s \mathbf{u}_s\left(h_s + \frac{1}{2}u_s^2 + U_{\text{pot}}\right)\right] + \nabla\cdot\mathbf{Q}_s = \left(\frac{\delta E_s}{\delta t}\right)_{\text{collision}}.
$$

(2.73)

The first term is the rate of change of the total energy density of species s. The second term takes into account the change of this energy density due to bulk transport of energy in or out of a volume; it also includes, by means of the *specific enthalpy* h_s, the energy gain or loss due to mechanical work associated with changes in the volume. The $\nabla\cdot\mathbf{Q}_s$ term is the heat conduction term that accounts for changes in the energy density associated with microscopic heat transport in or out of a volume. The right-hand side handles the local time rate of change of the energy density associated with collisional processes.

We can write the specific enthalpy, or enthalpy per unit mass, h_s, and the specific internal energy U_s as

$$
h_s = U_s + p_s/\rho_s = \frac{\gamma_s}{\gamma_s - 1}\frac{p_s}{\rho_s} \qquad U_s = \frac{f_s}{2}\frac{p_s}{\rho_s} = \frac{1}{\gamma_s - 1}\frac{p_s}{\rho_s},
$$

(2.74)

where f_s is the number of degrees of freedom of the gas species s. $f_s = 3$ for an ideal gas whose particles have no internal structure. $f_s = 5$ for ordinary air at room temperature (3 for translational degrees of freedom plus 2 for rotation). The ratio of specific heats for species s is equal to $\gamma_s = (f_s + 2)/f_s$. For an ideal monatomic gas $\gamma_s = 5/3$. $\gamma_s = 7/5$ for air and $\gamma_s = 1$ for a gas with only one degree of translational energy. The internal energy per particle is equal to the number of degrees of freedom multiplied by $k_B T_s/2$. The enthalpy is equal to the internal energy plus an extra term that accounts for mechanical energy gained or lost due to changes in the volume of a fluid parcel.

The potential energy term is often just the gravitational potential: $U_{\text{pot}} = U_{\text{gravity}}$. However, an energy relation for the plasma as a whole can be found by adding the energy equation (2.73) for each of the plasma species (i.e., the index s must include the electrons and all ion species) plus a relation for the electromagnetic energy. In this case, the potential energy term must also include the magnetic energy density, $B^2/2\mu_0$, and the electric field energy density $\varepsilon_0 E^2/2$. Other "electromagnetic terms" also appear throughout the combined plasma energy equation. These terms are explained in Section 7 of the Appendix, and the combined plasma energy equation in its conservative form will be given in Chapter 4.

$(\delta E_s/\delta t)_{\text{coll}}$ is the local change in energy density per unit time due to collisional processes, and it essentially represents the heating or cooling of species s due to collisions with other (colder or warmer) species. For instance, an ion gas can be heated by Coulomb collisions with hotter electrons, or the ion gas can be cooled by collisions with colder neutrals. Sometimes, for electrons, cooling due to inelastic collisions with neutral species must also be included. Banks and Kockarts (1973) have an extensive discussion of various heating and cooling terms. A relatively simple collisional energy term including only elastic collisions between species s and neutrals (subscript n) can be written as

$$\left(\frac{\delta E_s}{\delta t}\right)_{\text{coll}} = \mathbf{u}_s \cdot \left(\frac{\delta \mathbf{M}_s}{\delta t}\right)_{\text{coll}} + \mu_{sn} \nu_{sn} n_s [(3k_{\text{B}}/m_s)(T_n - T_s) + |\mathbf{u}_s - \mathbf{u}_n|^2]$$

$$- \left(\frac{1}{2} m_s u_s^2 - \frac{3}{2} k_{\text{B}} T_s\right) S_s, \tag{2.75}$$

where μ_{sn} is the reduced mass between species s and the neutral species, T_n is the temperature of the neutrals, $\mathbf{u}_n$ is the neutral flow velocity, and ν_{sn} is the collision frequency for species s and neutrals. Note that the momentum change due to collisions appears in this expression as does the net production rate of species s (S_s).

The vector $\mathbf{Q}_s$ is the heat flux of species s and represents the transport of heat from one location to another by "microscopic" processes (rather than by bulk flow, which was taken care of by other terms on the left-hand side of Equation (2.73)). Although the heat flux vector is quite complicated in general, if the collision frequency is sufficiently large and if the temperature gradient is small enough, the heat conduction expression can be simply approximated by

$$\mathbf{Q}_s = -K_s \nabla T_s. \tag{2.76}$$

Equation (2.76) states that heat flows in response to temperature gradients – heat flows from hot regions to cold regions. K_s is the conductivity coefficient, which is proportional to the collision mean free path λ_{mfp}. The conductivity coefficient can be written, to within a factor of order unity, as $K_s \approx (n_s k_{\text{B}} v_{s,\text{th}}) \lambda_{\text{mfp}}$, where $v_{s,\text{th}}$ is the thermal speed. The conductivity for Coulomb collisions in a fully ionized Maxwellian plasma (i.e., the *Spitzer conductivity* – see Banks and Kockarts again and Spitzer, 1962) is given by

$$K_s = CT^{5/2} \quad [\text{eV/m/s/K}] \tag{2.77}$$

with $C \approx 7.7 \times 10^7$ for most space plasmas.

Two other very useful forms of the energy equation can be derived from the energy conservation form by using the continuity and momentum equations (see Problems 2.10 and 2.11). One form of the energy equation is written in terms of the convective derivative of the pressure (i.e., thermal energy density) of species s:

$$\frac{D}{Dt}\left(\frac{1}{\gamma_s - 1} p_s\right) + \frac{1}{\gamma_s - 1} p_s \nabla \cdot \mathbf{u}_s + (\tilde{\mathbf{P}}_s \cdot \nabla) \cdot \mathbf{u}_s + \nabla \cdot \mathbf{Q}_s$$

$$= \left(\frac{\delta E_s}{\delta t}\right)_{\text{coll}} - \mathbf{u}_s \cdot \left(\frac{\delta \mathbf{M}_s}{\delta t}\right)_{\text{coll}} + \frac{1}{2} m_s u_s^2 S_s. \tag{2.78}$$

If species s were an ideal monatomic gas then the polytropic index would be $\gamma_s = 5/3$. The second and third terms account for the thermal energy changes due to compression or expansion of a fluid volume. For isotropic pressure, the third term is equal to $(p_s \nabla \cdot \mathbf{u}_s)$ and the second and third terms together become $[\gamma_s/(\gamma_s - 1)]p_s \nabla \cdot \mathbf{u}_s = h_s \nabla \cdot \mathbf{u}_s$. Yet another form of the energy equation can be written in terms of the convective derivative of the temperature:

$$\frac{3}{2}n_s k_{\mathrm{B}} \frac{DT_s}{Dt} + p_s \nabla \cdot \mathbf{u}_s + \nabla \cdot \mathbf{Q}_s$$

$$= \left(\frac{\delta E_s}{\delta t}\right)_{\mathrm{coll}} - \mathbf{u}_s \cdot \left(\frac{\delta \mathbf{M}_s}{\delta t}\right)_{\mathrm{coll}} + \left(\frac{1}{2}m_s u_s^2 - \frac{3}{2}k_{\mathrm{B}}T_s\right)S_s. \quad (2.79)$$

Isotropic pressure has been assumed here.

The above forms of the energy equation can often be simplified. Let us consider four possible simplifications that are often useful in space physics applications.

2.4.5.1 *Steady flow without heat sources or sinks: Bernoulli's equation*

In addition to assuming that the flow is steady $(\partial/\partial t = 0)$ and that there are no collisional sources or sinks of heat, let us assume that heat conduction is unimportant and that we can neglect the potential energy. The following expression, which is one form of *Bernoulli's equation*, can then be derived from Equations (2.73) and (2.74) plus the continuity equation for steady $(\partial/\partial t = 0)$ flow without sources or sinks (see Problem 2.12):

$$\frac{\gamma_s}{\gamma_s - 1} \frac{p_s}{\rho_s} + \frac{u_s^2}{2} = constant. \quad (2.80)$$

Equation (2.80) states that $h_s + u_s^2/2$ is constant along a streamline. For an ideal monatomic gas, Equation (2.80) states that $(5/2)k_{\mathrm{B}}T_s + u_s^2/2$ is a constant. For example, as a parcel of fluid moves from a region of fast flow to a region of slower flow, it heats up; kinetic energy of the bulk flow is converted into thermal kinetic energy. The volume of a fluid parcel decreases as it slows down, and from elementary thermodynamics we know that a gas that is *adiabatically* compressed (i.e., no heat transfer occurs into or out of the volume although mechanical work can be done on the gas) has its internal energy increased.

2.4.5.2 *Polytropic energy relation*

An even simpler energy relation can be found when the conductive heat transport and the collisional terms are unimportant:

$$p/\rho^\gamma = constant. \quad (2.81)$$

Relation (2.81) even applies to time-dependent situations but does not apply across discontinuities in the flow. This equation is applicable when *specific entropy* (i.e., entropy per unit mass) is conserved by the flow. This is called *isentropic* flow.

$\gamma = 1$ is appropriate for an isothermal gas and $\gamma = 5/3$ for an ideal monatomic gas (see Problem 2.13). $\gamma = 2$ is appropriate for a strongly magnetized plasma and for plasma motions perpendicular to the magnetic field.

2.4.5.3 Heat conduction equation

When the thermal speed is much greater than the fluid speed ($v_{\text{th},s} \gg u_s$), the "dynamical" terms and mass-loading terms can be neglected. The dominant processes are then heat conduction and local collisional heating and cooling. The energy equation (2.79) can then be reduced to the *heat conduction equation* for species s:

$$\frac{3}{2} n_s k_B \frac{\partial T_s}{\partial t} = -\nabla \cdot \mathbf{Q}_s + H_s - L_s(T_s). \tag{2.82}$$

The conductive heat flux was given by Equation (2.76). The collisional terms in Equation (2.79) have been reorganized into local heating and cooling rate terms, H_s and $L_s(T_s)$, respectively. The heating and cooling terms can be neglected for some situations, in which case the one-dimensional heat conduction equation simply becomes

$$\frac{\partial T_s}{\partial t} = \frac{2}{3 n_s k_B} \frac{\partial}{\partial z} \left(K_s \frac{\partial T_s}{\partial z} \right), \tag{2.83}$$

where z is the relevant spatial coordinate (e.g., distance along the magnetic field line). In Problem 2.14, Equation (2.83) is solved for an ionospheric electron gas.

2.4.5.4 Local collisional energy balance

At low altitudes in a planetary ionosphere where the neutral density (and thus the electron–neutral and ion–neutral collision frequencies) is high, local collisional energy transfer becomes more important in the energy balance relation than either convective or conductive heat transport. In this case, the simple energy equation (2.82) can be approximated by the following local heat balance equation:

$$H_s = L_s(T_s). \tag{2.84}$$

Equation (2.84) can sometimes be solved to give an analytic expression for T_s. Banks and Kockarts (1972) discussed the heating and cooling terms (H_s and L_s) appropriate for the ionosphere of Earth. The electron–neutral cooling rate can usually be expressed in the form $L_{e,en}(T_e) = b_{en} n_n n_e (T_e - T_n)$, where the parameter b_{en} is roughly independent of altitude z, although it does depend on the electron temperature T_e and on the neutral composition. For the terrestrial ionosphere between an altitude of about 120 km and 200 km, taking into account the neutral composition of this atmospheric region (N_2, O_2, and O), one finds that $b_{en} \approx 5 \times 10^{-13}$ [eV cm^3/(Ks)] if the electron and neutral densities (n_e and n_n, respectively) are given in units of cm^{-3}. Expressing the electron heating rate (due to collisions with suprathermal electrons – photoelectrons or auroral precipitating electrons) as

$H_e(z) = a_{ee}(z)n_e$, we can see that the electron temperature as a function of altitude is given, in this approximation, by

$$T_e = T_n + a_{ee}(z)/(b_{en}n_n). \qquad (2.85)$$

For the terrestrial daytime ionosphere, the function $a_{ee}(z)$ has a maximum value of about 0.05 eV/s at an altitude of about 160 km. Clearly, according to equation (2.85), the electron temperature is simply equal to the neutral temperature ($T_e = T_n$) at the lowest altitudes ($z < 130$ km) where the neutral density n_n is very large. However, at higher altitudes the electron temperature increases with altitude since n_n decreases with increasing altitude. Electron temperatures measured in the terrestrial ionosphere for altitudes above 200 km or so typically exceed the neutral temperature by several thousand degrees. However, at these higher altitudes vertical heat conduction becomes an important part of the electron energetics and Equation (2.85) is no longer valid.

2.5 Macroscopic sources for Maxwell's equations

The Vlasov equation, and the moment/conservation equations derived from it, include the electric and magnetic fields, $\mathbf{E}(\mathbf{x}, t)$ and $\mathbf{B}(\mathbf{x}, t)$, respectively, via the Lorentz force contribution to the average acceleration $\langle \mathbf{a} \rangle$. These fields are macroscopically averaged fields and do not include the very small-scale, microscopic fields associated with collisions. The fields can be found from Maxwell's equations (see the appendix) where the source terms are macroscopic (or average) quantities:

$$\nabla \cdot \mathbf{E} = \frac{\rho_c}{\varepsilon_0} \qquad (2.86)$$

$$\nabla \cdot \mathbf{B} = 0 \qquad (2.87)$$

$$\nabla \times \mathbf{E} = -\frac{\partial \mathbf{B}}{\partial t} \qquad (2.88)$$

$$\nabla \times \mathbf{B} = \mu_0 \mathbf{J} + \frac{1}{c^2} \frac{\partial \mathbf{E}}{\partial t}. \qquad (2.89)$$

The macroscopic source terms – charge density $\rho_c(\mathbf{x}, t)$ and the current density $\mathbf{J}(\mathbf{x}, t)$ – are expressed in terms of the densities and bulk flow velocities of all the charged particle species in the plasma:

$$\rho_c(\mathbf{x}, t) = \sum_s q_s n_s(\mathbf{x}, t), \qquad \mathbf{J}(\mathbf{x}, t) = \sum_s q_s n_s(\mathbf{x}, t)\mathbf{u}_s(\mathbf{x}, t). \qquad (2.90)$$

The sum over s must include both electrons and ions.

The continuity and momentum equations provide prescriptions for finding n_s and $\mathbf{u}_s$, respectively, and these equations in turn require $\mathbf{E}$ and $\mathbf{B}$, which are determined by Maxwell's equations using the source functions ρ_c and $\mathbf{J}$. In order to complete the set of self-consistent fluid equations, ρ_c and $\mathbf{J}$ are specified in terms of n_s and $\mathbf{u}_s$ by means of Equation (2.90).

The fluid equations that we have presented in this chapter will be used in Chapter 4 to develop a fluid theory useful for describing space plasmas. In the next section, we will use the material in this chapter to discuss a basic property of plasmas – Debye shielding.

2.6 Debye shielding and the plasma parameter

Plasmas are different from neutral gases in that they are composed of charged particles that can exert forces on each other through the electric and magnetic fields they create. In space plasmas, the number densities of electrons and ions are equal, on the average, and the plasma is said to be *quasi-neutral*. Slight departures from charge neutrality ($\rho_c = e(n_i - n_e) = 0$) can occur on large spatial scales for nonequilibrium plasmas but not for most equilibrium plasmas. However, significant departures from neutrality can exist on short spatial scales even for an equilibrium plasma. What we mean by "short" spatial scale will become clear from the following discussion of *Debye shielding* of a test charge in an equilibrium Maxwellian plasma.

2.6.1 Electrostatic potential of a test charge in a plasma

First let us consider the electric potential, $V(r)$, of a point test charge, q_T, located at the origin in a vacuum. $r = |\mathbf{x}|$ is the radial distance from the charge. Combining Gauss's law – Equation (2.86) – and the electrostatic relation, $\mathbf{E} = -\nabla V$, we obtain Poisson's equation,

$$\nabla^2 V = -\rho_c/\varepsilon_0. \tag{2.91}$$

For a point charge in a vacuum, the charge density is $\rho_c = 0$ for $r > 0$. The solution of Equation (2.91) for a point charge in a vacuum is

$$V(r) = \frac{1}{4\pi\varepsilon_0} \frac{q_T}{r}. \tag{2.92}$$

Now we suppose that at time $t = 0$, the test charge is immersed within an initially uniform plasma in which the electron and ion densities are equal to each other as well as to a reference density, $n_i = n_e = n_0$. The initial charge density is zero and the electric potential is still given by Equation (2.92). For positive values of q_T, the ions in the plasma are repelled by the test charge and the electrons are attracted; for negative q_T, the electrons are repelled and the ions attracted by the test charge. After a sufficiently long time, the electrons and ions rearrange themselves in response to the electrostatic forces on them, and the plasma eventually reaches a new equilibrium configuration that takes into account the existence of the test charge. The ions move much more slowly than the electrons, so that for an intermediate

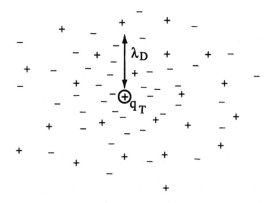

Figure 2.6. Shielding cloud surrounding a test charge.

time scale we assume that they remain motionless. The ion density then remains uniform: $n_i = n_0$. However, the density of electrons near the test charge increases so that $n_e > n_0$ (for $q_T > 0$) or $n_e < n_0$ (for $q_T < 0$). The charge density is no longer zero near the test charge because $n_i \neq n_e$ (Figure 2.6), and Poisson's equation becomes

$$\nabla^2 V = -e[n_i - n_e]/\varepsilon_0. \qquad (2.93)$$

For a Maxwell–Boltzmann distribution of electrons in an electrostatic field, we can use expression (2.20),

$$n_e(r) = n_0 \exp(+eV(r)/k_B T_e). \qquad (2.94)$$

This expression tells us that, in thermal equilibrium, the electron density is greatest at those locations where the electric potential V is the most positive – that is, n_e is higher in the vicinity of the test charge (for positive values of q_T). Here T_e is the electron temperature. The density variation is greater when the electron gas is cold than when the gas is hot. Substituting expression (2.94) into the spherical coordinate version of Equation (2.91), we find

$$\frac{1}{r^2}\frac{d}{dr}\left(r^2\frac{dV}{dr}\right) = -\frac{n_0 e}{\varepsilon_0}\left[1 - \exp\left(\frac{eV(r)}{k_B T_e}\right)\right], \quad r > 0, \qquad (2.95)$$

where we used the assumption that the ion density was uniform. For longer time scales, for which both the ions and electrons are in thermal equilibrium, another exponential term with the ion temperature would appear in the brackets on the right-hand side of (2.95) in place of the 1. Equation (2.95) is a complete equation for the potential V as function of r, subject to the condition that as $r \to 0$, the potential should look like that of a point charge, q_T.

We can approximate Equation (2.95) by restricting ourselves to radial distances that are large enough so that $|eV| \ll k_B T_e$. This condition means that a typical electron's kinetic energy is much greater than its potential energy. We can now expand the exponential function in Equation (2.95) using

$$e^x = 1 + x + \frac{x^2}{2!} + \cdots, \quad x \ll 1, \tag{2.96}$$

where $x = eV(r)/k_B T_e$. Poisson's equation then becomes

$$\frac{1}{r^2} \frac{d}{dr} \left(r^2 \frac{dV}{dr} \right) \cong \left[\frac{n_0 e^2}{\varepsilon_0 k_B T_e} \right] V(r) = \frac{1}{\lambda_D^2} V(r). \tag{2.97}$$

The constants contained within the brackets have been collected to form the parameter λ_D, which is called the *Debye length* or *Debye shielding length*. In SI units

$$\lambda_D = \sqrt{\frac{\varepsilon_0 k_B T_e}{n_0 e^2}} \quad [\text{m}], \tag{2.98a}$$

and in cgs units

$$\lambda_D = \sqrt{\frac{k_B T_e}{4\pi n_0 e^2}} \quad [\text{cm}]. \tag{2.98b}$$

The solution of Equation (2.97) is given by Problem (2.15):

$$V(r) = \frac{1}{4\pi\varepsilon_0} \frac{q_T}{r} \exp\left(-\frac{r}{\lambda_D} \right). \tag{2.99}$$

Clearly, as $r \to 0$, the potential is essentially that of a point charge in a vacuum, whereas for $r \gg \lambda_D$, Equation (2.99) demonstrates that $V(r) \to 0$ (as does the electric field) much faster than it does for a point charge. The vacuum Coulomb force is long range, but now in a plasma this force only extends a Debye length or so from the source, as a consequence of the Debye shielding cloud. For a positive test charge ($q_T > 0$) and positive potential, the shielding cloud contains an excess of electrons, whereas for a negative potential, the cloud has a deficit of electrons. It can easily be shown, using Gauss's law, that the total net charge contained in the shielding cloud, q_c, is equal and opposite to the test charge: $q_c = -q_T$. Another way of interpreting the Debye shielding phenomenon is that, although on small scales ($L \approx \lambda_D$) a plasma in thermal equilibrium can have significant departures from charge neutrality ($n_e \neq n_i$), for long spatial scales ($L \gg \lambda_D$) an equilibrium plasma must maintain charge neutrality. This property is called *quasi-neutrality*.

The size of the shielding cloud (λ_D) increases as the electron temperature increases because electrons with greater kinetic energy are better able to overcome the Coulomb attraction associated with the potential. And λ_D is smaller for a denser plasma because more electrons are available to populate the shielding cloud. We now consider the value of λ_D for two typical space plasmas.

Example 2.1 (Debye length in the ionosphere and in the solar wind) A numerical expression for the Debye length [m] is given by $\lambda_D = 69 \, [T_e/n_e]^{1/2}$ with electron temperature T_e in units of K and n_e in units of m^{-3}.

Typical temperature and density values in the topside terrestrial ionosphere are $T_e \approx 1000 \, K$ and $n_e \approx 10^{11} \, m^{-3}$, respectively, giving the following value for the Debye length:

$$\lambda_D = .007 \, m \, (\approx 1 \text{ cm}).$$

This Debye length is much less than either the vertical ($L \approx 300 \, km$) or horizontal ($L \approx 3,000 \, km$) extent of the ionosphere: That is, $\lambda_D \ll L$. Hence, the ionosphere can be considered to be quasi-neutral.

Typical parameters in the solar wind near 1 AU are $T_e \approx 10^5 \, K$ and $n_e \approx 10^7 \, m^{-3}$, giving

$$\lambda_D = 7 \, m.$$

Seven meters is much less than the macroscopic spatial scale of the solar wind ($L \approx 1 \, AU \approx 10^8 \, km$). Hence the solar wind can also be considered to be quasi-neutral. However, note that λ_D is greater than, or comparable to, the size of most spacecraft that have traversed the interplanetary medium. This must be taken into account when designing instruments to measure solar wind plasma properties.

2.6.2 The plasma parameter

Each particle in a plasma – be it an electron or an ion – can act as a "test charge" and carry its own shielding cloud. The concept of Debye shielding as it has just been developed requires the presence of a sufficiently large number of electrons and ions so that "density" can be defined in a statistically meaningful way. A useful parameter in this regard is the *number of particles in a Debye sphere*,

$$N_D = n_0 \left[4\pi \lambda_D^3 / 3 \right]. \tag{2.100}$$

N_D is approximately equal to the Λ parameter that appears in the Coulomb logarithm. The *plasma parameter* is defined by

$$g_{plasma} = 1/N_D. \tag{2.101}$$

A useful expression for N_D in SI units is $N_D = 138 \, T^{3/2}/n^{1/2}$. For example, in the solar wind and in the ionosphere, we have

solar wind: $N_D \approx 10^{10}$, $g_{plasma} \approx 10^{-10}$

ionosphere: $N_D \approx 10^5$, $g_{plasma} \approx 10^{-5}$.

In both of these plasmas, the number of particles in a Debye sphere is extremely large and the plasma parameter is very small, which indicates that Debye shielding is a statistically valid concept.

Another interpretation of the plasma parameter is that for $g_{plasma} \ll 1$, large-scale plasma phenomena are much more important than short-range Coulomb collisions (Nicholson, 1983). This is equivalent to the statement that for an equilibrium plasma for which $g_{plasma} \ll 1$, the average kinetic energy of a plasma particle ($\langle KE \rangle$) is much larger than its average potential energy ($\langle PE \rangle$). One can show (see Problem 2.17) that

$$g_{plasma} \approx \langle PE \rangle / \langle KE \rangle. \tag{2.102}$$

$\langle PE \rangle$ for an equilibrium plasma indicates the importance of the Coulomb collisions.

Three criteria are commonly used (see references listed at the end of this chapter) for determining whether or not a charged particle gas is a "good" plasma. "Good" is not meant to be an ethical judgment, but indicates that the plasma is quasi-neutral on important length scales and that collective, collisionless, and long-range phenomena such as plasma oscillations (discussed in Chapter 4) are more important than short-range collisional phenomena. The three criteria are:

(1) $\lambda_D \ll L$ (quasi-neutrality on length scales of interest),
(2) $g_{plasma} \ll 1$ (Coulomb collisions are not important and λ_D is defined),
(3) $\omega \tau_n \gg 1$ (other collisional processes are not important).

In criterion (3), ($\omega = 2\pi f$) denotes the angular frequency of the relevant plasma process (such as a wave mode) and τ_n is a collision time for electron or ion collisions with neutrals. The last criterion states that collisions with neutrals do not constitute an important process on time scales of interest. For example, for a plasma to sustain a large-scale oscillation (or waves) with wave frequency ω, then ω must be much greater than the collision frequency for ion–neutral or electron–neutral collisions ($\tau_n^{-1} \approx \nu_{in}$ or ν_{en}). This last criterion is not always met in an ionospheric plasma or in the solar atmosphere, although it usually is met in the solar wind or magnetosphere; in fact in the ionosphere, some of the most interesting phenomena are associated with charged particle collisions with neutrals.

In this chapter a fluid theory has been introduced that can be used to describe the statistical behavior of a collection of charged particles. This theory will be used in Chapter 4 to derive a more refined set of equations (e.g., the magnetohydrodynamic equations) that can be directly applied to space plasma problems. But first, in Chapter 3, we will study the behavior of individual (or single) particles in specified electric and magnetic fields. The study of single particle motion is often useful for understanding very low density plasmas, but it can also be useful for obtaining a physical understanding of the effects of fields on charged particles in fluids.

Problems

2.1   Find the number density associated with the "beam" distribution function given by Equation (2.13).

2.2   Demonstrate by substitution that the Maxwellian distribution, given by
       Equation (2.15), is a solution of the collisionless Boltzmann equation
       (2.11).

2.3   The equation of state for an ideal gas relates pressure to density and tem-
       perature: $p_s = n_s k_B T_s$.
           Show that for a Maxwellian gas the average kinetic energy of a particle
       is given by

$$\langle KE \rangle = (3/2) k_B T_s$$

       and that T_s is indeed the temperature appearing in the Boltzmann factor.
       Then show that for a Maxwellian distribution, Equation (2.31) yields the
       equation of state.

2.4   Find the scalar pressure p_s associated with the shell distribution function
       given by Equation (2.14).

2.5   In the derivation of the momentum equation from the Boltzmann equation
       several mathematical manipulations were undertaken. Derive the following
       expressions:

       (a) $\int v_i a_x \frac{\partial f_s}{\partial v_x} d^3 v = 0$ for $i = y, z$,
            where the acceleration $\mathbf{a}$ includes the Lorentz force.
       (b) $\int v_x a_x \frac{\partial f_s}{\partial v_x} d^3 v = -n_s \langle a_x \rangle$.

2.6   Start with the following "conservative" form of the momentum equation
       for species s:

$$\frac{\partial}{\partial t}(\rho_s \mathbf{u}_s) + \nabla \cdot [\rho_s \mathbf{u}_s \mathbf{u}_s] + \nabla p_s = \rho_s \langle \mathbf{a}_s \rangle + \left(\frac{\delta \mathbf{M}_s}{\delta t}\right)_{\text{collisions}}$$

       Use the continuity equation for species s to transform this form of the
       momentum equation into the following form:

$$\rho_s \left[\frac{\partial \mathbf{u}_s}{\partial t} + \mathbf{u}_s \cdot \nabla \mathbf{u}_s\right] + \nabla p_s = \rho_s \langle \mathbf{a}_s \rangle - m_s \mathbf{u}_s S_s + \left(\frac{\delta \mathbf{M}_s}{\delta t}\right)_{\text{collisions}},$$

       where m_s is the mass of a species s particle and where S_s is the net pro-
       duction rate of species s.
           Note that in component notation

$$\{\nabla \cdot [\rho_s \mathbf{u}_s \mathbf{u}_s]\}_i = \nabla \cdot [(\rho_s u_{s,i})\mathbf{u}_s] = \frac{\partial}{\partial x_j}(\rho_s u_{s,i} u_{s,j}).$$

2.7   Show that the units of the following collision term in the momentum equa-
       tion are $[N\,m^{-3}]$:

$$\left(\frac{\delta \mathbf{M}_s}{\delta t}\right)_{\text{collisions}} = -\rho_s \sum_{t \neq s} \nu_{st} (\mathbf{u}_s - \mathbf{u}_t).$$

Also show that the units of the collision frequency ν_{st} must be $[s^{-1}]$, starting from an expression for this collision frequency that includes the momentum transfer cross section.

2.8   For a slow-moving, uniform, charged particle gas (i.e., species s) with zero electric and magnetic field, moving through a stationary background neutral gas, the momentum equation simply becomes

$$\partial \mathbf{u}_s / \partial t = -\nu_{sn}\mathbf{u}_s.$$

Find the flow velocity as a function of time given an initial flow velocity of $\mathbf{u}_{s0}(t = 0)$.

2.9   Derive the Spitzer (i.e., Coulomb) heat conduction expression as given by Equation (2.77) (to within a factor of order unity) by using an expression for the Coulomb collision mean free path,

$$\lambda_{\mathrm{mfp}} = v_{\mathrm{th}}/\nu,$$

where v_{th} is the thermal speed and ν is the Coulomb collision frequency.

2.10   Derive the form of the energy equation given by Equation (2.78) from the "conservative" form of the energy equation (Equation (2.73)). Use the continuity and momentum equations. Assume that the electric and magnetic fields are zero.

2.11   Derive the form of the energy equation (2.79) from the form given by Equation (2.78), assuming an isotropic pressure.

2.12   Derive the simple energy relation, Equation (2.80), from the "conservative" form of the energy equation (Equation (2.73)) using Equation (2.74). Carefully consider the assumptions needed for this simple equation that were discussed in the text.

2.13   Derive the polytropic energy relation (2.81) for an ideal gas, starting from the energy equation (2.78) by neglecting the collision terms and heat conduction.

        Hint: You will also need the continuity equation without sources or sinks.

2.14   The electron temperature as a function of altitude z in a planetary ionosphere can frequently be described using a one-dimensional, steady-state, heat conduction equation without local heating and cooling:

$$K_e(\partial T_e/\partial z) = Q_{e0},$$

where Q_{e0} is the downward electron heat flux at the top of the ionosphere ($z = z_{\mathrm{top}}$). The constant Q_{e0} can often be equated to the integrated heating rate at higher altitudes associated with such processes as magnetospheric heating. The neutral density, and therefore electron–neutral collisional cooling rate, decreases sharply with increasing altitude in an ionosphere; hence the assumption of no heating and cooling is not unreasonable. The lower boundary can be placed at an altitude z_{bottom} where cooling first begins to be important. At this lower boundary, assume that the temperature is specified: $T_e(z = z_{bottom}) = T_{e0}$. It is usually appropriate to use the Coulomb heat conductivity coefficient.

(a) Derive the steady-state heat conduction equation given in this problem from Equation (2.83).

(b) Show that the solution, $T_e(z)$, of the steady-state heat conduction equation is given by

$$T_e^{7/2} = T_{e0}^{7/2} + (7/2)(Q_{e0}/C)(z - z_{bottom})$$

where C is the constant in Equation (2.77).

(c) Let $T_{e0} = 300\,K$ at $z_{bottom} = 100\,km$ and let $z_{top} = 1000\,km$. Also let the heat flux into the top of the ionosphere have a (typical) value of $Q_{e0} = 10^{13}\,eV/m^2/s$. Plot the electron temperature as a function of altitude for these conditions. What is T_e at the top of the ionosphere?

2.15   Demonstrate by substitution that the electrostatic potential given by Equation (2.99) is the solution of the differential Equation (2.97).

2.16   The Debye shielding length was found in Chapter 2 with the assumption that the ions were stationary. Repeat the analysis given in the chapter, but for mobile ions (which are in thermal equilibrium like the electrons) with temperature T_i, and find a new expression for the Debye length λ_D. How does this expression for λ_D compare with the expression given by Equation (2.98)?

2.17   Demonstrate that the plasma parameter is approximately (to within a factor of order unity) given by the ratio of the average potential energy of an electron in the plasma to the average kinetic energy:

$$g_{plasma} \approx \langle PE \rangle / \langle KE \rangle.$$

Assume that the relevant value of the test charge is $q_T = e$.

Bibliography

Banks, P. G. and G. Kockarts, *Aeronomy*, Academic Press, New York, 1973.

Bhatnagar, P. L., E. P. Gross, and M. Krook, A model for collision processes in gases, I, *Phys. Rev.*, **94**, 511, 1954.

Bittencourt, J. A., *Fundamentals of Plasma Physics*, Pergamon Press, New York, 1986.

Burgers, J. M., *Flow Equations for Composite Gases*, Academic Press, New York, 1969.

Gombosi, T. I., *Gaskinetic Theory*, Cambridge Univ. Press, Cambridge, UK, 1994.

Krall, N. A. and A. W. Trivelpiece, *Principles of Plasma Physics*, McGraw-Hill, New York, 1973.

Nicholson, D. R., *Introduction to Plasma Theory*, Wiley, New York, 1983.

Rees, M. H., *Physics and Chemistry of the Upper Atmosphere*, Cambridge Univ. Press, Cambridge, UK, 1989.

Spitzer, L., Jr., *Physics of Fully Ionized Gases*, Interscience Publishers (J. Wiley and Sons), New York, 1962.

3

Single particle motion and geomagnetically trapped particles

The motion of a single charged particle in specified electric and magnetic fields is described in this chapter. Charged particles trapped in the Earth's inner magnetosphere are then discussed.

3.1 Equation of motion

The trajectory of a single particle in a force field as a function of time and position can be determined by solving the equation of motion,

$$m\frac{d\mathbf{v}}{dt} = \mathbf{F}(\mathbf{x}, t), \tag{3.1}$$

where

$$\frac{d\mathbf{x}}{dt} = \mathbf{v}(\mathbf{x}, t).$$

$\mathbf{F}(\mathbf{x}, t)$ is the force (units of Newtons) on a particle with mass m, position vector $\mathbf{x}$, and velocity vector $\mathbf{v}$. If the particle has an electrical charge q, then one of the forces on the particle is the Lorentz force given by

$$\mathbf{F} = q(\mathbf{E} + \mathbf{v} \times \mathbf{B}). \tag{3.2}$$

Solving Equation (3.1) is easy in principle if $\mathbf{E}(\mathbf{x}, t)$ and $\mathbf{B}(\mathbf{x}, t)$ are completely specified. However, in most plasmas the electric field intensity $\mathbf{E}$ and the magnetic flux density $\mathbf{B}$ depend on the positions and velocities of all the other charges in the system and are very complicated functions. This complexity is the essence of plasma physics. Nonetheless, it is still instructive to consider the motion of a single charged particle in electric and magnetic fields that are independent of time and that are simple functions of position (e.g., uniform fields). Physical insight into important aspects of plasma physics can be gained in this manner, and single particle motion methods are applicable to a number of geophysical and astrophysical problems.

3.2 Uniform magnetic field (E = 0)

The equation of motion for a particle with charge q and mass m in a uniform magnetic field is

$$m\frac{d\mathbf{v}}{dt} = q(\mathbf{v} \times \mathbf{B}). \tag{3.3}$$

It is very useful to separate the velocity into its components parallel ($\mathbf{v}_\parallel$) and perpendicular ($\mathbf{v}_\perp$) to the magnetic field:

$$\mathbf{v} = \mathbf{v}_\parallel + \mathbf{v}_\perp. \tag{3.4}$$

The equation for $\mathbf{v}_\parallel$ can be found by taking the dot product of Equation (3.3) with the unit vector $\hat{\mathbf{b}}$ in the direction of the magnetic field $\mathbf{B}$. Since $(\mathbf{v} \times \mathbf{B}) \cdot \hat{\mathbf{b}} = 0$, we obtain

$$m\frac{d\mathbf{v}_\parallel}{dt} = 0. \tag{3.5}$$

The solution of Equation (3.5) is $\mathbf{v}_\parallel = $ constant; that is, the particle velocity along the field is independent of time.

The equation for $\mathbf{v}_\perp$ can be found by noting that $\mathbf{v}_\parallel \times \mathbf{B} = 0$ and by subtracting Equation (3.5) from (3.3); this gives

$$m\frac{d\mathbf{v}_\perp}{dt} = q(\mathbf{v}_\perp \times \mathbf{B}) = m\mathbf{v}_\perp \times \mathbf{\Omega}, \tag{3.6}$$

where

$$\mathbf{\Omega} = \Omega\hat{\mathbf{b}}$$

with magnitude $\Omega = qB/m$. Ω is the *gyrofrequency* (or *Larmor frequency*) and has the units of radians per second. In cgs units, $\Omega = qB/mc$.

Without any loss of generality, we can let the field be directed along $\hat{\mathbf{z}}$, so that $\mathbf{v}_\perp$ lies in the x–y plane; that is, $\mathbf{v}_\perp = v_x\hat{\mathbf{x}} + v_y\hat{\mathbf{y}}$. Using $\hat{\mathbf{x}} \times \hat{\mathbf{z}} = -\hat{\mathbf{y}}$ and $\hat{\mathbf{y}} \times \hat{\mathbf{z}} = \hat{\mathbf{x}}$, Equation (3.6) can be broken into its x and y components:

$$\frac{dv_x}{dt} = \Omega v_y, \tag{3.7}$$

$$\frac{dv_y}{dt} = -\Omega v_x. \tag{3.8}$$

Combining Equations (3.7) and (3.8), we obtain a second-order ordinary differential equation for v_x:

$$\frac{d^2v_x}{dt^2} + \Omega^2 v_x = 0. \tag{3.9}$$

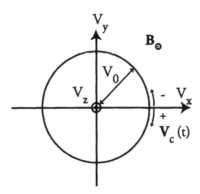

Figure 3.1. Particle trajectory in velocity space of a single charge in a uniform magnetic field directed out of the page. The + sign indicates the direction of positively charged particles, and the − sign indicates the direction of negatively charged particles.

The general solution to Equation (3.9) can be written as

$$v_x(t) = v_0 e^{i(|\Omega|t+\delta)}, \tag{3.10}$$

where v_0 is an amplitude and δ is a phase; these parameters are determined by the initial conditions. Using Equation (3.10) for v_x, we can find v_y from Equation (3.7):

$$v_y = \pm i v_0 e^{i(|\Omega|t+\delta)}. \tag{3.11}$$

The $\pm$ sign is just the sign of the charge q. The velocity components specified by Equations (3.10) and (3.11) are complex variables; the actual (i.e., measurable) velocity components are the real parts of these complex variables:

$$\begin{aligned} v_x &= v_0 \cos(|\Omega|t + \delta), \\ v_y &= \mp v_0 \sin(|\Omega|t + \delta). \end{aligned} \tag{3.12}$$

The total perpendicular velocity is independent of time and is given by

$$v_\perp^2 = v_x^2 + v_y^2 = v_0^2. \tag{3.13}$$

The particle follows a circular trajectory in v_x–v_y velocity space (Figure 3.1), with positively charged particles moving in a left-handed (LH) direction and negatively charged particles moving in a right-handed (RH) direction.

The *gyroperiod*, the period of time required for the particle to complete a cycle of its circular motion, is

$$T = \frac{2\pi}{|\Omega|} = 2\pi \frac{m}{|q|B}. \tag{3.14}$$

The trajectory of the charge in space, $\mathbf{x}(t)$, is determined by integration of the velocity $\mathbf{v}$ over time. In the parallel direction, this integration simply yields $z = z_0 + v_\parallel t$, where z_0 is the initial value of z. The trajectory is circular in the x–y

Table 3.1. *Approximate values of gyrofrequencies and gyroradii of protons (p)*
and electrons (e) in space plasmas

Location	v_p	v_e	B	Ω_p	Ω_e	$r_{L,p}$	$r_{L,e}$
solar wind	50 km/s[a]	1000 km/s[a]	5 nT	$0.5\,s^{-1}$	$10^3\,s^{-1}$	100 km	1 km
inner mag-netosphere at 3 R_E	4000 km/s[b]	5×10^4 km/s[c]	10^3 nT	$100\,s^{-1}$	$2 \times 10^5 s^{-1}$	40 km	300 m
ionosphere	5 km/s	200 km/s	3×10^4 nT	$3000\,s^{-1}$	$5 \times 10^6\,s^{-1}$	2 m	5 cm

[a]Typical thermal speed in solar wind reference frame.
[b]Speed of a proton with an energy of about 100 keV.
[c]Speed of an electron with an energy of about 10 keV.

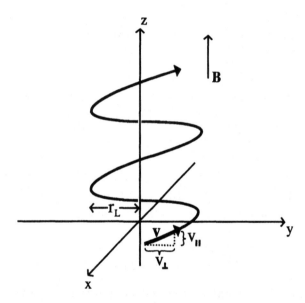

Figure 3.2. Helical trajectory of a negative charge in a uniform magnetic field. $v_\parallel$ and $v_\perp$ are the parallel and perpendicular components of the velocity **v**.

plane and this result is obtained in Problem 3.1. The complete trajectory is a helix wound about the field direction (Figure 3.2). The radius of this circle is called the *gyroradius* (or *cyclotron radius* or *Larmor radius*) and is given by

$$r_L = \frac{v_\perp}{|\Omega|} = \frac{mv_\perp}{qB},$$
(3.15)

where from now on Ω is understood to be positive in expressions like (3.14) or (3.15), whatever the sign of the charge q. Table 3.1 provides typical values of the gyrofrequency and the gyroradius for three space plasma environments.

The ratio of the perpendicular component of the velocity to the parallel component of the velocity remains constant along the trajectory. This ratio is the tangent of the angle α between the velocity vector and the field direction:

$$\tan \alpha = \frac{v_\perp}{v_\parallel} = \frac{v_{\perp 0}}{v_{\parallel 0}}. \tag{3.16}$$

α is called the *pitch angle* of the particle.

A quantity that will be useful later is the magnetic moment μ_m of the particle. The magnetic moment is the electrical current I in a current loop multiplied by the area A of the loop: $\mu_m = I A$. The current for the single particle can be identified as the charge q divided by the gyroperiod T. Using the area of the circle defined by the gyroradius, we get the following expression for the magnetic moment:

$$\mu_m = \frac{W_\perp}{B}, \tag{3.17}$$

where $W_\perp = (1/2)\, m v_\perp^2$ is the perpendicular energy of the particle.

3.3 Uniform magnetic field and uniform external force

Now we consider the motion of a charged particle in both a uniform magnetic field **B** and a uniform force field **F**, which does not depend on the particle position or velocity (i.e., an "external" force). The gravitational force and the electrostatic force are two examples of forces that can often be considered to be uniform, at least over some limited region of space.

The uniform external force can be decomposed into its parallel and perpendicular components:

$$\mathbf{F} = \mathbf{F}_\parallel + \mathbf{F}_\perp. \tag{3.18}$$

We can then consider separately the parallel and perpendicular parts of the equation of motion. The parallel part of the equation of motion (3.1) is simply

$$\frac{dv_\parallel}{dt} = \frac{F_\parallel}{m}. \tag{3.19}$$

The solution of Equation (3.19) is

$$v_\parallel(t) = v_{\parallel 0} + \left(\frac{F_\parallel}{m}\right) t, \tag{3.20}$$

where $v_{\parallel 0}$ is the initial parallel velocity. The particle acceleration in the direction of the magnetic field is $a_\parallel = F_\parallel / m$.

The equation of motion for the total perpendicular velocity $\mathbf{v}_{\perp t}$ can be written as

$$\frac{d\mathbf{v}_{\perp t}}{dt} = \mathbf{v}_{\perp t} \times \boldsymbol{\Omega} + \frac{1}{m}\mathbf{F}_\perp. \tag{3.21}$$

Equations (3.6) and (3.21) are almost the same, but Equation (3.21) has a $\mathbf{F}_\perp/m$ term added to the right-hand side. Now let us assume that we can decompose $\mathbf{v}_{\perp t}(t)$ into a gyrating part (as discussed in the last section), $\mathbf{v}_c(t)$, and a uniform drift velocity part, $\mathbf{v}_F$, with $d\mathbf{v}_F/dt = 0$:

$$\mathbf{v}_{\perp t}(t) = \mathbf{v}_c(t) + \mathbf{v}_F. \tag{3.22}$$

The validity of this assumption can be checked later by substitution of the final solution into Equation (3.21). Now let us substitute Equation (3.22) into (3.21) and recognize from the previous section that for simple gyromotion, $d\mathbf{v}_c/dt = \mathbf{v}_c \times \mathbf{\Omega}$. Then using $d\mathbf{v}_F/dt = 0$, we find that Equation (3.21) is reduced to

$$0 = \mathbf{v}_F \times \mathbf{\Omega} + \frac{1}{m}\mathbf{F}_\perp. \tag{3.23}$$

Because $\mathbf{v}_F$ is orthogonal to $\mathbf{\Omega}$ (i.e., to $\mathbf{B}$), we can see that $\mathbf{v}_F = -(\mathbf{v}_F \times \mathbf{\Omega}) \times \mathbf{\Omega}/\Omega^2$. Then we use expression (3.23) to substitute $\mathbf{F}_\perp/m$ for $\mathbf{v}_F \times \mathbf{\Omega}$, which gives

$$\mathbf{v}_F = -(\mathbf{v}_F \times \mathbf{\Omega}) \times \frac{\mathbf{\Omega}}{\Omega^2}$$
$$= \frac{\mathbf{F}_\perp \times \mathbf{\Omega}}{m\Omega^2}$$
$$= \frac{\mathbf{F} \times \mathbf{B}}{qB^2}. \tag{3.24}$$

The overall motion orthogonal to $\mathbf{B}$ consists of gyromotion plus a constant drift in a direction mutually orthogonal to both the force $\mathbf{F}$ and the field $\mathbf{B}$. For forces independent of the charge q, such as gravity, the drift motion is oppositely directed for positive and negative charges because of the sign of q. In a charge-neutral plasma, in which there equal number densities of positive and negative charges (e.g., ions and electrons), the application of a uniform force in a uniform magnetic field results in an electric current with current density

$$\mathbf{J}_F = \frac{n(\mathbf{F}_+ + \mathbf{F}_-) \times \mathbf{B}}{B^2} \quad \left[\frac{A}{m^2}\right], \tag{3.25}$$

where n is the number density of $+$ (or $-$) charged particles. $\mathbf{F}_+$ and $\mathbf{F}_-$ are the forces on positively and negatively charged particles, respectively. Equation (2.90) for the current density and Equation (3.24) for the drift velocity were used to derive Equation (3.25).

3.4 Uniform electric and magnetic fields

The most important example of a uniform force is the electrostatic force, $\mathbf{F} = q\mathbf{E}$, where $\mathbf{E}$ is the electric field intensity. We now consider the motion of charged particles in the presence of both a uniform electric field and a uniform magnetic field.

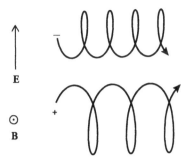

Figure 3.3. Trajectories of positive and negative particles in uniform electric and magnetic fields. The negatively charged particle is assumed to be less massive than the positive charge and therefore has a smaller gyroradius for comparable gyrovelocities ($v_\perp$). The $\mathbf{E} \times \mathbf{B}$ drift speeds of the two types of particles are the same.

For the parallel motion, we can use Equation (3.20) and the particle acceleration in the direction of $\mathbf{B}$, $a_\parallel = (q/m)E_\parallel = (q/m)\mathbf{E} \cdot \hat{\mathbf{b}}$. In order to determine the drift motion perpendicular to the magnetic field, we substitute $\mathbf{F} = q\mathbf{E}$ into the general expression (3.24) for the velocity and obtain

$$\mathbf{v}_E = \frac{\mathbf{E} \times \mathbf{B}}{B^2}. \tag{3.26}$$

$\mathbf{v}_E$ is known as the $\mathbf{E} \times \mathbf{B}$ *drift velocity* and is independent of the sign of the charge. Electrons and ions drift in the same direction and with the same drift speed. We see from Equation (3.25) that the current density due to $\mathbf{E} \times \mathbf{B}$ drift motion is zero. In cgs units, the right-hand side of Equation (3.26) must be multiplied by the speed of light c.

The complete motion perpendicular to the field consists of gyromotion plus the constant drift motion, just derived, that is mutually orthogonal to both $\mathbf{E}$ and $\mathbf{B}$ (Figure 3.3). Physically, this drift motion can be understood in terms of a gyroradius that varies along the trajectory. As a positively charged particle moves in the direction of $\mathbf{E}$, it gains energy and the total particle speed $|\mathbf{v}_t| = |\mathbf{v}_c + \mathbf{v}_E|$ increases; at this point in time, the "local" gyroradius is large and the particle moves a relatively large distance in the $\mathbf{E} \times \mathbf{B}$ direction. As the particle moves in a direction opposite to $\mathbf{E}$, it is decelerated and slows down, so that the local value of $v_{\perp t}$ is small, and the particle moves a relatively small distance opposite to the $\mathbf{E} \times \mathbf{B}$ direction. Hence, the net motion is in the $\mathbf{E} \times \mathbf{B}$ direction. Note that the total perpendicular velocity, $v_{\perp t}$, is a periodic function of time; therefore, the magnetic moment found using $v_{\perp t}$ varies with time.

It is useful to examine the particle motion in a frame of reference that moves with velocity $\mathbf{v}_E$. In this frame of reference, the perpendicular velocity consists solely of gyromotion (Equation 3.12), and the perpendicular speed $v_\perp$ is constant in time. The magnetic moment defined in this frame of reference is $\mu_m = mv_\perp^2/(2B)$, which is independent of time. As is shown in the appendix, the electric field in this moving frame of reference is zero, that is,

$$\mathbf{E}' = \mathbf{E} + \mathbf{v}_E \times \mathbf{B} = 0. \tag{3.27}$$

Example 3.1 (Motion of a charged particle initially at rest in uniform electric and magnetic fields) In this example, we determine the velocity and trajectory of a particle with positive charge $+q$ in a uniform electric and a uniform magnetic field. The initial velocity of the particle is assumed to be zero. Let $\mathbf{B} = B\hat{\mathbf{z}}$ and $\mathbf{E} = E\hat{\mathbf{y}}$; that is, $\hat{\mathbf{b}} = \hat{\mathbf{z}}$. The drift velocity, given by Equation (3.26), is

$$\mathbf{v}_E = \frac{\mathbf{E} \times \mathbf{B}}{B^2} = \frac{E}{B}\hat{\mathbf{x}}. \tag{3.28}$$

The gyromotion takes place in the x–y plane and the velocity of gyromotion, $\mathbf{v}_c(t)$, is given by Equation (3.12) and can be written

$$\mathbf{v}_c(t) = v_0[\hat{\mathbf{x}}\cos(\Omega t + \delta) - \hat{\mathbf{y}}\sin(\Omega t + \delta)] \tag{3.29}$$

with $\Omega = |qB/m|$. The amplitude v_0 remains unspecified for now. The total velocity includes both the gyromotion and the $\mathbf{E} \times \mathbf{B}$ drift:

$$\mathbf{v}(t) = \mathbf{v}_c(t) + \mathbf{v}_E = \hat{\mathbf{x}}\left[v_0\cos(\Omega t + \delta) + \frac{E}{B}\right] - \hat{\mathbf{y}}v_0\sin(\Omega t + \delta). \tag{3.30}$$

From the initial condition, $\mathbf{v} = 0$ at $t = 0$, we can determine δ and v_0:

$$\delta = \pi \text{ radians} \quad \text{and} \quad v_0 = E/B. \tag{3.31}$$

From $v_0 = E/B$, we can see that for this example the particle gyrates with a speed equal to its drift speed. The total velocity of the particle is zero when $t = nT$ (integer $n = 0, 1, 2, \ldots$), where T is the gyroperiod.

Assume that the particle starts at the origin ($\mathbf{x}(0) = 0$). We find the position of the particle by integrating Equation (3.30), obtaining

$$x(t) = \left(\frac{E}{B}\right)\left[t - \left(\frac{1}{\Omega}\right)\sin(\Omega t)\right],$$
$$y(t) = \frac{E}{\Omega B}[1 - \cos\Omega t]. \tag{3.32}$$

Note that the gyroradius is $r_L = E/(B\Omega)$. The trajectory is sketched in Figure 3.4, and such a curve is called a *cycloid*.

3.5 Nonuniform magnetic field

We now consider a magnetic field whose magnitude and/or direction is a function of position. The applied electric field is assumed to be zero. The kinetic energy

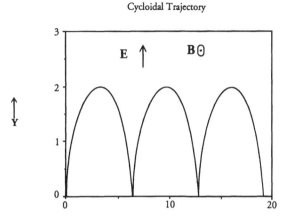

Figure 3.4. Cycloidal trajectory for a particle starting at rest in a uniform electric field and a uniform magnetic field. The coordinates are in units of gyroradii.

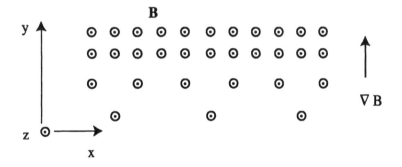

Figure 3.5. Magnetic field directed out of page with a field gradient present.

of a particle remains constant in such a time-independent field. To prove this, you can take the dot product of both sides of Equation (3.3) with **v** and show that the particle energy (or its speed) remains constant. The approach that will be taken in this section is called the *guiding center approximation*, in which deviations from gyromotion are assumed to be small. The motion of a particle is decomposed into gyromotion (i.e., the unperturbed orbit) plus some type of drift motion, which can be associated with the *guiding center*, about which the particle gyrates.

3.5.1 Gradient B drift

Consider a magnetic field that is in one direction only ($\mathbf{B}(\mathbf{x}) = B(\mathbf{x})\hat{\mathbf{z}}$). The field strength has a gradient, ∇B, that is perpendicular to **B** as shown in Figure 3.5. The local gyroradius, r_L, is large wherever B is small, and it is small wherever B

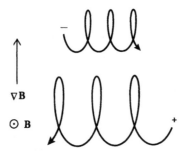

Figure 3.6. Grad-B drifts of a negatively charged particle and a positively charged particle.

is large. On physical grounds, we then expect a positive charge to drift to the left for the field geometry pictured in Figure 3.6. We now derive an expression for the grad-B drift velocity, $\mathbf{v}_{\nabla B}$.

If the particle gyroradius is very small with respect to the scale length L of the field gradient ($r_L \ll L = |\mathbf{B}/\nabla B|$), then the particle motion does not deviate significantly from gyromotion. The particle velocity can then be split into a gyration part and successively smaller corrections to the velocity, that is,

$$\mathbf{v} = \mathbf{v}_c + \mathbf{v}_1 + \cdots \qquad |\mathbf{v}_1| \ll |\mathbf{v}_c|. \tag{3.33}$$

Here $\mathbf{v}_c(t)$ is given either by Equations (3.10) and (3.11) or by (3.12) if we assume, without loss of generality, that $\mathbf{B}$ is in the z direction.

Similarly, one can expand the function $\mathbf{B}(\mathbf{x})$ about the origin ($\mathbf{x} = 0$) using a Taylor's series expansion:

$$\mathbf{B}(\mathbf{x}) = \mathbf{B}(\mathbf{x} = 0) + (\mathbf{x} \cdot \nabla)\mathbf{B} \ldots, \tag{3.34}$$

$$|(\mathbf{x} \cdot \nabla)\mathbf{B}| \ll |\mathbf{B}|,$$

where $\mathbf{x}$ is the position vector. Equations (3.33) and (3.34) are substituted into Equation (3.3), and second-order terms such as $q\mathbf{v}_1 \times [\mathbf{x} \cdot \nabla \mathbf{B}]$ are dropped. From the resulting equation, we then subtract Equation (3.6) (where $\mathbf{v}_\perp$ is identified as $\mathbf{v}_c$), thus obtaining

$$\frac{d\mathbf{v}_1}{dt} = \frac{q}{m}\mathbf{v}_1 \times \mathbf{B}(0) + \frac{q}{m}\mathbf{v}_c \times [(\mathbf{x} \cdot \nabla)\mathbf{B}], \tag{3.35}$$

where $\mathbf{B}(0) = \mathbf{B}(\mathbf{x} = 0)$. Motion parallel to $\mathbf{B}$ is not relevant and so $v_\parallel = 0$ has been assumed. The time average of $\mathbf{v}_1$ over one gyroperiod (or over an integral number of gyroperiods) is the drift velocity we wish to find,

$$\mathbf{v}_{\nabla B} = \langle \mathbf{v}_1 \rangle = \frac{1}{T}\int_0^T \mathbf{v}_1 \, dt. \tag{3.36}$$

Taking this average of expression (3.35) and recognizing that $\langle d\mathbf{v}_1/dt \rangle = 0$ (i.e., constant drift velocity), we find

$$-\mathbf{v}_{\nabla B} \times \mathbf{B}(0) = \langle \mathbf{v}_c \times [(\mathbf{x} \cdot \nabla)\mathbf{B}] \rangle, \tag{3.37}$$

which can be expressed as

$$\mathbf{v}_{\nabla B} = -\frac{1}{B(0)^2}\langle \mathbf{v}_c(\nabla B \cdot \mathbf{x}_c)\rangle \times \mathbf{B}(0), \qquad (3.38)$$

where we have used the fact that $\mathbf{B}$ is only in the z direction. The average is taken over an unperturbed particle orbit, and thus we identify $\mathbf{x}(t)$ as the gyromotion trajectory, $\mathbf{x}_c(t)$, which can be found by integrating the unperturbed velocity $\mathbf{v}_c$ given by Equations (3.10) and (3.11):

$$\mathbf{x}_c(t) = \frac{v_0}{|\Omega|}[\sin(|\Omega|t)\hat{\mathbf{x}} \pm \cos(|\Omega|t)\hat{\mathbf{y}}]. \qquad (3.39)$$

The constants of integration were chosen so that the guiding center starts at the origin, $\mathbf{x} = 0$. The remainder of the derivation requires recognizing that $\langle \sin(\Omega t) \cos(\Omega t)\rangle = 0$ and $\langle \cos^2(\Omega t)\rangle = 1/2$. Equation (3.38) is then averaged over a gyroperiod, using expression (3.39) for $\mathbf{x}_c(t)$, to give

$$\mathbf{v}_{\nabla B} = \pm\frac{v_\perp^2}{2|\Omega|B(0)}[-\hat{\mathbf{x}}(\nabla B \cdot \hat{\mathbf{y}}) + \hat{\mathbf{y}}(\nabla B \cdot \hat{\mathbf{x}})]. \qquad (3.40)$$

The term in brackets is just $\hat{\mathbf{b}} \times \nabla B$, where the unit vector in the direction of the magnetic field is given by $\hat{\mathbf{b}} = \hat{\mathbf{z}}$. Equation (3.40) becomes

$$\mathbf{v}_{\nabla B} = \pm\frac{v_\perp^2}{2|\Omega|B^2}\mathbf{B} \times \nabla B$$

$$= \pm\frac{1}{2}v_\perp r_L\frac{\mathbf{B} \times \nabla B}{B^2}, \qquad (3.41)$$

where B is understood to refer to $B(0)$. $\mathbf{v}_{\nabla B}$ is called the *grad-B drift velocity*. The ∇B drift motions of electrons and ions are oppositely directed, and consequently a current is associated with this type of drift. The upper and lower signs in Equations (3.40) or (3.41) are for positively and negatively charged particles, respectively. A quick dimensional analysis of Equation (3.41) confirms that $\mathbf{v}_{\nabla B}$ has units of velocity.

3.5.2 Curvature drift

A particle moving along a curved field line experiences an outward centrifugal force in the frame of reference moving with the parallel velocity $\mathbf{v}_\parallel$ (Figure 3.7). This force $\mathbf{F}_c$ is equal to

$$\mathbf{F}_c = \frac{mv_\parallel^2}{R_c}\hat{\mathbf{R}}_c, \qquad (3.42)$$

where $\mathbf{R}_c$ is a vector pointing radially outward and has a magnitude equal to the local radius of curvature of the field line.

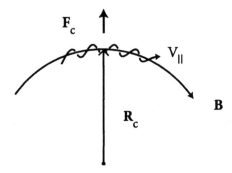

Figure 3.7. Curved magnetic field line and outward centrifugal force on a particle following this field line.

According to the general drift formula (3.24) for a uniform force, the centrifugal force gives rise to a drift motion, which is called the *curvature drift*:

$$\mathbf{v}_{cB} = \pm \frac{v_\parallel^2}{|\Omega| R_c} \hat{\mathbf{R}}_c \times \hat{\mathbf{b}}. \tag{3.43}$$

Curvature drift cannot be the only type of drift operating in a vacuum; grad-B drift must simultaneously be present because a curved magnetic field cannot have a constant magnitude or it violates Ampère's law for a vacuum, $\nabla \times \mathbf{B} = 0$. Taking the curl of $\mathbf{B} = B_\phi(r)\hat{\phi}$ in cylindrical coordinates, we can see that the gradient of B is just $\partial B_\phi / \partial r = -B_\phi / r$ (Problem 3.4). Identifying the distance from the axis, r, as R_c, this gradient can be used in the grad-B formula (3.41). Adding the grad-B and curvature drifts together, we find the following expression for the total drift velocity in a vacuum:

$$\mathbf{v}_B = \mathbf{v}_{\nabla B} + \mathbf{v}_{cB} = \pm \frac{\hat{\mathbf{R}}_c \times \hat{\mathbf{b}}}{|\Omega| R_c} \left(v_\parallel^2 + \frac{1}{2} v_\perp^2 \right). \tag{3.44}$$

Notice that $\mathbf{v}_B$ depends on the sign of the charge, so that for a gas with equal number densities of electrons and ions, a current density exists.

The drift motions and their associated currents are very important for charged particle motion in the *ring current* region (also known as the *radiation* or *van Allen* belts) of the Earth's magnetosphere. This will be discussed later in this chapter.

3.6 Adiabatic invariants – The first invariant

An important class of problems in classical mechanics is periodic motion (Symon, 1971). A quantity called the *action* remains invariant for slow changes in a system. The action is defined in terms of a *generalized coordinate*, q_{gen}, and its *conjugate momentum*, p_{gen}:

$$J = \oint p_{\text{gen}} \, dq_{\text{gen}}. \tag{3.45}$$

The integral is evaluated over one period of the motion. For charged particle motion in the geomagnetic field, three adiabatic invariants are possible. These are obtainable from Equation (3.45) if the generalized momentum is chosen as $\mathbf{p}_{\text{gen}} = \mathbf{p} + q\mathbf{A}$, where $\mathbf{p} = m\mathbf{v}$ is the ordinary momentum and $\mathbf{A}$ is the magnetic vector potential (Walt, 1994). Here we will obtain the first of these adiabatic invariants in a simpler fashion.

The gyromotion of a charged particle in a magnetic field is obviously periodic. A convenient choice of generalized coordinate is the azimuthal angle, $q_{\text{gen}} = \phi$; the conjugate momentum is then the angular momentum, $p_{\text{gen}} = \ell = mv_{\perp}r_{\text{L}}$. Note that in cylindrical coordinates, the gyromotion velocity vector is $\mathbf{v}_{\text{c}} = v_{\perp}\hat{\phi}$. The action then becomes

$$J = \int_0^{2\pi} mv_{\perp}r_{\text{L}}\, d\phi = 2\pi mv_{\perp}r_{\text{L}} = \frac{4\pi m}{q}\mu_{\text{m}}, \tag{3.46}$$

where it is safe to assume that r_{L} and $v_{\perp}$ are constant over a gyroperiod for a slowly changing magnetic field B. The action is just equal to the magnetic moment, μ_{m}, multiplied by a constant. μ_{m} is called the *first adiabatic invariant* of charged particle motion in a magnetic field.

Another way to demonstrate the invariance of μ_{m} is to simply evaluate its time derivative using the chain rule:

$$\frac{d\mu_{\text{m}}}{dt} = \frac{d}{dt}\left(\frac{mv_{\perp}^2}{2B}\right) = \frac{m}{2}\left(\frac{1}{B}\frac{dv_{\perp}^2}{dt} - \frac{v_{\perp}^2}{B}\frac{1}{B}\frac{dB}{dt}\right). \tag{3.47}$$

We now evaluate the terms on the right-hand side of this equation. First, we must consider energy conservation for charged particle motion. We start with the equation of motion (3.1) plus the Lorentz force, Equation (3.2). Taking the dot product of both sides of Equation (3.1) with $\mathbf{v}$ and recognizing that $\mathbf{v} \cdot (\mathbf{v} \times \mathbf{B}) = 0$, we find that

$$\frac{dW}{dt} = \frac{d}{dt}\left(\frac{1}{2}mv^2\right) = m\mathbf{v} \cdot \frac{d\mathbf{v}}{dt} = q\mathbf{v} \cdot \mathbf{E}, \tag{3.48}$$

where W is the total kinetic energy of the particle. The electric field appearing in Equation (3.48) generally includes a contribution from the choice of the reference frame. If the electric field is zero, the kinetic energy (and particle speed) remains constant, and only the direction of the velocity changes. One can also show that an equation identical to (3.48) holds in the frame of reference of the guiding center:

$$\frac{dW_{\perp}}{dt} = q\mathbf{v}_{\text{c}} \cdot \mathbf{E}, \tag{3.49}$$

where $W_{\perp} = (1/2)mv_{\perp}^2$ and $\mathbf{v}_{\text{c}} = v_{\perp}(t)\hat{\phi}(t)$ is just the velocity of gyromotion in cylindrical coordinates.

Now we return to evaluating Equation (3.47) for the time rate of change of the magnetic moment. We are not interested in (3.47) on time scales less than a gyroperiod, and thus we consider the parameters in it to be time averaged over a gyroperiod

($\langle \rangle$ as defined by Equation (3.36)). Taking the time average of Equation (3.49) we find that

$$\frac{d}{dt}\left\langle \frac{1}{2}mv_\perp^2 \right\rangle = \frac{q}{T}\int_0^T \mathbf{v}_c(t)\cdot \mathbf{E}\,dt. \tag{3.50}$$

The time integral can be converted to a closed path integral over the particle orbit using the incremental path length $d\mathbf{l} = \mp \mathbf{v}_c dt$, where the upper sign is for positive particles and the lower sign for negative particles. Equation (3.50) for positively charged particles becomes

$$\frac{d}{dt}\left\langle \frac{1}{2}mv_\perp^2 \right\rangle = -\frac{q}{T}\oint \mathbf{E}\cdot d\mathbf{l} = -\frac{|q|}{T}\int_S (\nabla \times \mathbf{E})\cdot d\mathbf{S}, \tag{3.51}$$

where the first integral is taken along the path of the particle orbit and the second integral is over the area of the orbit, S. The negative sign indicates that the direction of the line integral is opposite to the direction of motion of a positive charge. Stokes theorem was used to obtain the last part of Equation (3.51).

Using Faraday's law, $\nabla \times \mathbf{E} = -\partial \mathbf{B}/\partial t$ (see the appendix) to eliminate $\nabla \times \mathbf{E}$ from Equation (3.51), we find that

$$\frac{d}{dt}\left\langle \frac{1}{2}mv_\perp^2 \right\rangle = \frac{|q|}{T}\int_S \frac{d\mathbf{B}}{dt}\cdot d\mathbf{S} = \frac{|q|}{T}(\pi r_L^2)\frac{dB}{dt}, \tag{3.52}$$

where the area enclosed by the orbit is $S = \pi r_L^2$. We assumed that dB/dt is uniform over the area of the gyro-orbit S. In the frame of reference of the guiding center, we can write $\partial B/\partial t = dB/dt$. By substituting Equation (3.52) for $d/dt\langle(1/2)mv_\perp^2\rangle$ in the time average of expression (3.47) we find that

$$\frac{d\mu_m}{dt} = 0 \quad \text{or} \quad \mu_m = constant. \tag{3.53}$$

The constancy of the magnetic moment (or first adiabatic invariant) is valid for both spatially and temporally varying magnetic fields – as long as the change of B over either a gyro-orbit or a gyroperiod is small.

A corollary of Equation (3.53) is that the magnetic flux, Φ_m, through the area of a particle orbit is conserved for slow variations of B:

$$\Phi_m = \pi r_L^2 B = constant. \tag{3.54}$$

The derivation of Equation (3.54) from Equation (3.53) and the definition of μ_m is left as an exercise for the reader (Problem 3.5).

Example 3.2 (Example of betatron acceleration) What are the gyroperiod and gyroradius of a 1 keV proton in a 100 nT magnetic field, which is a typical field value in the "middle" magnetosphere? Assume that the parallel velocity is zero. What is the proton's energy after the magnetic field strength has been slowly increased to 200 nT? What is the new gyroradius?

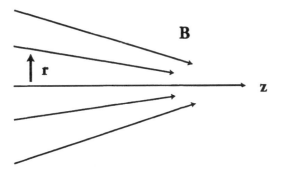

Figure 3.8. Magnetic mirror.

The speed of a 1 keV proton is $v_\perp = 440$ km/s.
The gyrofrequency is $\Omega = qB/m = 9.63\,\mathrm{s}^{-1}$.
The gyroperiod is $T = 2\pi/\Omega = 0.65$ s.
The gyroradius is $r_{\mathrm{L}} = v_\perp/\Omega = 45.7$ km.
The magnetic moment is $\mu_{\mathrm{m}} = W_\perp/B = 10^{-2}$ keV/nT.

As B doubles, μ_{m} remains constant, and so the energy $W_\perp$ must double. The new proton energy is thus 2 keV. μ_{m} is proportioned to $r_{\mathrm{L}}^2 B$ so that r_{L} must decrease by the factor $\sqrt{2}$ as B doubles, giving a new value of $r_{\mathrm{L}} = 32.3$ km. The proton has been accelerated and the energy comes from whatever powers the current creating the magnetic field. This type of acceleration is called *betatron acceleration*.

3.7 Magnetic mirrors and bottles

A common magnetic field configuration both in space and in laboratory plasmas is one in which the field lines are converging, as illustrated in Figure 3.8. The field pictured in Figure 3.8 is axisymmetric and can be represented in cylindrical coordinates as

$$\mathbf{B} = B_r \hat{\mathbf{r}} + B_z \hat{\mathbf{z}}. \tag{3.55}$$

The field is assumed to be primarily in the z direction ($B_r \ll B_z$) and B_z increases as z increases. The component of the field in the r direction can be found from Maxwell's equation:

$$\nabla \cdot \mathbf{B} = \frac{1}{r}\frac{d}{dr}(r B_r) + \frac{dB_z}{dz} = 0. \tag{3.56}$$

Expression (3.56) enables us to find B_r in terms of the gradient of $\mathbf{B}$ along the z direction; in the vicinity of the axis ($r = 0$) and for $B_r \ll B_z$ we have

$$B_r \cong -\frac{1}{2}r\frac{dB_z}{dz}. \tag{3.57}$$

The magnetic moment remains constant as a charged particle moves with parallel velocity $v_\parallel$ in the z direction. From the constancy of μ_m, given by Equation (3.53), and from the expression for μ_m, Equation (3.17), we see that as the field strength increases from B_0 to B, $v_\perp$ must increase from its "initial" value, $v_{\perp 0}$, according to the expression

$$v_\perp^2 = \left(\frac{B}{B_0}\right) v_{\perp 0}^2. \tag{3.58}$$

The total energy, parallel plus perpendicular, must remain constant for a static magnetic field configuration, that is,

$$v_\perp^2 + v_\parallel^2 = v_{\perp 0}^2 + v_{\parallel 0}^2, \tag{3.59}$$

where $v_{\perp 0}$ and $v_{\parallel 0}$ are the velocity components at some reference point. B_0 is the field strength at the reference point. As B increases, so does $v_\perp$, and Equation (3.59) indicates that $v_\parallel$ must decrease to preserve energy conservation. As B continues to increase with increasing distance z along the field, $v_\parallel$ eventually becomes zero, and at that point the particle is "reflected" and starts to move back toward the region of weaker field. A field configuration such as that shown in Figure 3.8 is called a *magnetic mirror*.

What is the physical explanation for the $v_\parallel$ decrease, the $v_\perp$ increase, and the particle reflection as a particle encounters a magnetic mirror? First let us explain the increase in the perpendicular velocity component by considering Equation (3.47) for the time derivative of μ_m. The electric field is zero for a static magnetic field and the total kinetic energy is conserved; however, in a frame of reference moving with parallel velocity $v_\parallel$ and at a distance from the axis of $r = r_L$, there exists a motional electric field, $E = -v_\parallel B_r$, directed along the direction of the gyromotion (for a positively charged particle). The force on a charged particle ($+$ or $-$ charge) due to this electric field is such that $v_\perp$ increases (see Equation (3.49)). We can also explain the increase of $v_\perp$ as B increases by using Equation (3.52) for the time derivative of $\langle (1/2)mv_\perp^2 \rangle$ and recognizing that $dB/dt = v_\parallel dB/dz$.

Alternatively, we can determine why $v_\parallel$ decreases, rather than why $v_\perp$ increases. The component of the Lorentz force in the z direction is

$$F_\parallel = q(\mathbf{v} \times \mathbf{B})_z = q v_\perp B_r = -\frac{1}{2} q v_\perp r_L \frac{dB_z}{dz}, \tag{3.60}$$

where B_r was taken from Equation (3.57) with $r = r_L$. Equation (3.60) for the force can also be written as

$$\mathbf{F} = -\mu_m \nabla B. \tag{3.61}$$

The force is directed along $\hat{\mathbf{z}}$, opposite to the gradient of B, and this force decelerates particles that are moving toward larger values of the magnetic field strength.

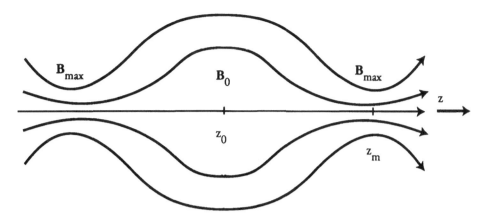

Figure 3.9. Magnetic bottle with maximum field B_{max} and minimum field B_0.

A magnetic bottle consists of two magnetic mirrors, as shown in Figure 3.9. Consider a particle whose parallel and perpendicular velocities are $v_{\parallel 0}$ and $v_{\perp 0}$, respectively, at a point in the bottle ($z = z_0$) where the field strength is at a minimum. We can use the energy conservation relation (3.59), the invariance of μ_m in the form (3.58), plus the definition of pitch angle to relate the pitch angle α to the local field magnitude and to the initial pitch angle, α_0:

$$\sin^2 \alpha = \frac{B}{B_0} \sin^2 \alpha_0. \qquad (3.62)$$

The details of this derivation are left to the reader as an exercise (Problem 3.6). The initial pitch angle α_0 is the pitch angle at $z = z_0$. As a particle travels in the bottle away from z_0, B increases and, from Equation (3.62), α must increase. The particle reflects at a location where B is large enough so that $\alpha = 90°$ (i.e., $v_{\parallel} = 0$). This happens where $B = B_0 / \sin^2 \alpha_0$. The field strength required to cause the particle to reflect increases as α_0 decreases. The maximum field strength of the bottle determines the minimum value, α_0, required for confinement. For a given value of the *mirror ratio*, which is defined by $R_{mir} = B_{max}/B_0$, the criterion for confinement in the bottle is (from (3.62))

$$\sin \alpha_0 > \frac{1}{R_{mir}^{1/2}} = \left(\frac{B_0}{B_{max}} \right)^{1/2}. \qquad (3.63)$$

All particles whose pitch angles at $z = z_0$ are less than $\sin^{-1}(R_{mir}^{-1/2})$ can escape from the bottle, whereas all particles whose pitch angles exceed this are confined to the bottle. Consequently, if we start out with an *isotropic distribution* of particles at $z = z_0$ (that is, equal particle fluxes in all directions), then a short time later the distribution will be missing all particles that do not satisfy Equation (3.63), as illustrated in Figure 3.10. The missing portion of the distribution is called the *loss cone*.

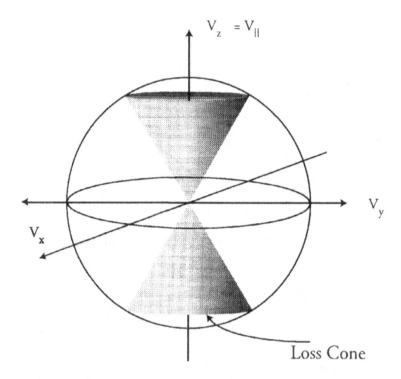

Figure 3.10. Schematic showing the loss cone where particles are missing from an otherwise isotropic distribution.

Particle distributions were introduced in Chapter 2 and will be discussed further in Section 10 of Chapter 3.

A particle confined in a magnetic bottle bounces back and forth with a *bounce period* T_b given by the following integral over a "bounce cycle":

$$T_b = \oint \frac{dz}{v_\|(z)} = 4 \int_{z_0}^{z_{max}} \frac{dz}{v \left[1 - \dfrac{B(z)}{B_0} \sin^2 \alpha_0 \right]^{1/2}}, \tag{3.64}$$

where v is the total particle speed. Equation (3.62) was used to obtain the second part of Equation (3.64). We can now define another invariant of the motion because the bounce motion is periodic. Let the generalized coordinate be the distance $q_{gen} = z$ along the field, in which case the relevant conjugate momentum is then $p_{gen} = mv_\|$. From the general equation for the action, (3.45), we can define the *second (or longitudinal) adiabatic invariant* as

$$J_L = m \oint v_\| dz = 4mv \int_{z_0}^{z_{max}} \left[1 - \frac{B(z)}{B_0} \sin^2 \alpha_0 \right]^{1/2} dz. \tag{3.65}$$

where the second part of the equation assumes a symmetric bottle. J_L is an adiabatic invariant if the shape of the magnetic bottle changes slowly (that is, adiabatically) in comparison with the bounce period T_b. In general, to keep J_L invariant as the size of the bottle is slowly changed, both v and α_0 must change; that is, the total kinetic energy of the particle does not remain constant. Why? The answer can be found by considering Equation (3.48) (Problem 3.7).

We now consider a simple example of bounce motion.

Example 3.3 (Simple bounce motion – Fermi acceleration) Suppose that B is constant along virtually the entire length of a magnetic bottle, except right at the ends. Charged particles are reflected at the end points of the bottle. The longitudinal invariant J_L from Equation (3.65) can be simply approximated in this case as

$$J_L \cong 4m v_{\|0} z_{\max}. \tag{3.66}$$

The length of the magnetic bottle is $2z_{\max}$. Suppose that the ends of the bottle slowly approach each other; that is, the length of the bottle shrinks. Equation (3.66) states that $v_{\|0}$ must increase as $2z_{\max}$ decreases, in such a way as to keep the longitudinal invariant constant. Since $v_{\perp 0}$ does not change owing to the constancy of μ_m, the particle energy $W = (1/2)m(v_{\|0}^2 + v_{\perp 0}^2)$ must increase. The longitudinal invariant explains why the parallel velocity increases in a shrinking magnetic bottle for time spans long in comparison with the bounce period. On a shorter time scale it can also be shown that a particle gains or loses energy from a single encounter with a moving magnetic mirror. If the magnetic mirror is moving at a velocity v_{mir}, then a particle after reflection has a parallel velocity $v_{\|0} = v_{\|0} + 2v_{\mathrm{mir}}$, where $v_{\|0}$ was the parallel velocity of the particle prior to its encounter (or "collision") with the moving mirror. The particle gains energy from each bounce if it is confined between two converging magnetic mirrors.

This method of accelerating a particle is called *Fermi acceleration*. Enrico Fermi suggested that cosmic rays are accelerated by this mechanism in the interstellar medium due to "collisions" with interstellar clouds. Of course, if the length of the magnetic bottle increases rather than decreases, then from Equation (3.66), $v_{\|0}$ decreases and the particle energy decreases. If we consider a medium containing a large number of scatterers (i.e., mirrors) moving in all directions, a particle will gain and lose energy in a stochastic manner. If an initial collection or distribution of particles starts out with a monoenergetic distribution, then over a period of time some particles will gain energy and some will lose energy, so that the particle distribution at later times will "spread out" or "diffuse" in energy. This mechanism is called *second-order Fermi acceleration*.

3.8 The dipole magnetic field

The most important example of a magnetic bottle field configuration in space physics is the Earth's magnetic field, which can be closely approximated as a magnetic dipole with its magnetic north pole located near the geographic south pole.

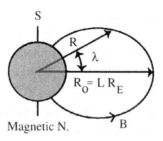

Dipole Magnetic Field Line

Figure 3.11. Dipole magnetic field line. The Earth's magnetic north pole is located in the southern polar cap at a geographic latitude of 78°S.

The components of the magnetic flux density of the terrestrial magnetic dipole in spherical coordinates are given by

$$B_R(R, \lambda) = -\frac{M_E}{R^3} 2 \sin \lambda,$$

$$B_\lambda(R, \lambda) = \frac{M_E}{R^3} \cos \lambda \tag{3.67}$$

Here R is the radial distance from the center of the Earth, and λ is *magnetic latitude*. The coordinate system used is essentially the same as spherical polar cooridnates but with polar angle (in radians) $\theta = \pi/2 - \lambda$. Magnetic latitude and geographic latitude typically differ by about 20°, depending on the longitude, because the Earth's magnetic dipole moment vector, M_E, is not exactly aligned with the Earth's rotation axis but is tilted by approximately 20° with respect to the Earth's rotation axis. $M_E = B_E R_E^3$ is the *dipole moment of the Earth*, where $B_E = 0.32 \times 10^{-4}$ T is the field strength at the equator at the Earth's surface where the radial distance is $R = R_E = 6{,}500$ km. The field strength associated with Equation (3.67) is

$$B(R, \lambda) = \frac{M_E}{R^3}(1 + 3 \sin^2 \lambda)^{1/2}. \tag{3.68}$$

The formula describing a magnetic field line is given by

$$R = R_0 \cos^2 \lambda, \tag{3.69}$$

where R_0 is the radial distance to the field line at the equator ($\lambda = 0$). It can be seen that R_0 identifies a particular field line (see Figure 3.11). The surface of rotation of a field line about the magnetic dipole axis is called an *L-shell* and is identified by the value $L = R_0/R_E$. The definition of the *McIlwain L-shell* is not quite this simple for the real geomagnetic field, which is not exactly dipolar.

The magnetic latitude where a field line encounters the Earth's surface is called the *invariant latitude*: $\Lambda = \lambda$ (at $R = R_E$). From Equation (3.69) and the definition of L-shell, we can see that an L-shell with value L intersects the Earth at an invariant latitude given by

$$\cos^2 \Lambda = \frac{1}{L}. \tag{3.70}$$

Thus $\Lambda = 45°$ for $L = 2$, $\Lambda = 60°$ for $L = 4$, and $\Lambda = 71.6°$ for $L = 10$. High L-shell values correspond to high invariant latitudes.

The incremental distance, ds, along a particular field line is given in terms of magnetic latitude by

$$\frac{ds}{d\lambda} = L R_E \cos \lambda (1 + 3 \sin^2 \lambda)^{1/2}, \tag{3.71}$$

and the field strength as a function of magnetic latitude along a given field line is given by

$$B(L, \lambda) = \frac{B_E}{L^3} \frac{(1 + 3 \sin^2 \lambda)^{1/2}}{\cos^6 \lambda}. \tag{3.72}$$

Clearly, B is a minimum at the magnetic equator and increases with increasing values of $|\lambda|$. That is, the geomagnetic field has a magnetic bottle configuration!

3.9 Charged particle motion in a dipole field

We can apply the understanding of charged particle motion in a magnetic field that we obtained earlier in this chapter to particle motion in the terrestrial dipolelike field. The gyroperiod (T_g) of a charged particle is always less than the time scale over which the magnetospheric magnetic field changes significantly, and thus the first adiabatic invariant, given by Equation (3.17), is indeed an invariant of the motion. Equation (3.62) can then be used together with Equation (3.72) to determine the pitch angle of a particle as a function of magnetic latitude λ:

$$\sin^2 \alpha = \sin^2 \alpha_0 \frac{(1 + 3 \sin^2 \lambda)^{1/2}}{\cos^6 \lambda}, \tag{3.73}$$

where α_0 is the particle's pitch angle at the magnetic equator. The mirror point occurs where $\sin \alpha = 1$, and expression (3.73) can be used to determine the magnetic latitude, λ_M, of this location in terms of α_0.

Charged particles are trapped in the magnetic bottle formed by the geomagnetic field. The bounce period (for a "round trip") can be found by writing Equation (3.64) in terms of λ instead of the distance along the field $s = z$:

$$T_b = 4 \int_0^{\lambda_m} \frac{1}{v_\parallel} \left(\frac{ds}{d\lambda} \right) d\lambda$$

$$\cong \frac{4 L R_E}{v} (1.30 - 0.56 \sin \alpha_0), \tag{3.74}$$

Table 3.2. *Time periods at $L = 4$ in the magnetosphere*
for charged particles with $\alpha_0 = \pi/2$ radians[a]

Particle	T_g(gyration)	T_b(bounce)	T_D(drift)
1 keV electrons	7.4×10^{-5} s	4.0 s	184 hrs
100 keV electrons[b]	8.8×10^{-5} s	0.46 s	1.5 hrs
1 keV protons	0.14 s	172 s	184 hrs
100 keV protons	0.14 s	17.2 s	1.8 hrs

[a] Adapted from Lyons and Williams (1984).
[b] These values were calculated using relativistically correct
formulae, which are slightly different than the nonrelativistic
formulae given in this chapter.

where expression (3.71) for $(ds/d\lambda)$ and an approximation to Equation (3.73) were
used by Hamlin et al. (1961) to derive the second part of Equation (3.74) (cf. Lyons
and Williams, 1984). Typical values of T_g and T_b for the terrestrial magnetosphere
are given in Table 3.2. Note that the bounce periods greatly exceed the gyroperiods
for both electrons and ions ($T_b \gg T_g$).

We saw earlier that charged particles also drift perpendicular to the magnetic
field if gradients are present in the field or if electric fields are present. However, if
the drift time is much less than a bounce period, then the longitudinal invariant J_L,
given by Equation (3.65), remains valid. In magnetospheric physics, an equivalent
invariant, I, is often used in place of J_L:

$$I = \frac{J_L}{2mv} = \int_{s1}^{s2} \left(1 - \frac{B}{B_M}\right)^{1/2} ds, \tag{3.75}$$

where $s1$ and $s2$ are the distances along the field line from the equator to the two
mirror points and where $B_M = B_0/\sin^2 \alpha_0$ is the field strength at the mirror points.
I remains constant as a charged particle drifts around the Earth, "defining" the
L-shell surface. Yet another adiabatic invariant, K, is sometimes used in space
physics and is closely related to I and J_L.

Now let us find the drift speed of a charged particle in the geomagnetic field. Both
$\mathbf{E} \times \mathbf{B}$ and gradient/curvature drifts are present in the magnetosphere, but $\mathbf{E} \times \mathbf{B}$ drifts
can be neglected for the energetic ring current – radiation belt – plasma in the inner
magnetosphere (i.e., for particle energies greater than about 1 keV). However, $\mathbf{E} \times \mathbf{B}$
motion is very important for colder plasma as will be described later in Chapter 8.
Equation (3.44) for the drift velocity can be transformed into an equation containing
the term $\mathbf{B} \times \nabla B$, the pitch angle, and the total kinetic energy $E_T = mv^2/2$:

$$\mathbf{v}_B = \frac{E_T}{qB^3}(1 + \cos^2 \alpha)\mathbf{B} \times \nabla B. \tag{3.76}$$

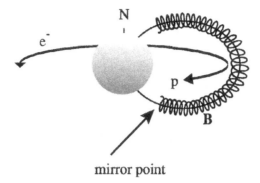

Figure 3.12. Particle motion in the geomagnetic field. Gyration, bounce, and drift motions are illustrated.

Note that the direction (i.e., the sign) of the drift velocity depends on the sign of the charge q. For a dipole field, we can write

$$\frac{\mathbf{B} \times \nabla B}{B^3} = -\hat{\phi} \frac{3 \cos^5 \lambda}{L R_E B_0} \frac{1 + \sin^2 \lambda}{(1 + 3 \sin^2 \lambda)^2}. \tag{3.77}$$

Positively charged particles such as protons drift westward azimuthally and negatively charged electrons drift eastward (Figure 3.12). Recall that B_0 is the field at the magnetic latitude $\lambda = 0$ and at a radial distance $R = R_0 = L R_E$. The drift speed clearly depends on the magnetic latitude of the particle, but it is often useful to average expressions (3.76) and (3.77) over a bounce period to obtain a bounce-averaged drift speed, $\langle v_B \rangle$, which is independent of the latitude. We will not carry out this average here but state the result: $\langle v_B \rangle \cong [6E_T/(q B_0 L R_E)] (0.35 + 0.15 \sin \alpha_0)$ (cf. Lyons and Williams, 1984).

The time required for a charged particle to drift once around the Earth is given by the circumference divided by the drift speed, or $T_D = 2\pi L R_E/|\langle v_B \rangle|$. Typical values of the drift period T_D are given in Table 3.2. A *third adiabatic invariant* is associated with this periodic azimuthal drift (Walt, 1994). You should note that there is a clean separation of the periods of the three types of periodic charged particle motion in the inner magnetosphere of the Earth (gyration, bounce, and drift motions):

$$T_g \ll T_b \ll T_D. \tag{3.78}$$

The same types of energetic particle motion are also known to be present in the inner magnetospheres of Jupiter, Saturn, Uranus, and Neptune.

3.10 The radiation belt and ring current – The inner magnetosphere

In this section, we apply the principles of single particle motion to the charged particle populations found in the inner magnetosphere. Other aspects of magnetospheric physics will be dealt with in Chapter 8. Several particle populations exist in the Earth's inner magnetosphere ($R < 10 R_E$):

(1) *Cold, thermal plasma*, with particle energies of 100 eV or less, is really the magneto-
    spheric extension of the ionosphere. The motion of particles in this plasma population
    is largely determined by $\mathbf{E} \times \mathbf{B}$ drift for directions orthogonal to the geomagnetic field
    and by diffusion for motions parallel to the field. The $\mathbf{E} \times \mathbf{B}$ motion in the innermost
    magnetosphere ($L < 5$ typically, although this "boundary" L-value varies with the
    level of geomagnetic activity) is usually such as to cause the cold plasma to corotate
    with the Earth in solid body – like fashion (as will be described in Chapter 8). How-
    ever, for greater radial distances ($L > 5$), the $\mathbf{E} \times \mathbf{B}$ drift motion associated with the
    magnetospheric *convection electric field* (again, this is considered in Chapter 8) carries
    this cold plasma to the dayside magnetopause where it is lost.

(2) *Ring Current plasma*, with particle energies ranging from about 100 eV up to several
    hundred keV, is "injected" into the inner magnetosphere from the magnetotail during
    magnetic storms. This particle population makes the major contribution to the particle
    energy density in the inner magnetosphere. The motion of these particles is primarily
    determined by gradient and curvature drifts associated with the geomagnetic field
    rather than by $\mathbf{E} \times \mathbf{B}$ drift motion. The ring current contains protons, O^+ ions, and
    He^+ ions.

(3) *Trapped radiation belt* particles are the high-energy extension of the ring current plasma
    population and have energies in excess of about 1 MeV for protons and in excess of
    tens of keV for electrons. The discovery of the radiation belt particle populations was
    made by James Van Allen and his University of Iowa research group in 1958 using
    Geiger tubes on board the *Explorer I* satellite (see the book by Van Allen cited in
    Chapter 1). The trapped radiation belt is often called the *Van Allen belt.*. The higher
    energy portion of the radiation belt population is produced by the *CRAND mechanism*
    (CRAND stands for Cosmic Ray Atmospheric Neutron Decay, as will be discussed
    later in this chapter).

Both ring current particles and the more energetic radiation belt particles are
trapped in the magnetic bottle created by the Earth's magnetic field configuration.
These particles also drift around the Earth due to grad-B and curvature drift motions.
Both particle populations have *loss cone* type distributions in pitch angle (see
Figure 3.10), because if a charged particle has a mirror point deep enough in the
atmosphere it is lost (as far as the magnetosphere is concerned) via collisions either
with atmospheric neutrals or with ionospheric ions and electrons. For example, at
$L = 4$ if all particles whose mirror points are below an altitude of about 1,000 km
(the exact altitude is not critical) are lost via collisions, then the distribution function
of these particles at the magnetic equator ($\lambda = 0$) has a loss cone starting at an
approximate pitch angle of $\alpha_{lc} = 7°$ (see Problem 3.12). Particles cannot be trapped
if the magnetic field lines they are associated with are "open," meaning connected
to the interplanetary magnetic field, in which case the magnetic field is *not* in a
"bottlelike" configuration. Trapping requires "closed" field lines that connect to
the Earth at both ends. Field lines at large L-shells (equivalent to large invariant
latitudes) are open during favorable solar wind conditions, as will be discussed
in Chapter 8. Furthermore, energetic particles that are nominally "trapped" on

field lines that are closed on the nightside do not remain trapped if the particles have drifts that carry them to the magnetopause on the dayside. This particle loss typically happens for $L > 10$ (which is no longer in the "inner" magnetosphere) for the energetic particles in the ring current and radiation belt populations, which mainly undergo curvature and grad-B drift, but this loss happens for $L > 5$ for the colder plasma, whose particle motion is controlled by $\mathbf{E} \times \mathbf{B}$ drift.

The sources and sinks of trapped particles in the inner magnetosphere will now be discussed. First, the loss processes for both ring current and radiation belt particles are discussed; then in separate subsections some characteristics of these two particle populations and their sources are discussed.

3.10.1 Loss processes for energetic particles in the inner magnetosphere

Several loss processes are relevant for trapped ring current or radiation belt particles. The four major ones are described next.

3.10.1.1 Loss due to collisions with neutrals

Charge transfer collisions between an energetic particle species A^+ and an atmospheric neutral species B is the most important process of this type and was represented by Equation (2.67): $A^+ + B \rightarrow A + B^+$. After the collision, virtually all the momentum of the fast particle A^+ is carried off by the neutral product A, leaving only a slow ion B^+. Thus, the energetic particle population suffers a loss. An important example of this process is the collision of a fast proton with an H atom found in the extended *hydrogen exosphere* or *geocorona* that surrounds the Earth:

$$H^+{}_{\text{fast}} + H \rightarrow H_{\text{fast}} + H^+. \tag{3.79}$$

The cross section for this reaction is roughly $\sigma_{\text{ct}} \approx 10^{-15} \text{ cm}^2$ for proton energies less than about 50 keV, but for $E > 50$ keV σ_{ct} decreases rapidly with increasing energy E, with $\sigma_{\text{ct}} \approx 10^{-19} \text{ cm}^2$ by $E = 300$ keV. The cross section for charge transfer collisions of protons with neutral O atoms is also about 10^{-15} cm^2 for $E < 50$ keV or so, as is the cross section for fast O^+ colliding with H or O.

The lifetime of a particle against loss by charge exchange or transfer collisions can be written as

$$\tau_{\text{ct}} = \frac{1}{n_n v \sigma_{\text{ct}}}, \tag{3.80}$$

where n_n is the density of the relevant neutral species and v is the particle speed. This collisional lifetime is obviously very short where n_n is high, as in the atmosphere itself. Particle losses in the atmosphere itself result in the formation of a loss cone. However, in this section we are interested in particle losses out in the magnetosphere itself, in which case the most abundant neutral species is the atomic hydrogen populating the geocorona. This loss also operates for particles not in the loss cone. Near a radial distance of a few R_E, the neutral H density is $n_H \approx 3 \times 10^2$ cm^{-3}. For ring current ions we find from Equation (3.80) that the charge exchange lifetime τ_{ct} is between several hours and one day. However, because σ_{ct} is so much smaller for the more energetic radiation belt protons, we find for this population that $\tau_{ct} \approx$ several years.

The energetic electrons in the radiation belt can also collide with neutrals, but the cross section is small at high energies and the lifetime is very long for particles outside the loss cone.

3.10.1.2 Loss via Coulomb collisions

A population of cold plasma (ions and electrons) coexists with the energetic particle populations in the inner magnetosphere (i.e., this cold plasma region, called the *plasmasphere*, will be discussed in Chapter 8). Coulomb collisions between fast and slow charged particles can lead to energy loss for the more energetic species. A formula for the collision frequency (ν_{st}) for Coulomb collisions between species s and t was given by Equation (2.68) as a function of the average relative speed between the species. For energetic particles colliding with cold particles the relative speed is well approximated as the speed of the energetic species; hence, in Equation (2.68), we can use $\langle g_{st}^3 \rangle = v^3$. The Coulomb collision time is the inverse of the collision frequency, $\tau_{Coul} = 1/\nu_{st}$, which is thus proportional to v^3. The collision time, τ_{Coul}, is inversely proportional to the number density n_t of the target species. The Coulomb collision lifetime increases strongly with energy due to the v^3 velocity dependence. For ring current plasma, however, charge exchange loss dominates except for ring current O$^+$ ions with energies less than 30 keV, for which Coulomb energy loss dominates. The Coulomb collision loss process is also more important than the charge exchange process for proton energies greater than a few hundred keV, because even though τ_{Coul} increases as v^3, τ_{ct} increases even more rapidly as a function of v at high enough energies. Coulomb collisional loss constitutes the main loss process for radiation belt protons.

The thermal plasma density in the plasmasphere (located typically where $L < 5$) is of the order of 10^3 cm^{-3}. Using this value for n_t we can find that τ_{Coul} is about several years for a 100 keV proton. Coulomb scattering is also an important loss process in the innermost magnetosphere ($L < 1.7$) for radiation belt electrons with MeV energies.

3.10.1.3 Loss via pitch-angle scattering into loss cone by wave–particle interactions

A charged particle whose pitch angle (at the magnetic equator) lies outside the loss cone is trapped, but if its pitch angle is somehow changed so that it is within the loss cone ($\alpha < \alpha_{lc}$), then the particle is lost to the distribution and from the magnetosphere. Collisions (or scattering) can suddenly change the direction of a particle's motion and thus its pitch angle. Note that the first adiabatic invariant (i.e., the magnetic moment μ_m) is violated during the scattering process. Coulomb collisions are not very efficient at changing the direction of an energetic particle by a large amount; instead, this process results in the gradual loss of energy. Charge exchange collisions entirely remove the energetic charged particle rather than simply changing the particle direction. However, charged particle motion can be affected by small-scale variations in the electric and magnetic fields. These field variations are usually associated with *wave* motion of some type in the background magnetospheric plasma. Small-scale in this case means scales greater than the microscopic scale associated with collisions, yet much smaller than the scale of the magnetosphere. For variations (or waves) in **E** and **B** to be able to alter the first adiabatic invariant of a charged particle, their scale size (i.e., wavelength λ) should be of the order of the gyroradius r_L or their time scale (i.e., wave period) should be of the order of the gyroperiod.

Wave–particle interactions can alter both the pitch angle and energy of a charged particle. An extreme example of this was discussed in Section 3.7 when second-order Fermi acceleration was introduced. Typically, the "scattering" of a charged particle by waves in the plasma happens in a random, or stochastic, manner, and results in *pitch-angle scattering or diffusion* and/or in *energy diffusion*. *Spatial diffusion* of a charged particle can also result from wave–particle interactions. We will discuss wave–particle interactions in a little more detail in the next section, but Schultz and Lanzerotti (1974) or Walt (1994) can be consulted for an in-depth discussion of this topic.

Wave–particle interactions leading to pitch-angle diffusion into the loss cone, and the consequent loss of particles from the inner magnetosphere, provide the most important loss process for radiation belt electrons. The most important plasma wave mode, with wavelengths of the order of a radiation-belt electron gyroradius (e.g., $r_L \approx 20$ km for an electron energy of ≈ 1 MeV) is the *whistler wave mode*. Whistler waves in the inner magnetosphere are electromagnetic waves that are produced by lightning and propagate along magnetic field lines (in what are called "ducts"), but these waves can also be produced locally in the magnetospheric plasma by a microscopic plasma instability associated with the loss cone distribution function (see Figure 3.10). Lyons and Williams (1984) have an extensive discussion of the effects of wave–particle interactions on particle distribution functions in the inner magnetosphere. The lifetime of a radiation belt particle for this process, τ_{waves}, is quite difficult to evaluate theoretically, but estimates were made from the measurements in the late 1950s and early 1960s of the decay of the artificial radiation belts created by high-altitude nuclear explosions, code-named *Starfish*, carried out by the United States and also by the explosions carried out by the Soviet Union.

Charged particles that were created both directly and indirectly (e.g., radioactive decay products) by these explosions populated different parts of the inner magnetosphere, as determined by the invariant latitude (i.e., L-shell) of each of the explosions. Significant enhancements of both the electron and proton components of the radiation belts were created by these explosions, and observations of the subsequent decay of the artificial components of the radiation belts provided estimates of the loss times as well as of spatial transport and diffusion times. Hess (1968) (see the bibliography at the end of the chapter) contains an excellent discussion of this topic and also of the radiation belts in general. The lifetime of a 2 MeV electron was shown to be $\approx$30 days at $L = 1.2$, but this lifetime rapidly increases with increasing L-value to a maximum of about 800 days (or $\approx$3 years) at $L = 1.5$. The lifetime then decreases again to $\approx$30 days by $L = 2$ and remains at that value out to $L \approx 4$. Theoretical work has demonstrated that Coulomb scattering can account for the lifetime in the region $L < 1.7$, but that wave–particle interactions must be invoked to explain the lifetime for larger L-values.

3.10.1.4 Loss via convection and drift

Particle loss from the magnetosphere due to particle drifts or convection into the magnetopause is not an important process for energetic particles in the inner magnetosphere. This loss process, however, is the most important one for cold ($E < 100$ eV) plasma, but we will defer consideration of it to Chapter 8. There does exist a loss of energetic particles associated with particle drift into the *South Atlantic Anomaly (SAA)*. The geomagnetic field is not a perfect magnetic dipole; in particular, the field strength in the vicinity of the South Atlantic is relatively low. Hence, particles mirror deeper in the atmosphere in this region and can be lost via collisions more readily. As particles drift around the Earth into this region the loss cone widens, and particles previously outside the loss cone find themselves within this *drift loss cone* and are lost from the radiation belts.

3.10.2 Radiation belt electrons and protons

Ions and electrons exist in the magnetosphere with energies ranging from a few electron volts up to hundreds of MeV. The most energetic protons (energies $E > 1$ MeV) and electrons ($E > 50$ keV) are created as a by-product of the interaction of galactic and solar cosmic rays with the terrestrial atmosphere. Energetic cosmic ray protons collide with the nuclei of air molecules and produce neutrons that travel out to the magnetosphere where they undergo β decay, producing energetic protons and electrons. The neutron, being neutral, is unaffected by the geomagnetic field and is able to freely traverse the magnetosphere, but its charged energetic decay products are trapped by the geomagnetic field and are the main source of the radiation belt population. This is called the CRAND mechanism. Energetic radiation belt particles undergo gyration, bounce, and drift motions, as discussed earlier in this chapter, and remain trapped until they are lost from the radiation belts due to one of the loss processes discussed earlier.

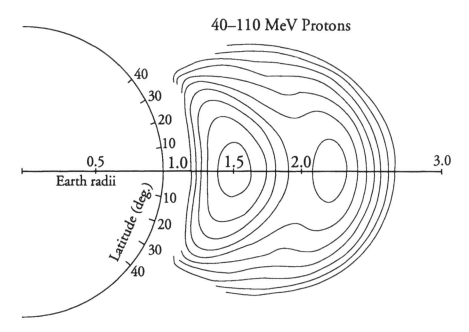

Figure 3.13. Contours of omnidirectional fluxes [J_0 has units of $\mathrm{cm^{-2}\,s^{-1}}$] of energetic protons with energies between 40 and 110 MeV as measured by the *Explorer 15* satellite in the inner magnetosphere. The outermost contour has $\log_{10}(J_0) = 2.3$, and the contours increase by 0.2 increments so the innermost contour near 1.5 R_E has $\log_{10}(J_0) = 4.1$. Adapted from McIlwaine (1961).

Figure 3.13 shows the measured spatial distribution of magnetospheric protons with energies in excess of 40 MeV. As discussed in the last subsection, protons with energies this high can survive in the inner magnetosphere for many years. The fluxes of the higher energy protons are most intense in two belts – the *inner belt* near $L \approx 1.5$ and the *outer belt* at $L > 2$. Typical fluxes of protons with energies in excess of 40 MeV are about $10^4 \, \mathrm{cm^{-2}\,s^{-1}}$ in the equatorial plane near $L = 1.5$. Proton fluxes at $L = 1.5$ for energies in excess of 4 MeV are about 200 times greater than this (for a very detailed description of the observations see the book by Hess cited in the bibliography at the end of the chapter). A figure is not included for radiation belt electrons (again see Hess if you are interested), but the omnidirectional flux of electrons with energies greater than 1.9 MeV is about $10^8 \, \mathrm{cm^{-2}\,s^{-1}}$ at $L \approx 1.5$, where the lifetime is greatest, and is $\approx 2 \times 10^6 \, \mathrm{cm^{-2}\,s^{-1}}$ for $L > 2$, where the lifetime is much shorter.

3.10.3 Ring current

The population of particles with energies less than a few hundred keV or greater than a few hundred eV is called the ring current. As we saw in the last section, the lifetime of ring current ions (a day or so) is much shorter than for radiation belt particles (i.e., months and years) mainly as a consequence of charge exchange

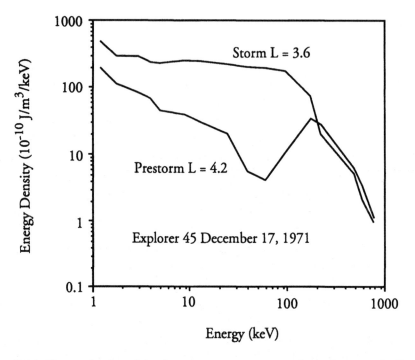

Figure 3.14. Energy density of the ring current versus energy for both magnetically quiet and disturbed times. Adapted from Lyons and Williams (1984).

collisions of ring current ions with exospheric hydrogen. The greater portion of the particle energy density is contained in the ring current population rather than in the colder thermal plasma or in the more energetic radiation belt population. However, most of the total ion number density is accounted for by the cold plasma that will be discussed in Chapter 8. Figure 3.14 shows the differential energy density of ring current plasma at $L \approx 4$ as a function of particle energy at two times – during a magnetically quiet time period (prestorm) and during a magnetic storm. The ring current is typically most intense near L-shells of about 4. Clearly, the energy density is greater during the storm period than during the quiet time period. The ring current plasma varies considerably with time unlike the stable radiation belt plasma. In fact, we can see in Figure 3.14 that the prestorm and storm energy densities are essentially the same for $E > 300$ keV (an energy close to radiation belt energies).

Ring current ions are "injected" into the inner magnetosphere from the *plasmasheet* region during geomagnetic storms. This region of the magnetosphere is located in the magnetotail near the magnetic equator beyond $L \approx 8$. The particle density and pressure are relatively high in this region, in comparison with the plasma found in *tail lobes*, which are located outside the plasmasheet. The ion thermal population in the plasmasheet has an average thermal energy of several keV. These magnetospheric regions will be discussed in Chapter 8. During magnetically active times the large-scale electric field in the magnetosphere (the *convection electric field*) becomes much larger, and the tail plasma $\mathbf{E} \times \mathbf{B}$ drifts toward the inner magnetosphere. If we suppose that a 5 keV proton initially starts at $L = 10$ and then is moved (or drifts) to $L = 5$ with its magnetic moment remaining invariant, then its perpendicular energy must increase by a factor of 8 because the magnetic field strength increases as L^3 for a dipole field. Hence the proton energy ends up as 40 keV, which is a ring current energy. The field at $L \approx 10$ is not quite dipolar due to the electrical currents flowing in the magnetosphere, but the principle remains the same. Once this 40 keV ring current ion has been created it is energetic enough for magnetic drifts to become more important than the $\mathbf{E} \times \mathbf{B}$ drift, and the $\mathbf{v}_B$ drift carries the particle in an almost circular drift "orbit" around the Earth. In this section, the motion of ring current ions will be described using the principles of single particle motion introduced earlier in this chapter.

Positive ions undergo gradient and curvature drift westward, whereas electrons drift eastward; hence, the plasma produces a westward-directed current of electrical current. The westward-directed "ring" of current circling the Earth in the inner magnetosphere is called the *ring current*. The plasma population most responsible for this current is itself often referred to simply as the ring current. Equations (3.76) and (3.77) can be used to determine the drift velocities of ring current electrons and ions. The current density associated with these drifts is given by $\mathbf{J}_D = e\{n_i \langle \mathbf{v}_{Bi} \rangle - n_e \langle \mathbf{v}_{Be} \rangle\}$, where n_j is the number density of ring current electrons ($j = e$) or ions ($j = i$) and where the drift velocities are averaged over the distribution functions of electrons and ions, respectively. The electron contribution to $\mathbf{J}_D$ is very small and will be neglected in the following derivation.

It is convenient to define parallel and perpendicular particle pressures (see Chapter 2 for a discussion of pressure):

$$p_\| = nm\langle v_\|^2 \rangle,$$
$$p_\perp = \tfrac{1}{2}nm\langle v_\perp^2 \rangle,$$

(3.81)

where n is the total ring current ion density and m is the ion mass. Strictly speaking, one should separately add the pressures due to ring current protons and ring current oxygen ions, whose relative densities depend on the level of geomagnetic activity, but we will not distinguish between ion species here. The averages in Equation (3.81) can be evaluated using the ring current ion distribution function. Using Equations (3.76) and (3.77) we find the current density to be

$$\mathbf{J}_D = \frac{\mathbf{B} \times \nabla B}{B^3}[p_\| + p_\perp].$$

(3.82)

Equation (3.81) is used to define the pressures in Equation (3.82). The bracketed term contains the total particle pressure, which is the same as the total ion energy density.

Expression (3.82) does *not* account for all the current density in the inner magnetosphere. Current density can be created – even for zero drift motion (that is, for stationary particle guiding centers) – by the gyromotion itself for a collection of charged particles. This current density is called the *magnetization current density* and is given by

$$\mathbf{J_M} = \nabla \times \mathbf{M}, \qquad (3.83)$$

where $\mathbf{M}$ is the magnetization vector of the plasma. This magnetization can be written as the average of the magnetization (or magnetic moment – see Equation (3.17)) of a single particle, $\mathbf{m}_B$, multiplied by the particle density, that is, with

$$\mathbf{m}_B = -\frac{mv_\perp^2}{2B}\hat{\mathbf{b}}, \qquad (3.84)$$

we have

$$\mathbf{M} = n\langle\mathbf{m}_B\rangle = -p_\perp\frac{\hat{\mathbf{b}}}{B}, \qquad (3.85)$$

and so

$$\mathbf{J_M} = \nabla \times \mathbf{M} = -\nabla \times \left[p_\perp\frac{\hat{\mathbf{b}}}{B}\right]. \qquad (3.86)$$

If the magnetic field is not strongly affected by the plasma (i.e., if the field from local currents is much less than the unperturbed geomagnetic field), then the vacuum field relation $\nabla \times \hat{\mathbf{b}} = \hat{\mathbf{b}} \times \nabla B/B$ (see Problem 3.9) can be used to rewrite expression (3.86) as

$$\mathbf{J_M} = -\frac{2p_\perp}{B^2}\hat{\mathbf{b}} \times \nabla B + \frac{\hat{\mathbf{b}} \times \nabla p_\perp}{B}. \qquad (3.87)$$

The first term of Equation (3.87) is the current associated with the charged particle orbits being crowded together on the inside of a curved field line. The second term of Equation (3.87) comes from gradients in the particle density (i.e., pressure), as illustrated in Figure 3.15. The total current density in the inner magnetosphere includes both the magnetization current density, given by Equation (3.87), and the drift current density, given by Equation (3.82):

$$\mathbf{J} = \frac{\hat{\mathbf{b}} \times \nabla p_\perp}{B} + (p_\parallel - p_\perp)\frac{\hat{\mathbf{b}} \times \nabla B}{B^2} \quad [\text{A/m}^2]. \qquad (3.88)$$

In Equation (3.88) a portion of the magnetization current cancels some of the current due to the gradient drift. For an isotropic particle distribution, $p_\parallel = p_\perp$, only the first term of Equation (3.88) contributes to the current density.

The ring current plasma is known to have a maximum energy density near $L = 3$–4 (that is, at a radial distance from the Earth of about 3 to 4 R_E), depending on the level of magnetic activity. Hence, $\nabla p_\perp$ points inward for L values

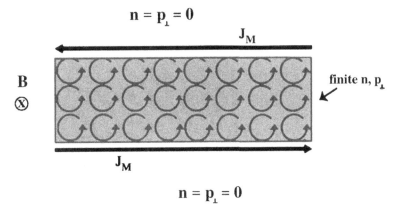

$$n = p_\perp = 0$$

Figure 3.15. Schematic of magnetization current density for a slab of plasma.

greater than about 4 and the current density from the first term of Equation (3.85) flows westward, whereas for lower L-values, $\nabla p_\perp$ points outward and this component of the current flows eastward. The total ring current is the current density integrated over the complete cross-sectional area of the inner magnetosphere. The westward component exceeds the eastward component so that the overall ring current is directed westward. The magnitude of the total ring current is typically several million amperes. The following example (and Problem 3.10) applies the above results to a simple "schematic" ring current.

Example 3.4 (Schematic ring current) Assume that the energetic plasma in the inner magnetosphere consists of monoenergetic 100 keV protons with a number density that is uniform (n_0) over a cross-sectional area defined by latitudes $\lambda = \pm\lambda_m$ with $\lambda_m = 0.5$ radians and by L-shells $L = 2$ and 4. The density is taken to be zero outside this area. Also assume that the proton distribution is isotropic ($p_\perp = p_\parallel$). The assumption of isotropy is actually inconsistent with the magnetic latitude limits just described, because the limited latitude extension implies the existence of a loss cone distribution with a growing cone angle as λ increases. However, we will overlook this inconsistency.

Only the first term of Equation (3.88) contributes to the current density in this case since $p_\parallel = p_\perp$ has been assumed. The current is produced by the density gradients at the inner and outer edges of our schematic ring current, and it is clear that current flows only in a thin sheet along the inner and outer edges of the populated region. The surface current density, given by Problem 3.10, is

$$K_L = \frac{\frac{1}{3}mv_0^2 n_0}{B(L, \lambda \approx 0)} \quad [\text{A/m}], \tag{3.89}$$

where v_0 is the speed of a 100 keV proton and where B was assumed to be always equal to its equatorial value at a given L-shell. For our simple scenario a surface current density appears at $L = 2$ and 4 (K_2 and K_4, respectively). As indicated in

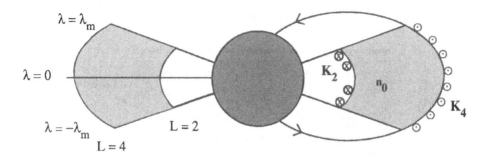

Figure 3.16. Schematic of a simplified ring current.

Figure 3.16, the current at $L = 2$ flows eastward and the current at $L = 4$ flows westward. Integrating K_L over magnetic latitude gives a total current flowing at the respective L-shell (also see Problem 3.10) of

$$I_L = \frac{2\lambda_m}{3B_E} mn_0 v_0^2 R_E L^4 \quad [\text{A}]. \tag{3.90}$$

The westward current (I_4) is a factor $4^4 = 256$ greater than the eastward current (I_2), and hence the total ring current ($I \cong I_4$) is westward. I_4 is equal to 5 MA for a ring current proton density of $n_0 = 10^7 \, \text{m}^{-3}$. Recall that $B_E = 0.32 \times 10^{-4} \, \text{T}$.

All electrical currents produce magnetic fields (see the appendix) as prescribed by Ampère's law. The magnetic field produced by a westward ring current near the vicinity of the Earth's surface is southward (and therefore directed opposite to the direction of the intrinsic geomagnetic field itself). The total field strength at the Earth's surface is thus reduced as a consequence of the ring current. The simple ring current postulated in the above example produces a geomagnetic field perturbation at the Earth's surface of about 130 nT, as can be demonstrated by the reader in Problem 3.10. This represents a reduction of about 0.4% in the equatorial surface field of the Earth. The field perturbation measured by magnetometers at the Earth's surface *near* the equator (but not right at the equator where an ionospheric current system called the *equatorial electrojet* can also produce magnetic field perturbations) is called the D_{st} index of geomagnetic activity. For our example, $D_{st} = -130$ nT (1 nT = 1 gamma), which is quite typical of geomagnetically active time periods, when the inner magnetosphere is filled with energetic plasma. During quiet times, the inner magnetosphere has a lower population of energetic particles and the D_{st} index might be only several nanotesla, positive as well as negative. Figure 3.17 displays the D_{st} index during a *geomagnetic storm* that took place on April 18, 1965. The D_{st} reached a value of -130 nT and took a couple of days to recover to its prestorm value. This recovery time is consistent with the charge-exchange loss time of ring current ions given by Equation (3.80) and discussed earlier. The few nanotesla positive excursion of the D_{st} that occurs at

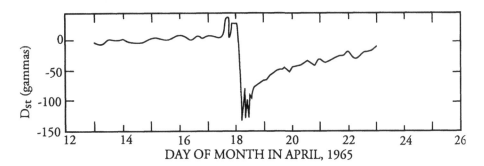

Figure 3.17. D_{st} index of geomagnetic activity during a geomagnetic storm on April 18, 1965. Adapted from Lyons and Williams (1984).

the start of the storm is called the *storm sudden commencement (SSC)* and will be explained in Chapter 7 as being caused by electrical currents flowing on the magnetopause. We will return to the subject of magnetic storms and geomagnetic activity later in the book.

3.11 Wave–particle interactions and quasi-linear diffusion

This section is rather outside the stated scope of this textbook in that it contains material beyond the simple fluid approach to space plasma physics. We will be concerned with how particle distribution functions respond to fluctuations and waves in the electric and magnetic fields, and this has several applications throughout the solar system including the evolution of radiation belt particle populations, as mentioned in the last section, and the transport of galactic cosmic rays in interplanetary space (taken up in Chapter 7). Furthermore, this topic goes beyond the motion of single particles and involves a consideration of a collection, or distribution, of particles, but the motion of a single particle in more complicated field configurations is an important part of this subject.

The concept of a distribution function of species s, $f_s(\mathbf{x}, \mathbf{v}; t)$, was introduced in Chapter 2, as was the Boltzmann equation, which can be solved for the distribution function. Examples of distribution functions in Chapter 2 included the beam, the shell, and the Maxwellian. And in the current chapter, we discussed the loss cone distribution. For a collisionless plasma, the Boltzmann equation becomes the Vlasov equation and no collision term exists. Nonetheless, "collisions" of a sort can be introduced into the Vlasov equation because of wave–particle interactions, as briefly discussed in Section 3.10.2 with respect to the loss of radiation belt electrons. One can convert the Vlasov equation to a diffusion equation, in which the diffusion coefficients depend on the wave electric and magnetic fields. It should be noted that these wave fields are still macroscopic, rather than being microscopic as in collisions, but still possess scale lengths or time periods that are small in comparison with the other macroscopic scales of the relevant region or phenomenon.

3.11.1 Quasi-linear theory

If the electric and magnetic fields vary only slowly, and if there are no collisions, then the adiabatic invariants of the motion introduced in this chapter remain valid even if the "single" particle is a member of a larger distribution of particles. However, these invariants can be violated by means of "real" collisions or by small-scale (or short time period) variations in the fields themselves. If these field irregularities are small in amplitude, then the departure of the particle motion from its "adiabatic motion" (e.g., gyration, bounce, and drift for the inner magnetosphere) also remains small and we can use a perturbative approach. This perturbative approach is called *quasi-linear theory*. In standard quasi-linear theory the linear theory of plasma waves (Krall and Trivelpiece, 1973) is employed to determine the growth rate and intensity levels of the plasma waves. However, an empirical approach is often taken in which one simply adopts parameters for the waves that agree with measurements. In either case, the particle distribution function is handled perturbatively, by splitting it into an unperturbed part (f_0), which would apply exactly if the waves did not exist, and a part (f_1) associated with the effects of plasma waves. Thus we have

$$f(\mathbf{x}, \mathbf{v}, t) = f_0(\mathbf{x}, \mathbf{v}, t) + f_1(\mathbf{x}, \mathbf{v}, t), \tag{3.91}$$

where $|f_1| \ll f_0$. Here we have dropped the explicit reference to the particle species s for simplicity.

The unperturbed distribution function depends on variables, related to $\mathbf{x}$ and $\mathbf{v}$, that are appropriate to the unperturbed single particle motion. For example, for particle motion in a smooth background magnetic field, $\mu = \cos \alpha$ and $v = |\mathbf{v}|$ are better variables than simply the Cartesian velocity vector $\mathbf{v}$. Another acceptable choice of variables would be $v_\parallel$ and $v_\perp$. Or, if the first adiabatic invariant μ_m is obeyed, one could also express f in terms of that variable rather than simply using $v_\perp$. Similarly, in the inner magnetosphere, if the longitudinal invariant K is obeyed, then an appropriate form of distribution function would be $f(L, \phi, \mu_m, K, t)$, where L and azimuthal angle ϕ were used instead of the Cartesian position vector $\mathbf{x}$ (and where one only needs to keep track of the distribution function at the magnetic equator). We also expand the electric and magnetic fields into smooth background parts, $\mathbf{E}_0$ and $\mathbf{B}_0$, and wave (or fluctuation) parts, $\mathbf{E}_1$ and $\mathbf{B}_1$:

$$\begin{aligned}
\mathbf{E}(\mathbf{x}, t) &= \mathbf{E}_0(\mathbf{x}) + \mathbf{E}_1(\mathbf{x}, t), \\
\mathbf{B}(\mathbf{x}, t) &= \mathbf{B}_0(\mathbf{x}) + \mathbf{B}_1(\mathbf{x}, t),
\end{aligned} \tag{3.92}$$

where we assume that $|E_1| \ll |E_0|$ and $|B_1| \ll |B_0|$. The $\mathbf{E}$ and $\mathbf{B}$ fields given by Equation (3.92) must obey Maxwell's equations, of course. The Lorentz force with fields given by Equation (3.92) can be split into two pieces: (1) an unperturbed part $q(\mathbf{E}_0 + \mathbf{v} \times \mathbf{B}_0)$ and (2) a smaller perturbation part that depends on $\mathbf{E}_1$ and $\mathbf{B}_1$.

Equation (3.91) and the Lorentz force with the fields given by Equation (3.92) can be substituted into the Vlasov equation (2.11). If we retain on the left-hand side only those terms that depend on the unperturbed quantities and gather on the right-hand side all other terms, we obtain a quite complicated equation. If we were to average this equation over a time scale long with respect to the wave motion but short with respect to the unperturbed motion, we would obtain the following equation for the time evolution of f_0:

$$\frac{\partial f_0}{\partial t} + \mathbf{v} \cdot \nabla f_0 + \frac{q}{m}(\mathbf{E}_0 + \mathbf{v} \times \mathbf{B}_0) \cdot \nabla_{\mathbf{v}} f_0 = \left(\frac{\delta f}{\delta t}\right)_{\text{wp}} + net\, source, \quad (3.93)$$

where a net source term has been added on so that we can include source and loss processes (including real collisions) if we wish. All the terms that involved f_1, E_1, or B_1 were incorporated (and time averaged) to obtain the wave–particle interaction term, $(\delta f/\delta t)_{\text{wp}}$, which describes how the unperturbed distribution function f_0 evolves in response to the waves. Equation (3.93) and the wave term $(\delta f/\delta t)_{\text{wp}}$ can take many different forms depending on the specific problem. In the next section, we consider a form of the wave term appropriate for *gyrotropic* (wherein particles can be found at all phase angles *of gyration with equal probability*) particle distributions. We will not repeat the derivation of these various forms of $(\delta f/\delta t)_{\text{wp}}$ here because they are lengthy and outside the scope of this text but will merely reproduce a few of the more important versions of this term.

3.11.2 Pitch-angle and energy diffusion

Earlier in this chapter we introduced the pitch angle of a particle (Section 3.2) and also mentioned the loss cone distribution function. In most space physics situations charged particle distributions are gyrotropic. In this case, we do not need to express the distribution function in terms of a full three-component velocity vector; $v_\parallel$ and $v_\perp$ alone are sufficient. Equivalently, we can use μ and v and express the distribution function as $f(s, \mu, v, t)$, where s is the distance along $\mathbf{B}_0$. We also identify f as being f_0. The quasi-linear diffusion equation (3.93) can also be transformed to a reference frame (i.e., to one moving "with" the plasma as will be discussed in Chapter 4) where the large-scale electric field is zero:

$$\frac{\partial f}{\partial t} + \mu v \frac{\partial f}{\partial s} + \frac{\mu_{\text{m}}}{mv} \frac{\partial B}{\partial s} \frac{\partial f}{\partial \mu} = \left(\frac{\delta f}{\delta t}\right)_{\text{wp}} + S - L, \quad (3.94)$$

where the net particle source as a function of μ, v, s, and t is denoted $S - L$ (production minus loss). The first two terms on the left-hand side of this equation comprise the convective derivation of f, where μv is just equal to $v_\parallel$. The terms proportional to $\partial B/\partial s$ represent the effect of a converging or diverging magnetic field (i.e., mirroring). Recall that the magnetic moment is μ_{m}.

A highly simplified form of the wave–particle interaction term for this type of distribution function is given by

$$\left(\frac{\delta f}{\delta t}\right)_{\text{wp}} = \frac{\partial}{\partial \mu}\left[D_{\mu\mu}\frac{\partial f}{\partial \mu}\right] + \frac{1}{v^2}\frac{\partial}{\partial v}\left[v^2 D_{vv}\frac{\partial f}{\partial v}\right], \qquad (3.95)$$

where $D_{\mu\mu}$ and D_{vv} are the *quasi-linear pitch-angle diffusion* and *energy diffusion coefficients*, respectively. These coefficients depend on the type of wave or fluctuation that are present and, in general, are quite complex. Nonetheless, they almost always can be put into the following form:

$$D_{\mu\mu} = G(\mu, v)\langle B_1^2\rangle \qquad (3.96)$$

and

$$D_{vv} = v_{\text{wave}}^2 D_{\mu\mu}, \qquad (3.97)$$

where $G(\mu, v)$ is a function, which we will leave unspecified for now, that depends on the details of the wave–particle interaction. The quasi-linear pitch-angle diffusion $D_{\mu\mu}$ is proportional to the average level of the fluctuations in the magnetic field, $\langle B_1^2\rangle$, but we could equally well express $D_{\mu\mu}$ in terms of the average level of the electric field fluctuations by using Maxwell's equations to convert B_1 to E_1. $D_{\mu\mu}$ has units of inverse time [s^{-1}]. The wave propagation speed is denoted v_{wave}, and the energy diffusion coefficient is proportional to v_{wave}^2. D_{vv} has units of [m^2 s^{-1}]. If the field fluctuations are stationary (i.e., nonpropagating, $v_{\text{wave}} = 0$), then $D_{vv} = 0$.

The effect of the $D_{\mu\mu}$ term in the quasi-linear equation – the combined Equations (3.94) and (3.95) – is to take an initial pitch-angle distribution and "smooth" it out in the variable μ. If $D_{\mu\mu}$ is large enough and enough time is allowed, then the limit of this "pitch-angle" diffusion is the isotropization of the distribution function f. One can see that if f is independent of μ, then $\partial f/\partial \mu = 0$, and the first term in Equation (3.95) is zero. Similarly, the effect of the D_{vv} term is to take an initial particle distribution and spread it out, or diffuse it, in the variable v. We can estimate how long these wave–particle processes take to operate by dimensionally analyzing the relevant terms in Equation (3.95), which provides us with the following *diffusion time* expressions:

$$\tau_{\mu\mu} \approx \frac{\Delta\mu^2}{D_{\mu\mu}} \approx \frac{1}{4D_{\mu\mu}} \qquad (3.98)$$

and

$$\tau_{vv} \approx \frac{\Delta v^2}{D_{vv}} \approx \frac{v^2}{v_{\text{wave}}^2 D_{\mu\mu}}, \qquad (3.99)$$

where a typical interval of μ we wish to diffuse over is $\Delta\mu \approx 0.5$ and a typical v interval is $\Delta v \approx v$. These diffusion times can be compared with other relevant time scales, such as loss times or the time for particle transport from one spatial region to another. Notice that the ratio of the energy diffusion time, τ_{vv}, to the pitch-angle diffusion time, $\tau_{\mu\mu}$, is $\tau_{vv}/\tau_{\mu\mu} = [v/v_{\text{wave}}]^2$. In many space physics applications, $v_{\text{wave}} \ll v$, and hence $\tau_{\mu\mu} \ll \tau_{vv}$; that is, energy diffusion is much slower than pitch-angle scattering.

Let us consider in a bit more detail the pitch-angle diffusion coefficient. Particles interact with some waves, or fluctuations, more easily than with others. There are many different wave modes (i.e., types of waves) that can exist in a plasma. The two most basic categories are *electrostatic* and *electromagnetic*. The propagation of high-frequency electromagnetic waves through a plasma is mentioned in the appendix. Electrostatic wave modes such as *Langmuir oscillations* and *ion-acoustic* waves will be discussed in Chapter 4. Very low frequency electromagnetic *MHD wave* modes will also be discussed in Chapter 4. The *dispersion relationship*, $\omega(\mathbf{k})$, characterizes the propagation properties of a specific wave mode, where ω is the angular frequency of the wave and $\mathbf{k}$ is the wavenumber with magnitude $k = 2\pi/\lambda$ (λ is the wavelength) and direction in the wave propagation direction. The phase speed of a wave is $V_{\text{phase}} = \omega/k$ and the group speed is given by $V_{\text{group}} = \partial\omega/\partial k$. For electromagnetic waves in a vacuum the dispersion relation is $\omega^2 = k^2 c^2$ and $V_{\text{phase}} = V_{\text{group}} = c$ (the speed of light). In a plasma, the dispersion relation for electromagnetic waves is $\omega^2 = k^2 c^2 + \omega_p^2$, where ω_p is the plasma frequency (see Chapters 2 and 4 and the appendix).

Waves of many different frequencies and wavenumbers can coexist in a plasma, as long as the appropriate dispersion relation is obeyed. In fact, in some situations the presence of highly non-Maxwellian particle distributions can lead to the "spontaneous" growth of waves (see Lyons and Williams, 1984, for a general discussion) whose energy is derived from the particle energy. This wave growth is called a *plasma instability*. The intensity of the waves as a function of the wavenumber (or frequency) and the nature of this function depend on the type of plasma instability operating. A convenient way of representing this functional dependence is with the *power spectrum* of the waves, or *power spectral density*. The power, which is usually denoted $P(k)$ and is a function of wavenumber, is the Fourier transform of a wave train versus distance (e.g., $B_1(\mathbf{x}, t)$) multiplied by its complex conjugate. $P(k)$ is normalized such that its integral over all k space is equal to the root mean square of the fluctuations, $B_{\text{rms}}^2 = \langle B_1^2 \rangle$.

Particles and waves interact most readily (either for wave–particle scattering or for wave generation by the plasma) when the following *resonance condition* is met (Lyons and Williams, 1984):

$$\omega - k_\parallel v_\parallel + n\Omega_s = 0. \tag{3.100}$$

$k_\parallel$ and $v_\parallel$ are the wavenumber and particle speed in the direction of the background magnetic field, respectively, Ω_s is the gyrofrequency of species s, and n is an integer. For $n = 0$, we have the *Landau resonance*, in which the parallel phase speed of the wave matches the parallel speed of the particle. The other values of n are associated with *cyclotron harmonic resonances*, with $n = -1$ being the *principle cyclotron resonance*. For the cyclotron resonance, a gyrating particle interacts with a rotating circularly polarized wave. A resonant particle experiences the same phase of the wave over a prolonged period of time, thus permitting a "strong interaction," whereas nonresonant particles experience a rapidly varying field so that the average force on the particle is small. The cyclotron resonances are relevant to electromagnetic waves and the Landau resonances to electrostatic waves. For a particle with given pitch angle and speed, the waves that contribute most to $D_{\mu\mu}$ are those that meet the resonance condition (3.100); $D_{\mu\mu}$ is proportional to the power of those resonant waves.

In Chapter 6, we will use the quasi-linear diffusion equation to describe cosmic ray propagation in the interplanetary medium. In the next section, we will consider the pitch-angle scattering of trapped radiation-belt particles.

3.11.3 Pitch-angle diffusion of trapped particles in the magnetosphere

Let us consider how the wave–particle interaction picture described in Section 3.11.2 applies to the inner magnetosphere. First, we are most interested in wave–particle interaction time scales that significantly exceed particle bounce times in the inner magnetosphere. Suppose that the spatial coordinate s, which is distance along a field line, is defined so that $s = 0$ refers to the magnetic equator. It makes sense in this case to worry just about the particle distribution function at the equator, $f(s = 0, L, \mu, v, t)$, where the L-shell has also been included as a variable. The distribution function at other positions along the field line, $f(s, L, \mu, v, t)$, can be simply determined from $f(s = 0, L, \mu, v, t)$, using Equation (3.94) without sources, sinks, wave–particle interaction term, and time derivative term. We assume for this purpose that all these neglected terms are only important for time scales long in comparison with the bounce period. Thus, we only need to know about the equatorial distribution function. For example, a distribution function that is fully isotropic (i.e., f is uniform in μ) at the equator remains isotropic everywhere along the field. And a distribution that is isotropic except for a loss cone remains isotropic outside the loss cone everywhere along the magnetic field with a loss cone angle that increases as s (and B) increases (you will demonstrate this in Problem 3.14).

We now transform the more general quasi-linear diffusion equation (3.94) to a diffusion equation for the equatorial distribution function alone:

$$\frac{\partial f}{\partial t} = \frac{\partial}{\partial \mu}\left[D_{\mu\mu}\frac{\partial f}{\partial \mu}\right] + S - L. \tag{3.101}$$

We have also neglected the energy diffusion term for simplicity, which is a reasonably good approximation in the inner magnetosphere. The diffusion coefficients and net source term in this equation are understood to be bounce averaged, and f is understood to be the equatorial distribution function. Often a cross-L-shell spatial diffusion term is relevant, but we consider this separately in the next section.

For the ring current or radiation belts, the source term must include all appropriate sources such as neutron decay (i.e., the CRAND mechanism), and the loss term should include the effects of charge transfer collisions or Coulomb collisions. These losses can be simply incorporated into Equation (3.101) by writing $L = -f/\tau_{\text{coll}}$, where τ_{coll} is the relevant collision time, as discussed earlier. If $D_{\mu\mu} = 0$ and $S = L = 0$, then the equatorial distribution function does not change in time and remains a loss cone distribution. However, if $D_{\mu\mu} \neq 0$, then loss of particles to the magnetosphere occurs even with $L = 0$ because of *diffusion into the loss cone*. For the outer planets, particle loss due to collisions with the natural satellites and rings of those planets should also be included in the loss term.

We can think of Equation (3.101) as a type of continuity equation in which the diffusion term appears as the divergence of a flux: $\partial f/\partial t = -\partial(Flux)/\partial\mu + S - L$. The particle flux (for the variable μ) is then given by

$$Flux = -D_{\mu\mu}\frac{\partial f}{\partial \mu}. \tag{3.102}$$

This flux is proportional to the gradient of f in the variable μ. This particle flux expression has the same form as the heat flux as given by Equation (2.76). Similarly, the pitch-angle diffusion equation has the same form as the heat conduction Equation (2.82). Both equations, and in fact all diffusion equations, are parabolic partial differential equations. We will also see other examples of diffusion equations in Chapters 4 and 5. Note that the first adiabatic invariant, μ_{m}, is violated for a particle undergoing a pitch-angle scattering event. Some space physicists prefer to work with diffusion in the variable μ_{m} rather than with pitch-angle diffusion.

For a loss cone distribution, the gradient $\partial f/\partial \mu$ is such as to produce a flux of particles into the loss cone. Particles in the loss cone are lost over a time period equal to about half the bounce period. *Particle precipitation* into the atmosphere results from diffusion into the loss cone, because particles with pitch angles inside the loss cone have mirror points down in the atmosphere. This loss process can be represented as

$$L = \begin{cases} -\dfrac{f}{\frac{1}{2}\tau_{\text{B}}} & (\alpha \text{ inside loss cone}) \\ 0 & (\alpha \text{ outside loss cone}), \end{cases} \tag{3.103}$$

where τ_{B} is the bounce period. Inside the loss cone means that $\alpha < \alpha_{\text{lc}}$ or $\alpha > |\pi - \alpha_{\text{lc}}|$. If the diffusion time given by Equation (3.98) is much greater than the bounce period, τ_{B}, then the lifetime of a particle for this process is of the order of the diffusion time itself. If pitch-angle diffusion is so fast as to ensure an isotropic distribution even at the loss cone (this is the so-called *strong pitch-angle diffusion limit*), then the particle lifetime is roughly the bounce period divided by the fraction of particles for an isotropic distribution that lie within the loss cone.

Example 3.5 (The effects of pitch-angle diffusion on the time evolution of a loss cone distribution) The purpose of this example is to demonstrate in a simple manner the qualitative effects of pitch-angle diffusion on a magnetospheric loss cone distribution function. The bounce-averaged pitch-angle diffusion equation was numerically solved for the equatorial distribution function $f(\mu, t)$, where the particle energy is implicit. The appropriate diffusion equation is Equation (3.101) with $S = 0$ and a loss rate L given by Equation (3.103). The bounce time appearing in Equation (3.103) was chosen to be $\tau_B = 2$ s and the loss cone pitch angle was $\alpha_{lc} = 10°$. Although these values were not chosen for any particular type or energy particle, they are reasonable choices for tens of keV electrons near an L of 3–4. The diffusion coefficient has the following typical form:

$$D_{\mu\mu} = D_0 \frac{1 - \mu^2}{\mu^2 + b^2} \qquad (3.104)$$

with the values $D_0 = [1 \text{ hr}]^{-1}$ and $b = .55$. This diffusion coefficient has a maximum near $\alpha = 90°$, which is qualitatively correct. The approximate pitch-angle diffusion time is $\tau_{\mu\mu} \approx 1/4D_{\mu\mu} \approx 1$ hour for this example, which is the time it takes for the distribution to evolve to a significant extent. Note that real electron pitch-angle diffusion loss times in the magnetosphere are of the order of days rather than an hour.

We assume that at time $t = 0$ we have a loss cone distribution (i.e., the distribution function is isotropic except for pitch angles inside the loss cone for which $f = 0$). The solution to the pitch-angle diffusion equation at several times is shown in Figure 3.18. Notice that f as a whole (and hence the average phase space density at a given particle energy) decreases with time with a lifetime of about 3,000 s (roughly 1 hour) because particles are being lost to the loss cone. Also notice that the distribution function becomes smoother as time progresses; in particular, f near the "edge" of the loss cone smooths out. All diffusion equations have the property that small scales (i.e., small values of $\delta\mu$ for this example) rapidly smooth out, so that the solution evolves into a smoother distribution. This can be shown analytically using Fourier analysis. We will discuss this further in Chapter 4 when magnetic diffusion is introduced.

How does pitch-angle diffusion operate in the inner magnetosphere? The quantitative details are rather complex, but the simple picture just presented should give you some qualitative feeling for what is thought to happen. If you desire more information read Lyons and Williams (1984), Roederer (1970), or Walt (1994). Energetic electrons in the radiation belt are able to be in resonance with relatively low frequency (a few hundred hertz) waves known as *whistler mode waves* or just *whistlers*. Lightning in the terrestrial atmosphere produces whistlers that propagate up to the inner magnetosphere, but the main source of magnetospheric whistlers is thought to be a plasma instability associated with the existence of loss cone

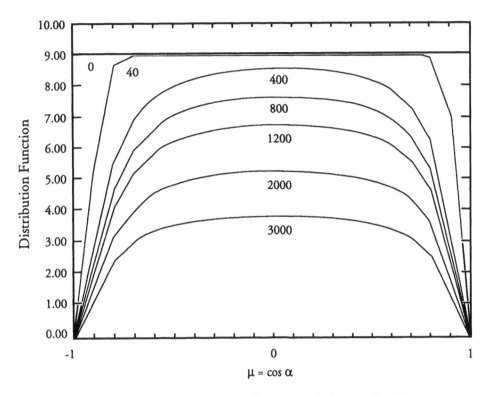

Figure 3.18. Distribution function versus cosine of the pitch angle for different times in units of seconds – from a numerical solution of a simple pitch-angle diffusion equation that includes particle loss in the loss cone. Courtesy of Dr. C. N. Keller.

distributions. Magnetic fluctuation levels associated with these waves are typically $B_{rms} \approx 3 \times 10^{-2}$ nT. The particle lifetime of 2 MeV electrons due to the pitch-angle diffusion into the loss cone associated with these waves is about 10 days near $L \approx 2.5$ but is longer at either lower or higher L-shells. Electron lifetimes for other energies are shown in Figures 5.25 and 5.26 of Lyons and Williams (1984).

What about wave–particle interactions for energetic ions in the inner magnetosphere? Just as energetic electrons can be in resonance with whistler mode waves, energetic ring current and radiation belt ions are able to be in resonance with a wave mode known as *ion cyclotron waves*. In fact, these waves are thought to be generated by a plasma instability associated with the nonisotropic nature of the ring current ion distribution function. But for ring current ions, either Coulomb collisions or charge exchange collisions, depending on the ion energy, are thought to be more important loss processes than pitch-angle diffusion.

3.11.4 Cross-L-shell diffusion in magnetospheres

Wave–particle interactions can lead to spatial diffusion as well as to pitch-angle or energy diffusion. Spatial diffusion along the magnetic field direction is important for applications such as cosmic ray propagation along interplanetary magnetic field lines. This type of spatial diffusion is associated with pitch-angle diffusion. A particle scattered in pitch angle by a magnetic fluctuation has had its parallel velocity changed and no longer "freely" streams along the field. A spatial diffusion coefficient can be determined by suitably averaging the pitch-angle diffusion coefficient over pitch angle. We will discuss this type of spatial diffusion again at the end of Chapter 6, in connection with cosmic ray propagation.

Spatial diffusion in directions perpendicular to the magnetic field can also result from wave–particle interactions. This is manifested in the inner and middle magnetospheres of the Earth as *cross-L shell diffusion*. The magnetic moment (μ_m) and longitudinal adiabatic invariant (I) are preserved during this type of diffusion but the *third invariant* is violated. The electric or magnetic fluctuations responsible for cross-L diffusion have resonant wave periods approximately equal to the appropriate particle drift periods. These wave periods are rather long in comparison with the waves responsible for pitch-angle scattering because particle drift times in the magnetosphere are much greater than the bounce times (see Table 3.2).

The rate of cross-L diffusion depends, not surprisingly, on the gradient of the distribution function in the variable L. The flux of particles at a given energy is proportional to $D_{LL}(\partial f/\partial L)$, where D_{LL} is the cross-L or *radial diffusion coefficient*. The time rate of change of f due to this type of diffusion is given by (cf. Lyons and Williams, 1984)

$$\frac{\partial f}{\partial t} = L^2 \frac{\partial}{\partial L}\left[D_{LL} L^{-2} \frac{\partial f}{\partial L}\right], \tag{3.105}$$

where f is understood to be the distribution function at the magnetic equator and the radial diffusion coefficient D_{LL} is assumed to be bounce averaged. Other terms such as sources, sinks, pitch-angle diffusion, etc. were omitted from Equation (3.105) but could be included if needed. Both electric and magnetic fluctuations with long wave periods contribute to D_{LL} in the Earth's inner magnetosphere, although the magnetic contribution is more important. The electric field fluctuations contribute only for energies less than about 500 keV and for lower L values. Nakada and Mead (1965) deduced the following expression for the magnetic part of D_{LL} for $L \approx 6$ using ground-based magnetometer data:

$$D_{LL} \approx 3 \times 10^{-10} L^{10} \quad \text{[units of } R_E^2/\text{day]}. \tag{3.106}$$

This expression is appropriate for magnetically quiet times and should be multiplied by a factor of 20 for magnetically active times. We see from the following example that radial diffusion is mainly important in the outer radiation belts ($L > 3$), where with diffusion times of tens to hundreds of days it is competitive with other loss processes.

Example 3.6 (Time scale for radial diffusion in the terrestrial magnetosphere)
A time scale for radial diffusion can be estimated using $\tau_{LL} \approx (\Delta L)^2/D_{LL}$. For an L

value of 6, D_{LL} from Equation (3.106) is roughly $3 \times 10^{-10}(6)^{10} \approx 0.02\,[R_E^2/\text{day}]$. Using $\Delta L \approx 2$, we find a diffusion time of

$$\tau_{LL} \approx (\Delta L)^2/D_{LL} \approx 2^2/0.02 \approx 200 \text{ days}.$$

This is for magnetically quiet conditions. For active conditions, we divide by 20 to get $\tau_{LL} \approx 10$ days. Repeating this for $L \approx 3$, we find much longer diffusion times ($2^{10} \approx 1,024$ times longer – diffusion times of several years even for active conditions).

Typically, in the outer part of the radiation belt, particles are injected from the outer magnetosphere during magnetically active time periods and then continue to diffuse inward radially. The radial diffusion is then balanced by other loss processes – mainly pitch-angle scattering into the loss cone for energetic electrons or loss due to Coulomb collisions for MeV protons. In the outer radiation belt one finds that $\tau_{\mu\mu} \approx \tau_{LL}$ (cf. Lyons and Williams, 1984). Particles spatially diffuse inwards in L and then are lost into the loss cone, thus precipitating into the atmosphere. Note that the magnetic moment remains constant (or invariant) during the spatial diffusion part of this process and that the particle energy increases as a consequence of this inward diffusive transport because the magnetic field strength increases as L decreases.

In the inner radiation belt, radial diffusion is not important and the main source of energertic protons is neutron decay, as discussed earlier. The lower energy ring current plasma mainly "convects" inward due to the large-scale magnetospheric electric field rather than diffusing inward like the more energetic radiation belt plasma. Furthermore, ring current ions have lifetimes of only a day or so, which is much less than the radial diffusion time.

Problems

3.1 For a uniform magnetic field in the z direction, $\mathbf{B} = B_0\hat{\mathbf{z}}$, find the trajectory in space of a particle with charge q, whose initial ($t = 0$) location is given by (x_0, y_0, z_0) and whose initial velocity vector is (v_{x0}, v_{y0}, v_{z0}). That is, find the position vector $\mathbf{x}(t)$.

3.2 A proton moves in uniform electric and magnetic fields given by $\mathbf{E} = 10$ V/m $\hat{\mathbf{x}}$ and $\mathbf{B} = 10^{-4}$ T $\hat{\mathbf{z}}$. At $t = 0$ and at the origin $\mathbf{x} = 0$, $\mathbf{v}(t = 0) = 2.414 \times 10^5\,\hat{\mathbf{x}} - 1.414 \times 10^5\,\hat{\mathbf{y}}$ [m/s].

(a) Find the gyrofrequency Ω.
(b) Find $\mathbf{v}(t)$ and $\mathbf{x}(t)$, and sketch these trajectories in velocity space and configuration space.
(c) Find the $\mathbf{E} \times \mathbf{B}$ drift speed $\mathbf{V}_E$.
(d) Find gyration speed v_0 (in the frame of reference of the drift motion).

(e) What is the gyration energy of the proton in units of eV?

(f) What is the magnetic moment of the proton $\mu_\mathrm{m} = mv_0{}^2/2B$?

3.3   A proton with the same initial position and velocity as in Problem (3.2) moves in a nonuniform magnetic field with $\mathbf{B} = [10^{-3} + 10^{-6}\,y]\,\hat{\mathbf{z}}$ [T]. That is, $\mathbf{B}$ increases with increasing y (units of m). Also, $\mathbf{E} = 0$.

(a) Find the gyrofrequency Ω at $t = 0$ and $y = 0$.

(b) Find the grad-B drift velocity, $\mathbf{v}_{\nabla B}$.

(c) Find $\mathbf{v}(t)$ and $\mathbf{x}(t)$, approximately.

(d) What electric field $\mathbf{E}$ would be needed to give a total drift speed almost equal to zero?

3.4   For a magnetic field in a vacuum, $\mathbf{B}(r) = B_\phi(r)\hat{\phi}$ (cylindrical coordinates), demonstrate that $\partial B_\phi/\partial r = -B_\phi/r$.

3.5   Demonstrate that the magnetic flux (Φ_m) through a particle orbit in a slowly varying magnetic field is independent of time. You can assume that the magnetic moment of the particle is an adiabatic invariant of the motion.

3.6   Derive the relation $\sin^2\alpha = [B/B_0]\sin^2\alpha_0$ (that is, Equation (3.62)) from the definition of pitch angle α and the invariance of the magnetic moment.

3.7   The kinetic energy of a charged particle increases with time in a shrinking magnetic bottle. Use Faraday's law to explain why this energy increase occurs.

3.8   Use Equations (3.76) and (3.77) to estimate the drift speeds v_B of electrons and ions at $L = 4$ in the terrestrial magnetosphere for the energies given in Table 3.2. Then verify the time scales listed in that table.

3.9   Demonstrate that in a vacuum $\nabla \times \hat{\mathbf{b}} = \hat{\mathbf{b}} \times \nabla B/B$, where $\hat{\mathbf{b}} = B/|\mathbf{B}|$.

3.10   (a) Use the information in Example 3.4 (and Figure 3.16) to derive the current density expression (3.89).

(b) Also derive Equation (3.90) for the total current I_L.

(c) Estimate the magnetic field perturbation produced at the surface of the Earth by the simple ring current of Example 3.4.

3.11   Estimate the current density $\mathbf{J}$ at $L=3$ in the vicinity of the magnetic equator ($\lambda = 0$) for a spatially uniform distribution of protons with density of $10^7\,\mathrm{m}^{-3}$, all with a pitch angle of $90°$ and with an average energy of 100 keV. What direction is $\mathbf{J}$ in? Repeat this calculation but with all the protons having a pitch angle of $45°$ at the magnetic equator.

3.12   Determine the loss cone pitch angles α_lc, for the Earth's magnetosphere at $L = 2, 3, 4, 5, 6$, and 10. What is the magnetic latitude λ of the mirror points for each of these cases if the loss cone mirror point is the altitude 1,000 km. For the $L = 4$ case, determine how α_lc varies along the magnetic field (i.e., versus λ).

3.13   Derive Equation (3.94) from Equation (3.93).

3.14   Determine the variation of $f(\mu, s)$ as a function of $\mu = \cos\alpha$ and distance s from the magnetic equator along a magnetic field line, if the equatorial distribution function is a loss cone distribution with loss cone pitch angle α_{lc0} at the magnetic equator. How does α_{lc} vary along the field line? Assume that pitch-angle diffusion, sources, and sinks are all negligible.

Bibliography

Chen, F., *Introduction to Plasma Physics*, Plenum Press, New York, 1974.

Hamlin, D. A., R. Karplus, R. C. Vik, and K. M. Watson, Mirror and azimuthal drift frequencies for geomagnetically trapped particles, *J. Geophys. Res.,* **66**, 1, 1961.

Hess, W. N., *The Radiation Belt and Magnetosphere*, Blaisdell Publ. Co., Waltham, MA, 1968.

Krall, N. A. and A. W. Trivelpiece, *Principles of Plasma Physics*, McGraw-Hill, New York, 1973.

Lyons, L. R. and D. J. Williams, *Quantitative Aspects of Magnetospheric Physics*, D. Reidel Publ. Co., Dordrecht, The Netherlands, 1984.

McIlwaine, C. E., Coordinates for mapping the distribution of magnetically trapped particles, *J. Geophys. Res.,* **66**, 3681, 1961.

Nakada, M. P. and G. D. Mead, Diffusion of protons in the outer radiation belt, *J. Geophys. Res.,* **70**, 4777, 1965.

Roederer, J. G., *Dynamics of Geomagnetically Trapped Radiation*, Physics and Chemistry in Space 2, Springer-Verlag, New York, 1970.

Schulz, M. and L. J. Lanzerotti, *Particle Diffusion in the Radiation Belts*, Springer-Verlag, New York, 1974.

Symon, K. R., *Mechanics*, 3rd. ed., Addison-Wesley, Reading, MA, 1971.

Walt, M., *Introduction to Geomagnetically Trapped Radiation*, Cambridge Univ. Press. Cambridge, UK, 1994.

4

Magnetohydrodynamics

In Chapter 3, we studied how single charged particles move in specified electric and magnetic fields, and we then applied our knowledge of single particle motion to the radiation belt and ring current plasma. However, the fields in some situations depend too much on the particle distributions to be readily specified and must be found self-consistently using the charged particle distribution functions. Often, it is not necessary to have complete information about the distribution functions in a system. In fact, it is usually sufficient to know only a few of the velocity moments of the distribution function, as derived in Chapter 2. In Chapter 4, we will adopt the "fluid" picture of a plasma, introduced in Chapter 2, and further refine it to obtain an analytical tool useful for studying space plasma phenomena. This analytical tool is called magnetohydrodynamics (or MHD for short). We cannot adequately cover in one chapter all the material that would be desirable to know about this subject and so the reader is encouraged to consult one or more of the references listed in the bibliography at the end of this chapter.

4.1 Two-fluid plasma

Let us consider a plasma consisting of two species: electrons (e) with mass m_e and a single ion species (i) with mass m_i. The continuity equations for electrons and ions are given by (see Equation (2.40))

$$\frac{\partial n_e}{\partial t} + \nabla \cdot (n_e \mathbf{u}_e) = S_e \qquad (4.1)$$

and

$$\frac{\partial n_i}{\partial t} + \nabla \cdot (n_i \mathbf{u}_i) = S_i, \qquad (4.2)$$

where $n_s, \mathbf{u}_s, S_s$ ($s = e, i$) refer to the number density, flow velocity, and net source, respectively, for species s. The respective mass densities can be expressed as $\rho_e = m_e n_e$ and $\rho_i = m_i n_i$.

The momentum equation for a plasma species s was given by Equation (2.61). We assume that both the electron and ion thermal pressure tensors are isotropic, with scalar pressures given by p_e and p_i, respectively. The electron and ion momentum equations are

$$n_e m_e \left[\frac{\partial}{\partial t} + \mathbf{u}_e \cdot \nabla \right] \mathbf{u}_e = -n_e e [\mathbf{E} + \mathbf{u}_e \times \mathbf{B}] - \nabla p_e + n_e m_e \mathbf{g}$$

$$- \sum_{t \neq e} n_e m_e \nu_{et} (\mathbf{u}_e - \mathbf{u}_t) \tag{4.3}$$

and

$$n_i m_i \left[\frac{\partial}{\partial t} + \mathbf{u}_i \cdot \nabla \right] \mathbf{u}_i = +n_i e [\mathbf{E} + \mathbf{u}_i \times \mathbf{B}] - \nabla p_i + n_i m_i \mathbf{g}$$

$$- \sum_{t \neq i} n_i m_i \nu_{it} (\mathbf{u}_i - \mathbf{u}_t) - P_i (m_i \mathbf{u}_i - m_n \mathbf{u}_n). \tag{4.4}$$

Recall from Chapter 2 that $\mathbf{g}$ is the acceleration due to gravity and ν_{st} is the effective momentum transfer collision frequency between species s and t. The "frictional" term includes a sum over all species but does not include self-collisions (i.e., $s \neq t$). For electrons, one must consider electron–ion and electron–neutral collisions. Equation (4.4) includes the "mass-loading" term, but the viscosity terms have been omitted. P_i is the production rate of ions due to the ionization of a neutral species with mass m_n and velocity $\mathbf{u}_n$.

The energy equations for electrons and ions can be found from Equation (2.73), (2.78), or (2.79) with $s = e$ or i. However, these equations will not be reproduced here but will be provided in convenient forms as they are required later in the chapter.

The $\mathbf{E}$ and $\mathbf{B}$ fields that are present in Equations (4.3) and (4.4) are macroscopic fields and can be found from Maxwell's equations with macroscopically defined source terms (see Chapter 2, Section 2.5).

The full two-fluid equations are difficult to use, although they have been occasionally used for space physics problems such as solar wind outflow from the Sun. Usually these equations are further approximated.

4.2 Plasma oscillations

4.2.1 Waves

Waves in plasmas can be studied using either the fluid equations or the Vlasov equation. However, the fluid approach is much easier to carry out and gives a better physical picture of the nature of wave propagation. We restrict ourselves to the fluid approach in this book, although it should be noted that wave growth or damping must be treated with the Vlasov equation approach. Before dealing with waves in plasmas, let us review some properties of waves in general.

Any general wave train (as specified by the wave part of the electric field $\mathbf{E}_1(\mathbf{x}, t)$, for example) can be represented by the sum of many plane waves with a range of frequencies. Formally, $\mathbf{E}_1(\mathbf{x}, t)$ is given by the inverse Fourier–Laplace transform of $\mathbf{E}_1(\mathbf{k}, \omega)$, where $\mathbf{k}$ is the *wave vector* and ω is the angular *frequency* of a single Fourier–Laplace component:

$$\mathbf{E}_1(\mathbf{x}, t) = \frac{1}{(2\pi)^4} \int \int d^3k \, d\omega \mathbf{E}_1(\mathbf{k}, \omega) e^{i[\mathbf{k}\cdot\mathbf{x}-\omega t]}. \qquad (4.5)$$

If a wave train has a very narrow range of frequencies and wavenumbers, ω and $\mathbf{k}$ (for example, a delta function), then we can write Equation (4.5) as

$$\mathbf{E}_1(\mathbf{x}, t) = \mathbf{E}_0 e^{i[\mathbf{k}\cdot\mathbf{x}-\omega t]}, \qquad (4.6)$$

where $\mathbf{E}_0$ is a constant amplitude vector, which can be complex (i.e., have real and imaginary parts). Equation (4.6) describes a plane wave. Actually, the measurable electric field for a plane wave is the real part of Equation (4.5) or (4.6). Recall that the exponential function of an imaginary argument is given by

$$e^{ix} = \cos x + i \sin x. \qquad (4.7)$$

Hence, if the wave amplitude $\mathbf{E}_0$ is real (although it does not have to be in general), we find that the electric field for our plane wave is

$$\mathbf{E}_1(\mathbf{x}, t) = \mathbf{E}_0 \cos(\mathbf{k} \cdot \mathbf{x} - \omega t). \qquad (4.8)$$

The wave vector $\mathbf{k}$ points in the direction of wave propagation. We write $\mathbf{k} = k\hat{\mathbf{x}}$ for a wave propagating in the x direction. The *wavenumber k* can be written in terms of the *wavelength λ* as

$$k = \frac{2\pi}{\lambda}. \qquad (4.9)$$

The angular frequency can be expressed as $\omega = 2\pi f$, where f is the wave *frequency* in *cycles/second* or *hertz* (Hz). For $\mathbf{k} = k\hat{\mathbf{x}}$, Equation (4.8) becomes

$$\mathbf{E}_1(x, t) = \mathbf{E}_0 \cos(kx - \omega t). \qquad (4.10)$$

$\mathbf{E}_1(x, t)$ varies sinusoidally in time with frequency ω for fixed x and varies sinusoidally in the spatial variable x for a fixed time. The *phase* of the wave can be written

$$phase = [kx - \omega t] = k\left[x - \left(\frac{\omega}{k}\right)t\right] = constant. \qquad (4.11)$$

The phase clearly remains constant for a reference frame moving with velocity (ω/k). The *phase speed* of the wave $(\Delta x/\Delta t)$ is thus given by

$$V_{\text{ph}} = \omega/k. \qquad (4.12)$$

A wave packet limited in spatial extent ($\infty \gg \Delta x \gg \lambda$) consists of the super-position of "many" plane waves for a range of wavenumbers (and frequencies) and is centered at (k, ω). The wave packet as a whole propagates at the *group velocity*

$$V_g = \frac{\partial \omega}{\partial k}, \tag{4.13}$$

where ω is expressed as a function of wavenumber k. The functional dependence of ω on k depends on the type of wave and is given by the *dispersion relation* for that type of wave:

$$\omega = \omega(k). \tag{4.14}$$

For example, the dispersion relation for electromagnetic wave propagation in a vacuum is

$$\omega^2 = k^2 c^2, \tag{4.15}$$

where c is the speed of light. In this case, we simply have $V_g = V_{ph} = \pm c$, which tells us that electromagnetic waves travel at the speed of light in a vacuum. The propagation of electromagnetic radiation in a plasma is considered in the appendix and in Problem 4.1. An electromagnetic wave does *not* travel at the speed of light in a plasma.

4.2.2 Plasma (Langmuir) oscillations

Plasmas support many different *wave modes* and *Langmuir oscillations/waves* comprise the most important mode. Langmuir oscillations are a high-frequency phenomena primarily involving electrons; ions are relatively massive and are slow to follow the wave motion. Hence, let us assume that the ions are motionless and that the ion density remains uniform: $n_i = n_0$. Let us further assume that the electron and ion fluids are both cold: $T_e = T_i = 0$. Let the magnetic field be zero and we also neglect collisions. The two-fluid equations are reduced to the following simple continuity and momentum equations for electrons:

$$\frac{\partial n_e}{\partial t} + \nabla \cdot (n_e \mathbf{u}_e) = 0 \tag{4.16}$$

and

$$m_e n_e \left[\frac{\partial \mathbf{u}_e}{\partial t} + (\mathbf{u}_e \cdot \nabla) \mathbf{u}_e \right] = -n_e e \mathbf{E}. \tag{4.17}$$

The electric field is given by Gauss's law with the charge density expressed in terms of the electron and ion densities:

$$\nabla \cdot \mathbf{E} = \frac{e}{\varepsilon_0}(n_i - n_e) = \frac{e}{\varepsilon_0}(n_0 - n_e). \tag{4.18}$$

We now separate all quantities into background (subscript "0") and wave parts (subscript "1") to obtain

$$n_e = n_{e0} + n_{e1}, \tag{4.19a}$$

$$\mathbf{u}_e = \mathbf{u}_{e0} + \mathbf{u}_{e1}, \tag{4.19b}$$

and

$$\mathbf{E} = \mathbf{E}_0 + \mathbf{E}_1. \tag{4.19c}$$

We take the background plasma to be quasi-neutral and uniform ($n_{e0} = n_i = n_0$). In this case, we must also have $\mathbf{u}_{e0} = 0$ and $\mathbf{E}_0 = 0$ for the background plasma in order to satisfy Equations (4.16)–(4.18). Note that $\nabla n_{e0} = 0$ and $\partial n_{e0}/\partial t = 0$.

Further progress can be achieved by assuming that the wave amplitude is small:

$$n_e = n_0 + n_{e1} \quad \text{with } |n_{e1}| \ll n_0. \tag{4.20}$$

All wave quantities, relative to the background quantities, are of the order of a parameter $\varepsilon = O(|n_{e1}|/n_0)$, which is assumed to be very small ($\varepsilon \ll 1$).

We can now write the electron continuity and momentum equations as

$$\frac{\partial n_{e1}}{\partial t} + \nabla \cdot [(n_0 + n_{e1})\mathbf{u}_{e1}] = 0 \tag{4.21}$$

and

$$m_e \left[\frac{\partial \mathbf{u}_{e1}}{\partial t} + \mathbf{u}_{e1} \cdot \nabla \mathbf{u}_{e1} \right] = -e\mathbf{E}_1 \tag{4.22}$$

$$\approx \varepsilon \qquad \approx \varepsilon^2 \text{ (neglect)} \qquad \approx \varepsilon.$$

The magnitudes of the terms in Equation (4.22) are indicated using the parameter ε.

Gauss's law becomes

$$\nabla \cdot \mathbf{E}_1 = \frac{e}{\varepsilon_0}[n_0 - (n_0 + n_{e1})] = -\frac{n_{e1}e}{\varepsilon_0}. \tag{4.23}$$

The net charge density is proportional to the wave density n_{e1}.

We now *linearize* Equations (4.21) and (4.22) by neglecting all terms of order higher than ε, such as the $\nabla \cdot (n_{e1}\mathbf{u}_{e1})$ term in Equation (4.21) and the $\mathbf{u}_{e1} \cdot \nabla \mathbf{u}_{e1}$ term in Equation (4.22). For example, the linearized continuity equation is

$$\frac{\partial n_{e1}}{\partial t} + n_0 \nabla \cdot \mathbf{u}_{e1} = 0. \tag{4.24}$$

Our next step is to assume that we have plane waves propagating in the x direction (i.e., $\mathbf{k} = k\hat{\mathbf{x}}$) (see Equation (4.6)). Then we have

$$n_{e1}(x, t) = \tilde{n}_{e1} \exp[i(kx - \omega t)], \tag{4.25}$$

$$\mathbf{u}_{e1}(x, t) = \tilde{u}_{e1}\hat{\mathbf{x}} \exp[i(kx - \omega t)], \tag{4.26}$$

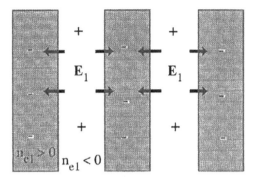

Figure 4.1. Schematic for plane-wave Langmuir wave showing slablike electron density perturbations.

and

$$\mathbf{E}_1(x, t) = \tilde{E}_1\hat{\mathbf{x}}\exp[i(kx - \omega t)], \tag{4.27}$$

where $\tilde{u}_{e1}, \tilde{n}_{e1}$, and $\tilde{E}_1$ are plane-wave amplitudes that are independent of x and t. Note that $\mathbf{E}_1$ is parallel to $\mathbf{k}$ because Langmuir waves are longitudinal. One-dimensional (slablike) perturbations in the density n_e (and therefore in the charge density) naturally lead to longitudinal wave electric fields (see Figure 4.1). In the x direction, positive and negative electron density perturbations (n_{e1}) alternate and give rise to negative and positive net charge densities, respectively. The slabs of charge produce a wave electric field ($\mathbf{E}_1$) that points from positive charge to negative charge. This electric field then accelerates the electrons away from regions of excess electron density.

The time derivative of an arbitrary plane-wave quantity, Q, gives

$$\frac{\partial}{\partial t}Q(x, t) = -i\omega Q(x, t). \tag{4.28}$$

And the gradient operator, when applied to plane waves propagating in the x direction, can be written

$$\nabla \Rightarrow +ik\hat{\mathbf{x}}. \tag{4.29}$$

We now substitute the plane-wave expressions for $n_{e1}(x, t)$, $\mathbf{u}_{e1}(x, t)$, and $\mathbf{E}_1(x, t)$ into the linearized versions of Equations (4.22)–(4.24).

The continuity equation becomes

$$-i\omega\tilde{n}_{e1} = -ik\tilde{u}_{e1}n_0, \tag{4.30}$$

the momentum equation becomes

$$-im_e\omega\tilde{u}_{e1} = -e\tilde{E}_1, \tag{4.31}$$

and Gauss's law becomes

$$ik\tilde{E}_1 = -\frac{e}{\varepsilon_0}\tilde{n}_{e1}. \tag{4.32}$$

We use Equations (4.30) and (4.31) to eliminate $\tilde{n}_{e1}$ from Equation (4.32), which becomes

$$ik\tilde{E}_1 = -\frac{e}{\varepsilon_0}\frac{kn_0}{\omega}\left(\frac{e}{im_e\omega}\right)\tilde{E}_1. \tag{4.33}$$

We now divide both sides by $\tilde{E}_1$, rearrange terms, and obtain the following dispersion relation for Langmuir waves in a cold plasma:

$$\omega^2 = \omega_{pe}^2, \tag{4.34}$$

where we define

$$\omega_{pe} \equiv \sqrt{\frac{n_0 e^2}{\varepsilon_0 m_e}}. \tag{4.35}$$

ω_{pe} is called the *electron plasma frequency*. In cgs units, the electron plasma frequency is given by

$$\omega_{pe} = \sqrt{\frac{4\pi n_0 e^2}{m_e}}. \tag{4.36}$$

Equation (4.34) describes how the wave frequency varies versus k

$$\omega(k) = \pm\omega_{pe}. \tag{4.37}$$

In fact, the function $\omega(k)$ is independent of k; and, hence, the group velocity ($V_g = \partial\omega/\partial k$) is zero. In this case, we actually have an oscillation – a *plasma oscillation* – rather than a propagating wave.

When ion motions are allowed, the relevant plasma frequency is given by

$$\omega_p = \sqrt{\omega_{pe}^2 + \omega_{pi}^2} \tag{4.38}$$

with

$$\omega_{pi} \equiv \sqrt{\frac{n_0 e^2}{\varepsilon_0 m_i}}$$

Because $m_i \gg m_e$ (unless we assume positrons rather than massive ions – see Problem 4.3), $\omega_p \cong \omega_{pe}$ is an excellent approximation.

Physically (see Figure 4.1), plasma oscillations occur because the moving electrons have inertia so that they overshoot their equilibrium position (where $\mathbf{E}_1 = 0$) and take a finite time to be decelerated. The deceleration occurs because of the electric field when the electrons pile up in a different location. Electron inertia "opposes" the electric force associated with the pile up of electrons. This is a collective phenomenon (i.e., involving large numbers of electrons acting together) in which a large-scale electron field ($\mathbf{E}_1$) is generated.

A useful expression (in SI units) for the frequency $f_p = \omega_p/2\pi$ is given by

$$f_p \cong 9\sqrt{n_0 \, (\mathrm{m}^{-3})} \quad [\mathrm{Hz}]. \tag{4.39}$$

Example 4.1 A typical *solar wind* electron density at 1 AU is

$$n_0 \approx 10^7 \, \mathrm{m}^{-3}.$$

Using this density we find from Equation (4.39) that a typical plasma frequency for the solar wind is

$$f_p \approx 3 \times 10^4 \, \mathrm{Hz} = 30 \, \mathrm{kHz}.$$

The average maximum *ionospheric* electron density at Earth is

$$n_0 \approx 10^{12} \, \mathrm{m}^{-3}.$$

The plasma frequency given by Equation (4.39) for the ionosphere is

$$f_p \approx 9 \, \mathrm{MHz}.$$

Electromagnetic waves (these are transverse waves, unlike Langmuir waves, which are longitudinal) cannot propagate in a plasma if their frequency is less than f_p (see the appendix and Problem 4.1). Instead, waves impinging on a plasma medium from the outside reflect at the location in the plasma where $f = f_p$. Recall that the AM (amplitude modulation) radio band is 0.5–1.6 MHz, and hence these waves are reflected from the terrestrial ionosphere. The FM (frequency modulation) radio band has $f > 88$ MHz – well above the maximum ionospheric plasma frequency – explaining why FM waves can propagate right through the ionosphere.

As we have just seen, a cold plasma can "support" Langmuir oscillations but not Langmuir waves that propagate. However, wave propagation is possible in a warm plasma. In a warm plasma ($T_e \neq 0$) the electron pressure gradient force term must be retained in the momentum equation, and the dispersion relation can be rederived (which you will do in Problem 4.2):

$$\omega^2 = \omega_{pe}^2 + 3k^2 v_{te}^2, \tag{4.40}$$

where the *electron thermal speed* is defined by

$$v_{te}^2 \equiv k_B T_e/m_e. \tag{4.41}$$

The frequency ω in Equation (4.40) (unlike in Equation (4.34)) now depends on the wavenumber, and thus the group velocity,

$$V_g = \frac{\partial \omega}{\partial k} = 3\frac{k}{\omega}v_{te}^2 = \frac{3v_{te}^2}{v_{ph}}, \tag{4.42}$$

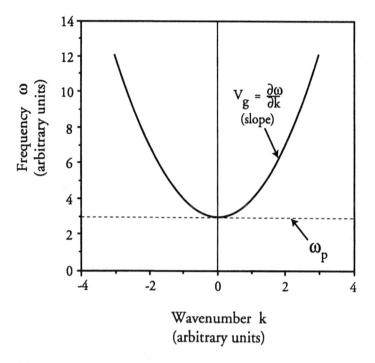

Figure 4.2. Plasma/Langmuir dispersion relation. The group velocity, V_g, is the first derivative of $\omega(k)$.

is nonzero and the phase speed, ω/k, is equal to

$$V_{\mathrm{ph}} = \frac{\omega}{k} = \frac{1}{k}\sqrt{\omega_{\mathrm{pe}}^2 + 3k^2 v_{\mathrm{te}}^2}$$

$$\approx \frac{\omega_{\mathrm{pe}}}{k} \quad \text{(for small } k\text{)}.$$

(4.43)

The dispersion relation, $\omega(k)$, is shown in Figure 4.2. The group velocity is just the slope of the function $\omega(k)$; we can clearly see that $V_g = 0$ at $k = 0$. And as $k \to \infty$, $V_{\mathrm{ph}} = V_g \to \sqrt{3}v_{\mathrm{te}}$. Plasma waves cannot exist with frequencies below ω_{pe}.

4.3 The single-fluid equations and the generalized Ohm's law

4.3.1 Quasi-neutrality and the generalized Ohm's law

If we are dealing with a plasma phenomenon that is both large scale (scale size $L \gg \lambda_{\mathrm{D}}$) and has a relatively low frequency ($\omega \ll \omega_{\mathrm{p}}$), then the plasma is quasi-neutral ($n_e \cong n_i$) on these length and time scales. Most interesting space plasma phenomena satisfy these two criteria. We can then assume that the electron density is equal to the total ion density:

$$n_e = n_i.$$

(4.44)

One consequence of Equation (4.44) is that we do not need separate continuity equations for the electron and ion gases; a single continuity equation suffices.

A difficulty arises if we adopt $n_e = n_i$: The *assumed* charge density is zero ($\rho_c = 0$), although the actual charge density must have very small deviations from zero. With zero charge density Gauss's law, Equation (4.18), simply becomes $\nabla \cdot \mathbf{E} = 0$ and is no longer useful for determining the electric field. An alternate means of finding $\mathbf{E}$ is required. The electron momentum Equation (4.3) can be used for this purpose with the assumption $n_e = n_i$. Solving Equation (4.3) for $\mathbf{E}$ we obtain a relation called the *generalized Ohm's law* (GOL):

$$\mathbf{E} = -\mathbf{u}_e \times \mathbf{B} - \frac{1}{n_e e}\nabla p_e + \frac{m_e}{e}\mathbf{g} + \frac{m_e}{e}\sum_{t \neq e}\nu_{et}(\mathbf{u}_e - \mathbf{u}_t)$$

$$-\frac{m_e}{e}\left[\frac{\partial \mathbf{u}_e}{\partial t} + \mathbf{u}_e \cdot \nabla\mathbf{u}_e\right]. \tag{4.45}$$

Equation (4.45) specifies the electric field required to maintain quasi-neutrality in a plasma. We can derive a simpler form of the GOL for a collisionless plasma by neglecting the friction term (i.e., the term that includes the collision frequency ν_{et}) and by neglecting gravity and the electron inertial terms (the latter is the last term), which are proportional to the very small electron mass:

$$\mathbf{E} = -\mathbf{u}_e \times \mathbf{B} - \frac{1}{n_e e}\nabla p_e. \tag{4.46}$$

$$\underbrace{\qquad\qquad}_{\substack{\text{motional} \\ \text{electric} \\ \text{field}}} \underbrace{\qquad\qquad}_{\substack{\text{ambipolar} \\ \text{electric field}}}$$

The *ambipolar electric field* (or *polarization electric field*) is proportional to the gradient of the electron pressure. The *motional electric field* (which is the first term on the right-hand side as indicated) is associated with the frame of reference of the electron gas. The electric field, $\mathbf{E}'$, in a reference frame moving at the electron flow velocity, $\mathbf{u}_e$, does not include this term:

$$\mathbf{E}' = \mathbf{E} + \mathbf{u}_e \times \mathbf{B} = -\frac{1}{n_e e}\nabla p_e. \tag{4.47}$$

Now consider the component of this electric field parallel to the magnetic field, $E_{\parallel} = \mathbf{E} \cdot \hat{\mathbf{b}}$, where $\hat{\mathbf{b}} = \mathbf{B}/|\mathbf{B}|$ is the unit vector parallel to $\mathbf{B}$. From Equation (4.46) we find

$$E_{\parallel} = -\frac{1}{n_e e}(\hat{\mathbf{b}} \cdot \nabla p_e) = -\frac{1}{n_e e}\frac{\partial p_e}{\partial s}, \tag{4.48}$$

where s is the distance along a magnetic field line. If we further assume that the electrons are isothermal with temperature T_e, and if we use the equation of state,

Equation (4.48) becomes

$$E_\parallel = -\frac{k_B T_e}{e} \frac{1}{n_e} \frac{\partial n_e}{\partial s}.$$ (4.49)

Equation (4.49) indicates that a polarization electric field exists as a consequence of gradients along **B** in the plasma density. The electric potential difference between two points along the magnetic field due to the polarization is determined by integrating Equation (4.49). If the electron temperature is independent of distance then this potential difference just depends on the variation of the electron density and is approximately $k_B T_e/e$ over a typical scale length for significant density changes. How can we determine the variation of the electron density n_e as a function of distance s? The electron density is just equal to the ion density n_i, by the assumption of quasi-neutrality, and n_i can be found by solving the fluid conservation equations for ions (we will return to this shortly).

Suppose the ion density as a function of s (i.e., $n_i(s)$) is specified. For example, suppose n_i increases as s increases (Figure 4.3). If no ambipolar field existed, and if we assumed strict charge neutrality initially, an unbalanced pressure gradient force on the electrons would exist and accelerate them so that the electrons would move from the high-density region to the low-density region. A small charge imbalance would then very quickly develop ($n_e \cong n_i$, but not exactly equal) as illustrated in Figure 4.3. When the electric field resulting from the very small, but nonzero, charge density just equaled the ambipolar field as specified by Equation (4.49) (or, actually, by Equation (4.45) if *all* forces on the electron gas are included) then an electron force balance would again be achieved with n_e almost equal to n_i (but not quite – the difference ($n_i - n_e$) is exceedingly small – see Problem 4.4). The ambipolar, or polarization, electric field thus holds the electron and ion gases together and maintains quasi-neutrality. Theoretically, this electric field could be found using Gauss's law; but, practically, the difference between n_e and n_i is so small (on length scales $L \gg \lambda_D$) relative to the magnitude of n_e that reliably calculating the charge density for real problems is almost impossible.

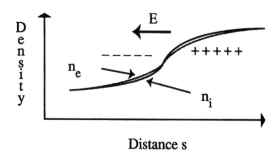

Figure 4.3. Ambipolar polarization electric field associated with electron pressure gradient force (i.e., with the density gradient for isothermal electrons). The Coulomb/electric force due to the minute charge imbalance counteracts the pressure gradient force.

4.3.2 Single-fluid equations

A multi-ion species plasma can be described with the set of Equations (4.1)–(4.4) plus the generalized Ohm's law. The generalized Ohm's law, Equation (4.45), tells us what $\mathbf{E}$ is in terms of the variables n_e, $\mathbf{u}_e$, and T_e. For *several* singly charged ion species, n_e is equal to the sum of the densities of all ion species (this is a generalization of Equation (4.44)):

$$n_e = n_i \equiv \sum_{\substack{\text{all ion} \\ \text{species } s}} n_s.$$ (4.50)

The summation in this equation is over all ion species in the plasma (e.g., H^+, O^+, ...). We can determine the density of each ion species, n_s, and the ion velocities, $\mathbf{u}_s$, from the appropriate continuity and momentum equations. We also need to know the temperature of each ion species, which can be found by using the appropriate energy equation. The electric field in all these fluid equations can be taken from the GOL. Our remaining unknown is the electron velocity $\mathbf{u}_e$, which cannot be found by using the electron momentum equation because that equation was "converted" into the GOL and was instead used to find $\mathbf{E}$. The electron flow velocity can be found from the electrical current density and the calculated ion velocities if the magnetic field is known.

At this point, we simplify the problem further by considering just a *single fluid* that combines the electron gas and all the ion species. We start with the definition of the total *mass density*:

$$\rho(x, t) \equiv n_e m_e + \sum_s n_s m_s,$$ (4.51)

where the sum is over all ion species. The *center of mass velocity* of the plasma is given by

$$\mathbf{u}(\mathbf{x}, t) \equiv \frac{1}{\rho} \left[n_e m_e \mathbf{u}_e + \sum_s n_s m_s \mathbf{u}_s \right].$$ (4.52)

For a single ion species, we can find expressions for the mass density ρ and the center of mass flow velocity $\mathbf{u}$, starting from Equations (4.51) and (4.52):

$$\rho = n_e m_e + n_i m_i = n_e(m_e + m_i) \cong n_e m_i,$$ (4.53)

$$\mathbf{u} = \frac{m_e \mathbf{u}_e + m_i \mathbf{u}_i}{m_e + m_i} \cong \mathbf{u}_i.$$ (4.54)

The approximate versions of Equations (4.53) and (4.54) are quite accurate because the electron mass is much less than the ion mass ($m_e/m_i \ll 1$). Our description of the plasma also requires knowledge of the current density $\mathbf{J}$:

$$\mathbf{J}(x, t) = \sum_s n_s Z_s e \mathbf{u}_s - n_e e \mathbf{u}_e,$$ (4.55)

where Z_s is the charge number on ion species s. For a single ion species (with $Z = 1$) Equation (4.55) becomes

$$\mathbf{J} = n_e e (\mathbf{u}_i - \mathbf{u}_e).$$ (4.56)

The *single-fluid mass continuity equation* is found by mass-weighting the electron and ion continuity equations (4.1) and (4.2) and using the definitions of ρ and $\mathbf{u}$ to obtain

$$\frac{\partial \rho}{\partial t} + \nabla \cdot (\rho \mathbf{u}) = m_i S_i.$$ (4.57)

A small source term, $m_e S_e$, was neglected in this equation.

The *single-fluid momentum equation* can be derived by adding together the electron and ion momentum equations; the electric field cancels out during this operation. Note that this is equivalent to substituting the electric field from the GOL, Equation (4.45), into the ion momentum equation (4.4). We have

$$\rho \left[\frac{\partial \mathbf{u}}{\partial t} + \mathbf{u} \cdot \nabla \mathbf{u} \right] = -\nabla p + \mathbf{J} \times \mathbf{B} + \rho \mathbf{g} - \rho \bar{\nu}(\mathbf{u} - \mathbf{u}_n) - P_i m_i (\mathbf{u} - \mathbf{u}_n).$$ (4.58)

Terms with the charge density ρ_c were neglected, as were terms of order m_e/m_i ($\ll 1$). In the mass-loading term at the end of this equation we have supposed that $m_i = m_n$. The average momentum transfer collision frequency of ions with neutrals is $\bar{\nu} \cong \nu_{in}$, and henceforth this will just be denoted ν. The total thermal pressure is $p = p_e + p_i$; the electron and ion pressures can be found using the appropriate energy equations discussed in Chapter 2.

A slightly more accurate version of the momentum equation (4.58) would also include a term $\rho_c \mathbf{E}$ on the right-hand side with $\mathbf{E}$ specified by the GOL. In this case, we would need the following *charge continuity equation* for ρ_c, which can be derived from Maxwell's equations (Problem 4.5):

$$\frac{\partial \rho_c}{\partial t} + \nabla \cdot \mathbf{J} = 0.$$ (4.59)

Typically, an excellent approximation to the charge continuity equation is given by

$$\nabla \cdot \mathbf{J} = 0.$$ (4.60)

The charge continuity equation (4.59) with suitable boundary conditions can be used to find the current density $\mathbf{J}$. The magnetic field must also be specified and we can use Ampère's law,

$$\nabla \times \mathbf{B} = \mu_0 \mathbf{J},$$ (4.61)

for this purpose. We have neglected the displacement current in Equation (4.61); this is a good approximation for phenomena with time scales τ such that $\tau^{-1} \ll \omega_p$.

There exists another way of finding the magnetic field, which is often much easier to use, as will be considered in the next section.

One difficulty remains – the electric field given by Equation (4.45) is written in terms of $\mathbf{u}_e$, but we really need it in terms of the variables $\mathbf{u}$ and $\mathbf{J}$. The conversion of Equation (4.45) into a new form of the generalized Ohm's law using $\mathbf{u}$ and $\mathbf{J}$ is messy but straightforward (this can be done as an exercise by the ambitious reader):

$$
\mathbf{E} = -\mathbf{u} \times \mathbf{B} + \frac{1}{n_e e}\mathbf{J} \times \mathbf{B} - \frac{1}{n_e e}\nabla p_e + \eta \mathbf{J} + \frac{m_e}{n_e e^2}\left\{ \frac{\partial \mathbf{J}}{\partial t} + \nabla \cdot [\mathbf{J}\mathbf{u} + \mathbf{u}\mathbf{J}] \right\}. \tag{4.62}
$$

with the labels: Hall term ($\frac{1}{n_e e}\mathbf{J} \times \mathbf{B}$), ambipolar polarization term ($-\frac{1}{n_e e}\nabla p_e$), motional field ($-\mathbf{u} \times \mathbf{B}$), Ohmic term ($\eta \mathbf{J}$), electron inertial term.

The commonly used names for the various terms are indicated.

The *Ohmic resistivity* is given by

$$
\eta = \frac{m_e \nu_e}{n_e e^2}, \tag{4.63}
$$

where ν_e is the total electron momentum transfer collision frequency (see Chapter 2), which for a single ion species is $\nu_e = \nu_{ei} + \nu_{en}$ (i.e., electron–ion Coulomb collision frequency and electron–neutral collision frequency).

The electron inertial term, typically being quite small relative to other terms in Equation (4.62), is often neglected. The Hall term results from using $\mathbf{u}$ rather than $\mathbf{u}_e$ in the motional electric field term; it is also often neglected. The ambipolar/polarization term was already discussed. In a collisionless plasma, the resistivity is zero ($\eta = 0$); equivalently, the *electrical conductivity* $\sigma = 1/\eta$ is infinite. For the opposite extreme of large η, with $\nabla p_e = 0$ and $\mathbf{B} = 0$, the generalized Ohm's law looks like the "ordinary" Ohm's law:

$$
\mathbf{J} = \frac{1}{\eta}\mathbf{E} = \sigma \mathbf{E}. \tag{4.64}
$$

If we retain the motional electric field but still neglect the rest of Equation (4.62) we can write Ohm's law as

$$
\mathbf{J} = \frac{1}{\eta}(\mathbf{E} + \mathbf{u} \times \mathbf{B}) = \frac{1}{\eta}\mathbf{E}', \tag{4.65}
$$

where $\mathbf{E}' = \mathbf{E} + \mathbf{u} \times \mathbf{B}$ is the electric field in the plasma frame of reference (also see Equation (4.47)).

The GOL has important implications for the time evolution of the magnetic field, as we will see in the next section.

4.4 Magnetic convection–diffusion ("freezing" law)

4.4.1 The magnetic induction equation (convection–diffusion equation)

The time evolution of the magnetic field is given by Faraday's law:

$$\frac{\partial \mathbf{B}}{\partial t} = -\nabla \times \mathbf{E}. \tag{4.66}$$

The time rate of change of $\mathbf{B}$ depends on the curl of the electric field. You will not be surprised to find that we will use the generalized Ohm's law to supply $\mathbf{E}$. Starting with Equation (4.62), we can again neglect the electron inertial terms. And even when the Hall and pressure gradient terms are not that small, the curls of these terms are usually small. We are left with

$$\mathbf{E} = -\mathbf{u} \times \mathbf{B} + \eta \mathbf{J}, \tag{4.67}$$

which is just Equation (4.65) rearranged.

Combining Equations (4.66) and (4.67) we obtain

$$\frac{\partial \mathbf{B}}{\partial t} = \nabla \times (\mathbf{u} \times \mathbf{B}) - \nabla \times (\eta \mathbf{J}). \tag{4.68}$$

Using Equation (4.61) (Ampère's law) to eliminate the current density in favor of the magnetic field we obtain the *magnetic convection–diffusion equation*:

$$\frac{\partial \mathbf{B}}{\partial t} = \underbrace{\nabla \times (\mathbf{u} \times \mathbf{B})}_{\substack{\text{magnetic} \\ \text{convection}}} - \underbrace{\nabla \times \left(D_B \nabla \times \mathbf{B}\right)}_{\substack{\text{magnetic} \\ \text{diffusion}}}, \tag{4.69}$$

where the *magnetic diffusion coefficient* is given by

$$D_B = \eta/\mu_0 = \frac{m_e \nu_e}{\mu_0 n_e e^2}. \tag{4.70}$$

Equation (4.63) for the resistivity was used in Equation (4.70). The two terms of the right-hand side of Equation (4.69) have been labeled "magnetic convection" and "magnetic diffusion" for reasons you will see below. We can simplify the magnetic diffusion term if $\nabla D_B \times (\nabla \times \mathbf{B}) = 0$ can be assumed, as is true for a medium of uniform resistivity. We can then employ a vector calculus identity and the Maxwell equation $\nabla \cdot \mathbf{B} = 0$ to get

$$\nabla \times \nabla \times \mathbf{B} = \underbrace{\nabla(\nabla \cdot \mathbf{B})}_{\text{equals 0}} - (\nabla \cdot \nabla)\mathbf{B}. \tag{4.71}$$

With the above simplifications, Equation (4.69) becomes

$$\frac{\partial \mathbf{B}}{\partial t} = \nabla \times (\mathbf{u} \times \mathbf{B}) + D_B \nabla^2 \mathbf{B}, \tag{4.72}$$

taking the form of a traditional diffusion-type equation in its last term.

4.4.2 Frozen-in magnetic flux

Let us compare the order of magnitude of the convection term (i.e., first term on the right-hand side) of Equation (4.72) to the order of magnitude of the diffusion term. That is, suppose L is the typical spatial scale over which the variables B and u vary significantly, and suppose that τ is the typical time constant for temporal variation of B. The "dimensional" analysis of Equations (4.69) or (4.72) gives us the following time constant (for $D_B = 0$):

$$\frac{B}{\tau} \sim \frac{u \cdot B}{L} \quad \text{or} \quad \tau \sim L/u. \tag{4.73}$$

For $u = 0$ (no convection), another simple dimensional analysis provides us with a typical magnetic diffusion time constant of

$$\tau \sim L^2/D_B. \tag{4.74}$$

Thus, for a convection-dominated situation, the time scale for evolution of B is directly proportional to the length scale L, whereas for a diffusion-dominated situation, the time scale varies as the square of L (as is typical for diffusion problems of all types).

How do we determine whether magnetic diffusion or convection is more important for a particular plasma regime? We compare the magnitude of the two terms to obtain the *magnetic Reynolds number* R_m:

$$R_m = \frac{(\text{convection term})}{(\text{diffusion term})}$$
$$\sim \frac{uB/L}{D_B B/L^2} \sim \frac{Lu}{D_B}. \tag{4.75}$$

The magnetic Reynolds number is analogous to the ordinary *Reynolds number*, which is the ratio of the viscosity term to the convection/advection term in the Navier–Stokes equation. The Navier–Stokes equation is a form of the momentum equation that includes viscosity effects (i.e., the viscosity term is the second term on the right-hand side of Equation (2.64)).

From Equation (4.75), we see that R_m is very large ($R_m \gg 1$) and convection dominates for small values of the magnetic diffusion coefficient D_B (i.e., for low resistivity/high electrical conductivity). For large values of D_B, the magnetic Reynolds number R_m is small ($R_m \ll 1$) in which case magnetic diffusion (i.e., Ohmic dissipation of currents) dominates the time evolution of the magnetic field. However, notice that even for small values of D_B (as is the case for most space plasmas) a small value of R_m results if the values of L and/or u are small enough. As we shall see in Chapter 8, this can happen in a narrow current sheet such as the one located at the Earth's magnetopause. For $R_m = 1$, magnetic convection and diffusion are of comparable importance.

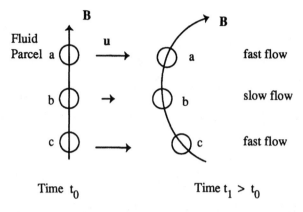

Figure 4.4. Schematic of frozen-in magnetic field. Field line stays attached to particular fluid parcels.

Magnetic flux is frozen into the plasma flow for very large values of the magnetic Reynolds number ($R_m \to \infty$). What is meant by "frozen-in magnetic flux"? The convection–diffusion equation for $D_B = 0$ simply becomes

$$\frac{\partial \mathbf{B}}{\partial t} = \nabla \times (\mathbf{u} \times \mathbf{B}). \tag{4.76}$$

If Equation (4.76) holds, then the following theorem also holds:

Theorem 4.1 *The magnetic flux through a closed loop within the (infinite conductivity) fluid, and moving with the fluid, remains constant over time.*

The proof of this theorem is not given here but can be found in many plasma physics books such as are listed in the bibliography at end of this chapter (e.g., Siscoe, 1982).

Consider the schematic example shown in Figure 4.4. A field line (or part of a field line) can be thought of as being "tied" to a particular parcel of fluid if $R_m \gg 1$. The field line gets distorted and stretched as the faster fluid with its piece of the field line outruns the slower fluid with its piece of the field line. The concepts of magnetic convection and diffusion can also be illustrated for a one-dimensional geometry, as we shall do in the next section.

4.4.3 One-dimensional convection–diffusion equation

A one-dimensional version of the magnetic convection–diffusion equation (4.72) is easier to understand than the full equation. Assume that the magnetic field is only in the x direction, $\mathbf{B} = B(z, t)\hat{\mathbf{x}}$, and that the flow is only in the z direction with a velocity $\mathbf{u} = u(z)\hat{\mathbf{z}}$ (see Figure 4.5).

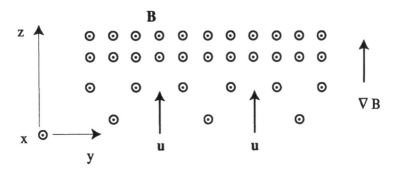

Figure 4.5. Simple one-dimensional geometry for magnetic convection and diffusion.

With this choice of coordinates, the convection–diffusion Equation (4.72) becomes

$$\frac{\partial B(z, t)}{\partial t} = -\frac{\partial}{\partial z}(uB) + D_B \frac{\partial^2 B}{\partial z^2}. \tag{4.77}$$

For $u = 0$ (and for $R_m = 0$), we have purely one-dimensional magnetic diffusion. A localized magnetic field enhancement with scale size L decays with the time constant given by Equation (4.74). In Problem 4.6, you are asked to solve Equation (4.77) (with $\mathbf{u} = 0$) and find the time evolution of B for a given initial profile. You will find that the small-scale variations of B decay most rapidly, as expected from Equation (4.74).

Now let us assume that $R_m \gg 1$, in which case we can neglect the magnetic diffusion term in Equation (4.77). Equation (4.77) then tells us that for a plasma where uB increases with z, the magnetic field at a fixed value of z decreases with time. In Figure 4.6 a steady-state scenario is shown. For this scenario, the plasma slows down and then speeds up as it moves from left to right. The steady-state ($\partial B / \partial t = 0$) solution of Equation (4.77) is obviously

$$u(z)B(z) = constant. \tag{4.78}$$

As the plasma slows down (smaller values of u), the frozen-in field lines "pile up" and B becomes larger. As the plasma speeds up (u increases), the field lines "spread out" and B becomes smaller. A traffic jam on a highway is analogous – as the speed of the traffic slows, cars pile up.

If the resistivity of a plasma is nonzero (i.e., $D_B \neq 0$), then magnetic flux can slip or "thaw." That is, field lines are no longer tightly tied to, or frozen to, particular fluid parcels. In this case, magnetic field enhancements tend to decrease due to diffusion of magnetic flux away from the enhancement. Equivalently, the field enhancement shown in Figure 4.5 is actually created by an electrical current in the y direction. Magnetic diffusion means that the electrical current undergoes Ohmic dissipation.

You should be aware of a complication that occurs for many space plasmas. The collisional resistivity in the solar wind and magnetosphere (but not in the

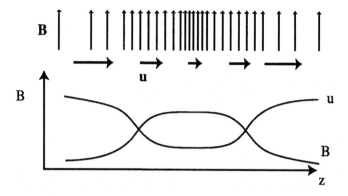

Figure 4.6. Schematic showing frozen-in magnetic field for one-dimensional plasma flow. A steady-state solution, $uB = constant$, is shown.

ionosphere) is virtually zero. And indeed we find that $R_m \gg 1$ *almost* everywhere in these plasma environments. However, near narrow current layers a phenomenon called *magnetic reconnection* takes place that requires $R_m \leq 1$ over a narrow region. However, the magnetic reconnection regions studied in space physics have turned out to be much broader than one would expect using the collisional form of the resistivity (i.e., Equation (4.70)). It seems that some *anomalous resistivity* is required and is thought to originate from microscopic (but still collisionless) plasma processes. A discussion of these processes is outside the scope of this text, but we can still use the convection–diffusion equation, as well as address the phenomenon of magnetic reconnection, by adopting suitable anomalous resistivity coefficients. We discuss magnetic reconnection at the end of this chapter and in Chapter 8.

4.5 The magnetohydrodynamic equations

Let us write again the single-fluid equations (Equations (4.57) and (4.58)) but include the diffusion–convection equation for the magnetic field:

$$\frac{\partial \rho}{\partial t} + \nabla \cdot (\rho \mathbf{u}) = m_i S_i, \tag{4.79}$$

$$\rho \frac{\partial \mathbf{u}}{\partial t} + \rho \mathbf{u} \cdot \nabla \mathbf{u} = -\nabla p + \mathbf{J} \times \mathbf{B} + \rho \mathbf{g} - \rho \nu (\mathbf{u} - \mathbf{u}_n)$$
$$- P_i m_i (\mathbf{u} - \mathbf{u}_n), \quad \text{with } p = p_e + p_i, \tag{4.80}$$

$$\frac{\partial \mathbf{B}}{\partial t} = \nabla \times (\mathbf{u} \times \mathbf{B}) - \nabla \times (D_B \nabla \times \mathbf{B}). \tag{4.81}$$

Alternatively, in place of (4.81) we could use Equation (4.60):

$$\nabla \cdot \mathbf{J} = 0. \tag{4.82}$$

Generally, we also need separate energy relations for electrons and ions, such as those given in Chapter 2. However, for now we merely write the polytropic relation (see Equation 2.81) in terms of some reference density and pressure, ρ_0 and p_0, respectively:

$$p/p_0 = (\rho/\rho_0)^\gamma, \tag{4.83}$$

where $\gamma = 5/3$ for an ideal monatomic gas and $\gamma = 1$ for an isothermal gas.

Equations (4.79)–(4.83) together comprise the *equations of magnetohydrodynamics* (i.e., *the MHD equations*). These equations can be used in many ways and with many different approximations. Next we introduce the concept of magnetic pressure.

4.5.1 Magnetic pressure

The term $\mathbf{J} \times \mathbf{B}$ in Equation (4.80) is the *"Maxwell" force (per unit volume)* on a magnetized plasma due to electrical currents. You might recall from your elementary physics course that the force on a length l of a straight wire, carrying current I, in magnetic field $\mathbf{B}$ is given by

$$\mathbf{F} = I\mathbf{l} \times \mathbf{B} \quad [N], \tag{4.84}$$

where the direction of $\mathbf{l}$ is the same as that of the current. This is just the "electric motor" force. $\mathbf{J} \times \mathbf{B}$ is the analogous force per unit volume on a plasma.

The current density $\mathbf{J}$ can be eliminated from $\mathbf{J} \times \mathbf{B}$ by using Ampère's law (minus the displacement current):

$$\mathbf{J} = \frac{1}{\mu_0} \nabla \times \mathbf{B}.$$

Thus

$$\mathbf{J} \times \mathbf{B} = \frac{1}{\mu_0}(\nabla \times \mathbf{B}) \times \mathbf{B}$$

$$= -\nabla\left(\frac{B^2}{2\mu_0}\right) + \frac{1}{\mu_0}\mathbf{B} \cdot \nabla\mathbf{B} \tag{4.85}$$

where a vector calculus identity, which can be found in most electromagnetics textbooks, was used in the final step. Using Equation (4.85) we can rewrite the momentum equation (4.80) as

$$\rho\frac{\partial \mathbf{u}}{\partial t} + \rho\mathbf{u} \cdot \nabla\mathbf{u} = -\nabla(p + B^2/2\mu_0) + \frac{1}{\mu_0}\mathbf{B} \cdot \nabla\mathbf{B}$$

$$+ \rho\mathbf{g} - \rho v(\mathbf{u} - \mathbf{u}_n) - P_i m_i(\mathbf{u} - \mathbf{u}_n). \tag{4.86}$$

It is apparent that the quantity $B^2/2\mu_0$ acts on the fluid in the same manner as the thermal pressure p; hence, this quantity is called *magnetic pressure*:

$$p_B = \frac{B^2}{2\mu_0} \quad \left[\text{units of } \frac{N}{m^2}\right]. \tag{4.87a}$$

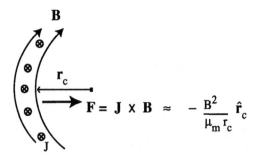

Figure 4.7. Tension force **F** (per unit mass) on palsma due to curved magnetic field lines. Curved field lines imply a current density **J**, as shown, due to Ampère's law. **J** × **B** is in the same direction as $-\hat{\mathbf{r}}_c$.

In cgs units the magnetic pressure takes the form

$$p_B = \frac{B^2}{8\pi} \quad [\text{dynes/cm}^2]. \tag{4.87b}$$

B has units of tesla in Equation (4.87a) and Gauss in (4.87b). A gradient in magnetic pressure gives rise to a force on the plasma. The magnetic pressure gradient requires a spatial variation of the magnitude of **B**, with the force pointing from the high-field region to the low-field region. The concept of magnetic pressure will become more clear to you shortly when we consider static force equilibrium for a magnetized plasma. The importance of the thermal pressure term relative to the magnetic pressure term in Equation (4.86) can be approximately judged by the ratio of these two terms, which is given the name of *plasma beta*:

$$\beta \equiv \frac{p}{p_B} = \frac{n_e k_B (T_e + T_i)}{B^2/2\mu_0}. \tag{4.88}$$

The second part of **J** × **B** as given by Equation (4.85) (or in Equation (4.86)) represents the force on the plasma due to curvature of field lines. This is the *magnetic tension force*, which we can write very approximately as

$$\frac{1}{\mu_0} \mathbf{B} \cdot \nabla \mathbf{B} \approx \frac{1}{\mu_0} \frac{B^2}{r_c}(-\hat{\mathbf{r}}_c), \tag{4.89}$$

where $\mathbf{r}_c$ is the radius of curvature of the field lines; see Figure 4.7. An analogy can be made to the force on a curved string under *tension*.

4.6 Static equilibrium

An important subclass of problems is one in which the plasma is stationary, or almost stationary: **u** = **0**. In this case, we can greatly simplify the momentum equations (4.80), or (4.86), and obtain the following *static force balance relation*:

$$\mathbf{0} = -\nabla p + \mathbf{J} \times \mathbf{B} + \rho \mathbf{g}. \tag{4.90}$$

We have further assumed that the neutral gas (if it exists) has zero velocity ($\mathbf{u}_n = \mathbf{0}$). Even Equation (4.90) can be further simplified using various approximations.

4.6.1 Hydrostatic balance

We first take the special case of an unmagnetized plasma ($\mathbf{B} = \mathbf{0}$), in which case Equation (4.90) becomes

$$\nabla p = \rho \mathbf{g}. \tag{4.91}$$

This is the *(hydro)static force balance relation*. This relation is especially useful for horizontally stratified atmospheres (e.g., the planets and the Sun), for which the acceleration due to gravity (downward) can be written as $\mathbf{g} = -g\hat{\mathbf{z}}$. We have $g = 9.88 \, \text{m/s}^2$ for the Earth. In this case, all quantities are functions only of z (altitude), and $\hat{\mathbf{z}}$ is a unit vector directed up. An equation of state can be used to express the pressure p in terms of the density ρ and temperature T: $p = \rho \tilde{R} T$. $\tilde{R}$ is the gas constant appropriate for the particular fluid/plasma under consideration. Equation (4.91) becomes

$$\frac{\partial p}{\partial z} = -\rho g \tag{4.92}$$

$$= -\frac{g}{\tilde{R} T} p.$$

The solution of (4.92) is simply

$$p(z) = p_0 \exp\left\{-\int_{z_0}^{z} \frac{g \, dz'}{\tilde{R} T(z')}\right\}, \tag{4.93}$$

where p_0 and z_0 are reference values of pressure and altitude, respectively. For an isothermal atmosphere ($T = \text{constant}$) and constant g (true for a limited altitude range), Equation (4.93) becomes

$$p(z) = p_0 e^{-(z-z_0)/H} \tag{4.94}$$

with *scale height*

$$H \equiv \frac{\tilde{R} T}{g}. \tag{4.95}$$

The pressure decreases with height in an exponential manner with an *e*-folding length called the *scale height H*. For an isothermal atmosphere, the density ρ is directly proportional to p and also decreases with increasing altitude in an exponential manner.

Example 4.2 (Hydrostatic balance in an isothermal neutral atmosphere) Consider a neutral atmosphere with uniform temperature $T = T_n$ and gas constant

$\tilde{R} = k_B/\bar{m}$, where $\bar{m}$ is the mean molecular mass. For air on Earth, $\tilde{R}$ is the ordinary gas constant and we can write Equation (4.95) to obtain the *neutral scale height*:

$$H_n = \frac{k_B T_n}{\bar{m} g}. \tag{4.96}$$

H_n obviously depends on the parameters T_n, $\bar{m}$, and g. For Earth, $T_n \cong 300$ K, $g = 9.88$ m/s^2, and $\bar{m} = 28.8$ amu (20% O_2 and 80% N_2), in which case

$$H_n = 8.7\,\text{km}.$$

For Jupiter, near the cloudtops, $T_n \cong 300$ K, $\bar{m} \cong 2$ amu (mostly H_2), and $g = 23$ m/s^2. The scale height in the Jovian atmosphere is then

$$H_n \cong 52\,\text{km}.$$

The pressure for a plasma must include both the electron and ion partial pressures: $p = p_e + p_i = n_e k_B (T_e + T_i)$, where we have used separate equations of state for electrons and ions plus we have assumed quasi-neutrality. Equations (4.92)–(4.95) still apply if we identify T as $(T_e + T_i)$ and use $\tilde{R} = 2k_B/(m_e + m_i)$. For isothermal electron and ion temperatures, we can easily show (see Problem 4.7) that for a horizontally stratified plasma near a planet (that is, for an ionosphere) the electron density varies with altitude exponentially:

$$n_e(z) = n_{e0} e^{-(z - z_0)/H_p} \tag{4.97}$$

with *plasma scale height*

$$H_p \cong \frac{k_B(T_e + T_i)}{m_i g}. \tag{4.98}$$

Note that this plasma scale height expression includes both the electron and ion temperatures, unlike the neutral scale height, which only included T_n. The presence of the electron temperature term is really due to the electron pressure gradient term in the generalized Ohm's law, Equation (4.45), that was used to eliminate $\mathbf{E}$ from the momentum equation.

Figure 4.8 shows an electron density profile measured in the mid-latitude terrestrial ionosphere (see the discussion in Rees, 1989). Hydrostatic equilibrium only applies for altitudes above about 300 km, where a maximum exists in the electron density profile. This ionospheric region above the maximum is called the *topside ionosphere*. In the lower ionosphere, on the other hand, chemistry is the controlling process. Note that the topside electron density has an exponential fall-off. The major ion species in the ionospheric "F-region" is O^+ ($m = 16$ amu). The temperatures in the ionosphere are $T_e \approx 3,500$ K and $T_i \approx 1,500$ K; hence we can calculate from

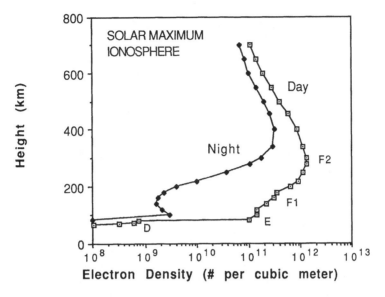

Figure 4.8. Electron density versus altitude in the terrestrial F-region ionosphere versus for solar maximum conditions. Typical day and night profiles are shown. The locations of the D, E, F1, and F2 regions are indicated. Note that the standard convention in the atmospheric sciences is to plot the dependent variable as the abscissa and altitude as the ordinate.

Equation (4.98) that the plasma scale height is

$$H_p \cong 260 \, \text{km}.$$

4.6.2 Static pressure balance in a planar geometry for a magnetized plasma

Now we consider another approximation to the static force balance relation (4.90). We neglect gravity ($g = 0$), so that the equation simply becomes $\mathbf{J} \times \mathbf{B} = \nabla p$. Furthermore, we assume a planar magnetic geometry (i.e., straight magnetic field lines). In this case, we can omit the second term on the right-hand side of expression (4.85) for $\mathbf{J} \times \mathbf{B}$, so that Equation (4.90) becomes

$$\nabla(p + B^2/2\mu_0) = 0, \tag{4.99}$$

which can be integrated to give

$$p + B^2/2\mu_0 = constant \text{ in a direction normal to } \mathbf{B} \tag{4.100}$$

or

$$p_e + p_i + p_B = p_{\text{total}} = constant.$$

$p_B = B^2/2\mu_0$ is obviously acting like pressure here. Equation (4.100) simply states that the total pressure remains constant; any increase in magnetic pressure must be

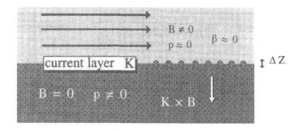

Figure 4.9. Static balance between thermal and magnetic pressure.

compensated for by a decrease in the thermal pressure or vice versa. Consider the following simple illustration of a static pressure balance at an interface between a magnetized, but very low density, plasma and a high-density unmagnetized plasma (Figure 4.9).

To satisfy Equation (4.100), the following condition must apply at the interface between the two media:

$$p = \frac{B^2}{2\mu_0}, \tag{4.101}$$

where p is the thermal pressure in the bottom layer and B is the field strength in the upper layer. However, you should not overlook that what is "really" balancing the thermal pressure at the interface is the $\mathbf{J} \times \mathbf{B}$ force integrated over the narrow extent of the interface. A narrow layer of electrical current, or a *current layer*, is located at the interface. Its current density (per unit length), given by Ampère's law, is

$$K = \int_{\Delta z} J \, dz = B/\mu_0, \tag{4.102}$$

where Δz is the thickness of the current layer. This current is called the *diamagnetic current*. The "magnetic" (i.e., "electric motor") force per unit area (or pressure) is given by

$$p_B = \int_{\Delta z} (\mathbf{J} \times \mathbf{B}) \cdot \hat{\mathbf{z}} \, dz = |\mathbf{K} \times \langle \mathbf{B} \rangle_{\Delta z}| = \frac{B}{\mu_0} \cdot \frac{B}{2} = \frac{B^2}{2\mu_0}. \tag{4.103}$$

This magnetic pressure must equal the thermal pressure p in order to keep the interface stationary (or static).

Example 4.3 (The Venus ionopause) The Venus ionopause provides a nice example of a static pressure balance between magnetic and thermal pressures. Figure 4.10 shows the magnetic field strength and the plasma density as functions of height above the surface of Venus for three orbital passes of the *Pioneer Venus Orbiter (PVO)*. *PVO* was launched by NASA and went into orbit around Venus in 1978. The orbit was highly elliptical and so during an orbital pass the ionosphere

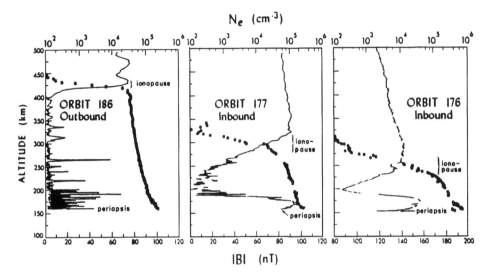

Figure 4.10. Measured magnetic field strength (*Pioneer Venus Orbiter* magnetometer – Russell and Vaisberg, 1983) and measured electron density (Langmuir probe) in the ionosphere of Venus for three orbits. The magnetic field is shown as a solid line and uses the bottom scale. The electron densities are shown as solid dots and use the upper scale. (From Russell and Vaisberg, 1983.)

was sampled above the *periapsis* (i.e., minimum) altitude, which was typically 145 km. Instruments on board measured plasma properties including electron and ion densities and temperatures as well as the magnetic field (Russell and Vaisberg, 1983). The *ionopause* is defined as the region where the ionospheric plasma stops, which is where the solar wind plasma begins.

The plasma just above the ionopause has very low density (and pressure) but is highly magnetized due to the compression of the interplanetary magnetic field. During Orbit 186 the solar wind "pressure" was low so that the magnetic field strength above the ionopause was only about 70 nT. In this case, the magnetic field strength in the ionosphere itself is essentially zero, on the average. However, for Orbits 176 and 177 the solar wind pressure was high, as was the magnetic field strength above the ionopause. Orbits 176 and 177 will not be discussed here, but we will return to them in Chapter 7. The solar wind interaction with Venus is briefly discussed in Chapter 7; what concerns us here is that a pressure balance between thermal and magnetic pressure has been experimentally demonstrated to exist at the Venus ionopause. You can also notice in Figure 4.10 that the ionospheric density (and the pressure, because the electron and ion temperatures are approximately independent of altitude) falls off exponentially as expected from hydrostatic equilibrium, except in the ionopause region.

Now we calculate for Orbit 186 the magnetic pressure just above the ionopause and the thermal pressure just below the ionopause and show that these are equal to each other – that is, we will now demonstrate that Equation (4.101) is satisfied. First, note that above the ionopause $B = 70$ nT so that the magnetic pressure is $B^2/2\mu_0 = 2 \times 10^{-9}$ N/m^2. Next note that the electron density in the ionosphere, just below the ionopause, is $n_e = 3 \times 10^4$ cm^{-3}. The measured electron and ion temperatures are $T_e = 3,800$ K and $T_i = 1,800$ K, respectively. We find that the thermal pressure is $p = n_e k_B (T_e + T_i) = 2.2 \times 10^{-9}$ N/m^2, which is equal to the magnetic pressure to within 10% (roughly the error of our estimates). Equation (4.101) is indeed satisfied at the Venus ionopause.

4.6.3 Force-free magnetic equilibrium

Now we consider yet another case of a static force balance in which we suppose that $\nabla p \approx 0$ and $g = 0$. Then the static force balance relation, (4.90), is simply

$$\mathbf{J} \times \mathbf{B} = 0. \tag{4.104}$$

This deceptively simple equation can be satisfied in one of two ways: (1) $\mathbf{J} = \mathbf{0}$, but this is not a particularly interesting case, and (2) $\mathbf{J} \parallel \mathbf{B}$; that is, $\mathbf{J}$ is everywhere parallel to $\mathbf{B}$, which results in a *force-free* magnetic structure. A force-free structure exists where the magnetic pressure gradient force part of $\mathbf{J} \times \mathbf{B}$ exactly counterbalances the tension force part of $\mathbf{J} \times \mathbf{B}$ (see Equation (4.85)).

Especially interesting are cylindrically symmetric force-free structures with magnetic field vector $\mathbf{B} = \mathbf{B}(r)$, where r is the radial distance from the axis. For this functional form of the magnetic field, the force relation $\mathbf{J} \times \mathbf{B} = 0$ is satisfied only if $\mathbf{J}$ is proportional to $\mathbf{B}$ multiplied by some scalar function of r; that is, the following relation must be satisfied:

$$\nabla \times \mathbf{B} = \alpha(r)\mathbf{B}, \tag{4.105}$$

where $\alpha(r)$ is some function only of r. In Problem 4.11 you are asked to find $\mathbf{B}(r)$ for the special case of $\alpha = constant$.

Force-free (or almost force-free) magnetic structures are present in many space plasma environments, such as in the solar corona, in the solar wind, and in the ionosphere of Venus. Appearing in the measured magnetic field profile in the ionosphere of Venus for Orbit 186 (Figure 4.10) are narrow spikes that are about 10 km across. A detailed analysis of the magnetic field vector in one of these spikes reveals a ropelike structure, as illustrated in Figure 4.11 (Russell and Elphic, 1979; Elphic et al., 1980). These structures have been given the name *magnetic flux ropes*. Analysis indicates that these ropes are essentially force free.

4.6.4 Stability

We have just finished a discussion of static equilibria. However, not all equilibria are stable. A circus performer balanced on a high wire is in a state of equilibrium,

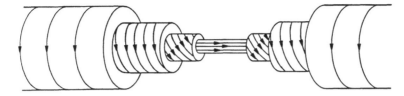

Figure 4.11. Magnetic flux rope structure as deduced from *Pioneer Venus Orbiter* magnetometer measurements. The flux rope is dissected so that its internal structure is visible. Reprinted with permission from *Nature* (Russell and Elphic, *Nature*, **279**, 616, 1979). Copyright (1979) Macmillan Magazines Limited.

albeit an unstable one. A slight unfortunate move to the left or right and (barring other action) the performer is no longer in a "static" equilibrium. Consider a ball in the bottom of a bowl; it is in a *stable* mechanical equilibrium, in that a slight departure from the point of equilibrium results in a restoring force (i.e., gravity in this example) that maintains equilibrium. In contrast, a ball balanced on the top of an overturned bowl is in an *unstable* equilibrium; a slight perturbation from the equilibrium configuration results in an acceleration away from the equilibrium. *Neutral stability* is also possible.

MHD equilibrium states can be stable or unstable. Let us consider the current layer configuration pictured in Figure 4.9. Without gravity, this configuration is neutrally stable. But suppose we introduce gravity. The equilibrium is stable because the low-β and low-density plasma is on top and the high-β, high-density plasma is on the bottom. The stability condition can be verified by displacing a small parcel of plasma from near the interface and determining whether the parcel seeks to return to its equilibrium position (i.e., stable) or not (i.e., unstable). If the gravitational acceleration **g** is directed downward in Figure 4.9, a parcel of nonmagnetized dense fluid that is moved up into the low-β, low-density region is denser than its new surroundings and, consequently, has negative buoyancy (see Problem 4.8 for a discussion of buoyancy). The force on this parcel is downward, thus restoring it to its original equilibrium position. Similarly, a parcel of magnetized plasma that is moved downward into the high-β region becomes less dense than the medium surrounding it and buoyancy results in an upward restoring force.

Suppose though that the denser fluid lies on top of the less dense fluid (i.e., invert the two regions in Figure 4.9 but keep **g** directed down). Then an upward (downward) displacement of a fluid parcel into the neighboring region results in positive (negative) buoyancy that leads to an acceleration of the parcel away from the equilibrium position (see Figure 4.12). This instability is called the *Rayleigh–Taylor instability* and it occurs whenever a lighter fluid supports a heavier fluid in a gravitational field. An example of a situation that is Rayleigh–Taylor unstable is a layer of oil supported by a layer of water. You can demonstrate this yourself with a

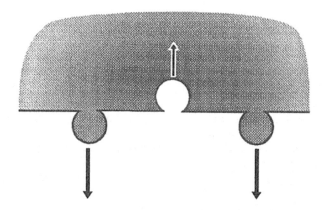

Figure 4.12. Rayleigh–Taylor instability of a heavy fluid supported by a lighter fluid.

container of oil-and-vinegar salad dressing. The Venus ionopause is stable against the Rayleigh–Taylor instability.

Other types of fluid instabilities exist, such as *streaming instabilities*, in whch plasma drifts can upset the equilibrium configuration of a plasma. An important example of a streaming instability is the *Kelvin–Helmholtz instability*. Without gravity, the diamagnetic boundary shown in Figure 4.9 is neutrally stable. But suppose the plasma on either side of the interface is drifting tangentially at different speeds; that is, suppose a *velocity shear* is present at the boundary. In this case, ripples that might be present at the boundary could grow via the Kelvin–Helmholtz instability.

The Rayleigh–Taylor and Kelvin–Helmholtz instabilities are examples of macroscopic, or fluid (or MHD), instabilities, in which the distribution functions f are Maxwellians or drifting Maxwellians. Another category of instabilities comprises *kinetic instabilities*, in which certain kinds of deviations of the distribution function from a Maxwellian distribution result in the growth of plasma perturbations or waves. This topic will not be addressed in this book (although it was briefly alluded to at the end of Chapter 3) and is left to more advanced books on plasma physics and space plasmas.

4.6.5 Diffusion in a partially ionized plasma

For a partially ionized plasma, we can simplify the momentum equation (4.86) (or, equivalently, Equation 4.80) by neglecting the left-hand side (i.e., the inertial terms). This approximation is not as extreme as the static equilibrium approximation made earlier in Section 4.6, because the friction and mass-loading terms, which contain the flow velocity **u**, are still being retained. The now simplified momentum equation can be algebraically manipulated to yield the following explicit expression for the plasma flow velocity:

$$\mathbf{u} = \mathbf{u}_n + \frac{1}{\rho\nu + P_i m_i}(-\nabla p + \mathbf{J} \times \mathbf{B} + \rho\mathbf{g}). \qquad (4.106)$$

Equation (4.106) in effect states that the ion–neutral friction force is in balance with the sum of all other forces on the plasma, including Maxwell stresses, thermal pressure gradient, and gravity.

Obviously, Equation (4.106) is applicable only to plasma environments in which the abundance of neutrals is sufficiently high such that a reasonably large friction term ($\rho\nu$) or mass-loading term ($P_i m_i$) exists. Furthermore, Equation (4.106) is also applicable only if the flow velocity ($\mathbf{u}$) is small enough such that the left-hand side of Equation (4.80) (i.e., the "inertial terms") is much smaller than the largest individual term on the right-hand side. Note that if $\mathbf{u} = \mathbf{u}_n$, then the "static" equilibrium expressed by Equation (4.90) is obtained. Often, $\mathbf{u}_n \approx 0$ is a good assumption because neutral flow velocities are almost always slow in comparison with typical plasma velocities.

Let us further simplify Equation (4.106) by (1) using Equations (4.85)–(4.87b) plus a planar geometry to convert $\mathbf{J} \times \mathbf{B}$ to $-\nabla p_B$, (2) assuming $P_i m_i \ll \rho\nu$ (neglecting mass-loading), and (3) assuming $\mathbf{u}_n = 0$ (stationary neutrals). We obtain

$$\mathbf{u} = -\frac{1}{\rho\nu}(\nabla p_{\text{tot}} - \rho\mathbf{g}), \qquad (4.107)$$

where $p_{\text{tot}} = p_B + p = B^2/2\mu_0 + n_e k_B(T_e + T_i)$. Equation (4.107) is one version of the *plasma diffusion equation*, and it is called the *ambipolar diffusion equation*. This equation states that plasma flows in response to gradients in the total pressure (plus gravity).

The diffusion equation is especially useful for determining the plasma flow in planetary ionospheres. In particular, we can easily convert Equation (4.107) to a form that is useful for plasma transport along a strong external magnetic field, such as is present in the terrestrial ionosphere. First, we assume that $\mathbf{g} = -g\hat{\mathbf{z}}$ (as in Section 4.6.1), where z is altitude. The magnetic field, $\mathbf{B} = B\hat{\mathbf{b}}$, is almost uniform within the relatively narrow ionosphere layer. The angle of inclination of the magnetic field with respect to the horizontal plane, θ_I, is given by (see Figure 4.13)

$$\sin\theta_I = +\hat{\mathbf{b}}\cdot\hat{\mathbf{z}}. \qquad (4.108)$$

Starting from Equation (4.107), we can derive an equation for the plasma flow speed (or diffusion velocity) along the magnetic field:

$$u_\parallel = -\frac{1}{m_i n_e \nu}\left\{\frac{\partial}{\partial s}[n_e k_B(T_e + T_i)] + n_e g m_i \sin\theta_I\right\}. \qquad (4.109)$$

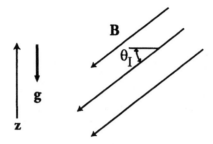

Figure 4.13. Plasma diffusion along a strong magnetic field in a planetary ionosphere. $u_\parallel$ is the plasma flow speed along **B**, and s is the distance along **B**.

A single ion species with mass m_i has been chosen for simplicity. Distance along the magnetic field is denoted s. The component of $\mathbf{u} = u_\parallel \hat{\mathbf{b}}$ in the z direction is the most important flow component for a stratified plasma medium such as a planetary ionosphere. Equation (4.109) can also be expressed (Problem 4.12) as

$$u_z = -D_a \sin^2 \theta_I \left[\frac{1}{n_e} \frac{\partial n_e}{\partial z} + \frac{1}{H_p} + \frac{1}{(T_e + T_i)} \frac{\partial}{\partial z}(T_e + T_i) \right], \qquad (4.110)$$

where the *ambipolar diffusion coefficient* D_a is given by

$$D_a = \frac{k_B(T_e + T_i)}{m_i \nu} \qquad (4.111)$$

and where H_p is the plasma scale height, given by Equation (4.98).

We can combine Equation (4.110) and a one-dimensional version of the continuity Equation (4.79) to find an expression for the electron density n_e:

$$\frac{\partial n_e}{\partial t} + \frac{\partial \phi_i}{\partial z} = S_i. \qquad (4.112)$$

The vertical plasma flux is $\phi_i = n_e u_z$ and S_i is the net plasma source.

Equation (4.112) combined with Equation (4.110) has the form of a standard diffusion equation for an isothermal gas if $\mathbf{g} = \mathbf{0}$ (Problem 4.12).

The ion–neutral collision frequency (ν) is proportional to the atmospheric neutral density, which decreases rapidly with increasing altitude. Consequently, D_a, which is inversely proportional to ν, is small at low altitudes (and u_z is small, even for large density gradients) and large at high altitudes. At high altitudes it is very often true that $|u_z/D_a| \ll 1/H_p$, in which case we can set $u_z = 0$ (this approximation is called *diffusive equilibrium*) and reobtain for an isothermal plasma the hydrostatic profile given by Equation (4.97).

The ambipolar diffusion equation will be discussed again in Chapter 7. Note that diffusion equations for individual ion species in a multi-ion species plasma can be also be found by starting with Equations (2.61) and (2.64), instead of the single-fluid momentum equation, by neglecting the inertial terms (i.e., left-hand side) and by solving for $\mathbf{u}_s$.

4.6.6 Dynamic pressure

The inertial terms in the momentum equation are important for fast flows. We now consider a steady-state ($\partial/\partial t = 0$), collisionless ($\nu = 0$), source-free ($S_i = P_i = 0$) plasma. We also consider flow only in one direction with $\mathbf{u} = u(x)\hat{\mathbf{x}}$ and with the magnetic field normal to the flow direction, $\mathbf{B} \perp \mathbf{u}$. With these assumptions, the x component of the momentum Equation (4.86) becomes

$$\rho u \frac{\partial u}{\partial x} = -\frac{\partial}{\partial x}(p + B^2/2\mu_0). \tag{4.113}$$

This equation can be further transformed (Problem 4.13) into

$$\frac{\partial}{\partial x}(\rho u^2 + p + B^2/2\mu_0) = 0$$

or

$$\rho u^2 + p + B^2/2\mu_0 = C_1, \tag{4.114}$$

where C_1 is a constant independent of x. That is, $\rho u^2 + p + B^2/2\mu_0$ is constant along a *streamline*.

If the total pressure $p + B^2/2\mu_0$ decreases along a streamline, then ρu^2 must increase as prescribed by Equation (4.113) (and vice versa). Thus, ρu^2 acts like pressure and is given the name *dynamic pressure*. Although the simple expression (4.113) is not strictly applicable to more complicated flow patterns, ρu^2 nonetheless provides a good estimate of the relative importance of the $\rho \mathbf{u} \cdot \nabla \mathbf{u}$ term to other terms in the MHD momentum equation. This term is important if the ratio of ρu^2 to the thermal pressure p is large; this ratio is closely related to the sonic Mach number, which we will discuss in Section 4.8.

4.7 MHD energy relations

Both electron and ion pressures appear in the MHD momentum equation (4.80), and thus a complete treatment of a plasma must include energy relations for both plasma species. Often, in place of a complete energy equation, a simple polytropic relation is used to relate the total thermal pressure p to the density ρ (see Equation 4.83) or to relate the partial pressure of an individual species, p_s, to its density ρ_s (see Equation 2.81). However, sometimes a more accurate treatment of the plasma energetics is required.

The energy equations we saw in Section 2.4.5 can be used separately for electrons and ions. The types of approximations that are reasonable can be quite different for electrons than for ions. For ionospheric plasmas, the heat conduction Equation (2.82) is usually appropriate for both species; even the local heat balance relation (2.84) often provides an adequate description of the energetics in the lower ionosphere (altitudes less than 200 km) where the neutral density is high. Local heat balance is often an especially good assumption for the ions if the ionospheric flow is slow (i.e., subsonic, $u \ll u_{i\,\mathrm{therm}}$). However, for fast flows ($u > u_{i\,\mathrm{therm}}$), it becomes necessary to use a more complete energy equation such as those given by Equations (2.73), (2.78), or (2.79).

An equation for the "total" energy (electron thermal energy and bulk kinetic energy, ion thermal energy and bulk kinetic energy, and electromagnetic field energy) is often convenient for MHD problems where we really don't wish to separate out individual plasma species but would rather just deal with the *total* thermal pressure p.

4.7.1 Combined MHD energy relation

We now combine the energy relation (2.73) as applied to both electrons ($s = e$) and ions ($s = i$). Energy density terms for electromagnetic fields (internally or externally generated) must be included. The total energy density (units of J/m^3) of the plasma is then given by

$$W = \sum_{s=e,i} \rho_s \left[U_s + \frac{1}{2} u_s^2 + U_{\mathrm{grav}} \right] + \frac{1}{2} \varepsilon_0 E^2 + B^2/2\mu_0, \qquad (4.115)$$

where you recall from Chapter 2 that the internal energy density for species s is $\rho_s U_s = [1/(\gamma_s - 1)]p_s$, U_{grav} is the gravitational potential, $(1/2)\varepsilon_0 E^2$ is the energy density associated with the electric field, and $B^2/2\mu_0$ is the energy density associated with the magnetic field (as well as being the magnetic pressure). Recognizing that the electron mass density, $\rho_e = m_e n_e$, is much less than the ion mass density, $\rho_i = n_i m_i$, we can write the total plasma energy density as

$$W \cong \frac{1}{\gamma_e - 1} p_e + \frac{1}{\gamma_i - 1} p_i + \frac{1}{2} \rho_i u_i^2 + \rho_i U_{\mathrm{grav}} + \frac{1}{2} \varepsilon_0 E^2 + B^2/2\mu_0$$

$$\cong \frac{1}{\gamma_i - 1} p + \frac{1}{2} \rho u^2 + \rho U_{\mathrm{grav}} + \frac{1}{2} \varepsilon_0 E^2 + B^2/2\mu_0, \qquad (4.116)$$

where $\rho \cong \rho_i$, $p = p_e + p_i$, $\gamma_e = \gamma_i$, and $u \cong u_i$ (see Equations (4.51)–(4.54)).

The first term in the combined energy relation should be $\partial W/\partial t$, where W is given by Equation (4.116). The next term is the combined electron and ion divergence term from Equation (2.73):

$$\nabla \cdot \sum_s \left[\rho_s \mathbf{u}_s \left(h_s + \frac{1}{2} u_s^2 + U_{\mathrm{grav}} \right) \right]$$

$$\cong \nabla \cdot \left(\frac{\gamma_e}{\gamma_e - 1} \mathbf{u}_e p_e + \frac{\gamma_i}{\gamma_i - 1} \mathbf{u}_i p_i + \frac{1}{2} \mathbf{u}_i \rho u_i^2 + \rho \mathbf{u}_i U_{\mathrm{grav}} \right), \quad (4.117)$$

where γ_e and γ_i are the ratios of specific heats for electrons and ions, respectively. A further simplification of Equation (4.117) is obtained if we take $\gamma_e = \gamma_i$ (calling it simply γ) and if $\mathbf{u}_e = \mathbf{u}_i = \mathbf{u}$:

$$\nabla \cdot \left\{ \rho \mathbf{u} \left[\frac{\gamma}{\gamma - 1} \frac{p}{\rho} + \frac{1}{2}u^2 + U_{\text{grav}} \right] \right\}. \tag{4.118}$$

The presence of the electromagnetic energy density in Equation (4.116) also necessitates the inclusion of a transport term for electromagnetic energy (i.e., $\nabla \cdot \mathbf{S}$ must be inlcuded, where $\mathbf{S}$ is the Poynting vector) and the inclusion of a source/sink of electromagnetic energy density, $-\mathbf{E} \cdot \mathbf{J}$. The appendix provides a review of this topic. Putting all the parts together, we find the following MHD energy relation:

$$\frac{\partial}{\partial t} \left[\frac{1}{2} \rho u^2 + \frac{p}{\gamma - 1} + \rho U_{\text{grav}} + \frac{1}{2} \varepsilon_0 E^2 + \frac{B^2}{2\mu_0} \right]$$

$$+ \nabla \cdot \left\{ \rho \mathbf{u} \left[\frac{\gamma}{\gamma - 1} \frac{p}{\rho} + \frac{1}{2}u^2 + U_{\text{grav}} \right] \right\} + \nabla \cdot \mathbf{S}$$

$$+ \nabla \cdot (\mathbf{Q}_e + \mathbf{Q}_i) = \left(\frac{\delta E_e}{\delta t} \right)_{\text{coll}} + \left(\frac{\delta E_i}{\delta t} \right)_{\text{coll}} - \mathbf{E} \cdot \mathbf{J}, \tag{4.119}$$

where $\mathbf{Q}_e$ and $\mathbf{Q}_i$ are conductive heat fluxes for electrons and ions, respectively. Electron and ion collisional energy terms are also included on the right-hand side of (4.119). More details on the ion collision term can be found in Chapter 2.

4.7.2 Bernoulli's equation

A simple energy relation (2.80) was found earlier for the case of steady flow with no collisional sources or sinks of heat and with no heat conduction. An analogous relation can be derived from Equation (4.119) for $\mathbf{u} \perp \mathbf{B}$:

$$\frac{\gamma}{\gamma - 1} \frac{p}{\rho} + \frac{1}{2}u^2 + U_{\text{grav}} + \frac{B^2}{\mu_0 \rho} = constant \text{ along a streamline}, \tag{4.120}$$

where the Poynting vector is approximated with only the motional electric field terms in the generalized Ohm's law. For $\mathbf{u}$ parallel to $\mathbf{B}$ we have the same expression but without the $B^2/\mu_0\rho$ term. Equation (4.120) is *Bernoulli's equation*, which has wide applicability (and is *not* just for one-dimensional flow as is the case for Equation (4.114)). But both Equation (4.114) and Bernoulli's equation (4.120) tell us that as a fluid parcel slows down its pressure (and temperature) increases; or, conversely, as a fluid parcel speeds up, its pressure decreases.

4.8 MHD waves

Earlier in this chapter we considered the topic of Langmuir waves (i.e., plasma oscillations and waves). The frequency for this wave mode was shown to be close to the plasma frequency ($\omega \approx \omega_p$). Plasma oscillations are associated with departures from quasi-neutrality. In the process of deriving the single-fluid MHD equations, we assumed plasma neutrality, and therefore the MHD equations (unlike the two-fluid equations) do not "contain" within them plasma oscillations. However, other wave modes that have much lower frequencies ($\omega \ll \omega_p$) can be found using the MHD equations (i.e., *the MHD wave modes*). These MHD modes can also be found using the two-fluid equations rather than with the MHD equations, as will be carried out in this Chapter, but this would require much more effort. MHD waves can be generated either by some external source or by an internal MHD instability. In this book, wave growth or damping are not treated, only wave propagation will be considered (i.e., only the real part of the dispersion relation for MHD wave modes will be studied here). We further simplify our task by emphasizing MHD waves that propagate either along **B** or at right angles to **B**, rather than the general case of propagation at an arbitrary angle with respect to the field. The method that we will use to determine the dispersion relations for the MHD wave modes is basically the same as the method used in Section 4.2 to determine the dispersion relation for Langmuir waves.

For all MHD wave modes, our starting point are the MHD equations (4.79–4.81), omitting collisional terms and also omitting the source terms and the gravity term:

$$\text{continuity} \qquad \frac{\partial \rho}{\partial t} + \nabla \cdot (\rho \mathbf{u}) = 0, \tag{4.121}$$

$$\text{momentum} \qquad \rho \frac{\partial \mathbf{u}}{\partial t} + \rho \mathbf{u} \cdot \nabla \mathbf{u} = -\nabla p - \nabla \frac{B^2}{2\mu_0} + \frac{1}{\mu_0} \mathbf{B} \cdot \nabla \mathbf{B}, \tag{4.122}$$

$$\text{magnetic induction} \qquad \frac{\partial \mathbf{B}}{\partial t} = \nabla \times (\mathbf{u} \times \mathbf{B}), \tag{4.123}$$

$$\text{energy} \qquad p = p_0 [\rho/\rho_0]^\gamma. \tag{4.124}$$

The polytropic energy relation has been included in our set of equations, in lieu of a more complicated energy relation.

4.8.1 Ion-acoustic waves (sound waves)

We assume that $\mathbf{B} = 0$ for this mode, although this mode also can exist in a magnetized plasma for waves propagating along **B** (i.e., for wave vector $\mathbf{k} \parallel \mathbf{B}$). The induction Equation (4.123) is not needed for this mode, and both terms containing **B** in Equation (4.122) can be omitted. In addition, we can use the polytropic relation (4.124) to express the pressure gradient as (see Problem 4.14)

$$\nabla p = \gamma \frac{p_0}{\rho_0} \left(\frac{\rho}{\rho_0} \right)^{\gamma-1} \nabla \rho. \tag{4.125}$$

Our set of MHD equations now simply consists of a continuity equation for ρ (in terms of ρ and $\mathbf{u}$) and a momentum equation for $\mathbf{u}$ (in terms of ρ and $\mathbf{u}$):

$$\frac{\partial \rho}{\partial t} + \nabla \cdot (\rho \mathbf{u}) = 0 \qquad (4.126)$$

and

$$\rho \frac{\partial \mathbf{u}}{\partial t} + \rho \mathbf{u} \cdot \nabla \mathbf{u} = \frac{\gamma p_0}{\rho_0} \left(\frac{\rho}{\rho_0} \right)^{\gamma-1} \nabla \rho. \qquad (4.127)$$

Now we write the density and the flow velocity in terms of a background part and a wave part:

$$\rho = \rho_0 + \rho_1 \quad \text{with } \nabla \rho_0 = 0 \quad \text{and } \rho_1 \ll \rho_0 \qquad (4.128)$$

and

$$\mathbf{u} = 0 + \mathbf{u}_1 \quad \text{with } |\mathbf{u}_1| \ll v_{\text{th}}, \qquad (4.129)$$

where v_{th} is the ion thermal speed.

The background plasma is assumed to be uniform and stationary; naturally, this is an equilibrium state. The wave parts of our fluid variables are assumed to be small so that we can substitute Equations (4.128) and (4.129) into Equations (4.126) and (4.127) and then *linearize* the resulting equations. Recall from Section 4.2 that linearization means retaining only terms linear (or less) in $|\rho_1|/|\rho_0|$ or $|\mathbf{u}_1|/|v_{\text{th}}|$. This procedure yields the following equations for variables ρ_1 and $\mathbf{u}_1$:

$$\text{continuity} \quad \frac{\partial \rho_1}{\partial t} + \rho_0 \nabla \cdot \mathbf{u}_1 = 0, \qquad (4.130)$$

$$\text{momentum} \quad \rho_0 \frac{\partial \mathbf{u}_1}{\partial t} = -\gamma \frac{p_0}{\rho_0} \nabla \rho_1. \qquad (4.131)$$

We have two ways to proceed from here. For the first way, we derive a wave equation. We start by taking the time derivative of Equation (4.130) and then we use Equation (4.131) to eliminate $\partial \mathbf{u}_1/\partial t$ from the resulting equation. This procedure yields the following wave equation for ρ_1:

$$\frac{\partial^2 \rho_1}{\partial t^2} - C_s^2 \nabla^2 \rho_1 = 0 \qquad (4.132)$$

$$\text{with} \quad C_s \equiv \sqrt{\frac{\gamma p_0}{\rho_0}}. \qquad (4.133)$$

C_s is the *ion-acoustic speed* or *sound speed*. One set of solutions of the wave Equation (4.132) are waves propagating in the x direction with the functional form

$$\rho_1(x, t) = \rho_1(x \pm C_s t). \qquad (4.134)$$

In Problem 4.15 you are asked to prove that this function is a general solution of the wave Equation (4.132) if $\rho_1(\zeta)$ is any function and if $\zeta = x \pm C_s t$. The $\pm$ sign

Figure 4.14. Density perturbation propagating at the speed of sound (or ion-acoustic speed).

denotes sound waves propagating in the negative/positive x direction at speed C_s. See Figure 4.14.

The second way to proceed with Equations (4.130) and (4.131) is to assume a plane wave propagating in the x direction with frequency ω and wave vector $\mathbf{k} = k\hat{x}$. Equation (4.131) then indicates that $\mathbf{u}_1$ must also point in the x direction. Our density and flow velocity perturbations are written in plane wave form as

$$\rho_1 = \tilde{\rho}_1 e^{i[kx - \omega t]} \tag{4.135}$$

and

$$u_1 = \tilde{u}_1 e^{i[kx - \omega t]}. \tag{4.136}$$

Substituting Equations (4.135) and (4.136) into Equations (4.130) and (4.131), we can find the following algebraic equations for the wave amplitudes $\tilde{\rho}_1$ and $\tilde{u}_1$:

$$-i\omega\tilde{\rho}_1 + \rho_0[ik\tilde{u}_1] = 0 \tag{4.137}$$

and

$$-i\omega\rho_0\tilde{u}_1 = -\gamma\frac{p_0}{\rho_0}ik\tilde{\rho}_1. \tag{4.138}$$

Both $\tilde{\rho}_1$ and $\tilde{u}_1$ can be eliminated from these two equations, leaving us with the relation

$$\omega^2 = k^2\left[\frac{\gamma p_0}{\rho_0}\right] = C_s^2 k^2. \tag{4.139}$$

Equation (4.130) is the *dispersion relation* for *ion-acoustic waves*. Both the phase velocity ($V_{ph} = \omega/k$) and group velocity ($V_g = \partial\omega/\partial k$) are equal to $\pm C_s$. The general wave equation solution, (4.134), can be obtained by superimposing many plane-wave solutions (this is called Fourier analysis).

The above analysis was for sound waves for a plasma, but an essentially identical analysis applied to a neutral gas would give ordinary sound waves if we were to substitute the neutral gas pressure $p = \rho R T_n$ and the neutral gas density ρ for the plasma pressure and plasma density, respectively. The sound speed for a neutral gas is given by $C_S = \sqrt{\gamma p/\rho} = \sqrt{\gamma R T_n}$. The polytropic index is equal to $\gamma = 7/5$ for air and $\gamma = 5/3$ for an ideal monatomic gas. T_n is the neutral temperature. Note that in the expression for C_S, p and ρ can be used in place of p_0 and ρ_0 because $p \cong p_0$ and $\rho \cong \rho_0$.

In a plasma the pressure must include both the electron and ion partial pressures ($p = p_e + p_i$), so that Equation (4.133) for the speed of ion-acoustic waves can be expressed as

$$C_S = \sqrt{\frac{\gamma(p_e + p_i)}{n_e m_i}} = \sqrt{\frac{\gamma k_B (T_e + T_i)}{m_i}}$$

$$= 91\sqrt{\frac{\gamma(T_e + T_i)}{A_i \,(\text{amu})}} \quad [\text{m/s}]. \tag{4.140}$$

Both T_e and T_i appear in the expression for C_S, although only the ion mass appears in the denominator. A_i is the ion mass in atomic mass units, or amu; that is, $m_i = A_i m_p$, where m_p is the proton mass. T_e and T_i are in units of Kelvin.

The key force underlying the ion-acoustic wave (the "restoring force") is the pressure gradient force, which includes both electron and ion contributions. Ion inertia opposes this restoring force; the electrons are too light to contribute to this inertial effect. The pressure gradients result from compressions and rarefactions of the fluid. Recall that the electron pressure gradient was brought into the MHD momentum equation via the electric field. Thus, an electric field is associated with ion-acoustic waves but obviously not with ordinary sound waves in a neutral gas. From the generalized Ohm's law, with suitable simplifications, we see that $\mathbf{E} \cong -(1/n_0 e)\nabla p_e$. This ambipolar field is in the same direction as the wave vector $\mathbf{k}$. This is the x direction for our simple plane wave as discussed above. Waves whose electric fields and wave vectors are in the same direction ($\mathbf{k} \parallel \mathbf{E}$) are called *longitudinal* waves.

For many space plasmas the electron temperature greatly exceeds the ion temperature ($T_e \gg T_i$) and the ion acoustic speed becomes

$$C_S \cong \sqrt{\frac{\gamma k_B T_e}{m_i}}. \tag{4.141}$$

When this mode propagates along a strong background magnetic field, the electrons in many space plasmas can be approximated as isothermal so that $\gamma = 1$ is often an appropriate choice of the polytropic index in Equation (4.141).

The *sonic Mach number* M_S is defined as the ratio of the bulk flow speed u of the plasma to the ion-acoustic speed C_S:

$$M_s \equiv \frac{u}{C_s}. \tag{4.142}$$

The Mach number squared, $M_s^2 = u^2/C_s^2 = \rho u^2/\gamma p$ is approximately equal to the ratio of the dynamic pressure to the thermal pressure. Dynamic pressure is more important than thermal pressure for highly *supersonic* flow, $M_s \gg 1$, whereas the opposite is true for highly *subsonic* flow, $M_s \ll 1$.

Example 4.4 (Ion-acoustic speed and sonic mach number for space plasma)

(1) *Solar Wind* The following choice of parameters is appropriate for the typical solar wind at a distance of 1 Astronomical Unit:

$$\text{polytropic index} \quad \gamma \approx 5/3$$
$$\text{protons} \quad T_p \cong 10^4 \, \text{K}$$
$$\text{electrons} \quad T_e \cong 10^5 \, \text{K}$$
$$\text{ion mass} \quad A_i \cong 1 \, \text{amu (protons)}.$$

The sound speed from Equation (4.140) then becomes $C_S = 39$ km/s. Typically in the solar wind, the flow speed is $u \approx 400$ km/s. Thus, the sonic Mach number is equal to

$$M_S \approx \frac{400 \, \text{km/s}}{39 \, \text{km/s}} \approx 10.$$

The solar wind flow is highly supersonic and its dynamic pressure is 100 times greater than its thermal pressure.

(2) *Ionosphere* An appropriate choice of parameters for the terrestrial ionospheric plasma is

$$\text{polytropic index} \quad \gamma \approx 5/3$$
$$\text{oxygen ions} \quad T_i \cong 1{,}000 \, \text{K} \ (A_i = 16)$$
$$\text{electrons} \quad T_e \cong 2{,}000 \, \text{K}.$$

With this choice of parameters, Equation (4.140) gives $C_s = 1.6$ km/s.

Ionospheric flow speeds range from 0 to 1 km/s, with lower values (e.g., $u \approx 100$ m/s) being typical at lower latitudes and higher values ($u \approx 1$ km/s) being typical in the polar ionosphere. Taking an "average" value of $u \approx 400$ m/s, we find that the sonic Mach number is

$$M_s \approx \frac{400 \, \text{m/s}}{1{,}600 \, \text{m/s}} \approx 0.25.$$

Ionospheric plasma flows are almost always subsonic.

4.8.2 Alfvén waves

Let us return to the simplified MHD Equations (4.121)–(4.124). We now introduce a uniform background magnetic field $\mathbf{B}_0 = B_0\hat{\mathbf{z}}$ (Figure 4.15), and we only consider waves propagating along this field; that is, we restrict ourselves to wave vectors of the form $\mathbf{k} = k\hat{\mathbf{z}}$. Ion-acoustic waves can also propagate along $\mathbf{B}_0$, but to exclude these waves we make the assumption that the pressure remains uniform so that $\nabla p = 0$ in Equation (4.122). From the polytropic relation, the density must also remain uniform ($\nabla \rho = 0$), and the continuity equation simply becomes $\nabla \cdot \mathbf{u} = 0$ (i.e., the flow is incompressible). The remaining two MHD equations are

$$\text{momentum} \quad \rho\frac{\partial \mathbf{u}}{\partial t} + \rho\mathbf{u} \cdot \nabla\mathbf{u} = -\nabla\frac{B^2}{2\mu_0} + \frac{1}{\mu_0}\mathbf{B} \cdot \nabla\mathbf{B}, \quad (4.143)$$

$$\text{magnetic induction} \quad \frac{\partial \mathbf{B}}{\partial t} = \nabla \times (\mathbf{u} \times \mathbf{B}). \quad (4.144)$$

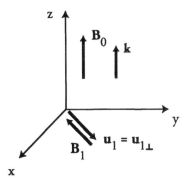

Figure 4.15. Geometry for Alfvén waves.

We now write the dependent variables **B** and **u** in terms of background plus wave parts,

$$\mathbf{B} = \mathbf{B}_0 + \mathbf{B}_1, \quad |\mathbf{B}_1| \ll |\mathbf{B}_0|$$

and $\hspace{6cm}$ (4.145)

$$\mathbf{u} = 0 + \mathbf{u}_1,$$

and we assume that the wave field is much less than the background field. Note that $\rho_1 = 0$. We substitute the expressions (4.145) into Equations (4.143) and (4.144) and we linearize the resulting equations. For the momentum equation we obtain

$$\text{momentum} \quad \rho_0 \frac{\partial \mathbf{u}_1}{\partial t} + \rho_1 \frac{\partial \mathbf{u}_1}{\partial t} + \rho_0 \mathbf{u}_1 \cdot \nabla \mathbf{u} = -\nabla \left[\frac{B_0^2 + 2\mathbf{B}_0 \cdot \mathbf{B}_1 + \mathbf{B}_1^2}{2\mu_0} \right]$$

$$\text{order } \varepsilon \quad \text{order } \varepsilon^2 \quad \text{order } \varepsilon^2 \quad \quad 0 \quad \quad \text{order } \varepsilon \quad \quad \text{order } \varepsilon^2$$

$$+ \frac{1}{\mu_0} \mathbf{B}_0 \cdot \nabla \mathbf{B}_1 + \frac{1}{\mu_0} \mathbf{B}_1 \cdot \nabla \mathbf{B}_0.$$

$$\text{order } \varepsilon \quad \text{order } \varepsilon$$

The order of magnitude, $\varepsilon \approx |\mathbf{B}_1|/|\mathbf{B}_0|$, of each term has been identified. Retaining only those terms that are of order ε and using the assumption that the background plasma is uniform, the linearized momentum equation becomes

$$\rho_0 \frac{\partial \mathbf{u}_1}{\partial t} = -\frac{1}{\mu_0} \nabla (\mathbf{B}_0 \cdot \mathbf{B}_1) + \frac{1}{\mu_0} \mathbf{B}_0 \cdot \nabla \mathbf{B}_1. \quad (4.146)$$

For the induction equation we find

$$\text{induction} \quad \frac{\partial \mathbf{B}_1}{\partial t} = \nabla \times (\mathbf{u}_1 \times \mathbf{B}_0) + \nabla \times (\mathbf{u}_1 \times \mathbf{B}_1). \quad (4.147)$$

$$\text{order } \varepsilon \quad \quad \text{order } \varepsilon^2$$

Only the first term on the right-hand side of Equation (4.147) is retained.

We have two vector equations for two unknown vector quantities, $\mathbf{u}_1$ and $\mathbf{B}_1$, from which we drop all terms of order ε^2 or higher. Next, we adopt plane-wave solutions

$$\mathbf{u}_1 = \tilde{\mathbf{u}}_1 \exp[i(kz - \omega t)] \tag{4.148}$$

and

$$\mathbf{B}_1 = \tilde{\mathbf{B}}_1 \exp[i(kz - \omega t)]. \tag{4.149}$$

Substitution of Equations (4.148) and (4.149) into Equations (4.146) and (4.147) leads to the following algebraic equations for the vector amplitudes $\tilde{\mathbf{u}}_1$ and $\tilde{\mathbf{B}}_1$:

$$-i\omega\rho_0\tilde{\mathbf{u}}_1 = -\frac{B_0}{\mu_0}ik\hat{\mathbf{z}}\tilde{B}_{1z} + \frac{\tilde{\mathbf{B}}_1}{\mu_0}i\mathbf{B}_0 \cdot \mathbf{k} \tag{4.150}$$

$$-i\omega\tilde{\mathbf{B}}_1 = i B_0 k \tilde{\mathbf{u}}_1 - \mathbf{B}_0(i\mathbf{k} \cdot \tilde{\mathbf{u}}_1). \tag{4.151}$$

We used $\mathbf{B}_0 \cdot \tilde{\mathbf{B}}_1 = B_0\tilde{B}_{1z}$ and $\mathbf{B}_0 \cdot \mathbf{k} = B_0 k$. Evaluating the z component of Equation (4.150), we can easily demonstrate that the right-hand side is zero; hence, $\tilde{u}_{1z} = 0$. This means that the plasma moves only in directions normal to $\mathbf{B}_0$ (and to $\mathbf{k}$) and $\tilde{\mathbf{u}}_1$ (and $\mathbf{u}_1$) must lie in the x–y plane; thus, $\tilde{\mathbf{u}}_1 = \tilde{\mathbf{u}}_{1\perp}$. Consequently, the second term on the right-hand side of Equation (4.151) is zero. We can now write Equations (4.150) and (4.151) as

$$-i\omega\rho_0\tilde{\mathbf{u}}_{1\perp} = \frac{i B_0 k}{\mu_0}\tilde{\mathbf{B}}_1 \quad \text{(from 4.150)},$$

which is just

$$\tilde{\mathbf{B}}_1 = -\frac{\omega}{k}\frac{\rho_0\mu_0}{B_0}\tilde{\mathbf{u}}_{1\perp} \tag{4.152}$$

and

$$\tilde{\mathbf{B}}_1 = -\frac{k}{\omega}B_0\tilde{\mathbf{u}}_{1\perp} \quad \text{(from 4.151)} \tag{4.153}$$

The vector $\tilde{\mathbf{B}}_1$ (and $\mathbf{B}_1$) is directed in the opposite direction from the plasma motion for positive k, and it is thus perpendicular to $\mathbf{B}_0$ (and also to $\mathbf{k}$). Combining Equations (4.152) and (4.153) and eliminating $\tilde{\mathbf{B}}_1$, we obtain the *dispersion relation for Alfvén waves*:

$$\omega^2 = \frac{B_0^2}{\mu_0\rho_0}k^2$$

$$= C_A^2 k^2$$

with

$$C_A \equiv \sqrt{\frac{B_0^2}{\mu_0\rho_0}} \tag{4.154}$$

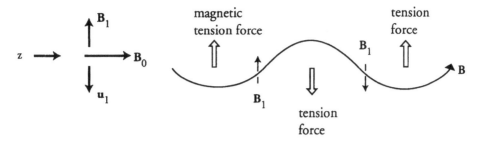

Figure 4.16. Schematic for Alfvén wave.

in SI units and

$$C_A \equiv \sqrt{\frac{B_0^2}{4\pi\rho_0}}$$

in cgs units. Both the phase velocity ($V_{ph} = \omega/k$) and the group velocity ($V_g = \partial\omega/\partial k$) of the Alfvén wave mode are equal to the *Alfvén speed* C_A.

The wave magnetic field is perpendicular to the wave vector ($\mathbf{k} \perp \mathbf{B}_1$). From Maxwell's equations, it can be proven that the wave electric field $\mathbf{E}_1$ is normal to both $\mathbf{B}_1$ and to $\mathbf{k}$ (Problem 4.16). Alfvén waves are thus *transverse* and *electromagnetic*. However, the energy density of the wave electric field is negligible in comparison with the energy density of the wave magnetic field.

Perturbations, or disturbances, in the magnetic field travel at the Alfvén speed along the background field as Alfvén waves. A better physical understanding of these waves can be realized by considering the analogy with mechanical waves on a string that is put under tension. Just as the restoring force for a wave on a string is the string tension, the restoring force in an Alfvén wave is magnetic tension; changes in the magnitude of $\mathbf{B}$ are second order in the parameter ε, and thus the magnetic pressure gradient force is not important for Alfvén waves. Magnetic tension accelerates the plasma and is opposed by ion inertia (see Figure 4.16). By analogy, the propagation speed of a wave on a string under tension T is $C_{string} = \sqrt{T/\lambda_s}$, where λ_s is the mass density per unit length of the string. The "wave" in this case is the displacement of the string up or down from the equilibrium straight configuration. The magnetic analogy to the string tension is the quantity B^2/μ_0.

The Alfvénic Mach number, M_A, is given by

$$M_A \equiv \frac{u}{C_A}. \tag{4.155}$$

The following example gives typical space plasma values of C_A and M_A.

Example 4.5 (Alfvén speed and Alfvénic Mach number in space plasmas) A numerical expression for C_A in units of km/s is

$$C_A = 2.19 \times 10^4 \sqrt{\frac{B_0^2 \,(\text{nT})}{n_0 \,(\text{m}^{-3}) A_i \,(\text{amu})}} \quad \left[\frac{\text{km}}{\text{s}}\right].$$

First we apply this expression to the solar wind plasma. A reasonable choice of parameters for the solar wind near 1 AU is

$$B_0 \cong 5\,\text{nT}$$

$$A_i \cong 1\,\text{amu (protons)}$$

$$n_e \cong 5\,\text{cm}^{-3} \;(= 5 \times 10^6\,\text{m}^{-3})$$

$$u \cong 400\,\text{km/s}.$$

The Alfvén speed and Alfvénic Mach number then become

$$C_A \cong 50\,\text{km/s}$$

$$M_A \cong 8 \;\text{(highly super-Alfvénic)}.$$

The solar wind is super-Alfvénic as well as supersonic (see Example 4.4); that is, the solar wind flows faster than either magnetic disturbances or pressure disturbances (i.e., sound) can propagate along the magnetic field.

We now apply Equations (4.154) and (4.155) to the terrestrial ionospheric plasma. A reasonable choice of parameters is

$$B_0 \cong 3 \times 10^4\,\text{nT} \quad \text{(this is the surface geomagnetic field strength)}$$

$$A_i \cong 16\,\text{amu (O}^+\text{ions)}$$

$$n_0 \cong 10^5\,\text{cm}^{-3} \;(= 10^{11}\,\text{m}^{-3})$$

$$u \cong 400\,\text{m/s}.$$

The Alfvén speed and Alfvénic Mach number then become

$$C_A \cong 140\,\text{km/s}$$

$$M_A \cong 10^{-3} \;\text{(highly sub-Alfvénic)}.$$

Ionospheric plasma flows are always sub-Alfvénic.

For the plasmasheet in the terrestrial magnetosphere (distances from the Earth of $\approx 10 R_E$), a reasonable choice of parameters is

$$B_0 \cong 20\,\text{nT}$$

$$A_i \cong 1\,\text{amu (protons)}$$

$$n_0 \cong 10^7\,\text{m}^{-3}$$

$$u \approx 50\,\text{km /s (although flow speeds are quite variable).}$$

The Alfvén speed and Alfvénic Mach number then become

$$C_A \cong 140 \, \text{km/s}$$

$$M_A \cong 0.3 \, \text{(marginally sub-Alfvénic)}.$$

Typical plasma flows in the terrestrial plasmasheet at this distance are marginally sub-Alfvénic.

4.8.3 Magnetosonic waves

We now consider *magnetosonic waves* in a magnetized plasma that propagate at right angles with respect to the background magnetic field ($\mathbf{k} \perp \mathbf{B}_0$). Our starting point is again the set of Equations (4.121)–(4.124). In Problem 4.17, you are asked to derive the dispersion relation for magnetosonic waves in the standard way. For magnetosonic waves both the pressure gradient force and the magnetic pressure gradient force need to be included, and it turns out that the magnetic tension force is negligible. The *dispersion relation for magnetosonic waves*, which you will hopefully derive yourself, is given by

$$\omega^2 = C_{MS}^2 k^2$$

where

$$C_{MS}^2 = C_s^2 + C_A^2$$

$$= \frac{\gamma p_0}{\rho_0} + \frac{B_0^2}{\mu_0 \rho_0}. \tag{4.156}$$

C_{MS} is the *magnetosonic speed*. Both the phase and group velocities for this MHD wave mode are equal to $\pm C_{MS}$.

The magnetosonic mode is similar to the ion-acoustic wave mode in that ion-acoustic waves are driven by the thermal pressure gradient force and magnetosonic waves are driven by the combined thermal and magnetic pressure gradient forces. Ion inertia again counteracts the restoring force. The restoring force and the velocity perturbations, $\mathbf{u}_1$, are in the same direction (or opposite direction) as $\mathbf{k}$. The motion of the plasma for this wave mode results in compressions and rarefactions of the frozen-in magnetic field (see Figure 4.17), which then results in magnetic pressure gradient forces.

The continuity and magnetic induction equations act in a similar fashion for this wave mode, as do the thermal and magnetic pressures. The similar roles of thermal and magnetic pressure can be illustrated by assuming that the plasma has a polytropic index of 2, in which case we can see that the wave speed is given by

$$C_{MS}^2 = \frac{1}{\rho}\left(\gamma p + 2\frac{B^2}{2\mu_0}\right) = \frac{2}{\rho} p_{\text{tot}}. \tag{4.157}$$

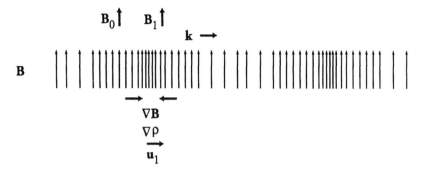

Figure 4.17. Schematic for magnetosonic wave.

We have dropped the subscript "0" from the background fluid variables in Equation (4.157). The total pressure appearing in Equation (4.157) was introduced in Equation (4.100).

The *magnetosonic Mach number* for a plasma with flow speed u is defined by

$$M_{\mathrm{MS}} \equiv \frac{u}{C_{\mathrm{MS}}}. \tag{4.158}$$

In the solar wind near the orbit of the Earth, $C_{\mathrm{MS}} \approx \sqrt{C_s^2 + C_A^2} \approx [40^2 + 50^2]^{1/2}$ km/s ≈ 64 km/s. And the magnetosonic Mach number of the solar wind is $M_{\mathrm{MS}} \cong 6.3$. C_S and C_A values for the solar wind were given in Examples 4.4 and 4.5.

A more general treatment of MHD waves allows the waves to propagate at any angle θ with respect to the background field. We will state the results here but will not derive them (see Siscoe, 1983, for a more complete treatment). Three wave modes are possible: the *intermediate mode* (i), the *slow mode* (s), and the *fast mode* (f). The intermediate mode is basically the Alfvén mode with a phase velocity of

$$\mathbf{V}_{\mathrm{ph}} = C_A \cos\theta \hat{\mathbf{k}}. \tag{4.159}$$

For $\mathbf{k} \parallel \mathbf{B}_0 (\theta = 0)$, $V_{\mathrm{ph}} = C_A$, but for $\mathbf{k} \perp \mathbf{B}_0 (\theta = \pi/2)$, $V_{\mathrm{ph}} = 0$ (that is, intermediate mode waves – Alfvén waves – do not exist for purely perpendicular propagation). The phase velocities for the slow and fast modes are

$$[V_{\mathrm{ph}}^2]_{\mathrm{f,s}} = \frac{1}{2}\left[(C_s^2 + C_A^2) \pm \sqrt{(C_s^2 + C_A^2)^2 - 4C_s^2 C_A^2 \cos^2\theta}\right] \tag{4.160}$$

with one sign giving the fast mode speed and the other sign giving the slow mode speed. The two phase velocities resulting from Equation (4.160) for parallel propagation ($\theta = 0$) are $\pm C_s$ and $\pm C_A$, and the fast mode velocity is identified as the one having the greater numerical value. That is, for parallel propagation only two MHD wave modes exist, one purely ion-acoustic and the other purely Alfvénic, with the latter also being the same as the intermediate mode. For $\theta = \pi/2$, only the fast mode exists with $V_{\mathrm{ph}}^2 = C_s^2 + C_A^2$. Thus, for perpendicular propagation, the fast mode is the same as the magnetosonic mode. The slow mode cannot be present for perpendicular propagation; it can only exist for oblique propagation.

4.9 MHD discontinuities (shock waves)

4.9.1 Discontinuities

Small disturbances in a magnetized plasma propagate through this medium in the form of one of the MHD wave modes (slow, intermediate, or fast mode – or for certain propagation angles, ion-acoustic, Alfvén, or magnetosonic modes). These wave modes were derived by linearizing the MHD equations, and this procedure applies only to very small amplitude waves. Large-amplitude variations in fluid quantities must be handled differently. A *MHD discontinuity* is a surface across which fluid quantities change discontinuously. The thickness of the discontinuity is much less than other length scales in the system; typically, for a collisional gas, the thickness δ of the discontinuity layer is of the order of the collision mean free path, λ_{mfp}.

A relatively simple treatment of MHD discontinuities will be given here; for a more complete treatment see Siscoe (1983) or other more advanced plasma texts. We perform the analysis in the frame of reference of the discontinuity. The flow velocity of the fluid **u** in this reference frame is given by

$$\mathbf{u} = \mathbf{u}(\text{original}) - \mathbf{u}_{\text{discon}}, \tag{4.161}$$

where **u** on the right-hand side is the original fluid velocity and $\mathbf{u}_{\text{discon}}$ is the velocity of the discontinuity. In the rest frame of the discontinuity, the flow is assumed to be steady state.

Fluid and field parameters change discontinuously across a discontinuity in a way that depends on the type of discontinuity. We introduce the following difference operation, $[\![\ ]\!]$, for the change of a quantity Q across a discontinuity:

$$[\![Q]\!] \equiv Q_1 - Q_2. \tag{4.162}$$

Q could represent density, temperature, or a velocity component. If the fluid is moving in a direction normal to the surface, then region 1 is considered to be the upstream region and region 2 to be the downstream region. (See Figure 4.18.)

Let us consider the field quantities, **E** and **B**, before looking at fluid quantities. Both **E** and **B** (in fact, all vectors) can be written in terms of normal (i.e., B_n) and

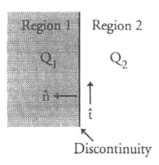

Figure 4.18. The change in quantity Q across the discontinuity is denoted $[\![Q]\!] = Q_1 - Q_2$. Unit vectors normal ($\hat{\mathbf{n}}$) and tangential ($\hat{\mathbf{t}}$) to the surface are shown.

tangential (i.e., B_t) components: $\mathbf{B} = B_n\hat{\mathbf{n}} + B_t\hat{\mathbf{t}}$. Maxwell's equation for $\mathbf{B}$ is

$$\nabla \cdot \mathbf{B} = 0. \tag{4.163}$$

Applying Gauss's integral theorem to Equation (4.163) we have that a small volume contained within a closed surface that straddles the discontinuity yields the equation

$$[\![B_n]\!] = 0. \tag{4.164}$$

This relation is true for all MHD discontinuities.

Faraday's law for steady-state situations is just $\nabla \times \mathbf{E} = \mathbf{0}$. Applying Stoke's theorem for a rectangular path aligned with the surface gives

$$[\![E_t]\!] = 0. \tag{4.165}$$

For a collisionless plasma, $\mathbf{E} = -\mathbf{u} \times \mathbf{B}$ (see Equation (4.67); this is consistent with the induction Equation (4.123)), and we find that Equation (4.165) can be written as

$$[\![u_n B_t - u_t B_n]\!] = 0, \tag{4.166}$$

where u_n and u_t are the normal and tangential components, respectively, of the flow velocity ($\mathbf{u} = u_n\hat{\mathbf{n}} + u_t\hat{\mathbf{t}}$).

We now address how the fluid variables change across the discontinuity by using the fluid conservation equations. For a very narrow discontinuity, sources can be neglected, as can friction (for a collisional plasma) and gravity. The steady-state, source-free form of the single-fluid continuity Equation (4.79) is just

$$\nabla \cdot (\rho\mathbf{u}) = 0. \tag{4.167}$$

The most convenient form of momentum equation for our current purposes is the steady-state, single-species version of Equation (2.59) with magnetic stress terms as given by Equation (4.85) rewritten in the form of a divergence:

$$\nabla \cdot \left(\rho\mathbf{u}\mathbf{u} + \tilde{\mathbf{P}} - \frac{\mathbf{B}\mathbf{B}}{\mu_0} + \frac{B^2}{2\mu_0}\tilde{\mathbf{I}} \right) = 0. \tag{4.168}$$

Here $\tilde{\mathbf{I}}$ is the unit tensor and $\tilde{\mathbf{P}}$ is the pressure tensor. Note that all the forces have been written in terms of the divergence of a stress tensor. The most convenient form of the energy equation is given by Equation (4.119); with suitable approximations (steady-state, zero heat flux and zero collision terms) this becomes

$$\nabla \cdot \left[\left(\frac{1}{2}\rho u^2 + \frac{\gamma}{\gamma - 1}p \right)\mathbf{u} + \mathbf{S} \right] = 0, \tag{4.169}$$

where p is the scalar pressure and $\mathbf{S}$ is the electromagnetic Poynting vector.

Application of Gauss's integral theorem to Equations (4.167), (4.168) and (4.169) for a small volume straddling the discontinuity gives the following relations:

$$[[\rho u_n]] = 0 \qquad (4.170)$$

$$\left[\!\left[\rho \mathbf{u} u_n + p\hat{\mathbf{n}} - \frac{\mathbf{B}}{\mu_0} B_n + \frac{B^2}{2\mu_0}\hat{\mathbf{n}}\right]\!\right] = \mathbf{0} \qquad (4.171)$$

$$\left[\!\left[\left(\frac{1}{2}\rho u^2 + \frac{\gamma}{\gamma - 1}p\right)u_n + S_n\right]\!\right] = 0. \qquad (4.172)$$

Multiplying Equation (4.171) by $\hat{\mathbf{n}}$ and $\hat{\mathbf{t}}$ yields the following two equations:

$$\left[\!\left[\rho u_n^2 + p + \frac{B_t^2}{2\mu_0}\right]\!\right] = 0 \qquad (4.173)$$

and

$$\left[\!\left[\rho u_n u_t - \frac{B_n B_t}{\mu_0}\right]\!\right] = 0. \qquad (4.174)$$

(Note: Siscoe (1983) performs this same analysis using anisotropic pressures, $p_\perp$ and $p_\parallel$; here, we restrict ourselves to an isotropic pressure with $p_\perp = p_\parallel$.)

We now have conservation relations that can be used to study several types of MHD discontinuities:

(1) **Contact Discontinuity ($u_n = 0, B_n \neq 0$)** All quantities except the density and temperature are continuous across a contact discontinuity. Such discontinuities are not interesting for space plasmas; if formed they rapidly dissipate due to heat conduction and particle diffusion across them.

(2) **Tangential Discontinuity ($u_n = 0, B_n = 0$)** All the conservation relations except Equation (4.173) become trivial for $u_n = 0$ and $B_n = 0$:

$$\left[\!\left[p + B_t^2/2\mu_0\right]\!\right] = 0 \qquad (4.175)$$

or

$$p + B_t^2/2\mu_0 = constant \text{ across the tangential discontinuity.}$$

Equation (4.175) just restates again in the context of a MHD discontinuity the static force balance relation (4.100). Since the discussion in Section 4.6.3 also applies to tangential discontinuities, we will not discuss this any further here.

(3) **Shock Waves ($u_n \neq 0$)** Shock waves are discontinuous slowdowns of gas flow at which the gas is compressed. Shocks appear in many space plasma situations, as will be discussed later. Four general types of MHD shocks are possible:

(1) ordinary (non-MHD) shock waves ($\mathbf{B} = \mathbf{0}$),
(2) perpendicular shocks ($B_n = 0, B_t \neq 0$),
(3) parallel shocks ($B_n \neq 0, B_t = 0$), and
(4) oblique shocks ($B_n \neq 0$ and $B_t \neq 0$; both slow and fast mode shocks are possible for oblique shocks).

4.9.2 Shock waves

A quantitative analysis will be presented only for ordinary gas dynamic shocks for which $\mathbf{B} = 0$. This analysis will show that a gas goes from being supersonic upstream of the shock to subsonic downstream of the shock. Equations (4.170), (4.171), and (4.172), with both components of $\mathbf{B}$ set equal to zero, are needed for this analysis. Equations (4.170) and (4.174) together indicate that $[\![u_t]\!] = 0$; that is, u_t does not change across the shock and is thus uninteresting. We assume that $u_t = 0$ and thus have $u = u_n$. We now have the following conservation relations:

$$[\![\rho u]\!] = 0 \qquad (4.176)$$

$$[\![\rho u^2 + p]\!] = 0 \qquad (4.177)$$

$$\left[\!\!\left[\left(\frac{1}{2}\rho u^2 + \frac{\gamma}{\gamma - 1}p\right) u \right]\!\!\right] = 0. \qquad (4.178)$$

Equation (4.176) is equivalent to

$$\rho_1 u_1 = \rho_2 u_2 \quad \text{or} \quad \frac{u_2}{u_1} = \frac{\rho_1}{\rho_2} = 1/Z_s, \qquad (4.179)$$

where we have introduced the ratio $Z_s \equiv \rho_2/\rho_1$, which is the density "jump" or "*shock jump*" across the shock. The *velocity jump* is inversely proportional to Z_s. Equations (4.177) and (4.178) can also be written in terms of the upstream and downstream values of the fluid variables (see Figure 4.19):

$$\rho_1 u_1^2 + p_1 = \rho_1 u_2^2 + p_2 \qquad (4.180)$$

$$\left(\frac{1}{2}\rho_1 u_1^2 + \frac{\gamma}{\gamma - 1}p_1\right) u_1 = \left(\frac{1}{2}\rho_2 u_2^2 + \frac{\gamma}{\gamma - 1}p_2\right) u_2. \qquad (4.181)$$

Let us suppose that all the upstream variables, ρ_1, u_1, and p_1, are known. Then Equations (4.179)–(4.181) constitute three equations that can be used to solve for the three unknowns, ρ_2, u_2, and p_2. In fact, ρ_2, u_2, and p_2 can all be found in terms of the upstream sonic Mach number M_1. From Equations (4.133) and (4.142), we have

$$M_1^2 = \frac{\rho_1 u_1^2}{\gamma p_1} \quad \text{and} \quad M_2^2 = \frac{\rho_2 u_2^2}{\gamma p_2}. \qquad (4.182)$$

Equations (4.179)–(4.181) can be solved to obtain (see Problem 4.18)

$$\frac{u_2}{u_1} = \frac{\gamma - 1}{\gamma + 1} + \frac{2}{(\gamma + 1)M_1^2}. \qquad (4.183)$$

Hence, from Equation (4.179) the shock jump Z_s is

$$Z_s = \frac{\rho_2}{\rho_1} = \frac{u_1}{u_2} = \frac{\gamma + 1}{\gamma - 1 + \dfrac{2}{M_1^2}}. \qquad (4.184)$$

Shock

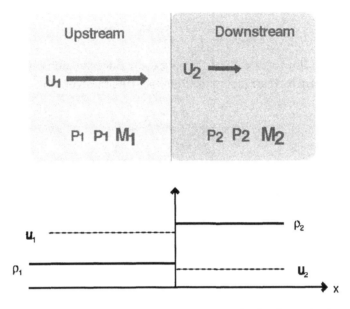

Figure 4.19. Schematic of shock wave showing the jump in density and a drop in velocity.

We can also use these results to find the pressure jump, which is

$$\frac{p_2}{p_1} = \frac{2\gamma M_1^2 - (\gamma - 1)}{\gamma + 1}. \tag{4.185}$$

Equations (4.183)–(4.185) are called the *Rankine–Hugoniot relations*.

For $M_1 = 1$, we have $u_2/u_1 = \rho_2/\rho_1 = p_2/p_1 = 1$; there is no shock and the flow stays sonic ($u = C_s$). We cannot have $M_1^2 < 1$. But for $M_1 > 1$ we have $Z_s > 1$, indicating that the density increases and the flow speed decreases across the shock (see Figure 4.19). Naturally, the mass flux must remain constant across a steady-state shock in order to prevent mass build-up (or loss) at the discontinuity surface. Because the flow decelerates at the shock there is compression. Compression and slowdown are associated with an increase in pressure (and temperature), as discussed in Section 4.7.2 on Bernoulli's relation. The temperature jump across the shock can be derived from Equation (4.185) and a suitable equation of state. For ordinary air, $p = \rho \tilde{R} T_n$, but for a plasma $p = p_e + p_i = \rho \frac{k_B}{m_i}(T_e + T_i)$, where m_i is the ion mass. Furthermore, for a plasma the Mach number must be determined using the ion-acoustic speed, which includes both T_e and T_i.

The *hypersonic limits* of the Rankine–Hugoniot relations can be found by taking the limit $M_1^2 \to \infty$:

$$\lim_{M_1^2 \to \infty} Z_s = \frac{\rho_2}{\rho_1} = \frac{u_1}{u_2} \to \frac{\gamma + 1}{\gamma - 1} \quad (= 4 \text{ for } \gamma = 5/3). \tag{4.186}$$

The maximum shock jump for $\gamma = 5/3$ is 4, but the maximum pressure jump is infinite:

$$\lim_{M_1^2 \to \infty} \frac{p_2}{p_1} = \frac{2\gamma}{\gamma + 1} M_1^2 \to \infty. \tag{4.187}$$

For supersonic flow upstream of the shock, the downstream flow must be subsonic, as we can see by rearranging the Rankine–Hugoniot relation (Problem 4.18) to get

$$M_2^2 = \frac{(\gamma - 1)M_1^2 + 2}{2\gamma M_1^2 - (\gamma - 1)}. \tag{4.188}$$

Clearly, this equation shows that the downstream gas flow is subsonic ($M_2 < 1$) for upstream gas flow that is supersonic ($M_1 > 1$). The hypersonic limit of the downstream Mach number is $M_2^2 \to (\gamma - 1)/(2\gamma)$, which is equal to 1/5 for an ideal monatomic gas with $\gamma = 5/3$.

Gas flow across a shock is thermodynamically irreversible; that is, the net change of entropy is positive and nonzero. The quantity p/ρ^γ is related to the entropy per unit mass (or *specific entropy*). Specific entropy is a constant (as in Equation (2.81)) for small perturbations such as typical sound waves. The flow is then said to be *isentropic*. However, the specific entropy increases across a shock (Problem 4.19):

$$\frac{p_2}{\rho_2^\gamma} > \frac{p_1}{\rho_1^\gamma}. \tag{4.189}$$

Irreversible "dissipation" of bulk kinetic energy (ρu^2) into thermal energy (p) takes place inside the shock discontinuity. This dissipation is related to collisions in an ordinary shock wave, but in space plasmas, the shocks are collisionless, and the nature of the dissipation mechanism becomes tricky. Nonetheless, shocks do exist in space and thus dissipation must be present, albeit associated with microscopic plasma instabilities and waves (i.e., small-scale structure in the plasma and fields) rather than ordinary collisions.

We have just analyzed ordinary gas shocks, but what about MHD shocks in general? The conservation relations are much messier if we retain the magnetic field terms. However, for *parallel shocks* ($\mathbf{B} \parallel \mathbf{u}$) the field again drops out of the equations and we reobtain the Rankine–Hugoniot relations for an ordinary shock. But, as just discussed, the dissipation mechanism for parallel, collisionless shocks in a space plasma is problematic (and not very efficient). Parallel shocks observed in space are not really discontinuities but appear as quite thick layers that have considerable plasma turbulence associated with them, as required for the dissipation.

Dissipation for collisionless *perpendicular shocks* ($\mathbf{B} \perp \mathbf{u}$) is more efficient than for parallel shocks and is associated with ion gyration. The shock thickness for this category of shock is roughly equal to an ion gyroradius. The MHD version of the Rankine–Hugoniot relations can be found from the appropriate conservation relations but will not be shown here; see Siscoe (1983) or Tidman and Krall

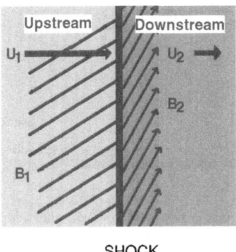

Figure 4.20. Oblique MHD fast-mode shock.

(1971). However, just as for ordinary shocks, the density increases and the velocity decreases across the shock. The change in the magnetic field can be found from Equation (4.166) with $B_n = 0$ (and $u = u_n$, $B = B_t$):

$$u_1 B_1 = u_2 B_2 \quad \text{or} \quad \frac{B_2}{B_1} = \frac{u_1}{u_2} = \frac{\rho_2}{\rho_1} = Z_s. \tag{4.190}$$

The magnetic field jump is the same as the density jump. The general case is not simple but the hypersonic limit is the same as for ordinary shocks, $Z_s = (\gamma + 1)/(\gamma - 1)$.

Oblique MHD shocks (in which the magnetic field is neither parallel to nor perpendicular to the flow) are even more complicated than perpendicular shocks. There are even two types of oblique shocks, one associated with slow-mode MHD waves and one associated with fast-mode waves. (Perpendicular MHD shocks are associated only with the fast/magnetosonic mode.) For the oblique fast-mode shock wave, the density increases and the flow speed decreases across the shock, as before. The conservation relations indicate that B_n stays constant and B_t increases across an oblique fast-mode shock; thus, the direction of **B** must change across the shock front (see Figure 4.20).

4.9.3 Role of shock waves

Why are shocks present in space, or in our atmosphere, for that matter? Often a high-amplitude sound wave will "steepen" into a moving shock front, as illustrated in Figure 4.21. Small-amplitude disturbances propagate at the sound speed and

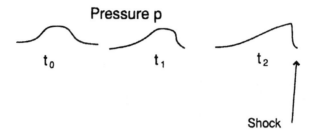

Figure 4.21. Schematic of nonlinear steepening of a pressure pulse into a traveling shock wave. A time sequence of pressure profiles is shown.

preserve their shape. But large-amplitude waves exhibit dispersion because the speed of propagation is larger for higher pressures (and temperatures) than for lower pressures (i.e., examine Equation (4.133) to convince yourself of this), allowing the middle of a disturbance to catch up with the leading edge. A traveling shock discontinuity forms once this process is complete.

Other examples of shock formation are the shock waves associated with super-sonic jets or with a fast piston moving down a long metal tube (i.e., a "shock tube" sometimes used to create hot gas and plasmas in the laboratory). Why are shocks re-quired in these examples? Dynamical information, such as a pressure disturbance, travels at the speed of sound. An object (or obstacle) moving at subsonic speed communicates its presence to the surrounding gas via sound waves, allowing the gas to adjust itself to the object, such as allowing the fluid to flow around the ob-stacle. A supersonically moving object outruns its sound waves – the gas upstream of this object is "unaware" of the object's existence. If we move to the frame of reference of the object, then the upstream gas is moving at supersonic speed toward the object, which acts as an obstacle to the flow. In order to be able to adjust itself (i.e., alter its density, pressure, and flow velocity) to the presence of the obstacle and flow around it, the gas must first become subsonic. The transition from supersonic flow to subsonic flow requires a shock wave. (See Figure 4.22.)

Many examples of shocks in space plasmas exist:

(1) Collisionless fast-mode MHD shocks in the solar wind flow called *planetary bow shocks* have been observed by spacecraft at all the planets in the solar system except Pluto.
(2) Shocks called *interplanetary shocks* have been observed in the solar wind. These are not associated with planets but with transient solar phenomena or with interaction of slow and fast "streams" in the solar wind.
(3) A shock called the *heliosphere termination shock* is thought to exist at the outer bound-ary of the *heliosphere*, where the solar wind runs up against the interstellar medium.
(4) Slow-mode MHD shocks are thought to be present in the Earth's magnetotail.

Some of these will be examined more carefully in later chapters.

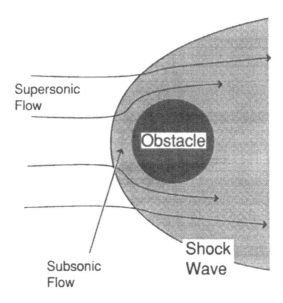

Figure 4.22. Sketch of fluid flow around a solid obstacle. Streamlines are shown.

4.10 Magnetic reconnection

Magnetic reconnection was alluded to in Section 4.4 and is related to the concept of frozen-in magnetic flux. Almost all space plasmas have a magnetic Reynolds number much greater than unity, $R_m \gg 1$, in which case a frozen-in magnetic field is a very good approximation as is the relation $\mathbf{E} = -\mathbf{u} \times \mathbf{B}$ (see Equations (4.66) and (4.75)). However, near the boundary between two different regions (such as the magnetopause boundary layer between the magnetosheath and magnetospheric plasma regions at Earth), narrow current layers exist in which R_m is small. Recall that R_m is proportional to the length scale L, and thus if a layer is narrow enough the value of R_m associated with it can be made small, even if the resistivity is quite small. These current layers can often be characterized as tangential discontinuities. At those places in the current layer where $R_m < 1$, magnetic field lines lose their identity (with a specific fluid parcel), and the topology of the magnetic field lines can be rearranged (i.e., magnetic "reconnection" or "merging" can occur).

As we found out in Section 4.4.2, R_m is inversely proportional to the resistivity of the plasma. The classical resistivity is virtually zero for most (collisionless) space plasmas. Therefore, collisional (fluid) magnetic reconnection is very inefficient; however, plasma instabilities are often able to develop in collisionless plasmas and can lead to what is called *anomalous resistivity*. Furthermore, the nature of single particle trajectories subject to small-scale electric and magnetic perturbations is such that magnetic field lines lose their identity and "collisionless magnetic reconnection" can take place and appear remarkably similar to fluid reconnection. In this book, we will stick with the fluid description of this phenomenon, but for a more advanced treatment, the reader is referred to the bibliography at the end of the chapter. This section will draw heavily from the introductory portions of the material by Cowley (1985) and by Axford (1984). Specific applications of magnetic reconnection will not be emphasized in this chapter but will be discussed later in the book as appropriate.

Magnetic reconnection is thought to occur in the following space plasma phenomena:

(1) Magnetic merging at the terrestrial dayside magnetopause and at the magnetopauses of other planets with large intrinsic magnetic fields. We will discuss this in Chapter 8.
(2) Magnetic merging in the magnetotails of the Earth and other planets, including maybe comets. *Magnetic substorms* are undoubtedly associated with magnetic reconnection (see Chapter 8).
(3) As the energy source for *solar flares*, and perhaps the energy source for the solar corona in general. This topic is discussed briefly in Chapter 5.

4.10.1 Neutral sheet for stationary plasma

Let us consider a very narrow tangential discontinuity (or *current sheet*) separating two regions with oppositely directed magnetic field. The field strength becomes zero right at the interface and is thus called a *neutral sheet*. For the geometry shown in Figure 4.23, the convection–diffusion equation (4.72) can be written as

$$\frac{\partial B_y}{\partial t} = -\frac{\partial}{\partial x}(u B_y) + \frac{\eta}{\mu_0}\frac{\partial^2 B_y}{\partial x^2}. \qquad (4.191)$$

If the plasma is not flowing ($\mathbf{u} = \mathbf{0}$), then a steady-state solution to Equation (4.191) is not possible. Formally, in this case, the magnetic Reynolds number is zero. We then have a pure magnetic diffusion problem (as in Problem 4.6) in which an initial magnetic field profile "decays." The initial magnetic field is equal and opposite in direction ($\pm\mathbf{B}_0$) on opposing sides of the layer. For uniform resistivity and for an infinitesimally thin initial current layer (see Cowley, 1985, and Axford, 1984) the

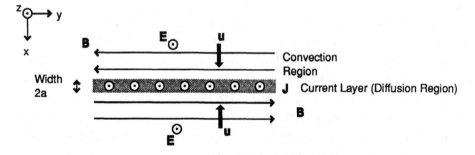

Figure 4.23. Schematic of current layer where field annihilation is occurring.

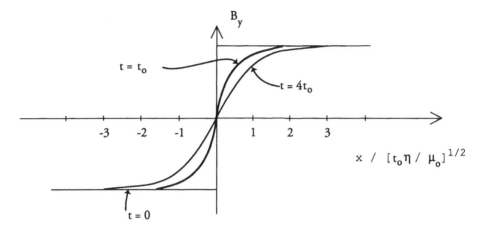

Figure 4.24. Solution for current layer evolution at two times expressed in terms of an arbitrary time scale t_0.

following "self-similar" solution can be obtained:

$$B_y = B_0 \operatorname{erf}(\zeta) \quad \text{with } \zeta = \left(\frac{\mu_0}{\eta t}\right)^{1/2} x. \tag{4.192}$$

Recall that $\operatorname{erf}(\zeta) \equiv (2/\sqrt{\pi}) \int_0^\zeta e^{-s^2} ds$ is the error function. This solution is shown in Figure 4.24. The width a of the current layer increases with time as

$$a \approx \left(\frac{\eta t}{\mu_0}\right)^{1/2}. \tag{4.193}$$

This expression can be found by setting $\zeta = 1$ in Equation (4.192). The current density $J_z(x, t)$ associated with this magnetic field profile is a Gaussian centered at $x = 0$, whose width increases in time as $t^{1/2}$.

Magnetic field is being "annihilated" by the magnetic diffusion process in the *diffusion region*, which is located near the neutral sheet. More precisely, the currents responsible for **B** are being dissipated by Ohmic resistivity. The total magnetic energy per unit area of the system is

$$W_B = \int w_B \, dx = \int \frac{B^2}{2\mu_0} \, dx. \tag{4.194}$$

The convection–diffusion equation plus Ampère's law can be used to find the following expression (Problem 4.20):

$$\frac{\partial W_B}{\partial t} = -\int \eta J_z^2 \, dx = -\int E_z J_z \, dx = -\int \mathbf{E} \cdot \mathbf{J} \, dx. \tag{4.195}$$

Equation (4.195) is consistent with the electromagnetic energy relation given by expression (A.45) in the appendix. Magnetic energy decreases with time and is converted to heat.

In the above example, the electric field is zero ($\mathbf{E} = \mathbf{0}$) outside the diffusion region but is large inside it; hence $\partial E_z/\partial x \neq 0$ and from Equation (4.192) we see that $\partial B_y/\partial t \neq 0$ (i.e., we have a non-steady-state situation). To achieve a steady-state situation we require a finite plasma flow speed $\mathbf{u}$, such that magnetic flux is carried by the plasma into the diffusion region to replace the magnetic flux annihilated there. Another difficulty with the static ($\mathbf{u} = \mathbf{0}$) scenario is that the magnetic pressure has a minimum at $x = 0$, and thus on both sides of the neutral sheet the magnetic pressure gradient force is directed inwards toward the sheet so that the plasma has a tendency to move toward the sheet (which would violate the assumption of $\mathbf{u} = \mathbf{0}$).

4.10.2 Steady-state current sheet

We now consider the scenario shown in Figure 4.23 in which the plasma flows toward the layer on both sides of the neutral sheet with flow velocity, $\mathbf{u} = u(x)\hat{\mathbf{x}}$, with $u(-x) = -u(x)$ and with $u(0) = 0$. For large $|x|$, we have $u = \pm u_0$ ($-$ for $x > 0$; $+$ for $x < 0$). The electric field from the generalized Ohm's law (4.67) can be written as

$$E_z = -u_x(x)B_z(x) + \underbrace{\frac{\eta}{\mu_0}\frac{\partial B_y}{\partial x}}_{\text{constant}} = +u_0 B_0. \tag{4.196}$$

According to Faraday's law for steady-state conditions ($\partial B_0/\partial t = 0$), E_z must be a constant. Far outside the diffusion region, we have $R_m \gg 1$, and the magnetic field is frozen into the plasma flow and $E_z = +u_0 B_0$. However, near the current sheet at $x = 0$, we have $\mathbf{u} = \mathbf{0}$ and $R_m = 0$, in which case $E_z = (\eta/\mu_0)\partial B_y/\partial x = +u_0 B_0$. The extent of the diffusion region can be defined as where R_m is about unity or less ($R_m < 1$). Using the width of the layer a for the length scale L in the magnetic Reynolds number expression (4.75) and letting R_m become unity, we can define the layer width as

$$R_m = \frac{\mu_0 u_0 a}{\eta} \to 1$$
$$a \approx \frac{\eta}{\mu_0 u_0}. \tag{4.197}$$

The example below describes how the magnetic field varies in this scenario for a particular choice of the velocity profile.

Example 4.6 (Steady-state magnetic field pattern at a current layer with inflowing plasma) Given the simple velocity profile shown in Figure 4.25a, the steady-state magnetic field profile shown in Figure 4.25b can be found by using Equation (4.191) (see Problem 4.21), assuming that $B = 0$ at $x = 0$. A qualitatively correct solution is shown here; you will find a better solution in the problem. The

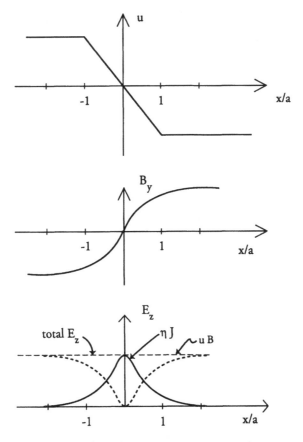

Figure 4.25. (Top) Plasma flow speed normal to the current layer (arbitrary units) (Middle) Magnetic field strength versus distance x. (Bottom) Electric field versus x plus contributions to this field from the motional electric field *(uB)* (short dashed line) and from the field associated with ohmic dissipation *(ηJ)* (solid line).

motional electric field, the "Ohmic" part of the electric field, and the total electric field (E_z) are separately plotted in Figure 4.25c. Notice that $E_z = constant$, as it must be for steady-state conditions, but the Ohmic contribution is a maximum in the center of the current layer.

The velocity profile shown in Figure 4.25a is arbitrary, albeit not totally inconsistent with the momentum equation in that the flow is toward the neutral sheet. A problem still exists with the above scenario: What happens to all the plasma that flows into the diffusion region? The continuity equation tells us that the density near $x = 0$ must increase with time unless electron–ion recombination can neutralize all the plasma being transported into that layer. Such recombination might actually be possible for some dense, cold plasmas (e.g., perhaps in the lower ionosphere of Mars), but the great majority of space plasmas are collisionless. Of course, a density build-up at $x = 0$, as well as not being consistent with a steady-state scenario, is associated with a pressure build-up that can alter the dynamics. We must conclude that a steady-state, one-dimensional scenario with collisionless plasma flow into a neutral sheet is difficult to achieve.

4.10.3 X-line magnetic reconnection

Constructing a steady-state scenario is made possible with a two-dimensional scenario. Inflow in one direction (x) can then be balanced by outflow in the other direction (y), even for an incompressible fluid. In this case, magnetic field is annihilated in one region but not in other regions, and we find a pattern such as that shown in Figure 4.26.

Many different models of magnetic reconnection exist in the space physics literature (cf. Cowley, 1985), but in all these models, magnetic field lines are convected (in the *convection region*) toward $x = 0$. The field lines then lose their identity in the *diffusion region* (where an *x-line* exists – note that the system is assumed to extend infinitely in the z direction). The "x" refers to the cross configuration of two field lines meeting in the middle of the diffusion region. The incoming field lines are "broken" in the diffusion region and are "reconnected" to field lines from the opposite side of the current sheet. After reconnection, the field lines are convected out the sides of the diffusion region. This process is called *magnetic reconnection* or *magnetic merging*.

Several steady-state, fluid reconnection models have been developed over the past 30 years. In the Sweet–Parker model (Parker, 1957, and Sweet, 1958) the length of the diffusion region is equal to the length of the system, and all plasma

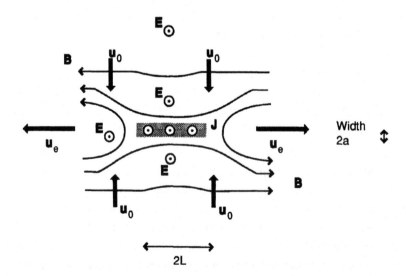

Figure 4.26. Schematic for steady-state magnetic reconnection. The hatched area with length $2L$ and width $2a$ is the magnetic diffusion region where $R_m < 1$. The x-line is a point at the center of this region and goes in and out of the page.

must flow through the diffusion region. In the Petschek/Sonnerup (P/S for short) class of models (cf. Cowley, 1985; Axford, 1984), the diffusion region length L is much less than the system length L_{syst}, although it is longer than the width of the diffusion region a (i.e., $L > a$). Most of the plasma in the P/S class flows outside the diffusion region; however, a piece of each field line must go through the diffusion region so that it can be reconnected. The various individual P/S models differ in their details. For example, in the original Petschek model (Petschek, 1964), the flow does not smoothly "turn" outside the corners of the diffusion region, as shown in Figure 4.26, but sharply changes direction at a slow-mode shock discontinuity. In this book, I will present a "generic" reconnection model and will not worry about the details; nonetheless, the results will remain generally consistent with more detailed treatments of steady-state reconnection.

For steady-state conditions and a two-dimensional configuration, the electric field must be constant everywhere:

$$E_y = -(\mathbf{u} \times \mathbf{B})_y + \eta J_y$$
$$= u_0 B_0. \tag{4.198}$$

Outside the diffusion region we have very low resistivity ($\eta/\mu_0 \approx 0$) and very large R_m, so that only the motional electric field is important. The incoming plasma has flow speed u_0 and field strength B_0. The outflow (exit) speed is u_e and the field strength is B_e. From Equation (4.198), we require for steady-state flow that

$$u_e B_e = u_0 B_0. \tag{4.199}$$

The *reconnection rate* is proportional to $E_y = u_0 B_0$.

The continuity equation applied in integral fashion to the diffusion region becomes (for incompressible flow, for which $\nabla \cdot \mathbf{u} = 0$)

$$u_0 L = u_e a \quad \text{or} \quad \frac{u_0}{u_e} = \frac{a}{L}. \tag{4.200}$$

The width of the diffusion region is still given by Equation (4.197); hence, defining R_m^o as the magnetic Reynolds number for the outflow convection region, we obtain

$$u_0 = u_e \frac{a}{L} = \frac{u_e}{u_0} \frac{\eta}{\mu_0 L} = \frac{u_e^2}{u_0} \frac{1}{R_m^o} \quad \text{or} \quad u_0 = u_e / \sqrt{R_m^o}, \tag{4.201}$$

with

$$R_m^o = \mu_0 u_e L / \eta.$$

The resistivity is assumed to be constant and uniform. The magnetic field in the outflow region, B_e, can be determined from the constancy of $|E_y| = |u_e B_e| = |u_0 B_0|$:

$$\frac{B_e}{B_0} = \frac{u_0}{u_e} = \frac{a}{L}. \tag{4.202}$$

For all reconnection models $a/L \ll 1$ and hence $B_e \ll B_0$, and $u_0 \ll u_e$. What is the speed of the outflowing plasma, u_e (and therefore what are u_0 and B_e)? We need to consider the energy balance to determine u_e.

We assume that not only is the plasma density uniform but so is the pressure. The energy equation (4.119) can be applied to this problem after numerous simplifications (see Figure 4.27). The energy per unit length transported into the diffusion region (from both sides) is given by $4Lu_0[(1/2)\rho u_0^2 + B_0^2/\mu_0]$, where the pressure term (i.e., thermal energy) has been neglected (this implicitly assumes that the plasma just outside the reconnection region has low β). The magnetic term comes from the divergence of the Poynting vector ($\nabla \cdot \mathbf{S}$) term in (4.119). The electromagnetic energy conversion rate per unit length in the diffusion "volume" can be found by integrating the right-hand side of Equation (4.119) over the volume of a unit length of the diffusion region:

$$\int\limits_{\substack{\text{diffusion} \\ \text{region}}} \mathbf{E} \cdot \mathbf{J}\, d^3x \approx 4LaE^2/\eta$$

$$= \frac{4La}{\eta}u_0^2 B_0^2 \approx 4u_0 L\frac{B_0^2}{\mu_0}, \qquad (4.203)$$

where the definition of the width a from Equation (4.197) was used in the second line. Notice that the rate of magnetic energy destruction in the diffusion region equals the flux of magnetic energy into the diffusion region. This implies that not only is the thermal pressure small compared to the magnetic pressure (or energy density) in the immediately incoming plasma but so is the dynamic pressure (i.e., u_0 is very small). However, we can do somewhat better than this with a slightly more careful analysis.

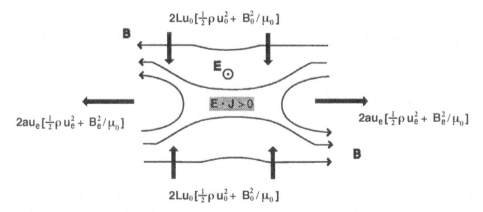

Figure 4.27. Energy balance for diffusion region surrounding the x-line. Energy input and outflow rates are indicated. Magnetic energy is converted into kinetic energy in the diffusion region. Thermal energy is neglected for simplicity.

The total energy flux leaving the diffusion region out of both sides is $4a\,u_e[(1/2)\,\rho u_e^2 + B_e^2/\mu_0]$, and this must equal the total energy flux into the region:

$$4Lu_0\left[\frac{1}{2}\rho u_0^2 + B_0^2/\mu_0\right] = 4au_e\left[\frac{1}{2}\rho u_e^2 + B_e^2/\mu_0\right]. \qquad (4.204)$$

Using Equations (4.200) and (4.204), we find

$$\frac{1}{2}u_0^2 + B_0^2/\rho\mu_0 = \frac{1}{2}u_e^2 + B_e^2/\rho\mu_0$$

or

$$\frac{1}{2}u_0^2 + C_{A0}^2 = \frac{1}{2}u_e^2 + \frac{B_e^2}{B_0^2}C_{A0}^2. \qquad (4.205)$$

C_{A0} is the Alfvén speed of the incoming flow, and $B_e/B_0 = a/L = u_0/u_e$ from Equation (4.202). Equation (4.205) can be rewritten as

$$u_e^2 = u_0^2 + 2C_{A0}^2\left[1 - \frac{u_0^2}{u_e^2}\right]. \qquad (4.206)$$

The solution of Equation (4.206) for u_e is either $u_e = u_0$, which requires $a = L$ (implying a minuscule diffusion region), or is

$$u_e = \sqrt{2}C_{A0}. \qquad (4.207)$$

Equation (4.207) states that the outflow speed is approximately equal to the Alfvén speed of the incoming flow. If the incoming flow is quite sub-Alfvénic ($M_{A0}^2 \ll 1$ or $u_0^2 \ll C_{A0}^2$), then a large amount of magnetic energy must be converted to plasma kinetic energy inside the diffusion region in order to accelerate the flow from its initial speed of u_0 to its final outgoing speed of approximately C_{A0}.

From Equations (4.202) and (4.207), we can see that the inflow speed must be equal to

$$u_0 = \sqrt{2}C_{A0}\left(\frac{a}{L}\right) = \sqrt{2}C_{A0}\frac{1}{\sqrt{R_m^o}}, \qquad (4.208)$$

where the external region (outflow) magnetic Reynolds number can be rewritten approximately as

$$R_m^o = \mu_0 C_{A0}L/\eta. \qquad (4.209)$$

The "total" reconnection rate is equal to the length of the system (on both sides), $4L_{\text{syst}}$, multiplied by $E_y = u_0 B_0$. And the inflow speed is given by Equation (4.208).

We can now use some results from this "generic" reconnection model to study the predictions of a couple of specific models (again see Cowley for details). In the Sweet–Parker model, the length of the diffusion region equals the length of the system ($L_{\text{syst}} = L$), and using Equation (4.208) we find that the total reconnection rate is about equal to (neglecting small $2^{1/2}$ type factors)

$$4Lu_0B_0 \approx 4C_{A0}aB_0 \approx 4LC_{A0}B_0\frac{1}{\sqrt{R_{\text{m}}^{\text{o}}}}. \tag{4.210}$$

In this case, the total rate of magnetic annihilation given by Equation (4.203) is equal, in the Sweet–Parker model, to the total rate of magnetic energy flux into the system. All plasma must pass through the diffusion region in this particular reconnection model.

In the Petschek/Sonnerup type reconnection model (unlike the Sweet–Parker model just discussed), the diffusion region is shorter than the length of the entire system, $L < L_{\text{syst}}$. The total reconnection rate equals $4L_{\text{syst}}u_0B_0 \approx 4C_{A0}aB_0$ (L_{syst}/L). Using Equation (4.208), we can also express the total reconnection rate as $\approx 4L_{\text{syst}}C_{A0}B_0(a/L) = 4L_{\text{syst}}C_{A0}B_0/\sqrt{R_{\text{m}}^{\text{o}}}$, where R_{m}^{o} is defined in terms of L. In contrast, if we define the magnetic Reynolds number in terms of L_{syst} instead of in terms of L (now denoted R_{m}^{os}), then the total reconnection rate is equal to

$$\left[4L_{\text{syst}}C_{A0}B_0/\sqrt{R_{\text{m}}^{\text{os}}}\right]\sqrt{\frac{L_{\text{syst}}}{L}}. \tag{4.211}$$

This rate is the same as the total reconnection rate for the Sweet–Parker model but with the addition of a $\sqrt{L_{\text{syst}}/L}$ factor that can be much greater than unity if the external conditions "driving" the reconnection require it. Thus, magnetic reconnection is more efficient for the P/S type model than for the Sweet–Parker model. Not all magnetic flux being "reconnected" is "processed" through the diffusion region in the P/S model – most of the magnetic flux avoids this region if $L \ll L_{\text{syst}}$.

The total reconnection rate in the Sweet–Parker model is quite low (i.e., Equation (4.210)) because R_{m}^{o} is typically very large for space plasmas. Since $R_{\text{m}} \approx 10^6$–$10^8$ in the solar wind the reconnection rate per unit area is proportional to $10^{-4}C_{A0}B_0$. This limits the magnetic flux throughput (and must also limit u_0). However, in the P/S models reconnection can be made efficient enough to accommodate much larger values of u_0. However, the quantitative value of $\sqrt{L_{\text{syst}}/L}$ in Equation (4.211) depends on the particular P/S model under consideration. Detailed analysis (which will not be presented here, but see the discussion in Cowley (1985)) indicates that the maximum inflow speed for a P/S type model is given by

$$u_{0\,\text{max}} \approx \frac{\pi}{8}\frac{C_{A0}}{\ln\left(R_{\text{m}}^{\text{os}}\right)}. \tag{4.212}$$

The maximum possible reconnection rate (per unit area) found using Equation (4.212) roughly equals $0.02\,C_{A0}B_0$ for a solar wind value of R_{m}^{os}, whereas the Sweet–Parker value of the reconnection rate was $\approx 10^{-4}C_{A0}B_0$. Note that $u_{0\text{max}}$ depends only on the log of the magnetic Reynolds number of the external plasma. Another way of expressing Equation (4.212) is that the maximum possible Alfvénic

Mach number, $M_{A0} = u_0/C_{A0}$, for the incoming flow is $M_{A0,max} = \pi/[8 \ln R_m^{os}]$. One can then deduce that $L/L_{syst} \approx 10^{-4}$–$10^{-5}$.

4.10.4 Collisionless magnetic reconnection – Anomalous resistivity

In the fluid treatment of magnetic reconnection given in Section 4.10.3 we assumed a collisional resistivity η. But most space plasmas are essentially collisionless ($\eta \approx 0$). However, a collisionless picture(s) of magnetic reconnection can be constructed using the theory of single particle motion near neutral sheets and/or using plasma instabilities to provide an effective, or *anomalous, resistivity*. Because this topic is beyond the scope of this text (the interested student is again referred to Cowley (1985)), in this section we will limit ourselves to a simplistic view and will look for a "quick fix" to the fluid picture of magnetic reconnection.

It is difficult to create a current sheet in a collisionless plasma that is much narrower than the ion gyroradius. For simplicity, and without further justification, let us assume that the width of the current layer (and also of the diffusion region) is equal to the ion gyroradius appropriate for the "general vicinity of" the current layer:

$$a \approx r_L. \tag{4.213}$$

Using Equation (4.213) and the fluid definition of the layer width a from Equation (4.197), we can define an anomalous (or effective) resistivity as

$$\eta_{eff} = r_L \mu_0 u_0. \tag{4.214}$$

The magnetic Reynolds number for the incoming flow region determined using η_{eff} is $R_m \approx L_{syst}/r_L$ ($\approx 10^6$ for the solar wind). The magnetic Reynolds number discussed earlier, R_m^{os}, is of the same order as this R_m.

Detailed studies of single particle motion (cf. Cowley, 1985) have demonstrated that the plasma speed of the outflow from the diffusion region is about equal to the Alfvén speed of the incoming plasma C_{A0} – just as in the fluid picture. The inflow speed given by Equation (4.208) can now be rewritten as $u_0 \approx (a/L)C_{A0} = (r_L/L)C_{A0}$, where L is the length of the diffusion region. The Mach number of the incoming flow can now be given by $M_{A0} = r_L/L$. If M_{A0} is known, then the length of the diffusion region is given by

$$L \approx \frac{r_L}{M_{A0}}. \tag{4.215}$$

For a P/S-type model (but using η_{eff}), the maximum possible value of M_{A0} is $\pi/[8 \ln R_m^{os}] \approx 0.02$, and hence from Equation (4.215) the minimum length of the diffusion region is given by

$$L_{min} \approx \frac{r_L}{\pi} 8 \ln R_m^{os} \approx 50 r_L. \tag{4.216}$$

(for solar wind conditions)

Note that the incoming flow must be sub-Alfvénic ($M_{A0} < 1$).

You should also be aware that magnetic reconnection does not necessarily have to take place in a steady-state manner, as we have assumed in this chapter. For example, observational evidence suggests that reconnection is probably occurring sporadically (i.e., in a "bursty" fashion) at the terrestrial dayside magnetopause.

Problems

4.1   Derive the following dispersion relation for electromagnetic wave propagation in an unmagnetized plasma:

$$\omega^2 = \omega_{pe}^2 + k^2 c^2,$$

where the plasma frequency ω_{pe} is given by Equation (4.35). Start from Maxwell's equations with zero charge density but with current density equal to $\mathbf{J} = n_e e \mathbf{u}_e$, where n_e is the electron density and $\mathbf{u}_e$ is the bulk flow speed of the electron gas. You will also need the following approximate electron momentum equation for a collisionless plasma:

$$m_e n_e \frac{\partial \mathbf{u}_e}{\partial t} \cong -n_e e \mathbf{E},$$

where $\mathbf{E}$ is the wave electric field and m_e is the electron mass.

4.2   Derive the dispersion relation, Equation (4.40), for Langmuir waves in a warm plasma in an analogous manner to the derivation carried out in the text for a cold plasma. You will need to retain the pressure gradient term in the electron momentum equation.

4.3   Derive the plasma frequency for a cold ($T_e = T_+ = 0$) plasma consisting of electrons and positrons with background densities $n_e = n_+ = n$.
      (*Hint:* Start with the two-fluid equations, linearize them, assume plane wave solutions, and derive a dispersion relation for Langmuir oscillations.) How is the plasma frequency for this plasma different from that in a plasma made of electrons and ions?

4.4   Consider a typical terrestrial ionospheric plasma, in which the electron density varies with altitude z as

$$n_e(z) = 10^5 \exp[-z/H_p] \quad [\text{cm}^{-3}],$$

where the plasma scale height is approximately $H_p \approx 300\,\text{km}$. The electron temperature is approximately $T_e \approx 3,500\,\text{K}$. Estimate the ambipolar electric field associated with this density profile and also estimate the associated departure from charge neutrality in the plasma: $|n_e - n_i|/n_e$. n_i is the ion density.

4.5   Derive the charge conservation Equation, (4.59), solely from Maxwell's equations.

4.6 Consider a static ($\mathbf{u} = \mathbf{0}$) magnetized plasma in a slablike region $0 \leq x \leq L/2$. The resistivity of the plasma is equal to η. The magnetic field in the plasma is directed only in the z direction and is a function only of x and t:

$$\mathbf{B}(x, t) = B(x, t)\hat{z}.$$

The magnetic field at the boundaries is kept at the value B_0. And the field strength at time $t = 0$ is given by the expression

$$B(x, t = 0) = B_0 + a_0(0) \sin[2\pi x/L] + a_1(0) \sin[4\pi x/L],$$

where $a_0(0)$ and $a_1(0)$ are arbitrary constants.

(a) **Estimate** the time it takes for the field to decay significantly due to Ohmic dissipation of the currents, in terms of the quantities η and L. (Note that this is a one- or two-line solution.)
(b) Find $B(x, t)$ as a function of x and t.
(c) Assume now that the plasma density is $n_e = 10^5 \, \text{cm}^{-3}$, the electron temperature is $T_e = 1,000 \, \text{K}$, and the neutral density is $n_n = 5 \times 10^9 \, \text{cm}^{-3}$. Find the resistivity of the plasma given that the electron–ion and electron–neutral collision frequencies are given by $\nu_{ei} = 50 n_e/T_e^{3/2}$ and $\nu_{en} = 10^{-8} n_n$ (where the densities in these expressions are in cgs units).

The plasma is confined to a region of length $L/2 = 100 \, \text{km}$ but has infinite extent in the other directions. For the choice of initial field parameters, $B_0 = 100 \, \text{nT}$, $a_0(0) = 50 \, \text{nT}$, and $a_1(0) = 50 \, \text{nT}$, find the solution $B(x, t)$ using the general solution from part (b). Plot this solution as a function of x at $t = 0$ and at three other later times.

4.7 Derive Equations (4.97) and (4.98) from Equation (4.92). For the topside terrestrial ionosphere (i.e., for altitudes above about 300 km) where $T_e \approx 3,500 \, \text{K}$ and $T_i \approx 1,500 \, \text{K}$, determine the plasma scale height H_p. Assume that the major ion species is O^+.

4.8 A balloon made with a massless, but strong, membrane has an arbitrary shape and has a volume V. This balloon is evacuated (i.e., a vacuum is created) and is placed in a planetary neutral atmosphere at an altitude where the pressure is p and the temperature is T. The mean molecular mass on this planet is M and the gravitational acceleration is g.

(a) Find a general expression for the upward force on the balloon, using the law of hydrostatic balance for a neutral atmosphere. In particular, demonstrate that this upward buoyant force is equal to the weight of the displaced fluid. (You are finding a mathematical expression for *Archimede's Law*.)
(b) Evaluate the force for a spherical balloon with a radius of 1 m in an atmosphere whose pressure (at the location of the balloon) is 50 bars and whose temperature is 500 K. The gravitational acceleration is 9 m/s^2 and the mean molecular mass is 44. (1 bar = 10^5 Pa.)

4.9 Determine the pressure profile as a function of radial distance r in the solar corona (i.e., the outer solar atmosphere) from hydrostatic balance for an isothermal corona with temperature $T_0 = 10^6$ K. We will discuss this topic in Chapters 5 and 6, but the only information you need here to handle this problem is that the acceleration due to gravity at the Sun's surface is 274 m/s^2 and that above the surface this acceleration varies inversely as the square of the radial distance from the center of the Sun. We will show in Chapter 6 that hydrostatic balance is *not* a good description of the outer solar corona, but for now we will over-look this.

4.10 The solar wind dynamic pressure is approximately "converted" into magnetic pressure in the magnetic barrier located just outside the ionopause of Venus:

$$\rho_{sw} u_{sw}^2 \cong B^2/8\pi \quad \text{(in cgs units),}$$

where ρ_{sw} is the mass density of the solar wind upstream of Venus, u_{sw} is its flow speed, and B is the field strength in the magnetic barrier.

Derive an expression for the ionopause height z_{ip} as a function of $\rho_{sw} u_{sw}^2$. Assume that the electron density at a reference altitude, $z_0 = 150$ km, is equal to a constant n_0. Assume that the electron and ion temperatures are independent of altitude z and that the major ion species is O$^+$. Assume that the ionospheric plasma is static.

(*Hint:* You will need an expression for n_e as a function of altitude. Ignore chemistry in getting this expression.)

4.11 Determine the magnetic field of a cylindrically symmetric configuration as a function of distance from the axis (r): $\mathbf{B}(r) = B_z(r)\hat{\mathbf{z}} + B_\phi(r)\hat{\phi}$. Assume a force-free field configuration of

$$\nabla \times \mathbf{B} = \alpha \mathbf{B},$$

where we will assume here that the function α is a constant, $\alpha(r) = \alpha_0$. *Hint:* Brush up on your ordinary differential equations and on Bessel functions.

4.12 (a) Derive Equation (4.110) for ambipolar diffusion from Equation (4.109).

(b) Demonstrate that Equation (4.110) plus Equation (4.112) have the following "standard" form for a diffusion equation if the plasma is isothermal and if we neglect gravity (which are actually not good assumptions in practice for real ionospheres):

$$\frac{\partial n_e}{\partial t} = \sin^2 \theta_1 D_a \frac{\partial^2 n_e}{\partial z^2}.$$

4.13 Transform Equation (4.113) into Equation (4.114). Suppose something slows down the solar wind to almost zero speed. Estimate the stagnation pressure of this solar wind by using Equation (4.114) with $B = 0$.

4.14   Find Equation (4.125) from the polytropic energy relation (4.124).

4.15   Prove that any function of the form $\rho_1(x \pm C_s t)$ is a solution of the wave Equation (4.132) for sound waves. Note that in working in only one direction (i.e., x) the Laplacian, ∇^2, just becomes $\partial^2/\partial x^2$.

4.16   Use Maxwell's equations to prove that for any electromagnetic wave, including Alfvén waves, the wave electric and magnetic fields are mutually orthogonal to each other and to the wave vector $\mathbf{k}$. In other words, electromagnetic waves are transverse. Derive an expression for the electric field amplitude of an Alfvén wave in terms of the magnetic field amplitude. Demonstrate that the energy density of the wave electric field is much less than the energy density of the wave magnetic field for the solar wind.

4.17   Consider a stationary ($\mathbf{u} = \mathbf{0}$) uniform magnetized plasma with magnetic field $\mathbf{B}_0 = B_0 \hat{\mathbf{z}}$. Assume that the background pressure and mass density are p_0 and ρ_0, respectively. Derive a dispersion relation for waves propagating perpendicular to the background field direction; that is, find the dispersion relation for *magnetosonic waves*. Assume that the wave amplitude is small (so that you can linearize the relevant fluid equations). What are the phase and group velocities for this wave mode? Ignore collisions and use the polytropic relation in lieu of the full energy equation. Start from the MHD Equations (4.121)–(4.124).

4.18   Use Equations (4.174)–(4.181) to find the Rankine–Hugoniot relations (4.183)–(4.185). Then find expression (4.188) for the downstream Mach number M_2, in terms of the upstream Mach number M_1.

4.19   Demonstrate that the specific entropy increases across a shock by verifying with the Rankine–Hugoniot relations the inequality given by Equation (4.189).

4.20   Derive Equation (4.195) from Ampère's law plus the magnetic convection–diffusion equation. Estimate $\partial W_B/\partial t$ as a function of time.

4.21   Find the magnetic field profile appropriate for the velocity shown in Figure 4.25a as discussed in Example 4.6. The profile you find will not be identical to the approximate one shown in Figure 4.25b. What difficulties might arise from the momentum balance associated with this scenario?

Bibliography

Axford, W. I., "Magnetic field reconnection," p. 1 in *Magnetic Reconnection in Space and Laboratory Plasmas*, Geophys. Monograph 30, American Geophysical Union, Washington, DC, 1984.

Bittencourt, J. A., *Fundamentals of Plasma Physics*, Pergamon Press, Oxford, UK, 1986.

Chen, F. F., *Introduction to Plasma Physics*, Plenum Press, New York, 1974.

Cowley, S. W. H., "Magnetic Reconnection," p. 121 in *Solar System Magnetic Fields*, ed. E. R. Priest, D. Reidel Publ. Co., Dordrecht, The Netherlands, 1985.

Elphic, R. C., C. T. Russell, J. A. Slavin, and L. H. Brace, Observations of the dayside ionopause and ionosphere of Venus, *J. Geophys. Res.*, **85**, 7670, 1980.

Parker, E. N., Sweet's mechanism for merging magnetic fields in conducting fluids, *J. Geophys. Res.*, **62**, 509, 1957.

Petschek, H. E., "Magnetic field annihilation," p. 425 in *AAS-NASA Symposium on The Physics of Solar Flares, NASA Spec. Publ. SP-50*, 1964.

Priest, E. R., *Solar MHD*, D. Reidel Publ. Co., Dordrecht, The Netherlands, 1982.

Rees, M. H., *Physics and Chemistry of the Upper Atmosphere*, Cambridge Univ. Press, Cambridge, UK, 1989.

Russell, C. T. and R. C. Elphic, Observations of flux ropes in the Venus ionosphere, *Nature*, **279**, 616, 1979.

Russell, C. T. and O. Vaisberg, "The interaction of the solar wind with Venus," p. 873 in *Venus*, ed. D. M. Hunten, L. Colin, T. M. Donahue, and V. I. Moroz, Univ. Arizona Press, Tucson, 1983.

Siscoe, G. L., "Solar System Magnetohydrodynamics," p. 11 in *Solar-Terrestrial Physics: Principles and Theoretical Foundation*, ed. R. L. Carovillano and J. M. Forbes, D. Reidel Publ. Co., Dordrecht, The Netherlands, 1983.

Sweet, P. A., "The neutral point theory of solar flares," in *Electromagnetic Phenomena in Cosmical Physics*, ed. B. Lehnert, Cambridge Univ. Press, Cambridge, UK, 1958.

Tidman, D. A. and N. A. Krall, *Shock Waves in Collisionless Plasmas*, Wiley-Interscience, New York, 1971.

5

Solar physics

The Sun is a star. As stars go, the Sun is rather cool and small and has the gross characteristics listed in Table 5.1. The Sun is the source of virtually all energy in our solar system, including the Earth. Solar radiation heats our atmosphere and provides the light needed to sustain life on our planet. The Sun is also the source of space plasmas throughout the solar system. For example, solar *extreme ultraviolet (EUV) radiation* is largely responsible for the existence of planetary ionospheres via the photoionization of atoms and molecules in the upper atmospheres of the planets. The solar wind plasma is really an extension of the solar corona out into interplanetary space. The Sun is also, naturally, the source of *solar activity*. Solar activity refers to both short-term and long-term temporal variations in the solar atmosphere (and hence in the solar wind) that create changes in the Earth's plasma environment (i.e., *geomagnetic activity*). We will deal with the effects of solar activity on the Earth later.

The field of *solar physics* has advanced dramatically during the past few decades, due to observations made by increasingly sophisticated ground- and space-based observatories, including NASA's *OGO*, *Skylab*, and *Solar Maximum* missions and the NASA/ESA *SOHO* (*Solar and Heliospheric Observatory*) mission, and due to theoretical developments in the areas of stellar nuclear physics, stellar radiative transfer, and solar MHD. We cannot possibly properly cover such a scientific discipline in one chapter; thus, only a brief introduction will be provided here. You are encouraged to read one of the references listed at the end of this chapter, such as Foukal (1990) or Zirin (1988), for a more complete introduction to solar physics.

5.1 Solar interior

5.1.1 Regions of the solar interior

The interior of the Sun can be divided into four zones as shown in Figure 5.1 and as listed below:

Table 5.1. *Properties of the Sun*

Radius	$R_\odot$	$= 6.96 \times 10^5$ km
Mass	$M_\odot$	$= 1.99 \times 10^{30}$ kg
Average mass density	$\langle \rho_\odot \rangle$	$= 1410$ kg/m^3
Luminosity (total energy output)	$L_\odot$	$= 3.86 \times 10^{26}$ W

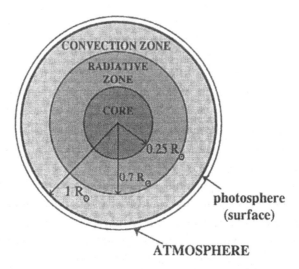

Figure 5.1. Zones of the solar interior and the solar atmosphere. The photosphere is about 500 km thick and the chromosphere is about 2500 km in extent. The corona extends many solar radii into space.

(1) **The core** The core is the high-density, high-temperature (15 million K) gas at the center of the Sun, in which thermonuclear reactions occur. One half of the solar mass is contained in the core, although its radius is only one fourth of the solar radius. Virtually all of the Sun's energy output (or *solar luminosity*) is generated in the core.

(2) **The radiative zone** The energy produced by nuclear reactions in the core must somehow make its way to the surface where it can be radiated away into space. Throughout most of the solar interior (i.e., the core and radiative zone), the energy gets transported via diffusion of gamma-ray photons. The photons are scattered, absorbed, and reemitted over and over again, gradually making their way toward the surface.

(3) **The convection zone** This zone is located in the uppermost 30% of the solar interior. The solar gas in this region is *convectively unstable* – that is, buoyancy forces result in up and down motions (see Problem 4.8). Convection cells of many different sizes exist in the convection zone and transport energy up to the solar surface where evidence of this convective activity can be seen in the observed small-scale structure in the solar atmosphere.

(4) **The atmosphere** The *solar atmosphere* is the region of the Sun that can be observed from the outside and from which solar energy is radiated into space. A small amount

of energy is also lost from the atmosphere in the form of kinetic energy of particles: the solar wind and solar cosmic rays. The atmosphere can itself be divided into four regions: *the photosphere, the chromosphere, the transition region,* and *the corona.*

5.1.2 Thermonuclear energy generation

The interior of a star must be very hot (10 million degrees or more) and dense (about 10^5 kg/m^3) in order to have a sufficiently high pressure to withstand the tremendously large gravitational forces exerted on the gas. These high temperatures and densities also provide the conditions required for nuclear reactions to take place, thus producing the energy needed to sustain the high temperature in the core over geological time scales. In Section 5.1.3, we discuss the density and temperature structure inside the Sun. Here we briefly review the most basic nuclear physics relevant to the Sun.

The solar interior is fully ionized and mostly consists of protons and electrons. Two protons (designated H^1, where the "1" stands for an atomic mass number of unity) sometimes approach close enough to react to form a deuterium nucleus (D^2 – one proton and one neutron) plus a positron (β^+) and an electron neutrino (ν_e):

$$H^1 + H^1 \rightarrow D^2 + \beta^+ + \nu_e \quad (+1.44 \, \text{MeV}). \tag{5.1}$$

Reaction (5.1) produces an energy of 1.44 MeV. This reaction proceeds very slowly. High temperatures are required to make this reaction go at all because of the large electrostatic repulsion between the two positively charged protons. In Problem 5.1 you are asked to find the distance of closest approach between two protons with a kinetic energy equivalent to 15×10^6 K.

Reaction (5.1) is followed by the reactions

$$D^2 + H^1 \rightarrow He^3 + \gamma \quad (+5.5 \, \text{MeV}) \tag{5.2}$$

and

$$He^3 + He^3 \rightarrow H^1 + H^1 + He^4 \quad (+12.9 \, \text{MeV}). \tag{5.3}$$

In reaction (5.2), a deuterium nucleus and a proton react to produce a light helium nucleus (He^3 – two protons and one neutron) plus a gamma-ray photon (γ), which carries off an energy of approximately 5.5 MeV. In reaction (5.3), two helium nuclei react to produce a stable helium nucleus plus 12.9 MeV of kinetic energy of the product nuclei.

The chain of reactions (5.1)–(5.3) has the net result of converting four protons into one helium nucleus:

$$4H^1 \rightarrow He^4$$

with an overall mass change of

$$\Delta m = (4m_{\text{H}} - m_{\text{He}}) = 0.029 \, \text{amu}. \tag{5.4}$$

Note that mass is lost in the above nuclear process; the rest masses of four protons (four m_{H}) exceed the mass of one helium nucleus (m_{He}) by 0.029 atomic mass units (amu). Energy conservation demands that this missing rest mass turn up as another form of energy/mass, such as the energy of the gamma-ray photon and the neutrino and kinetic energy of the various reaction products. The total amount of all this energy for a single "cycle" of reactions (5.1)–(5.3) is

$$\Delta E = \Delta mc^2 = (0.029 \, \text{amu})c^2 = 27 \, \text{MeV}. \tag{5.5}$$

This energy is just the sum of the energies listed for reactions (5.1)–(5.3), and this energy multiplied by the frequency of reactions throughout the solar core is responsible for the solar luminosity (see Problem 5.2).

Reactions other than (5.1)–(5.3) also take place in the solar interior but are not as important. For example, light helium nuclei can be converted into beryllium ($_4\text{Be}^7$) and boron ($_5\text{B}^8$), and the boron nucleus decays into the beryllium isotope $_4\text{Be}^8$ plus a 14 MeV neutrino. These neutrinos plus the lower energy neutrinos from reaction (5.1) can directly escape from the solar interior because neutrinos interact only very weakly (via the *weak force*) with matter. The neutrinos from the beryllium decay have been detected (with extremely low efficiency) at the Earth, but there is a discrepancy between the observed neutrino flux and the flux predicted by theoretical stellar evolution/structure calculations. The observed neutrino flux is approximately a factor of two less than the predicted flux (Haxton, 1995).

5.1.3 Structure of the interior and energy transport

Fluid equations very similar to those discussed in Chapter 4 can be employed to describe the density and temperature structure in the solar interior. The vertical momentum balance is very nicely approximated by the hydrostatic equilibrium relation given by Equation (4.92). Identifying the pressure gradient as $\nabla p = \hat{\mathbf{r}} \partial p / \partial r$, this relation becomes

$$\frac{\partial p}{\partial r} = -g\rho, \tag{5.6}$$

where r is the radial distance from the center of the Sun and the acceleration due to gravity at r is given by $g(r) = GM(r)/r^2$. G is the gravitational constant and $M(r)$ is the mass contained within a sphere of radius r,

$$M(r) = \int_0^r 4\pi r'^2 \rho(r') \, dr'. \tag{5.7}$$

The pressure p and density ρ also obey the following equation of state:

$$p \cong (n_{\text{H}} + n_{\text{He}} + n_e)k_{\text{B}}T, \tag{5.8}$$

where n_H, n_{He}, and n_e are the hydrogen (i.e., protons), helium (i.e., alpha particles), and electron number densities, respectively, and T is a common temperature. Note that $n_{He}/n_H \approx 0.08$ throughout the Sun; hence the helium abundance is 30% by mass. The electron density is related via charge neutrality to the ion number densities by $n_e = n_H + 2n_{He} + 8n_O + \cdots \cong n_H + 2n_{He} \cong 1.16 n_H$, where n_O is the oxygen number density; although all other constituents should also really be included we have included only the most important ones. We can rewrite Equation (5.8) as

$$p = \rho \frac{k_B}{\bar{m}} T \qquad \text{with } \bar{m} = \frac{m_H[1 + 4 \times 0.08 + \cdots] + m_e}{2 + 0.16 + \cdots} \cong 0.6 \, m_H, \qquad (5.9)$$

where $\bar{m}$ is the average mass. The solution to Equation (5.6) was given in Chapter 4 (Equation (4.93)) as

$$p(r) = p_0 \exp\left[-\int_0^r \frac{dr'}{H(r')}\right] \qquad \text{with } H(r) = \frac{k_B T(r)}{\bar{m} g(r)}. \qquad (5.10)$$

$H(r)$ is the scale height as a function of radial distance and p_0 is the pressure at the center of the Sun.

Of course, in order to determine the pressure (and density) using Equation (5.10), one needs to know the temperature as a function of radial distance, $T(r)$. The temperature is a strong function of r, decreasing from a peak value in the center of the core of 15×10^6 K (at $r = 0$) down to $T = 5,700$ K at the surface of the Sun ($r = R_\odot$). The temperature can be calculated using an energy equation that includes both particle and radiation (or photon) parts and that also includes energy transport and local heating by thermonuclear reactions. However, here we will merely adopt a temperature profile, $T(r)$, determined from a complex theoretical/numerical calculation (see Figure 5.2). Given $T(r)$, Equations (5.7)–(5.10) can be used to determine the pressure and density (also shown in Figure 5.2). Problem 5.3 asks you to do just this.

The main means of transporting energy throughout most of the solar interior is via radiation. More specifically, the energy is transported by the diffusion of gamma-ray photons. The photons are absorbed, reemitted, and scattered over and over again. The full energy equation for both particles and radiation must not only include the radiative energy transport but the thermonuclear energy generation in the core. Energy in the core is generated at a total rate of $L_\odot$, the solar luminosity. This energy must make its way via photon diffusion from the core up to the convection zone.

The mean free path for a photon depends on the number density of targets (i.e., absorbers and scatterers) n, which is related to the mass density by $\rho = \bar{m}n$, as well as on the photon–particle interaction cross section σ:

$$\lambda_{\text{mfp}} = \frac{1}{n\sigma} = \frac{1}{\kappa\rho}. \qquad (5.11)$$

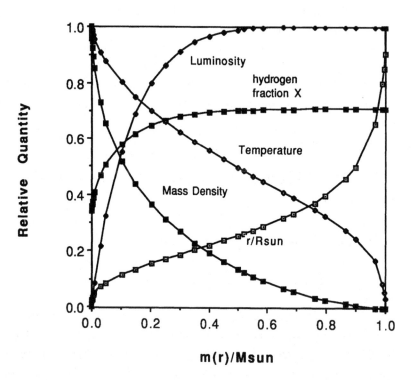

Figure 5.2. Density, temperature, and luminosity versus total mass enclosed within radial distance *r* relative to the solar mass in the solar interior from a theoretical model by Bahcall and Ulrich (1988). The radial distance is plotted versus this fractional mass enclosed. The density and temperature are relative to values at the solar center: density $\rho = 1.48 \times 10^5$ kg/m^3; temperature $T = 1.56 \times 10^7$ K. The luminosity is relative to the solar luminosity. Also shown is hydrogen fraction by mass. Adapted from Table 6.3 of Foukal (1990). Also see Figure 6.1 of Foukal (1990).

The mean free path λ_{mfp} can also be written in terms of the opacity κ and the mass density. Parameters σ or κ must include many processes and are functions of r. Near the outer part of the core $\kappa \approx 4$ cm^2/g, where $\rho \approx 10$ g/cm^3, and Equation (5.11) indicates that $\lambda_{\mathrm{mfp}} \approx 2 \times 10^{-4}$ m. Obviously, $\lambda_{\mathrm{mfp}} \ll R_{\odot}$ and an incredibly large number of scattering events must take place before a photon can escape from the interior of the Sun. The energy transport process is well approximated in the radiative zone by a diffusion equation; the diffusive photon flux at frequency v is given by

$$F_v = -D_v \nabla n_v, \tag{5.12}$$

where D_v is the diffusion coefficient of photons at frequency v, and n_v is the differential (per unit frequency) number density of photons at frequency v. We can multiply Equation (5.12) by the energy of a photon, hv, and then integrate over all frequencies. Assuming spherical symmetry in the Sun, we then find an equation for the flux of photon energy in the radial direction (units of J/m^2/s or W/m^2):

$$F_{ph} = -D_{ph}\partial\varepsilon_{ph}/\partial r. \tag{5.13}$$

Here F_{ph} is the photon energy flux and $\varepsilon_{ph} = \int n_\nu h\nu\, d\nu$ is the total photon energy density. The diffusion coefficient can be written (accurate to a factor of order unity) as a product of the characteristic velocity and the mean free path:

$$D_{ph} = \frac{1}{3}c\lambda_{mfp} = \frac{c}{3\kappa\rho}. \tag{5.14}$$

The speed of light is designated by c, and D_{ph} is an average over the complete photon spectrum. To an extremely high degree of accuracy the radiation field in the solar interior is really *blackbody radiation*. The frequency-integrated energy flux from the surface of a blackbody is $\sigma_{sb}T^4$, where σ_{sb} is the *Stefan–Boltzmann constant*. The energy density is then given by $\varepsilon_{ph} = (4/c)\sigma_{sb}T^4$ (see Problem 5.4), so that the energy flux from Equation (5.13) can be rewritten as

$$F_{ph} = -D_{ph}\frac{\partial}{\partial r}\left[\frac{4}{c}\sigma_{sb}T^4\right] = -\frac{4}{3}\frac{\sigma_{sb}}{\kappa\rho}\frac{\partial}{\partial r}(T^4). \tag{5.15}$$

Outside the core, local heat production is very small and the total energy flux through the area of a sphere of radius r must be independent of r and must equal the solar luminosity: $4\pi r^2 F_{ph} = L_\odot$. Hence, with this relation and Equation (5.15), we can obtain an expression for the temperature gradient in the solar interior due to radiation:

$$\frac{\partial T}{\partial r} = -\frac{3}{16\sigma_{sb}}\frac{\kappa\rho}{T^3}\frac{L_\odot}{4\pi r^2}. \tag{5.16}$$

In principle, Equation (5.16) can be integrated over r to give $T(r)$ from the core boundary out to the bottom edge of the convection zone, but $\rho(r)$ must also be solved for simultaneously using Equations (5.9) and (5.10), and the opacity κ must also be found versus r. Nonetheless, it is plausible to deduce from (5.16) (also see Figure 5.2) that as r increases, the temperature gradient, $\partial T/\partial r$, becomes more negative since ρ decreases but the opacity increases.

If the temperature gradient possible from radiation is too small to account for the solar luminosity, then a more efficient mode of energy transport takes over – that is, the bulk motion of the gas up and down associated with macroscopic-sized convection cells can transport energy radially in the solar interior. In fact, *convective transport* is the dominant mode of energy transport in the *solar convection zone*. Using *Bernoulli's relation* (Equation (4.120) with $u \approx 0$ (low-speed flow) and with no magnetic field ($B = 0$)) we find that the adiabatic variation of the temperature for a gas undergoing convection is

$$\frac{\gamma}{\gamma - 1}\frac{p}{\rho} + U_{grav} = constant \text{ in radial direction.} \tag{5.17}$$

This relation applies to fluid parcels slowly (and "adiabatically," so heat does not enter or leave the parcel) moving up or down. U_{grav} is the gravitational potential. Taking the derivative of Equation (5.17) with respect to the radial distance yields an expression for the *adiabatic temperature gradient*:

$$\left(\frac{dT}{dr}\right)_{\text{ad}} = -\frac{\gamma-1}{\gamma}\frac{1}{\tilde{R}}\frac{dU_{\text{grav}}}{dr}, \tag{5.18}$$

where the gas constant is $\tilde{R} = k_{\text{B}}/\tilde{m}$. The derivative of the gravitational potential is just the "local" gravitational acceleration: $g(r) = GM(r)/r^2$. The term $(dT/dr)_{\text{ad}}$ is also called the *adiabatic lapse rate*.

Another example of convective motion closer to home than solar convection is convective motion in the troposphere (i.e., lower atmosphere) of the Earth. In this case, we can simply write the gravitational potential as $U_{\text{grav}} = g(r - R_{\text{E}}) = gz$, where z is altitude and the gravitational acceleration for Earth is $g = 9.88$ m/s^2. The "adiabatic" temperature gradient from Equation (5.18) then becomes

$$\left(\frac{dT}{dr}\right)_{\text{ad}} = -\frac{\gamma-1}{\gamma}\frac{g}{\tilde{R}} = -\frac{g}{C_{\text{p}}}. \tag{5.19}$$

The temperature gradient, $(dT/dr)_{\text{ad}}$, is known as the *dry adiabatic lapse rate* and is equal to ≈ -10 K/km. A convectively mixed atmosphere becomes colder with increasing altitude at the rate of 10 K per kilometer. The quantity $C_{\text{p}} = \gamma\tilde{R}/(\gamma-1)$ is the *specific heat at constant pressure*. Actually, if the air is saturated with water vapor, such as in a cloud, then the relevant lapse rate is the *moist adiabatic lapse rate*, which is approximately -6.5 K/km; the latent heat of vaporization of condensing or evaporating water accounts for the difference between the dry and moist adiabatic lapse rates. If the *actual* atmospheric lapse rate, due to the radiation balance in the atmosphere or to other meteorological processes (e.g., weather fronts), happens to be more negative than the adiabatic lapse rate ($dT/dr < (dT/dr)_{\text{ad}}$), then the atmosphere is *convectively unstable* and convection cells of various sizes form and carry air (and heat) up and down; the lapse rate in the convectively moving air then becomes the adiabatic lapse rate. Thunderstorms are a dramatic example of what can happen when the atmosphere becomes convectively unstable.

We can invoke the buoyancy force to explain convective instability when the background atmosphere has an unstable lapse rate. If an air parcel is moved up, then the air in it cools at the adiabatic lapse rate, whereas the outside air at the same altitude is colder (due to the more negative lapse rate). Hence, the parcel is warmer (and less dense) than the surrounding air and moves up due to a positive buoyancy force. The air parcel then continues to move upwards like a hot air balloon. But if the parcel is moved down, then it becomes colder (and denser) than the surrounding air, and it sinks further due to negative buoyancy. In contrast, if the actual lapse rate is less negative than $(dT/dr)_{\text{ad}}$ (i.e., $dT/dr > (dT/dr)_{\text{ad}}$), then the atmosphere is stable and there is very little up and down air motion. In fact, if the lapse rate is

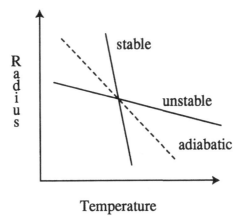

Figure 5.3. Schematic of temperature profiles for a stable and an unstable situation.

positive, then we have what is called an *atmospheric inversion*, which makes the atmosphere extremely stable.

A similar scenario holds true in the solar interior. If the radial temperature gradient (or lapse rate) due to radiation is more negative than the adiabatic lapse rate (for the relevant location in the Sun) then convective motions take place and the temperature gradient becomes equal to the adiabatic gradient. This happens in the convection zone. However, in the radiation zone the radiation-determined temperature gradient (as given by Equation (5.16)) is such that the gas is convectively stable in this region. Assuming negative temperature gradients, we can simply state the convection criterion as

$$\left| \frac{dT}{dr} \right| > \left| \left(\frac{dT}{dr} \right)_{ad} \right| \quad \text{unstable}$$

and

$$\left| \frac{dT}{dr} \right| < \left| \left(\frac{dT}{dr} \right)_{ad} \right| \quad \text{stable.} \tag{5.20}$$

This condition for convective stability is known in solar physics as the *Schwarzschild condition*. Figure 5.3 illustrates the convective stability criterion schematically. The convection zone of the Sun has convection cells (with up and down fluid motions) with size scales ranging from about 1,000 km (e.g., *granules*) up to a third the size of the Sun (e.g., *giant cells*) (see Figure 5.4). Evidence for convection cells of various sizes appears in the form of structures that have been observed on the surface of the Sun.

5.1.4 Solar oscillations

Line emission observed from the solar photosphere is Doppler shifted. The observed Doppler shifts have a periodic pattern changing from red to blue and back again

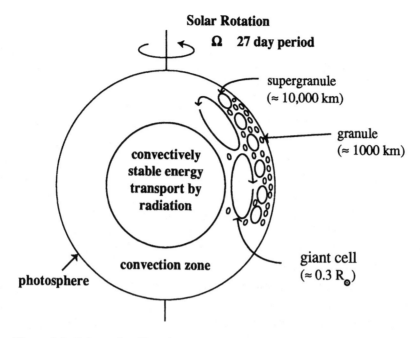

Figure 5.4. Schematic of interior of Sun and convection cells of various sizes.

indicating that the photospheric gas is periodically moving away from and toward the observer (i.e., the Earth). The photosphere oscillates radially, and the velocity amplitude depends on the particular emission feature being observed. Atomic lines formed deeper in the photosphere exhibit less Doppler shifting than lines formed higher up. This indicates that the velocity amplitudes of the observed oscillations increase with altitude. *Solar oscillations* are actually rather complicated and consist of a wide range of wave periods and horizontal wavelengths; however, the dominant oscillation period is *5 minutes*. Both *acoustic waves* and *internal gravity waves* can exist in the Sun, but acoustic waves dominate the wave spectrum observed at the solar surface. Much has been learned about the structure of the solar interior by studying these solar oscillations, just as much has been learned about the Earth's interior by studying the seismic waves generated by earthquakes. The study and interpretation of solar oscillations, or waves, is called *helioseismology* (cf. Gough et al., 1996).

The simplest type of oscillation in any gravitationally bound, stratified fluid is associated with buoyancy. If a fluid parcel is adiabatically displaced vertically from an equilibrium position in a stably stratified medium by a distance ζ, then it experiences a restoring force due to buoyancy. In this case, specific entropy is conserved and we can use the polytropic energy relation. Hydrostatic force balance applies to this type of fluid. Recall that the condition for stability was given by Equation (5.20) – for an atmosphere to be stable, (dT/dr) must be less negative than $(dT/dr)_{ad}$. For negative temperature gradients, this means that the absolute value of the temperature gradient of the surrounding medium (that is, the atmospheric lapse rate) must be less than the absolute value of the adiabatic lapse rate. In Problem (5.6) you are asked to apply Newton's law to the motion of this fluid parcel; note that the fluid parcel also has mass and thus inertia. You will hopefully be able to derive this second-order differential equation for the displacement: $d^2\zeta/dt^2 + N_{BV}^2\zeta = 0$. This is just the ordinary differential equation for a harmonic oscillator with oscillation frequency N_{BV}. The solution to this differential equation is the real part of $\zeta_0 \exp[-i N_{BV}t]$, where ζ_0 is the amplitude of the atmospheric oscillation. The natural vertical oscillation frequency in a stratified fluid such as is found in the Sun or in the Earth's atmosphere is called the *Brunt–Väisälä (BV) frequency* and is given by

$$N_{BV}^2 = \frac{g}{T}\left(\left(\frac{\partial T}{\partial r}\right) - \left(\frac{dT}{dr}\right)_{ad}\right),\qquad(5.21)$$

where the gas constant is $\tilde{R}$, g is the gravitational acceleration, and $(dT/dr)_{ad}$ is the adiabatic lapse rate. If $\tilde{R}$ varies with radial distance r, as is the case if the degree of ionization changes with r, then a more general form of Equation (5.21) must be used (see Zirin, 1988).

Note that N_{BV} is real (that is, $N_{BV}^2 > 0$) only when an atmosphere is stable as determined by the criterion given in Equation (5.20). For an unstable atmosphere, (dT/dr) is more negative than $(dT/dr)_{ad}$ and Equation (5.21) indicates that $N_{BV}^2 < 0$, so that N_{BV} is imaginary. An imaginary value of N_{BV} results in fluid motion in which the fluid parcel displacement is not oscillatory but exponentially grows in time. We can see that this is physically reasonable because a fluid parcel displaced vertically in an unstable atmosphere is indeed going to "unstably" rise or sink.

The Brunt–Väisälä frequency is real in the Sun's core, in the radiative zone, and in the solar atmosphere but is imaginary in the convection zone, which is convectively unstable. Simple buoyancy-related oscillations (called *internal gravity waves*) are not possible in the convection zone, although more general acoustic/buoyancy wave modes are possible.

The BV frequency takes on a particularly simple form for an isothermal background atmosphere $(dT/dr = 0)$. Substituting Equation (5.19) for the adiabatic lapse rate and using the definition of the sound speed C_s from Chapter 4, we can simplify Equation (5.21) to

$$N_{BV}^2 = \frac{g^2}{C_s^2}(\gamma - 1),\qquad(5.22)$$

where γ is the polytropic index.

Example 5.1 (Brunt–Väisälä frequency in the solar photosphere and in the lower atmosphere of the Earth) We can get some idea of the magnitude of the BV frequency in an atmosphere by using the simple expression (5.22) for an isothermal atmosphere. Note that an isothermal atmosphere must be convectively stable. For the solar photosphere, the acceleration due to gravity is $g = 274$ m/s^2. And for T $\approx$ 5,000 K we have for the sound speed $C_s^2 = \gamma \tilde{R} T \approx (8.3$ km/s$)^2$, in which case we have from Equation (5.22),

$$N_{BV} \approx 0.03 \quad \text{radians s}^{-1}.$$

The oscillation period $(2\pi/N_{BV})$ is thus about 4 minutes, which is close to the "nominal" 5-minute oscillation period..

In the troposphere of the Earth during conditions for which it is isothermal, we can also use Equation (5.22), but with $g = 9.88$ m/s^2 and with $C_s^2 \approx (0.4$ km/s$)^2$. We find that

$$N_{BV} \approx 0.02 \quad \text{radians s}^{-1}.$$

The oscillation period is again about 3 minutes.

The solar oscillations observed at the Sun are not simply buoyancy oscillations but are "combined" acoustic (or sound) and buoyancy modes that can have frequencies different from the BV frequency. The wave behavior depends on the vertical and horizontal wavenumbers, k_r and k_h, respectively. The horizontal wave vector can be further broken down into meridional and azimuthal parts. This behavior is characterized by the dispersion relation, which can be found by performing a wave analysis (i.e., linearization, assumption of plane waves, etc., as done in Chapter 4 for MHD waves) starting from the appropriate fluid conservation equations and with the appropriate background atmosphere in hydrostatic equilibrium. The dispersion relation that results from this type of analysis can be found in Zirin (1988) and is

$$(\omega^2 - \omega_{ac}^2)\omega^2 - (k_h^2 + k_r^2)C_s^2\omega^2 + k_h^2 C_s^2 N_{BV}^2 = 0, \tag{5.23}$$

where the *acoustic cutoff frequency* is given by

$$\omega_{ac} = \frac{\gamma g}{2C_s}. \tag{5.24}$$

For an isothermal atmosphere, we have $\omega_{ac}/N_{BV} = 0.5\gamma/(\gamma - 1)^{1/2} = 1.0206$ for an ideal gas. The acoustic cutoff frequency is almost equal to the Brunt–Väisälä frequency.

Waves with $\omega > N_{BV}$ are called *acoustic waves* and/or *modified sound waves*, whereas waves with $\omega < N_{BV}$ are known as *internal gravity waves*. Acoustic and gravity waves also occur in planetary atmospheres. For $\omega \gg N_{BV}$, the dispersion relation (5.23) reduces to the dispersion relation for a *simple sound wave*:

$$\omega^2 \cong k^2 C_s^2 + \omega_{ac}^2 \cong k^2 C_s^2, \tag{5.25}$$

where $k^2 = k_h^2 + k_r^2$. If we take the other extreme case, $\omega \ll N_{BV}$, we can isolate the gravity-wave mode. In the solar interior, gravity-wave modes cannot propagate through the convection zone since N_{BV} is imaginary in that region; consequently, the acoustic mode is needed to explain the observed solar oscillations. These waves are excited by convective motions: Only waves with certain wavenumbers are able to propagate into the photosphere and be observed and only some waves can propagate down into the solar radiative zone. In general, waves with larger meridional wavelengths can penetrate deep into the Sun; shorter wavelength waves are restricted to regions closer to the Sun's surface.

5.1.5 Dynamo theory

Magnetic fields are present throughout the Universe including the solar system. Both the Sun and the Earth possess sizable magnetic fields. How are these magnetic fields generated? They are created by electrical currents as prescribed by Ampére's law, but how are the currents sustained? Large-scale motions of electrically conducting fluids can act as the source of these magnetic fields. The magnetic induction equation discussed in Chapter 4 can be used to find the magnetic field $\mathbf{B}(\mathbf{x}, t)$:

$$\frac{\partial \mathbf{B}}{\partial t} = \nabla \times (\mathbf{u} \times \mathbf{B}) - \nabla \times \left(\frac{\eta}{\mu_0} \nabla \times \mathbf{B} \right), \tag{5.26}$$

where $\mathbf{u}$ is the plasma/fluid flow velocity. For a fluid with high electrical conductivity, such as in the solar interior, the electrical resistivity η is very small, and the second term on the right-hand side of Equation (5.26) can be neglected. As discussed in Chapter 4, the magnetic Reynold's number is very large in this case and the magnetic field is said to be frozen into the flow. Fluid motions are able to alter $\mathbf{B}$ and can amplify the field in certain regions. For instance, a small "seed" field can evolve into a large magnetic field. The process described by Equation (5.26) is called *dynamo theory*. If the flow velocity $\mathbf{u}(x, t)$ is merely adopted, or assumed, rather than self-consistently calculated with the MHD equations, then we have what is called *kinematic dynamo theory*. This theory is in general very complicated, and here we will only look schematically at some results. For a more detailed look at dynamo theory, see Parker (1983).

The Sun and the Earth both rotate and both contain electrically conducting fluids (the Earth has a molten iron core). The most important fluid motion is differential rotation, with an angular frequency that is a function of radial distance r and latitude λ: $\Omega = \Omega(r, \lambda)$. At the Sun, Ω is somewhat greater in the core than at the surface and is greater at the equator than at other latitudes. The rotation period of the solar surface at the equator is 27 days. Convective motions also play an essential role in solar and terrestrial dynamo theory. Let us consider what these motions do to an entrained magnetic field. Obviously, magnetic field lines are twisted, but how? At the solar surface, the rotation frequency has been observed to be equal to (Hathaway, 1996)

$$\Omega(\text{surface}) = 13.1 - 1.39 \sin^2 \lambda - 2.43 \sin^4 \lambda \ [\text{degrees/day}]. \tag{5.27}$$

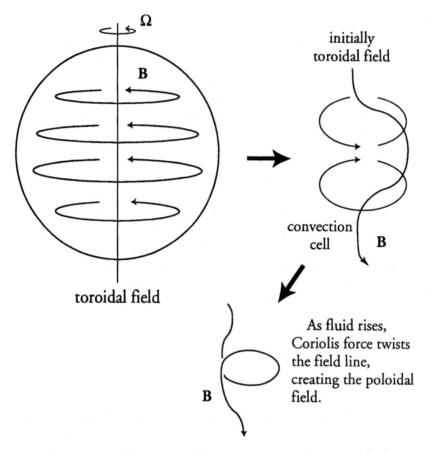

Figure 5.5. Sketch of dynamo-created toroidal and poloidal magnetic fields.

The dominant rotational motion acts to stretch any seed field in the azimuthal direction, thus creating a *toroidal field* (that is, a zonal/east–west directed field). You can picture this for yourself by sketching some magnetic field lines (Figure 5.5). Convection eddies or cells have motions that stretch an initially toroidal magnetic field as pictured in Figure 5.5. A rising fluid parcel carries up and distorts the magnetic field as shown. The Coriolis force then twists the fluid parcel and its associated field line. This results in the creation of a *poloidal field* (that is, a north–south directed field). This is called the $\alpha\omega$ *dynamo effect*. The Coriolis force is a force that acts on a moving parcel in a rotating (i.e., noninertial) frame of reference. For example, the frame of reference attached to the Earth's surface is rotating, and the Coriolis force plays a key role in meteorology. In the solar convection zone both toroidal and poloidal magnetic fields are present. Field strengths of ≈ 1 T (i.e., several thousand gauss) are possible. Convection cells of different sizes, and the magnetic fields associated with them, also manifest themselves on the Sun's surface. For example, sunspots are thought to be related to supergranule-scale convection; the magnetic field strength in a typical sunspot is about 0.1 T, which is 1,000 times greater than the average surface field strength.

The solar field has an average, large-scale, global-scale character that is approximately dipole. The polar regions of the Sun usually have fields with rather uniform polarity, with the two hemispheres having opposite polarity. The field structure at mid- and low latitudes is usually quite complex and shows evidence of the various size convection cells. The solar magnetic field has temporal as well as spatial structure. Smaller-scale structures are more short-lived than larger structures, with the time scale being roughly related to a convection cell "overturning" time. For instance, a typical sunspot persists for about a month. The solar magnetic field also has a very long period variation with a period of 22 years, associated with the largest-scale convection cell overturning time in the convection zone. This is the *solar cycle* time scale. Approximately every 11 years, the polarity of the "average" solar magnetic dipole reverses; the overall period of this solar cycle is 22 years. We will discuss the solar cycle more later.

5.2 The solar atmosphere

5.2.1 Structure of the solar atmosphere

The solar atmosphere has the density and temperature variation with radial distance shown in Figure 5.6. Solar physicists and astronomers have divided the solar atmosphere into four regions:

(1) The *photosphere* is the visible "surface" of the Sun and has an effective temperature of 5,800 K. The visible light that we see originates in the photosphere and accounts for the great majority of the total solar luminosity. This radiation is primarily in the form of *thermal blackbody emission*. The thickness of the photosphere is about 500 km, and the gas is mostly neutral with a fractional ionization (electron to neutral density ratio) of $n_e/n_n \approx 10^{-4}$.

(2) The *chromosphere* lies above the photosphere and has a thickness of 2,000–3000 km. The temperature is 4,300 K at the bottom of the chromosphere and increases to about 10^4 K at the top. The gas in the chromosphere is almost transparent to visible (continuum) radiation but is optically thick in certain spectrally narrow atomic transition lines such as Hα and for lines in the ultraviolet spectra (that is, wavelengths between 0.1 and 0.3 μm) of species like CIV, SiIII, and OV. Note that CIV denotes C^{+++}, CI means neutral carbon, CII means C^+, etc.

(3) The *transition layer* extends another 15,000 km above the chromosphere. The temperature in this layer increases from 10^4 K at the bottom up to about 10^6 K at the top (i.e., the base of the corona).

(4) The *solar corona* is the uppermost region of the solar atmosphere, and it extends many solar radii out into space where it gradually becomes the *solar wind* and *interplanetary medium. The corona is the source of the solar wind!* The corona is hot ($T \approx$ 1–2 million K) and is tenuous, so that it can only be observed in visible light during a full solar

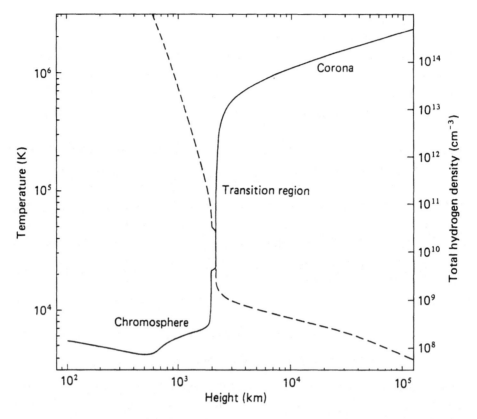

Figure 5.6. Temperature (solid line) and density (dashed line) versus height in the solar atmosphere. The height $h = 0$ occurs at the level of unit optical depth for visible radiation. (From Withbroe and Noyes, 1977. Reproduced, with permission, from the *Annual Review of Astronomy and Astrophysics*, vol. 15, © 1977, by Annual Reviews Inc.) Also see Noyes and Avrett (1987).

eclipse when the much more intense light from the photosphere is blocked out. The coronal gas is fully ionized and emits radiation primarily in the form of line emission from highly ionized species (e.g., FeXVI, SIX, HeII). This radiation is in the extreme ultraviolet and X-ray portions of the spectrum (wavelengths $< 0.1 \mu m$). Solar extreme ultraviolet radiation (EUV) can photoionize atoms and molecules, such as those found in the upper atmosphere of the Earth and other planets (e.g., $h\nu + N_2 \rightarrow N_2^+ + e$), and is thus primarily responsible for the existence of the partially ionized plasmas found in planetary ionospheres.

5.2.2 *Radiative transfer*

Unlike the interior, the solar atmosphere can be observed from the outside. In other words, photons created in the atmosphere have a high probability of escaping and thus being detected by telescope, human eye, etc. The study of the emission, absorption, and transport of radiation (i.e., photons) in an atmosphere such as

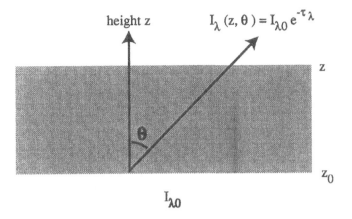

Figure 5.7. Geometry for radiation intensity in a vertically stratified atmosphere.

the Sun's is called *radiative transfer*. Both continuum and line emission as well as absorption are important in the solar atmosphere. *Atomic physics* also is important in the study of the solar atmosphere, in that cross sections for photon–atom, electron–atom, ion–atom, and atom–atom collisions are required. A detailed treatment of this subject is outside the scope of this book; if interested you are referred to the references listed at the end of this chapter.

The flux of photons within a narrow cone of solid angle about some direction and in some wavelength interval is called the *intensity* $I_\lambda(z, \theta)$ [units of $\text{cm}^{-2}\,\text{s}^{-1}\,\text{sr}^{-1}\,\text{cm}^{-1}$]. The unit of solid angle is the steradian (sr). The intensity is a function of height in the atmosphere, z, as well as direction (i.e., angle from the radial, or zenith angle, θ) and wavelength λ. If the intensity at some reference height, $z = z_0$, is equal to $I_{\lambda 0}$, then the intensity at another height, in the absence of sources, is given by

$$I_\lambda(z, \theta) = I_{\lambda 0} e^{-\tau_\lambda}. \tag{5.28}$$

τ_λ is the optical depth at wavelength λ for a path in a vertically stratified atmosphere as shown in Figure 5.7 and is given by

$$\tau_\lambda = \frac{1}{\cos\theta} \int_{z_0}^{z} n(z')\sigma_\lambda\, dz', \tag{5.29}$$

where $n(z)$ is the number density of absorbers or scatterers (e.g., H atoms or other species; strictly speaking (5.29) should contain a sum over all species of absorbers). The photon–atom interaction cross section at wavelength λ is denoted σ_λ. This cross section is a complicated function of wavelength and includes both continuum and discrete atomic line contributions. Often, $z = \infty$ is assumed, for external observers. The $1/\cos\theta$ factor is the ratio of path length increment to vertical height increment.

Obviously, in order to have a flux of photons leave the atmosphere, a source of photons must be present at some altitudes. Equation (5.28) is incomplete in that it does not include such a source. We now include a source of radiation that is a function of altitude, wavelength, and direction: $S_\lambda(z, \theta)$. The following *equation of radiative transfer* then describes the intensity $I_\lambda(z, \theta)$:

$$\frac{dI_\lambda(z, \theta)}{d\tau_\lambda} = -I_\lambda(z, \theta) + S_\lambda(z, \theta). \tag{5.30}$$

Integration of Equation (5.30) from the bottom to the top of the atmosphere yields the observable photon intensity at the top. Equation (5.30) can also be applied to planetary atmospheres.

In the chromosphere and corona, atomic and ionic line emission dominate the source function; that is, S_λ is small except in narrow spectral regions $\Delta\lambda$ surrounding spectral lines (λ_j) associated with atomic transitions. Emission results from both radiative and collisional excitation of the atomic transitions responsible for a particular spectral line. In the photosphere, the source function is primarily blackbody radiation, $S_\lambda(z, \theta) = B_\lambda(z)$, where the function B_λ is given by *Planck's formula*,

$$B_\lambda(z) = \frac{2c/\lambda^4}{\exp\left[\dfrac{hc}{\lambda k_B T}\right] - 1} \quad \text{[photons/cm}^2\text{/s/sr/cm]}, \tag{5.31}$$

where c is the speed of light and h is Planck's constant. B_λ depends on temperature, and the temperature is a function of z. For a given temperature, B_λ has a maximum at a wavelength $\lambda_{max} = 2{,}898 \, [\mu\text{m}]/T \, [\text{K}]$. The value of λ_{max} for the photospheric temperature of $\approx 6{,}000$ K is 550 nm, which is in the middle of the visible spectrum. The total energy flux (over all wavelengths) for a blackbody spectrum is proportional to T^4 – a fact that was used earlier in this chapter when we considered the temperature gradient in the solar interior.

The spectrum of radiation actually observed from the photosphere is determined by solving the equation of radiative transfer using (5.31) as the source function. Most of the radiation observed originates from altitudes at which the optical depth is approximately unity ($\tau = 1$). Hence, the emitted radiation observed above the atmosphere is approximately a blackbody spectrum with an effective temperature of that atmospheric level ($T \approx 5{,}800$ K). In addition, there is some line absorption of radiation that takes place from the overlying gas; these absorption features are known as *Fraunhofer lines*. The solar spectrum is shown in Figure 5.8. The far and extreme ultraviolet parts of the solar spectrum are dominated by line emission rather than by continuum blackbody radiation.

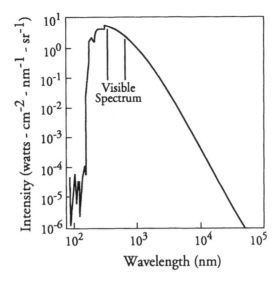

Figure 5.8. Solar spectrum. Intensity versus wavelength. The visible portion of the spectrum is indicated.

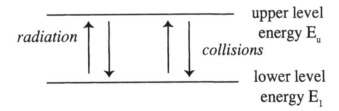

Figure 5.9. Two-level atom.

5.2.3 Excitation and ionization state of the gas in the solar atmosphere (the Saha equation)

The emission rate of a spectral line depends on the rate of excitation of the relevant excited state of the atom or ion. Let us consider an idealized, simple two-level atom in which the lower level is the ground state (Figure 5.9). Real atoms have many more than two energy levels. The upper level (or excited state) can be populated by collisions (with particles) or by radiation ("collisions" with photons), which cause excitation from the ground state to the excited state. The upper level can also be deexcited by collisions or by radiation (i.e., induced emission). For excitation, the required photon energy is given by $h\nu = E_u - E_l$, where E_u and E_l are the energies of the upper and lower levels, respectively, and h is Planck's constant. The photon frequency is $\nu = c/\lambda$. For collisional excitation, the particle energy must exceed the energy difference between the levels, which is called the *energy threshold*. For example, for collisional excitation of an atom by electron impact, the electron energy E_e must exceed $(E_u - E_l)$; that is, $E_e > (E_u - E_l)$. The rate of photon emission is the product of the probability of a single excited atom spontaneously deexciting via photon emission (this probability is known as the *Einstein A coefficient*) and the density (or population) of excited atoms. The population of atoms in the upper level can be found by solving a "rate" equation that takes into account all types of excitation and deexcitation.

In the ground state (i.e., the lowest energy level) of an atom, all the "orbital" electrons are "bound to the nucleus" and have negative total energy; that is, the potential energy of electrostatic attraction of the electrons to the nucleus is negative and exceeds the kinetic energy of the electrons. However, if an orbital electron in an atom somehow acquires enough energy so that the total energy becomes positive, then the electron is "free" and the atom has been ionized. The upper level for this situation is not discrete but lies in the *ionization continuum*. The minimum energy required to achieve this is called the *ionization potential*, χ. Starting with a neutral atomic species, designated M (or MI), an ionic species, M^+ (or MII), and an ionic species, M^{++} (or MIII), etc., can be created by photoionization:

$$
\begin{aligned}
h\nu + M &\rightarrow M^+ + e & h\nu &> \chi(\text{MI}) \\
h\nu + M^+ &\rightarrow M^{++} + e & h\nu &> \chi(\text{MII}) \\
h\nu + M^{++} &\rightarrow M^{+++} + e & h\nu &> \chi(\text{MIII}).
\end{aligned}
\tag{5.32}
$$

MI represents any neutral species, such as H, He, O, C, Fe, etc. MII represents a singly ionized species, and MIII a doubly ionized species. For example, CIII stands for C^{++}. Examples of the ionization potential are: HI (13.6 eV), HeI (24.6 eV), OI (13.6 eV), FeX (235 eV), and FeXV (390 eV). For photoionization to take place the photon energy must exceed χ; that is, the condition $h\nu > \chi$ must be satisfied. Particle collisions can also result in ionization; in fact, in the solar atmosphere, this is the most important cause of ionization. There is also the possibility that a collision between an electron and an ion can result in *electron–ion recombination*: $M^+ + e \rightarrow M + h\nu$. A photon is emitted during this process, which is called *radiative recombination*.

For an equilibrium to exist the number of ionization events in the gas must equal the number of recombination events for each ionization stage. For a gas in thermal equilibrium, the particle distribution functions are described by Maxwellians, the radiation field is blackbody, and ratio of the densities (or populations) of an ion stage $k + 1$ (n_{k+1}) to an ion stage k (n_k) is specified in terms of partition functions Z_s (for species s) as

$$
\frac{n_{k+1}n_e}{n_k} = \frac{Z_{k+1}Z_e}{Z_k} = 4.83 \times 10^{15} T^{3/2} \frac{g_{k+1}}{g_k} \exp[-\chi_k/k_B T],
\tag{5.33}
$$

where the densities are in units of [cm^{-3}]. This is known as the *Saha equation*. See Zirin (1988) for a more complete discussion of this important equation. Here e denotes electrons; n_e is the electron number density. χ_k is the ionization potential of stage k, and g_k is the *statistical weight* of the ground state of the ion. The *partition function* is the sum over all states of the Boltzmann factor multiplied by the statistical weight of each state. The Saha equation is applicable to the solar interior and to the lower part of the photosphere; these are regions that are close to being in thermal equilibrium. In Problem 5.7 you are asked to demonstrate that the ratio of H^+ to neutral H is roughly 10^{-5} in the photosphere where $n_0 \approx 5 \times 10^{16}\,\text{cm}^{-3}$ and $T \approx 5,000$ K and to demonstrate that the ratio is much greater than unity in the solar interior. At a radial distance $r \approx 0.5 R_\odot$, the density is $n_0 \approx 10^{24}\,\text{cm}^{-3}$ and the temperature is $T \approx 10^7$ K, in which case from Equation (5.33) we can estimate that neutral H is $\approx 10^2$ less abundant than H^+. The photosphere is primarily neutral hydrogen (hence, it is a partially ionized plasma), whereas the interior is almost fully ionized.

5.2.4 Coronal properties

The Saha equation is not applicable to the solar atmosphere above the photosphere because that gas is optically thin ($\tau \ll 1$) and is not in thermal equilibrium. In order to determine the relative abundances of the various ion stages (and the excited levels in each stage) one must perform a detailed balance analysis that takes into account all ionization and recombination processes – and this is a very difficult task. Collisional interactions, not radiative ones, are important in the solar corona. However, the relative abundance, n_{k+1}/n_k, still tends to be proportional to the factor $\exp[-\chi_k/k_B T]$, and thus as the temperature increases, higher ionization stages are populated. Figure 5.10 shows some of the ion species in the solar corona. These were found by a detailed balance and are consistent with the EUV observations of the corona. The radiation emitted from transitions in highly ionized species tends

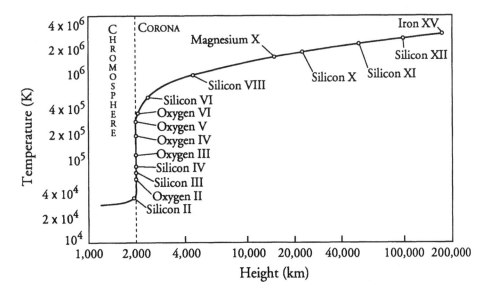

Figure 5.10. Coronal and transition region temperatures plus location of various ionization stages. Adapted from Goldberg (1969). Also see Gibson (1973).

Table 5.2. *Coronal properties*

$R/R_\odot$	n [m^{-3}]	T [K]	λ_D [m]	λ_{coll} [km]^a	C_A [km/s]	C_S [km/s]	f_p	B [T]
1.01	10^{15}	10^6	7×10^{-4}	10	30	150	0.3 GHz	10^{-4}
1.04	10^{14}	1.5×10^6	10^{-2}	2000	300	200	90 MHz	10^{-4}
3.00	4×10^{11}	10^6	0.1	$0.2R_\odot$	400	150	6 MHz	10^{-5}
10.0	3×10^9	5×10^5	1	$10R_\odot$	500	100	0.5 MHz	10^{-6}
214 (1 AU)	10^7	10^5	7	≈ 1 AU	30	30	30 kHz	10^{-8}
	$N_D \gg 1$		$R_m \gg 1$					

^{a}Electron–electron or electron–ion collisions.

to be in the EUV and soft X-ray parts of the solar spectrum ($\lambda < 100$ nm). Such spectra are shown by Zirin (1988).

Coronal properties have been determined from several types of observations, including EUV observations, X-ray observations, and observations of white light. White light from the corona can only be observed during solar eclipses or by using solar coronagraphs. The solar corona emits virtually no visible radiation, although it is bright in the EUV and X-ray parts of the electromagnetic spectrum. The white light observed from the corona is caused by the scattering by coronal electrons or dust of photons originating in the photosphere. Light scattered by electrons (*Thomson scattering*) is responsible for the *K corona* ($r < 3R_\odot$), and light scattered by the small dust particles causes the *F corona* ($r > 2R_\odot$). The K corona is especially important because observations of light scattered by electrons have been used to deduce the electron density in the corona. Some approximate information on average coronal properties are summarized in Table 5.2 (i.e., density n, temperature T, Debye length λ_D, collisional mean free path λ_{mfp}, Alfvén speed C_A, sound speed C_S, and plasma frequency f_p). The table also includes entries for radial distances of 10 and 200 solar radii (these pertain more to the solar wind than to the corona proper, as will be discussed in Chapter 6). The corona actually is quite variable both temporally and spatially, and consequently "average" properties must not be taken too seriously.

The distinguishing feature of the solar corona is its very high, million-degree, temperature. Energy is lost from the corona by the emission of EUV and X-ray radiation and by bulk gas outflow (i.e., the solar wind). The source of energy for the solar corona is still not known! It cannot be due to absorption of radiation because the corona is optically thin. Furthermore, radiative processes tend to transfer energy from high temperature regions to low temperature regions; and hence radiation acts as an energy sink for the corona rather than as an energy source. The coronal energy source is thought to be somehow associated with magnetic reconnection and, therefore, must be understood in terms of MHD processes. Magnetic reconnection is also thought to provide the power for solar flares – a very important solar phenomenon, and the same mechanism could be operating for the average corona on a

much smaller scale and much more frequently. We will briefly consider solar flares
in Section 5.4.

5.3 Small-scale solar structure

5.3.1 Overview

Giant cells, supergranules, and granules have already been discussed in connection
with the solar convection zone. Pictures of the solar atmosphere in various parts
of the spectrum show a great variety of structures ("solar dermatology"), such as
sunspots, quiescent and active filaments and prominences, coronal holes and rays,
plages, spicules, flares, white-light granulation, etc. An unspectacular but important type of structure in the solar photosphere is *granulation*. Granulation appears
at visible wavelengths and covers most of the Sun's surface like bubbles on the
surface of boiling water. Each granulation "cell" is about 1,000 km across and has a
characteristic lifetime of 10 minutes. The *supergranulation* in the convection zone
also manifests itself in the photosphere, although it is more difficult to distinguish
than the granulation. The *supergranulation network* can be observed most easily by
(a) observing the magnetic fields using the Zeeman effect, (b) looking at motions
of the gas using the Doppler effect, or (c) looking for manifestations of supergranulation in the chromosphere, such as structures called *spicules*. Supergranulation
cells are about 20,000 km across and have lifetimes of 1–2 days.

   The following types of small-scale structure are more striking than (albeit related
to) the granulation and supergranulation, although they generally cover a lower
percentage of the solar surface:

(1) *Sunspots* have a horizontal extent comparable to that of supergranulation (i.e., $\approx$20,000
    km). They appear as dark (i.e., cooler) spots on the Sun's surface. *Groups* of sunspots
    are associated with *active regions* in the overlying chromosphere. The magnetic field
    in sunspots is approximately 0.1 T, whereas the average surface field strength is only
    $\approx$$10^{-4}$ T (1 gauss). See Figure 5.11 for an Hα photograph of a sunspot group. (Hα is
    emission from the $n = 3$ to the $n = 2$ quantum levels of atomic hydrogen, where n is
    the principal quantum number.)

(2) *Filaments and prominences* extend from the chromosphere up into corona, appearing
    as archlike structures with thicknesses of $\approx$6,000 km, lengths of $\approx$100,000 km, and
    heights above the surface of $\approx$50,000 km. The gas in a *quiescent prominence* is typically
    about 300 times colder and denser than the surrounding coronal gas. The magnetic field
    in these structures is about 10^{-3} T and a magnetic cushion is thought to support them.
    See Figure 5.12.

(3) *Flares* are *explosively eruptive prominences* and are associated with large releases of
    energy in many parts of the spectrum. Magnetic energy released via magnetic recon-
    nection is thought to provide the power for flares.

(4) *Coronal holes* are very large in horizontal extent ($\approx$0.1–0.5$R_{\odot}$) and extend radially far
    out into the corona. They appear as less bright (i.e., cooler) regions in the soft X-ray

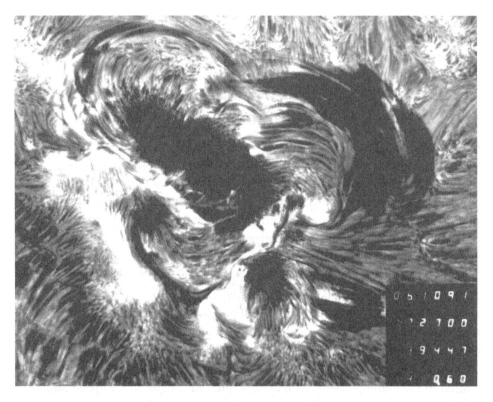

Figure 5.11. Photograph in hydrogen-alpha light of a sunspot group in June 1991. The bright areas are called plages and the dark ribbons separating plages are filaments. Fibrillike dark arches trace the magnetic field lines between the two polarities. Courtesy of H. Zirin. (Big Bear Solar Observatory, California Institute of Technology.)

part of the spectrum, and they are associated with "open" magnetic field lines and with *high-speed solar wind streams*. See Figure 1.2.

(5) *Coronal helmet streamers and coronal mass ejections (CMEs)* are large-scale (≈ 0.5–$1 R_\odot$) structures observed in the K corona that appear bright in the visible spectrum during solar eclipses. Helmet streamers extend from the inner corona, where they are associated with closed field lines, to the outer corona, where the field lines open up. They typically appear near the Sun's magnetic equator. The plasma density is relatively high in helmet streamers. CMEs are relatively high density regions appearing bright in visible light and often associated with helmet streamers that leave the Sun and move into interplanetary space (Figure 5.13 is a picture of a CME).

5.3.2 Sunspots

Sunspots are locally cool areas on the photosphere. Temperatures in the *umbra* (see Figure 5.14) are about 3,800 K – that is, 2,000 K colder than the surrounding photospheric gas. As mentioned earlier, the magnetic field in a sunspot ($B \approx 0.1$ T) is about 1,000 times greater than the average photospheric field. Sunspots usually

Figure 5.12. Photograph of a prominence off the limb photographed in hydrogen-alpha light on March 31, 1971. (Big Bear Solar Observatory, California Institute of Technology.)

appear in groups and the field in a sunspot group can be unipolar, bipolar, or very complex. Sunspots mainly appear in two zones of heliographic latitude, one each in the northern and southern hemispheres. The total number of sunspots is typically a couple of dozen but varies with a period of about 11.2 years (i.e., *the solar cycle*). The average latitude of the sunspots in a hemisphere progresses from mid-latitudes down to the equatorial region during the course of a solar cycle.

Our current understanding of why sunspots form is now briefly summarized. Regions of locally intensified B field generated by the dynamo mechanism in the convection zone are thought to be buoyant and thus rise to the surface, where they appear as sunspots (Figure 5.15). The simplest case consists of a pair of sunspots with opposite polarity fields. Why should a parcel of gas in the convection zone with enhanced magnetic field become buoyant? Because the fluid in such a parcel is less dense than the surrounding fluid. Pressure balance must exist at the interface between the magnetized and the external (relatively) unmagnetized material. Suppose that the external gas has the following density, pressure, and magnetic field strength: ρ_e, p_e, and B_e (which is ≈ 0), respectively. The same parameters for the magnetized parcel are denoted ρ, p, and B. Pressure balance at the interface gives

$$p_e = p + B^2/2\mu_0. \tag{5.34}$$

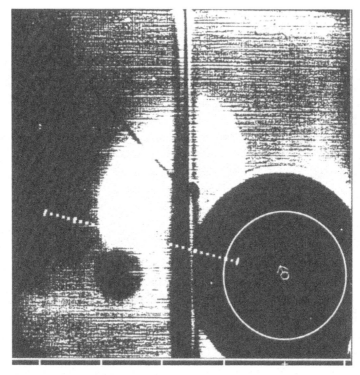

Figure 5.13. March 1989 photograph of a coronal mass ejection event in visible light. (From Feynman and Hundhausen, 1994.)

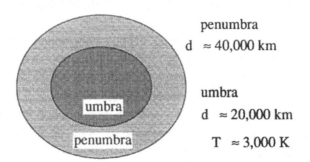

penumbra

d ≈ 40,000 km

umbra

d ≈ 20,000 km

T ≈ 3,000 K

Figure 5.14. Sketch of a sunspot showing the umbra and penumbra.

Equation (5.34) indicates that $p < p_e$, suggesting that $\rho < \rho_e$, if the temperatures inside and outside the parcel are about the same. The lower density of the parcel makes the parcel buoyant; it thus moves upward. However, as the flux tube/parcel rises, it cools adiabatically (see Equation (5.19)) and also might cool relative to the surrounding gas due to inhibition of heat transport into the parcel by the strong magnetic field. Hence, the gas in the flux tube that finally emerges above the surface (i.e., a sunspot) is cooler than the surrounding photospheric gas. However, if the cooling becomes too extreme, this results in the loss of buoyancy. The strong magnetic fields associated with sunspots and sunspot groups in the photosphere extend up into the

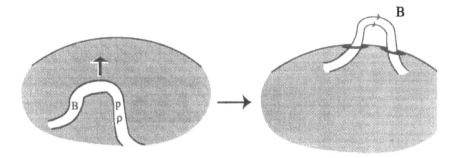

Figure 5.15. (Left) Magnetic flux tube beneath the surface. (Right) Emergence of flux tube: two sunspots.

chromosphere and corona and are observed as *active regions*. Solar prominences and solar flares are found in active regions. The closed loops of field lines over sunspot groups also appear as the brighter (i.e., hotter) parts of the corona.

5.3.3 Solar activity and the solar cycle

As we discussed earlier, the total number of sunspots observed on the Sun's surface varies with a period of about 11 years – the solar cycle. Figure 5.16 shows a plot of sunspot number versus year. The maximum number of sunspots in a given cycle varies considerably. During the course of a solar cycle the average location of the sunspots moves toward the equator. Sunspot pairs usually have opposite magnetic polarity, and the leading (in the direction of solar rotation) sunspot of a pair has opposite polarity in the northern and southern hemispheres. And this polarity switches every solar cycle. Furthermore, the polar regions of the Sun have weak magnetic fields but are all of one polarity (opposite in the two hemispheres). The average solar field on a very large scale is thus roughly dipolar. The dipole axis is in general tilted with respect to the solar rotation axis. The orientation of the dipole switches every cycle, and thus the true period of the magnetic field is 22 years rather than 11 years. The reversal of the polar fields occurs during solar maximum (i.e., at sunspot maximum).

Active regions of the Sun, which overlie sunspot groups, contain many magnetic "loop" structures. Phenomena that involve magnetic loops or arches include prominences and solar flares. Hotter coronal gas is associated with active regions, which are more common and extensive during solar maximum than during solar minimum. Because the "average" corona is hottest during solar cycle maximum, the solar EUV and X-ray fluxes are greatest during this portion of the solar cycle by about a factor of 3 or more. Shorter period variations are also observed; in particular, a 27-day period is apparent at the Earth in the measured EUV flux and is due to the solar rotation. Some parts of the Sun have more sunspots (active areas) than others, and the active areas are carried around by solar rotation to face the Earth every 27

YEARLY MEAN SUNSPOT NUMBERS

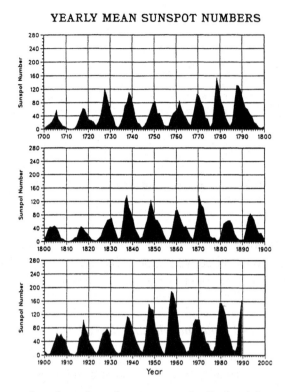

Figure 5.16. Observed total number of sunspots on the Sun's surface versus time. (From Gorney, 1990.)

days. Radio emission also varies with coronal properties and shows a solar cycle variation. For example, the decametric radio flux at a wavelength of 10.7 cm (the so-called F10.7 flux) correlates very well with the EUV flux, exhibiting a similar solar cycle dependance and a similar 27-day periodicity. F10.7 is more easily monitored than the EUV flux, which has to be measured with rocket- or satellite-borne instrumentation due to the strong atmospheric absorption of EUV photons. Thus F10.7 flux is used to estimate the EUV fluxes during time periods when the EUV flux itself is not being measured.

5.3.4 Solar flares, eruptive prominences, and coronal mass ejections

A large variety of magnetic field configurations are possible in the chromosphere and corona. Often these field configurations are associated with active regions, but they are also found in quiet regions. For example, quiescent prominences, mentioned earlier, are large filaments of cooler and denser gas supported by a magnetic cushion. The field underlying the prominence must be such that the magnetic pressure, $B^2/2\mu_0$, equals the weight per unit area of the overlying prominence material in a quasi-static manner. However, not all structures require large changes in the magnetic pressure. Force-free magnetic structures are also common in the solar corona. Recall from Chapter 4 that a force-free structure is one in which $\mathbf{J} \times \mathbf{B} = \mathbf{0}$, which can be achieved if $\mathbf{J}$ is parallel to $\mathbf{B}$. Helmet streamers are also present in the corona and are associated with current sheets; these structures are not force free but require thermal pressure gradients to stay in balance.

A static balance is not always possible in the solar corona. For example, many quiescent prominences do hang around for time periods of weeks or longer, but sometimes a coronal filament does not remain "static" and turns into an *explosively eruptive prominence*. Sometimes, coronal structures such as helmet streamers (or part of a helmet steamer) become unstable and are "ejected from" the Sun as a coronal mass ejection (CME). The cause of CMEs is still not understood.

Sometimes, but not always, the nonstatic readjustment of magnetic flux in the corona results in a *solar flare*. During the course of their 1–2 hour lifetimes, flares typically release $\approx 10^{25}$ J of energy in the form of X-ray emission and particle energy. Flare events are rather complicated both spatially and temporally, and they remain poorly understood, in spite of the large research effort that has been devoted to studying them. The consensus view of the evolution of a "two-ribbon" solar flare will be briefly summarized in a "cartoon" fashion.

All current theories of flare formation invoke as the power source the release of magnetic energy by magnetic reconnection (Priest, 1985). Let us suppose that a quiescent prominence is not only supported by a magnetic cushion but also has a magnetic arcade overlying it (see Figure 5.17a). Then, if the prominence slowly rises due to some imbalance of the forces on it, the field lines of the arcade become increasingly stretched until at some point in time, magnetic reconnection starts to take place underneath the prominence (Figure 5.17b). Once magnetic reconnection starts, the prominence no longer slowly evolves but changes in an explosive manner. Reconnection converts magnetic energy into kinetic (i.e., particle) energy, as discussed at the end of Chapter 4. Recall that the material coming "out" of the reconnection zone does so at the Alfvén speed. For the configuration shown in Figure 5.17, "out" corresponds to both downward and to upward. Thus, the prominence material is rapidly thrust upward, causing it to extend further and further into the solar corona. On the other side of the reconnection zone, material gets pushed down into the chromosphere. Energetic particles are shot down to the base of the reconnecting field lines where they collide with the atoms in the denser photosphere and chromosphere. The "footprints" of these field lines appear as two lines, or ribbons, of intense Hα emission (and other emission including other visible radiation, radio emission, and X rays from Bremstrahlung). Hence, the term *two-ribbon flare* is used to describe this important type of flare.

Other types of solar flares exist, but two-ribbon flares are the most important. The most active phase of a typical flare lasts about an hour. The energetic particles associated with flares include electrons with energies up to 10 MeV and nucleons with energies up to several hundred MeV. These most energetic flare particles are called *solar cosmic rays*.

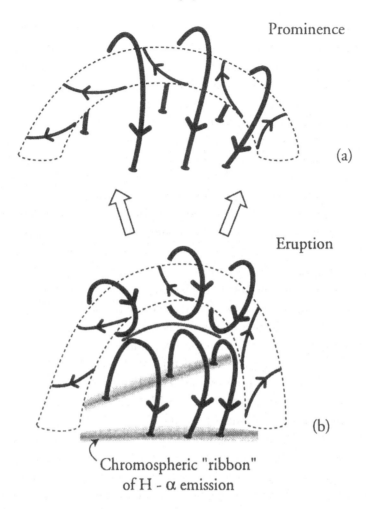

Prominence

(a)

Eruption

(b)

Chromospheric "ribbon"
of H - α emission

Figure 5.17. Schematic of a two-ribbon flare event. (a) Prominence with overlying magnetic arcade. (b) Two "ribbons" of chromospheric Hα emission. Magnetic field lines are shown as the solid lines.

Until a few years ago, the commonly held viewpoint was that there was a one-to-one correlation between solar flares and coronal mass ejections. In a CME event, a large chunk of the corona associated with a solar prominence or a helmet streamer moves out into interplanetary space. During the past few years, it has become apparent that flare and CME occurrences are *not* well correlated (Gosling, 1993). That is, CMEs often originate from prominences without a flare event and flares occur without an associated CME. In fact, Hundausen (1995) recently demonstrated that CMEs often *precede* flares. Some of the energetic particles detected during times of high solar activity are not produced at flares but are accelerated by shock waves moving ahead of some CMEs in the interplanetary medium. Both CMEs and flares are being actively researched in order to improve our understanding of these phenomena.

Problems

5.1 What is the average kinetic energy of a proton in a 15 million K gas, such as is found in the core of the Sun? What is the distance of closest approach during a binary collision between two protons with this average kinetic energy? How many proton radii is this?

5.2 The chain of reactions from Equation (5.1) to (5.3) provides almost all the power of the Sun (i.e., the solar luminosity). Throughout the whole Sun, how many times per second must this chain of reactions occur in order to account for the solar luminosity? Low-energy neutrinos are produced by the reaction (5.1). What is the flux of low-energy solar neutrinos at the Earth's surface? Note that matter is almost transparent to the passage of neutrinos.

5.3 Given the temperature profile as a function of radial distance r in Figure 5.2, determine the radial pressure and density profiles in the solar interior. You will need to carry out a numerical integration to accomplish this.

5.4 Show that the energy density associated with blackbody radiation is $\varepsilon_{\mathrm{ph}} = (4/c)\,\sigma_{\mathrm{sb}}T^4$ [J/m^3], where σ_{sb} is the Stephan–Boltzmann constant.

5.5 Derive the adiabatic lapse rate expression (5.19) in a way that is different than what was used in the text, by "adiabatically" displacing a parcel of fluid vertically and finding the resulting temperature. Assume that the parcel remains in pressure equilibrium with the surrounding fluid, which is in hydrostatic balance itself, and also use the polytropic energy relation (that is, specific entropy is conserved for adiabatic motions). Evaluate the lapse rate for the case of the terrestrial atmosphere near the ground and also estimate this lapse rate for the middle of the solar convection zone.

5.6 Derive Equation (5.21) for the Brunt–Väisälä (BV) frequency N_{BV}. Start by finding an expression for the buoyancy force of a parcel displaced vertically adiabatically. From Newton's laws you should then be able to find a second-order ordinary differential equation like that of a harmonic oscillator. Evaluate the BV frequency for an isothermal atmosphere at the Earth and also at the bottom of the solar photosphere.

5.7 Use the Saha equation to determine the density ratio of HII to HI in

(a) the bottom of the photosphere ($n_0 \approx 5 \times 10^{16}$ cm^{-3} and $T \approx 5{,}000$ K),
(b) the top of the photosphere ($n_0 \approx 5 \times 10^{15}$ cm^{-3} and $T \approx 4{,}200$ K),
(c) the solar interior at $r = 0.5R_\odot$ ($n_0 \approx 10^{24}$ cm^{-3} and $T \approx 10^7$ K).

5.8 The cross section for scattering of photons off of electrons (Thomson scattering) is $\sigma_T \approx 3 \times 10^{-25}$ cm^2. Estimate the relative intensity of white light scattered from the solar K corona (I_{scatt}) to the white-light intensity of the photosphere (that is, the solar disk), $I_\odot$. Show that $I_{\mathrm{scatt}}/I_\odot \approx 10^{-5}$. Use values of the electron density found in Table 5.2 plus reasonable interpolations.

5.9   A typical total energy for radiation (mostly X rays) and fast particles gener-
       ated by a moderate solar flare is 10^{25} J. Assuming 100% efficiency for the
       conversion of magnetic energy into X rays and particle kinetic energy via
       the reconnection process, find the coronal volume needed to obtain this en-
       ergy. If the height of the flare is comparable to its horizontal extent, what is
       the horizontal extent of the flare? Assume a typical magnetic field strength
       for an active region. Repeat these estimates for a 1% energy conversion
       efficiency.

Bibliography

Bahcall, J. and R. Ulrich, Solar models, neutrino experiments, and helioseismology, *Rev. Modern Physics*, **60**, 297, 1988.

Feynman, J. and A. J. Hundhausen, Coronal mass ejections and major solar flares: The great active center of March 1989, *J. Geophys. Res.*, **99**, 8451, 1994.

Foukal, P., *Solar Astrophysics*, Wiley, New York, 1990.

Gibson, E. G., *The Quiet Sun*, NASA SP-303, Scientific and Technical Office, Washington, DC, 1973.

Goldberg, L., Ultraviolet astronomy, *Sci. Am.*, **220**, 92, 1969.

Gorney, D. J., Solar cycle effects on the near-Earth space environment, *Rev. Geophys.*, **28**, 3, 1990.

Gosling, J.T., The solar flare myth, *J. Geophys. Res.*, **98**, 18937, 1993.

Gough, D. O., J. W. Leibacher, P. H. Scherrer, and J. Toomre, Perspectives in helioseismology, *Science*, **272**, 1281, 1996.

Hathaway, D. H., GONG observations of solar surface flows, *Science*, **272**, 1306, 1996.

Haxton, W. C., The solar neutrino problem, *Annu. Rev. Astron. Astrophys.*, **33**, 459, 1995.

Hundhausen, A. J., Coronal mass ejections: A summary of SMM observations from 1980 and 1984–1989, in *The Many Faces of the Sun*, ed. K.T. Strong, J. Saba, and B. Haisch, Springer-Verlag, New York, in press, 1995.

Noyes, R. W. and E. H. Avrett, The solar chromosphere, chap. 5 in *Spectroscopy of Astrophysical Plasmas*, ed. A. Dalgarno and D. Layzer, Cambridge Univ. Press, Cambridge, UK 1987.

Parker, E. N., Generation of solar magnetic fields I (p. 101); Generation of solar magnetic fields II (p. 113), in *Solar-Terrestrial Physics: Principles and Theoretical Foundation*, ed. R. L. Carovillano and J. M. Forbes, D. Reidel, Dordrecht, The Netherlands, 1983.

Priest, E. R., Introduction to Solar activity, p. 1 in *Solar System Magnetic Fields*, ed. E. R. Priest, D. Reidel Publ. Co., Dordrecht, Holland, 1985.

Ulrich, R. K., The influence of partial ionization and scattering states on the solar interior structure, *Astrophys. J.*, **258**, 404, 1982.

Withbroe, G. L. and R. W. Noyes, Mass and energy flow in the solar chromosphere and corona, *Annu. Rev. Astron. Astrophys.*, **15**, 363, 1977.

Zirin, H., *Astrophysics of the Sun*, Cambridge Univ. Press, Cambridge, UK 1988.

6

The solar wind

The solar wind is an extension of the solar corona out into interplanetary space. In this chapter we will answer the question of why there is a solar "wind." Why isn't the outer solar corona and interplanetary medium just static? In Section 6.1 we will show why an extended solar corona in hydrostatic balance is *not* possible, and in Section 6.2, we will derive the equations describing the supersonic solar wind outflow from the Sun. The interplanetary magnetic field is considered in Section 6.3, and solar wind structures such as sector structure and streams are discussed in Section 6.4. Section 6.5 contains a very brief discussion of other topics such as solar wind electrons, cosmic-ray modulation, and the heliopause.

6.1 Hydrostatic balance in the solar corona

6.1.1 Hydrostatic equilibrium

We derived a formula for hydrostatic force balance as a special case of the momentum equation in Chapter 4. Recalling Equation (4.92) or (5.6), and using a gravitational acceleration appropriate for the heliocentric distance r, we find that

$$\frac{\partial p}{\partial r} = -g\rho = -\frac{GM_\odot}{r^2}\rho, \tag{6.1}$$

where $M_\odot$ is the solar mass. The equation of state can be combined with Equation (6.1) (as was done in Chapter 5) to find the following equation for pressure as a function of radial distance r:

$$p(r) = p_\odot \exp\left(-\frac{GM_\odot}{\tilde{R}} \int_{R_\odot}^{r} \frac{dr'}{r'^2 T(r')}\right), \tag{6.2}$$

where $M_\odot$ is the solar mass, $\tilde{R} \approx 2k_B/m_p$ is the gas constant appropriate for the corona, $p_\odot$ is the gas (i.e., plasma) pressure at the base of the corona, and $T(r)$ is the coronal temperature profile. We assume that $T = T_e = T_i$, which is a reasonable assumption in the lower corona but a bad one at larger radial distances.

A temperature profile, $T(r)$, is required to evaluate Equation (6.2), and this then requires a consideration of the energetics. Here, a simple temperature profile will be adopted in which T varies as a simple power of the radial distance:

$$T(r) = T_\odot \left(\frac{R_\odot}{r} \right)^n, \tag{6.3}$$

where $T_\odot = T_c$ is the temperature at the base of the corona ($r = R_\odot$). The power-law exponent n is not necessarily an integer. Using Equation (6.3) in the evaluation of the pressure, one can find (Problem 6.1) the following expression for the gas pressure in the corona and interplanetary space:

$$p(r) = \begin{cases} p_\odot \exp\left[-\dfrac{GM_\odot}{\tilde{R} T_\odot R_\odot^n} \dfrac{1}{n-1} \left(r^{n-1} - R_\odot^{n-1} \right) \right] & n \neq 1 \\[3mm] p_\odot \left(\dfrac{R_\odot}{r} \right)^{[GM_\odot/(\tilde{R} T_\odot R_\odot)]} & n = 1. \end{cases} \tag{6.4}$$

First, let us consider the case of an isothermal corona ($T = T_\odot$ and exponent $n = 0$), in which case we find from (6.4) that

$$p(r) = p_\odot \exp\left[-\frac{GM_\odot}{\tilde{R} T_\odot} \left(\frac{1}{R_\odot} - \frac{1}{r} \right) \right]. \tag{6.5}$$

From Equation (6.5), we can see that the thermal pressure at $r = \infty$ does not go to zero but is instead equal to

$$p(\infty) = p_\odot \exp\left(-\frac{GM_\odot}{\tilde{R} T_\odot R_\odot} \right). \tag{6.6}$$

From Table 5.2 we see that the pressure at the bottom of the corona is approximately $p_\odot \approx 0.3$ N/m². Using $T_\odot = 1.5 \times 10^6$ K, Equation (6.6) indicates that at $r = \infty$, the pressure is $p(\infty) \approx 10^{-8}$ N/m². You might ask, so what? The total pressure of the local interstellar medium, p_{ISM}, is estimated to be only about 10^{-13} N/m² so that $p(\infty)$ greatly exceeds p_{ISM}. Thus, with these assumptions the solar pressure at very large distances exceeds the interstellar pressure, and the boundary surface separating the plasma of solar origin from the interstellar gas could not be in pressure balance. Hence, the coronal gas cannot remain static in this case but has a tendency to move outward. But can we devise another temperature profile to alleviate this? The pressure at $r = \infty$ remains finite for all temperature profiles that have $n < 1$. However, for $n = 1$ the pressure at $r = \infty$ is zero according to Equation (6.4).

What might a realistic profile temperature look like? If we are assuming a static gas, heat conduction is the most important mode of heat transport. At large enough radial distances, where the gas density is very low, collisional heat sources and sinks should be unimportant. An examination of the heat conduction equation, which we will undertake in the next section, indicates that a temperature profile similar to the one given by Equation (6.3) is not unrealistic if we keep the index n

smaller than 2/7. However, the pressure expression (6.4) with $n = 2/7$ still yields a finite pressure at infinity, and thus we must conclude that a static solution to the momentum equation for the corona is *not* realistic. What we need is a *dynamic* solution of the momentum equation – one which gives an outflow or "wind." Note that the heat conduction temperature solution is not entirely unrealistic for the electron gas – even for a nonstatic solar wind – but is unrealistic for the ion gas.

6.1.2 Heat conduction in the solar corona

In this section we will assume that heat conduction is the dominant physical process for the energetics (or at least for the electron energetics) and we will find an expression for the temperature as a function of radial distance. The heat conduction equation (2.82) for steady-state conditions and in the absence of sources and sinks is $\nabla \cdot \mathbf{Q} = 0$, where the heat flux vector for radial outflow is just $\mathbf{Q} = Q_r \hat{\mathbf{r}}$. The radial component of the heat flow can be found from Equation (2.76):

$$Q_r = -K \frac{dT}{dr}, \tag{6.7}$$

where for a fully ionized plasma the conductivity coefficient is $K = CT^{5/2}$ [eV/m/s/K] with $C = 7.7 \times 10^7$ from (2.77). The energy equation in spherical coordinates becomes, after integrating once over r,

$$r^2 K \frac{dT}{dr} = R_\odot^2 Q_{r\odot} = R_\odot^2 C T_\odot^{5/2} \left(\frac{dT}{dr}\right)_{r=R_\odot}, \tag{6.8}$$

where $Q_{r\odot}$ is the heat flux at the bottom of the corona. In Problem (6.2) you are asked to find the following expression for $T(r)$ starting from Equation (6.8) by integrating again:

$$T^{7/2} = T_\odot^{7/2} + \frac{7}{2C} R_\odot^2 Q_{r\odot} \left(\frac{1}{r} - \frac{1}{R_\odot}\right)$$

or

$$T^{7/2} = T_\odot^{7/2} \left[1 + \frac{7}{2}\zeta \left(\frac{R_\odot}{r} - 1\right)\right]$$

with ζ defined by

$$\left(\frac{dT}{dr}\right)_{r=R_\odot} \equiv \zeta \frac{T_\odot}{R_\odot}. \tag{6.9}$$

The temperature given by Equation (6.9) attains an asymptotic value at $r = \infty$ of $T_\infty = T(r = \infty)$ as given by

$$T_\infty^{7/2} = T_\odot^{7/2} \left(1 - \frac{7}{2}\zeta\right). \tag{6.10}$$

Equation (6.10) was found using Equations (6.8) and (6.9). The value of zero heat flux ($\zeta = 0$) gives an isothermal temperature profile. Note that to maintain a positive temperature at infinity, an upper limit exists on the heat flux; this limit can be expressed as $\zeta \leq 2/7$. Note that for $\zeta = 2/7$ then $T_\infty = 0$ and the temperature varies as a power of r:

$$T(r) = T_\odot \left(\frac{R_\odot}{r} \right)^{2/7}.$$

(6.11)

This equation is the same as Equation (6.3) with $n = 2/7$.

Heat conduction is not a bad assumption for the electron gas – even in a nonstatic corona. However, for the ion component of the plasma the dynamical terms of the energy equation are much more important than heat conduction, and for steady-state conditions, Bernoulli's law provides a better description of the energetics than the heat conduction equation.

6.1.3 Bernoulli's equation revisited

A simple steady-state energy equation without sources or sinks as given by Bernoulli's equation was discussed in both Chapters 2 and 4. For the solar corona, Bernoulli's equation can be written as

$$\frac{\gamma}{\gamma - 1} \tilde{R} T + \frac{1}{2} u^2 - \frac{GM_\odot}{r} = E_c$$

$$E_c = \frac{\gamma}{\gamma - 1} \tilde{R} T_c - \frac{GM_\odot}{R_c}.$$

(6.12)

We have assumed in the definition of the specific coronal energy, E_c, that the gas near the base of the corona is stationary ($u_c \approx 0$). An approximate numerical expression for E_c is

$$E_c \cong \frac{\gamma}{\gamma - 1} 1.65 \times 10^4 \, T_c - 1.9 \times 10^{11} \frac{R_\odot}{R_c} \quad \text{[J/kg]}.$$

(6.13)

R_c and T_c are the radial distance and temperature, respectively, at the base of the corona. We find from Equation (6.12) that the asymptotic flow speed at very large radial distances (i.e., the solar wind speed) must be equal to

$$u_\infty = \sqrt{2E_c}.$$

(6.14)

This speed was found by assuming that the temperature becomes zero as $r \to \infty$.

Example 6.1 (Solar wind speed from Bernoulli's equation) Let the coronal radius be $R_c = R_\odot$, and the coronal temperature be $T_c = 2 \times 10^6$ K. We also assume an ideal monatomic gas with $\gamma = 5/3$. Then we find from Equation (6.13) that E_c is negative! Solar wind outflow is impossible in this case: The coronal gas is "gravitationally bound." However, if we place the "base" of the corona at $R_c = 3R_\odot$, and keep the other parameters the same, then $E_c = 1.9 \times 10^{10}$ J/kg, and from Equation (6.14), $u_\infty = 140$ km/s, a value that is somewhat smaller than the typical solar wind speed but at least is positive.

Letting $R_c = R_\odot$ and $T_c = 2 \times 10^6$ K, but assuming a low value of the polytropic index, $\gamma = 1.15$, we find $E_c = 6.2 \times 10^{10}$ J/kg and $u_\infty = 355$ km/s. With $\gamma = 1.1$ the asymptotic flow speed is $u_\infty = 420$ km/s, which is indeed a realistic value for the solar wind flow speed. But what does a polytropic index of 1.1 mean?

Neither the heat conduction approach considered in Section 6.1.2 nor the Bernoulli's equation approach considered in this section are really appropriate. Heat conduction is more applicable to the electron gas and Bernoulli's equation to the proton gas. We should really use the two-fluid equations rather than single-fluid equations, at least for the energetics. Nonetheless, we can still obtain a reasonable description of solar wind outflow by using the one-fluid version of Bernoulli's law, but with a polytropic index that is "adjusted" so that in some sense it reflects other terms in the energy equation such as heat conduction in the electron gas. Although this approach is not really self-consistent, it is a simple one that enables us to make significant simplifications as well as to obtain reasonable results.

A more complete treatment of solar wind outflow requires a consideration of the continuity and momentum equations as well as the energy equation. In the next section, we will study steady, one-dimensional gas flow in general, and then we will take up again the solar wind outflow problem.

6.2 Solar wind outflow

6.2.1 General one-dimensional, steady gas flow

We first review one-dimensional, steady gas dynamics. Consider gas flow through a "tube" oriented in the radial direction with cross-sectional area $A(r)$, as shown in Figure 6.1. The area function, $A(r)$, is specified according to the geometry of the particular problem that one is dealing with. If the area function changes only gradually with r, then the non-r components of the flow velocity are small and to a good approximation the flow velocity is

$$\mathbf{u}(r) \cong u(r)\hat{\mathbf{r}}. \tag{6.15}$$

We wish to derive an expression for the flow speed – or better yet, the Mach number – as a function of the radial distance r. To do this we need to combine the continuity, momentum, and energy equations in a suitable way.

The single-fluid, steady-state continuity equation in the absence of sources and sinks is

$$\nabla \cdot [\rho \mathbf{u}] = 0. \tag{6.16}$$

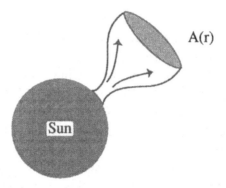

A(r)

Sun

Figure 6.1. Gas flow in the r direction through a tube with cross-sectional area $A(r)$.

The continuity equation for the geometry considered here can also be written in the following two forms (see Problem 6.3):

$$\rho(r)u(r)A(r) = constant$$

or

$$\frac{d\rho}{\rho} + \frac{du}{u} + \frac{dA}{A} = 0. \tag{6.17}$$

We can also assume that magnetic stress can be neglected ($\mathbf{J} \times \mathbf{B} = \mathbf{0}$), which is true for either a high-β plasma or for low-β plasma flow *along* the magnetic field. Also, we neglect friction and other collision terms, but we include the gravitational acceleration (which is taken to be for the Sun but which can be easily generalized). The r component of the momentum equation, (4.80), now becomes

$$\rho u \frac{du}{dr} + \frac{dp}{dr} + \frac{GM_\odot}{r^2}\rho = 0$$

or

$$u\,du + \frac{dp}{\rho} + \frac{GM_\odot}{r^2}\,dr = 0. \tag{6.18}$$

The single-fluid pressure is p. Equation (6.12) can be used for the energy equation; taking the differential of this equation and using the equation of state $\tilde{R}T = p/\rho$, we have

$$u\,du + \frac{\gamma}{\gamma - 1}\frac{dp}{\rho} - \frac{\gamma}{\gamma - 1}\frac{p\,d\rho}{\rho\,\rho} + \frac{GM_\odot}{r^2}\,dr = 0. \tag{6.19}$$

Let us assume that changes in the dependent variables in the flow direction (i.e., in the r direction) are not discontinuous; that is, we exclude shocks from our consideration. In this case the flow is *isentropic* – that is, the *specific entropy* is conserved along a streamline. We subtract equation (6.18) from (6.19) to obtain

$$\frac{\gamma}{\gamma - 1}\frac{dp}{\rho} - \frac{\gamma}{\gamma - 1}\frac{p\,d\rho}{\rho\,\rho} - \frac{dp}{\rho} = 0,$$

which can also be written

$$dp = \gamma \frac{p}{\rho} d\rho. \tag{6.20}$$

Integrating this equation once yields the familiar polytropic energy relation (recall Equations (4.83) and (2.81)):

$$p = p_0 \left(\frac{\rho}{\rho_0} \right)^{\gamma}. \tag{6.21}$$

Equation (6.21) can be used as the energy equation as long as no discontinuities are present in the flow.

It is convenient to work with the Mach number, $M = u/C_s$. Recall that C_s is the speed of sound and is equal to $[\gamma p/\rho]^{1/2}$. The momentum equation (6.18) can be partially expressed in terms of the Mach number by dividing by C_s^2, using the definition of Mach number, and by using Equation (6.20) recognizing that $dp = C_s^2 d\rho$. We find that

$$M^2 \frac{du}{u} + \frac{d\rho}{\rho} + \frac{GM_{\odot}}{C_s^2 r^2} dr = 0. \tag{6.22}$$

Obviously, we have not yet finished converting this equation to one solely in terms of the Mach number; we must find an expression for du/u in terms of dM/M. We must also eliminate any explicit reference to the dependent variable ρ in the second term.

Starting with the definition of M, we take its derivative to get

$$\frac{dM}{M} = \frac{du}{u} - \frac{dC_s}{C_s}. \tag{6.23}$$

Using the definition of the sound speed, we find

$$\begin{aligned}
\frac{dC_s}{C_s} &= \frac{d[\gamma p/\rho]^{1/2}}{[\gamma p/\rho]^{1/2}} = \frac{dp^{1/2}}{p^{1/2}} + \frac{d\rho^{-1/2}}{\rho^{-1/2}} = \frac{1}{2} \left(\frac{dp}{p} - \frac{d\rho}{\rho} \right) \\
&= \frac{\gamma - 1}{2} \frac{d\rho}{\rho} \\
&= -\frac{\gamma - 1}{2} \left(\frac{du}{u} + \frac{dA}{A} \right),
\end{aligned} \tag{6.24}$$

where the second line required that we recall $dp/p = \gamma d\rho/\rho$, and the third line needed the continuity equation, $d\rho/\rho = -du/u - dA/A$. Putting Equations (6.23) and (6.24) together yields

$$\frac{dM}{M} = \frac{du}{u} + \frac{\gamma - 1}{2} \frac{du}{u} + \frac{\gamma - 1}{2} \frac{dA}{A} = \frac{\gamma + 1}{2} \frac{du}{u} + \frac{\gamma - 1}{2} \frac{dA}{A}. \tag{6.25}$$

We can now use Equation (6.25) to eliminate du/u in favor of dM/M in the momentum equation (6.22),

$$\underbrace{M^2\frac{du}{u}}_{\left[\dfrac{2}{\gamma+1}\dfrac{dM}{M}-\dfrac{\gamma-1}{\gamma+1}\dfrac{dA}{A}\right]} + \underbrace{\frac{d\rho}{\rho}}_{\left[-\dfrac{2}{\gamma+1}\dfrac{dM}{M}-\dfrac{2}{\gamma+1}\dfrac{dA}{A}\right]} + \frac{GM_\odot}{C_s^2 r^2}\,dr = 0.$$

(6.26)

The expression for $d\rho/\rho$ comes from the continuity equation plus Equation (6.25). Equation (6.26) multiplied through by $(\gamma+1)/2$ becomes

$$(M^2-1)\frac{dM}{M} - \left(M^2\frac{\gamma-1}{2}+1\right)\frac{dA}{A} + \frac{\gamma+1}{2}\frac{GM_\odot}{C_s^2 r^2}\,dr = 0. \tag{6.27}$$

Rearranging this equation, we obtain the steady, one-dimensional, gas dynamic equation:

$$\frac{dM}{M} = \frac{\tilde{\Lambda}}{M^2-1}, \tag{6.28}$$

where the function $\tilde{\Lambda}$ is given by the expression

$$\tilde{\Lambda}(r) = \left[M^2\left(\frac{\gamma-1}{2}\right)+1\right]\frac{dA}{A} - \frac{\gamma+1}{2}\frac{GM_\odot}{C_s^2 r^2}\,dr. \tag{6.29}$$

An alternate form of Equation (6.28) is

$$\frac{1}{M}\frac{M}{dr} = \frac{\Lambda}{M^2-1}, \tag{6.30}$$

where

$$\Lambda(r) = \left[M^2\left(\frac{\gamma-1}{2}\right)+1\right]\frac{1}{A}\frac{dA}{dr} - \frac{\gamma+1}{2}\frac{GM_\odot}{C_s^2 r^2}. \tag{6.31}$$

Equations (6.28)–(6.31) describe the flow of gas through a tube with cross-sectional area A. The change of the Mach number with increasing distance r depends on the change of the area function. The sign of the change of M also depends on whether the flow is subsonic ($M < 1$) or supersonic ($M > 1$). And note that Equations (6.28) and (6.30) indicate that something strange happens whenever the Mach number is unity. Before applying these equations to the solar wind outflow problem, we first consider gas flow through a wind tunnel, in order to understand this type of flow better.

6.2.2 *One-dimensional gas flow through a tube (the Laval nozzle)*

Let us assume that we have gas flow through a wind tunnel, as pictured in Figure 6.1. Equation (6.28) describes the variation of the Mach number. If we neglect gravity,

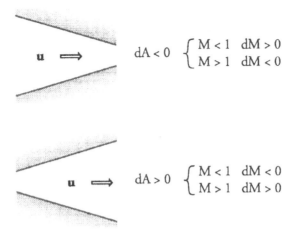

Figure 6.2. Nozzles and diffusers.

then Equation (6.29) simplifies to

$$\tilde{\Lambda}(r) = \left[M^2 \left(\frac{\gamma - 1}{2} \right) + 1 \right] \frac{dA}{A}. \tag{6.32}$$

The variation of $A(r)$ determines the variation of the Mach number (and flow speed). Suppose we have a converging flow tube in the flow direction (i.e., $dA < 0$), then (6.32) tells us that $\tilde{\Lambda}$ is negative. From Equation (6.28) we see that the Mach number increases ($dM > 0$) for subsonic flow ($M < 1$) and the Mach number decreases ($dM < 0$) for supersonic flow ($M > 1$). The former case is called a *nozzle* and the latter case is called a *diffuser* (Kundu, 1990). In contrast, for a diverging tube, the nozzle ($dM > 0$) is obtained for supersonic flow and the diffuser ($dM < 0$) for subsonic flow. Figure 6.2 summarizes this information. Note that, for the nozzle, the velocity increases and the pressure decreases ($dp < 0$) and, for the diffuser, the velocity decreases and the pressure increases ($dp > 0$). This follows from the steady-state momentum equation (6.18) if gravity is neglected.

Let us consider the converging tube case ($dA < 0$). For subsonic flow, the Mach number increases, so that at some point along the tube one finds $M = 1$ (Figure 6.3). In general for steady flow, the velocity (or Mach number) variation becomes ill-defined at the $M = 1$ location as the denominator of Equation (6.28) becomes zero. This situation is called *choking* in gas dynamics. However, suppose that just where M becomes 1, the tube changes from converging to diverging, as illustrated in Figure 6.4. That is, dA changes from negative to positive and is zero ($dA = 0$) right where $M = 1$. Note that the function $\tilde{\Lambda}$ becomes zero at this value of r. Equation (6.28) then has a finite limit given by

$$\frac{dM}{M} = \frac{\left(\frac{\gamma - 1}{2} M^2 + 1 \right) \frac{dA}{A}}{M^2 - 1} \rightarrow \frac{\frac{\gamma + 1}{2} 0}{0} = \text{finite}. \tag{6.33}$$

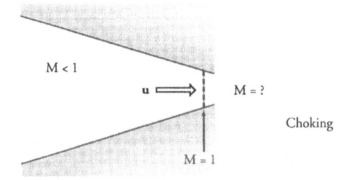

Figure 6.3. Choking flow where $M = 1$.

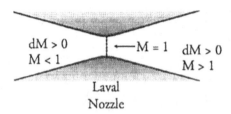

Figure 6.4. The Laval nozzle for the case where $M = 1$ at the throat of the nozzle.

Beyond the *transonic point* ($M = 1$), since $dA > 0$ and $M > 1$, we still have a nozzle and the flow continues to accelerate ($dM > 0$).

The converging/diverging flow tube is called a *Laval nozzle* (Kundu, 1990).The throat of the nozzle, where $dA = 0$ and where M goes through unity, is called the *critical point* of the flow. One can find $M(r)$ by integrating either Equation (6.28) or (6.30). Several categories of solutions are possible depending on the boundary condition existing at some initial value of $r = r_0$, as is described in Example 6.2 for one particular choice of area function $A(r)$.

Example 6.2 (The Laval nozzle) Consider a flow tube that has an area as a function of distance r given by $A(r) = r^2 + 1$, as shown in Figure 6.5. Gravity and other effects are neglected and an ideal monotomic gas ($\gamma = 5/3$) is assumed.

For this simple area function, Equations (6.28) and (6.32) together become

$$\frac{dM}{M} = \frac{\left(\frac{1}{3}M^2 + 1\right)\frac{dA}{A}}{M^2 - 1} = \frac{\left(\frac{1}{3}M^2 + 1\right)\frac{2r\,dr}{r^2 + 1}}{M^2 - 1}. \tag{6.34}$$

The critical point, where $dA = 0$, is located at $r = 0$ for this example. For transonic flow, M must equal 1 at $r = 0$. Equation (6.34) can be rearranged with all the M-dependence on one side and the r-dependence on the other side. Both sides

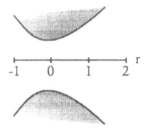

Figure 6.5. Dimensions of a flow tube (e.g., a wind tunnel) with constant width (into the page) and with the area function $A(r) = r^2 + 1$.

can then be integrated from an initial point $r = -1$ for the right-hand side and for an initial Mach number of M_0 for the left-hand side, with the result that

$$(M^2 + 3)^2 = QM,$$

where

$$Q = \frac{r^2 + 1}{2} \frac{(M_0^2 + 3)^2}{M_0}. \tag{6.35}$$

Q is a function of both r and the Mach number at the boundary (M_0), where for this example the boundary $r = r_0 = -1$. This algebraic equation is then solved to obtain $M(r)$ for $-1 < r < 2$ and for a range of M_0 values. As shown in Figure 6.6, several types of solution appear. For $M_0 < .28$, purely subsonic solutions are present, for which $M(r)$ increases to a subsonic maximum at the critical point $(r = 0)$ and then decreases again. For $M_0 > 2.7$, purely supersonic solutions are present, for which $M(r)$ decreases to a supersonic minimum at the critical point and then increases again. Especially interesting are the two transonic solutions, starting at $M_0 = .28$ and 2.7. Both these solutions reach $M = 1$ exactly at the critical point. The range of initial values, $.28 < M_0 < 2.7$, does not give any smooth steady solution over the complete range of r; the flow chokes for these initial conditions.

Which solution is the correct one depends on the boundary conditions. The pressure is important in this regard. In Problem 6.4 you are asked to derive the following expression for the density as a function of Mach number if gravity can be neglected:

$$\frac{\rho}{\rho_0} = \left(\frac{M_0}{M} \frac{A_0}{A} \right)^{\frac{2}{\gamma+1}}, \tag{6.36}$$

where ρ_0 is the density at the inner boundary ($r = -1$ for our simple example). The pressure can be found from the polytropic relation, (6.21). For our simple example, the pressure becomes

$$\frac{p}{p_0} = \left(\frac{M_0}{M} \frac{A_0}{A} \right)^{\frac{2\gamma}{\gamma+1}} = \left[\frac{2M_0}{M(r)(r^2 + 1)} \right]^{\frac{5}{4}}, \tag{6.37}$$

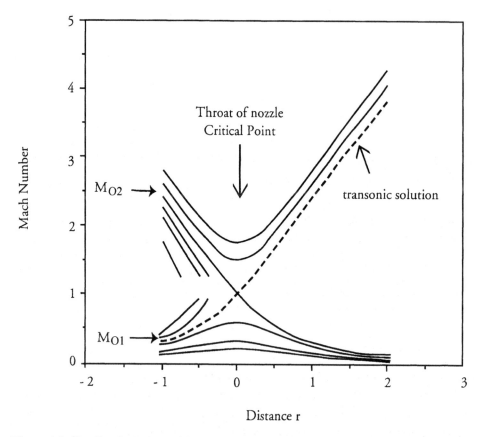

Figure 6.6. Family of solutions of the steady-state gas dynamic equation for different boundary conditions. Purely subsonic flow is present for M_0 (the Mach number at $r = -1$) less
than $M_{01} = 0.28$ and purely supersonic solutions for M_0 greater than $M_{02} = 2.7$. The two
transonic solutions start at $M_0 = M_{01}$ or M_{02}. The dashed line traces the transonic solution
that starts out subsonic and becomes supersonic.

where p_0 is the pressure at $r = -1$ and where the function $M(r)$ is shown in
Figure 6.6.

What is the pressure at large distances, $r \to \infty$? First, let us find $M(r)$ as $r \to \infty$.
For the subsonic branch, we find (Problem 6.4) that

$$\lim_{r \to \infty} M(r) = \frac{18 M_0}{(r^2 + 1)(M_0^2 + 3)^2}. \tag{6.38}$$

Note that for the subsonic branch, $M_0 < .28$. The Mach number goes to zero as
$r \to \infty$. This limit also applies to the transonic solution that ends up subsonic at
large r. Combining Equations (6.37) and (6.38), we find a finite expression for the
pressure at large r:

$$\lim_{r \to \infty} \frac{p}{p_0} = \left(\frac{1}{3} M_0^2 + 1\right)^{5/2}. \tag{6.39}$$

The pressure remains finite as $r \to \infty$ for this type of solution.

For the supersonic branch, which starts with $M_0 > 2.7$, we find (Problem 6.4) that the Mach number varies as

$$\lim_{r \to \infty} M(r) = r^{2/3} \frac{(M_0^2 + 3)^{2/3}}{(2M_0)^{1/3}}. \tag{6.40}$$

The Mach number becomes infinite as $r \to \infty$. This limit also applies to the transonic solution that ends up supersonic. The pressure limit associated with Equation (6.40) is

$$\lim_{r \to \infty} \frac{p}{p_0} = \frac{(2M_0)^{5/3}}{r^{10/3}(M_0^2 + 3)^{5/6}} \to 0. \tag{6.41}$$

The pressure has a zero limit as $r \to \infty$ for these supersonic solutions.

The type of solution that is applicable depends on the initial condition (M_0) and on the pressure at large values of r. Suppose the flow starts off subsonic with $M_0 < .28$ at $r = -1$ and that the (external) pressure at large distances is finite with $p \approx p_0$. Then the gas flow would behave in the manner described by a purely subsonic type solution. However, in the event that the pressure at large distances is extremely small and the flow starts out subsonic, then the only solution that applies is the transonic (accelerating) solution! As will be discussed in the next section, this is the solution relevant to solar wind flow.

The only force on the gas in Example 6.1, the classic Laval nozzle, was the pressure gradient force. The more general equations, (6.28) and (6.29), must be used when gravity is included. However, we can think of many other physical processes besides gravity that could affect the flow, including friction, mass addition, chemical effects, and heating and cooling (as in a flow tube in which combustion is taking place). All these effects can be represented by Equation (6.28) for steady, one-dimensional gas flow if a sufficiently general function $\tilde{\Lambda}(r)$ is defined by

$$\tilde{\Lambda}(r) = \left[M^2 \left(\frac{\gamma - 1}{2} \right) + 1 \right] \frac{dA}{A} - \frac{\gamma + 1}{2} \left(\frac{GM_\odot}{C_s^2 r^2} + \text{other terms} \right) dr, \tag{6.42}$$

where "*other terms*" takes into account the various physical processes just mentioned, which might be relevant for a particular flow. For example, if the gas under consideration is flowing through another, stationary, background gas, then the frictional force per unit volume is equal to $-\rho \nu u$, where ν is the collision frequency of the gas molecules with the background gas molecules. In this case, the "*other terms*" in (6.42) would appear as $[\nu u]/C_s^2$ or $\nu M/C_s$. This type of problem is discussed in fluid dynamics texts as "generalized, one-dimensional, steady gas flow" (cf. Kundu, 1990).

Whatever the "*other terms*" are for a particular situation, a transonic flow must exist if the flow starts out subsonic and if the pressure at large distances is very small (as indicated by Example 6.1). Where $M = 1$, at the critical point $(r = r_c)$, the denominator of the expression for dM/M given by Equation (6.28) becomes zero. The numerator, $\tilde{\Lambda}(r)$, must also be zero at the critical point. The function $\tilde{\Lambda}$ can be expressed as an "effective" area function if one wishes. The critical point for the generalized flow can be determined from the equation

$$\tilde{\Lambda}(r_c) = 0. \tag{6.43}$$

The critical point for Example 6.1 was $r_c = 0$. If the $\tilde{\Lambda}(r)$ function does not have a zero, then smooth, steady flow is not possible and one must revert to the more general form of the gas dynamics equations to find out how the flow behaves.

Some important examples in space physics where steady, one-dimensional gas flow serves as a reasonable approximation are:

Solar wind outflow This is the primary example we wish to consider in this chapter, and it will be discussed in more detail in the next section. The area function by itself does not have a minimum, and hence the gravity term must be included for $\tilde{\Lambda}(r)$ to have a zero; that is, gravity is needed to create the throat of the "nozzle" in the *effective* area function.

Polar wind outflow The magnetic field lines emanating from the polar cap of the Earth are "open" and do not connect back to the other hemisphere of the Earth, at least not anywhere in the near vicinity of the planet. The ionospheric plasma, mainly H^+ at high altitudes, is thus free to flow upwards along the magnetic field lines into a region of extremely low pressure. Steady, one-dimensional gas flow provides a useful description of this flow. Both gravity and ion–neutral friction are important.

Dusty cometary gas outflow The nucleus of a typical comet comprises a few-kilometer-diameter chunk of dust, rock, and ice. As it approaches the Sun, a cometary nucleus heats up and water vapor flows out into space, carrying the dust with it. This dusty gas flow can be approximated as a steady, one-dimensional flow. Gravity is negligible, but both the area function and gas–dust friction must be included in $\tilde{\Lambda}(r)$. See Problem 6.5.

6.2.3 Solar wind outflow

We now apply the formalism of generalized, steady, one-dimensional gas flow to the problem of solar wind outflow from the corona. The relevant function $\tilde{\Lambda}(r)$ has already been derived: Equation (6.29) or Equation (6.30) for $\Lambda(r)$, in which both area change and gravity are included. We have to choose a realistic area function, $A(r)$. A schematic of the plasma outflow in the lower corona is shown in Figure 6.1. The flow in the inner corona is confined, or channeled, by the magnetic field lines, and thus the cross-sectional area of a flow tube depends on the details of the field configuration. Along a flow tube the magnetic flux is a constant: $B(r)A(r) = constant$, where B is the field strength and A is the cross-sectional area of a flux (and flow) tube. The validity of this statement is examined more carefully in Section 6.3. For purely radial field lines, the area function is simply proportional to r^2, whereas for a magnetic dipole, $B(r)$ varies as r^{-3} so that $A(r)$ must vary as r^3. In general, we can write

$$A(r) = A_0(r/r_0)^s$$

giving

$$\frac{1}{A}\frac{dA}{dr} = \frac{s}{r}, \tag{6.44}$$

where s is a parameter. For example, $s = 3$ for a magnetic dipole. The actual field structure in the solar corona is complex, although near the polar regions and in coronal holes $s = 3$ is not a bad approximation.

We can use Equation (6.44) in our expressions for $\tilde{\Lambda}(r)$ or $\Lambda(r)$. Equations (6.30) and (6.31) can then be used to determine $M(r)$ given suitable boundary conditions. The types of solutions that are possible have already been discussed (see Figures 6.4 and 6.6). The particular solution that interests us is the transonic solution because of the relevant boundary conditions. We want the flow to start out subsonic (we know that the gas is almost stationary at the base of the corona) and to become supersonic far from the Sun so that the pressure far from the Sun is low (we know that the pressure of the interstellar gas is very low). The Mach number must equal unity at the critical point, $r = r_c$, where $\tilde{\Lambda}(r) = 0$ or $\Lambda(r) = 0$ (Equation (6.43)). The location of the critical point can be found from this condition and from expression (6.31):

$$\Lambda(r_c) = \left[M^2\left(\frac{\gamma-1}{2}\right)+1\right]\frac{1}{A}\frac{dA}{dr} - \frac{\gamma+1}{2}\frac{GM_\odot}{C_s^2 r_c^2} = 0$$

$$\text{with} \quad M = 1 \quad \text{and} \quad \frac{1}{A}\frac{dA}{dr} = \frac{s}{r}. \tag{6.45}$$

Equation (6.45) simplifies to

$$\frac{s}{r_c} = \frac{GM_\odot}{C_s^2 r_c^2}. \tag{6.46}$$

Before trying to solve Equation (6.46) for r_c, we should recall that $C_s^2 = C_s^2(r_c) = \gamma \tilde{R} T(r_c)$ is also a function of r_c, in general. But first let us deal with an isothermal corona ($\gamma \to 1$ in the polytropic energy relation, and $C_s^2 = constant$). Equation (6.46) in this case simply becomes

$$r_c = \frac{GM_\odot}{s\tilde{R}T_c}, \tag{6.47}$$

where T_c is the temperature of the corona.

Employing the value $s = 2$ (radial outflow) and substituting numerical values for the physical constants and parameters, such as G and $M_\odot$, we find that

$$\frac{r_c}{R_\odot} = \frac{5.6 \times 10^6 \text{ K}}{T_c(\text{K})}, \tag{6.48}$$

where $R_\odot$ is the solar radius. Typical coronal temperatures yield the following values of the critical radius:

$$\begin{aligned} T_c &= 1 \times 10^6 \text{ K} & r_c &= 5.6 R_\odot, \\ T_c &= 2 \times 10^6 \text{ K} & r_c &= 2.8 R_\odot. \end{aligned} \tag{6.49}$$

For an isothermal corona, Equation (6.49) indicates that the gas becomes supersonic at a distance of several solar radii. Outflow with $s = 3$ rather than 2 (i.e., flow along dipole field lines) has critical radii that are about $2/3$ of the Equation (6.49) values; that is, the flow accelerates more rapidly. Isothermal conditions imply a large conductive heat flux from the Sun out into the accelerating solar wind (Section 6.1). Next, let us consider the more general, nonisothermal case.

To find $C_s^2(r)$ we need to determine the coronal temperature as a function of distance. Bernoulli's equation provides the information we need (i.e., Equations (6.12) and (6.13)). Before using these equations, we must first divide through by $\gamma \tilde{R} T$ in order to convert the flow speed u into a Mach number M; this gives

$$\frac{1}{\gamma - 1} + \frac{1}{2} M^2 - \frac{GM_\odot}{C_s^2 r} = \frac{E_c}{C_s^2}. \tag{6.50}$$

Rearranging Equation (6.50), we find

$$C_s^2(r) = \frac{E_c + \dfrac{GM_\odot}{r}}{\dfrac{1}{\gamma - 1} + \dfrac{1}{2} M(r)^2}. \tag{6.51}$$

At $r = r_c$ (note that $r_c \neq R_c$ in general), we use $M = 1$ in Equation (6.51) and obtain

$$C_s^2(r_c) = 2\frac{\gamma - 1}{\gamma + 1}\left(E_c + \frac{GM_\odot}{r_c}\right). \tag{6.52}$$

By using Equations (6.46) and (6.52), we can find the following expression for the critical radius:

$$r_c = \frac{GM_\odot}{2s E_c}\left(\frac{\gamma + 1}{\gamma - 1} - 2s\right), \tag{6.53}$$

where E_c is the specific energy at the base of the corona as specified by Equation (6.13). More specifically, for $s = 2$ (radial outflow) and $s = 3$, expression

(6.53) becomes

$$s = 2 \qquad r_c = \frac{3}{4} \frac{GM_\odot}{E_c} \left(\frac{5/3 - \gamma}{\gamma - 1} \right) \qquad (6.54)$$

and

$$s = 3 \qquad r_c = \frac{5}{6} \frac{GM_\odot}{E_c} \left(\frac{7/5 - \gamma}{\gamma - 1} \right). \qquad (6.55)$$

The coronal energy E_c also depends on the polytropic index in these expressions. In Problem 6.6 you are asked to prove that Equations (6.48) and (6.54) are consistent with each other.

Example 6.3 (Nonisothermal flow in the solar corona) We use the coronal temperature chosen for Example 6.1 (i.e., $T_c = 2 \times 10^6$ K). For $R_c = R_\odot$ and $T_c = T_\odot$, then $E_c = 0.8 \times 10^{10}$ J/kg for $\gamma = 1.2$, $E_c = 6.3 \times 10^{10}$ J/kg for $\gamma = 1.15$ and $E_c = 8.82 \times 10^{10}$ J/kg for $\gamma = 1.1$. From Equation (6.14), we can determine that the asymptotic flow speeds (that is, the solar wind speed) $u_\infty = \sqrt{2E_c}$ for these values of E_c are 127 km/s, 355 km/s, and 420 km/s, respectively. Noting that $GM_\odot/R_\odot = 1.9 \times 10^{11}$ J, Equations (6.54) and (6.55) give us the following critical radii for these cases:

$$\begin{array}{llll}
s = 2 & \gamma = 1.2 & r_c/R_\odot = 41.6 & \\
& \gamma = 1.15 & r_c/R_\odot = 7.79 & \\
& \gamma = 1.1 & r_c/R_\odot = 9.15 & (6.56) \\
s = 3 & \gamma = 1.15 & r_c/R_\odot = 4.19. &
\end{array}$$

Clearly, the numerical value of r_c is quite sensitive to a number of parameters such as the geometry, the coronal temperature, and the polytropic index. For an ideal monotomic gas, the polytropic index is equal to 5/3. Recall the question posed at the end of Example 6.1, "But what does a polytropic index of 1.1 mean?" In a crude sense using this effective index allows the single-fluid equations to mimic the more complex two-fluid (electrons and protons) equations.

Asymptotic flow speeds (i.e., solar wind speeds) were given in Example 6.3. An expression for the Mach number as a function of heliocentric distance r can be found by integrating Equation (6.30) with the function $\Lambda(r)$ given by Equation (6.31). Combining the expression for $M(r)$ with an expression for the speed of sound as given by Equation (6.51), one can find the flow speed as a function of r. This task must be carried out numerically, which we will not do. Qualitatively, the desired $M(r)$ solution looks very much like the transonic solution shown for the simple Laval nozzle (Example 6.2).

Figure 6.7 shows some numerically calculated hydrodynamic solar wind flow speeds from the book by Parker (1963).

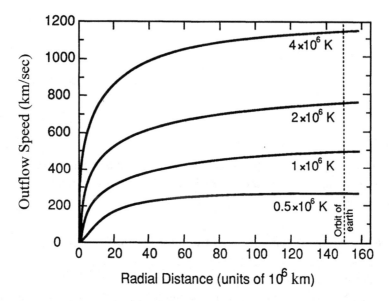

Figure 6.7. Outflow speed versus heliocentric distance for coronal gas from a numerical calculation by Parker (1963). Each curve is labeled with the coronal temperature. Adapted from Parker (1963).

It is mathematically simpler to merely consider the behavior of the solar wind flow at large distances from the Sun. We know from our simple example 6.2 that the Mach number M becomes large at large distances. The asymptotic form of Equations (6.30) and (6.44) as $r \to \infty$ is

$$\frac{dM}{M} \approx \frac{\gamma - 1}{2} \frac{s}{r} \, dr. \tag{6.57}$$

Equation (6.57) can be integrated to give an expression for $M(r)$:

$$M(r) \to Cr^{\gamma - 1}, \tag{6.58}$$

where $s = 2$ was adopted and where C is a constant depending on the solution at smaller values of r. Similarly, the asymptotic form of Equation (6.51) for C_s^2 can be found to be

$$C_s^2 \to \frac{2E_c}{[M(r)]^2}. \tag{6.59}$$

Two facts are apparent about Equation (6.59). First, since the definition of Mach number is $M(r) = u(r)/C_s(r)$, the asymptotic limit (6.59) also can be written as $u_\infty = \sqrt{2E_c}$. The solar wind speed, $u_{sw} = u_\infty$, is independent of r as $r \to \infty$, whatever the r-dependence of $M(r)$ or $C_s(r)$. This asymptotic value is reached for r values not too much larger than the critical radius, which is only a few solar radii; hence, long before the solar wind has even reached the orbit of Mercury, the flow speed has become approximately constant. This can also be seen from the numerical solutions (Figure 6.7). Second, together with Equation (6.58), Equation (6.59) indicates that the asymptotic form of C_s^2 is

$$C_s^2 \rightarrow \frac{1}{r^{2(\gamma-1)}}. \tag{6.60}$$

Note that the temperature is proportional to C_s^2. In the case of isothermal flow, $\gamma = 1$, and the asymptotic expressions indicate that both $M(r)$ and $C_s(r)$ are independent of r.

How do the density and pressure of the outflowing gas vary as a function of r? Given the flow speed $u(r)$ and the area function $A(r)$, then the continuity equation (6.17) can be used to obtain the following expression for the density:

$$\rho(r) = \rho_c \frac{C_s(r_c)}{u(r)} \left(\frac{r_c}{r}\right)^2. \tag{6.61}$$

We chose to normalize this expression at the critical point where $u = C_s$, and we assumed radial expansion ($s = 2$). Note that at large distances from the Sun, where $u \rightarrow u_\infty = u_{sw}$, the density varies as $1/r^2$. This has been confirmed by measurements made by spacecraft traversing interplanetary space at various heliocentric distances. See Figure 6.8 for numerical values of the solar wind density versus heliocentric distance (again adapted from Parker's book), as well as some observational data from an article by Bagenal and Gibson (1991). Note that close to the Sun the density decreases faster than $1/r^2$, but that at larger values of r the density does decrease in the expected asymptotic manner.

The pressure asymptotically becomes extremely small for accelerating transonic flow, according to our Laval nozzle analogy; this is not surprising considering that both the density and the temperature decrease as r increases. Putting together the relevant asymptotic formulae, we find that as $r \rightarrow \infty$ the pressure varies as

$$p(r) \rightarrow \frac{C}{r^{2\gamma}}, \tag{6.62}$$

where C is a constant depending on the flow at smaller values of r. Clearly, the pressure goes to zero as $r \rightarrow \infty$ for any reasonable value of γ.

Many issues concerning solar wind outflow have not been addressed, but these are beyond the scope of an introductory textbook. Suggestions for further reading are provided at the end of this chapter. We will further allude to the issue of the geometry of the outflow (i.e., the area function, $A(r)$) later in this chapter because it involves the structure of the magnetic field in the solar corona.

6.3 The interplanetary magnetic field

The magnetic field, and the force on the plasma associated with it, was neglected in our analysis of solar wind outflow. Implicit in this analysis was the assumption

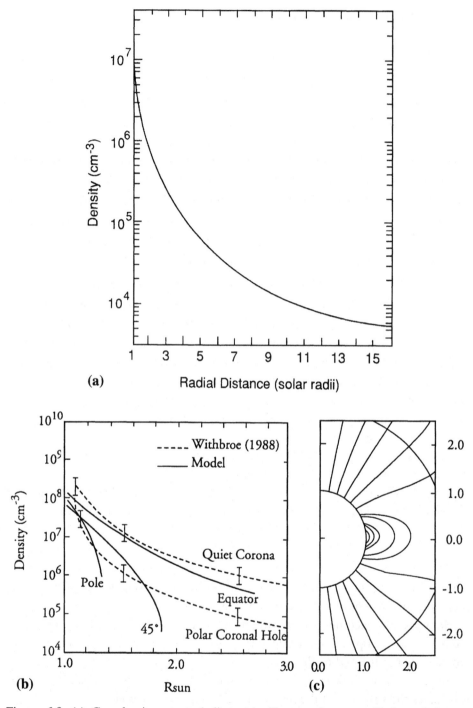

Figure 6.8. (a) Gas density versus heliocentric distance for coronal/solar wind gas. Adopted from Parker (1963). (b) Solid lines are coronal electron density profiles for both the equatorial plane and a polar coronal hole from the model of Bagenal and Gibson (1991). The model consists of an optimization of magnetostatic balance and corono-graph white-light K-corona data. The data are from the review of Withbroe (1988). (From Bagenal and Gibson 1991.) (c) Field lines from the Bagenal and Gibson (1991) model. (From Bagenal and Gibson, 1991.)

that the coronal plasma flows along the field, with the field lines defining a flow tube with some cross-sectional area. Now we must examine the validity of this assumption. We will also investigate the coronal magnetic field and the interplanetary magnetic field. As discussed in Chapter 5, the photospheric magnetic field pattern can be quite complex, with strong fields in sunspots and weaker fields elsewhere. The magnetic field is frozen into the plasma in the solar atmosphere and in the solar wind because the magnetic Reynolds number is very large; that is, $R_m \gg 1$ everywhere. This means that plasma can "slip" freely along a field line, but that otherwise a fluid parcel and a field line are attached to one another. However, this still does not tell us whether or not the plasma flow controls the field or the field controls the plasma flow. In this section we will examine the MHD momentum equation and address this question. In Section 6.3.2, the coronal magnetic field will be described using scalar potential theory, and in Section 6.3.3 an expression for the average interplanetary magnetic field will be derived.

6.3.1 Forces on the coronal plasma and the source surface

The single-fluid MHD momentum equation, (4.80), is the starting point for our discussion. The steady-state collisionless version of this equation becomes

$$
\begin{array}{cccccc}
\rho \mathbf{u} \cdot \nabla \mathbf{u} & = & -\nabla p & + & \mathbf{J} \times \mathbf{B} & + & \rho \mathbf{g} \\
\rho u^2 / L & & p/L & & p_B/L & & \rho g \\
\rho u^2 / p_B & & p/p_B & & 1 & & \rho g L / p_B \\
M_A^2 & & \beta \approx C_s^2 / C_A^2 & & 1 & & U_{\text{gravity}} / C_A^2.
\end{array}
\tag{6.63}
$$

The rough magnitude of each term is shown in the second line of (6.63). The third line consists of dimensionless numbers and was obtained by multiplying each term in the second line by L/p_B, and the fourth line is just the third line rewritten in terms of common physical quantities. L is a typical coronal length scale. p and p_B are the thermal and magnetic pressures, respectively. $M_A = u/C_A$ is the Alfvénic Mach number, and C_A and C_s are the Alfvén and sound speeds of the plasma, respectively. β is the plasma beta, and U_{gravity} (or U_g) is the gravitational potential. You should review Chapter 4 and convince yourself that the dimensionless terms given in Equation (6.63) are correct. Where M_A^2 is large, the dynamic pressure, ρu^2 (i.e., the inertial forces), exceeds the magnetic pressure; in this case the dynamic pressure is more important than the $\mathbf{J} \times \mathbf{B}$ force in determining the flow pattern. Conversely, wherever $M_A^2 \ll 1$, the $\mathbf{J} \times \mathbf{B}$ force exceeds the inertial force and the magnetic field can constrain the flow, rather than the field lines being "carried along" with the flow. Similarly, wherever $\beta \gg 1$, the pressure gradient force is more important than the $\mathbf{J} \times \mathbf{B}$ force in controlling the plasma.

Table 6.1. *Comparison of terms in MHD momentum
equation for the solar corona*

$r/R_\odot$	M_A^2	β	$U_{gravity}/C_A^2$
1.01	10^{-3}	10	10
1.04	10^{-3}	0.5	0.5
3	0.3	0.1	0.1
10	1	0.05	0.05

The importance of gravity relative to the $\mathbf{J} \times \mathbf{B}$ force is indicated by U_g/C_A^2. In the solar corona, the order of magnitude of the gravitational potential is given by $U_g \approx GM_\odot/R_\odot$.

The coronal physical variables listed in Table 5.2 can be used to estimate the dimensionless numbers in the last line of Equation (6.63) (Table 6.1). Values of the flow speed from Figure 6.7 are also used.

Both the mass density and thermal pressure in the solar photosphere are extremely large: $\beta \gg 1$ and $M_A^2 \gg 1$. In fact, in the vertical direction, as you recall, hydrostatic equilibrium is a very good approximation, and the momentum balance is primarily between the pressure gradient force and gravity. The $\mathbf{J} \times \mathbf{B}$ force is not important in comparison with the other forces, including the inertial force (since $M_A^2 \gg 1$), and as the gas moves around the solar surface due to convective motion driven from below, the magnetic field tends to go along for the ride.

The density in the solar atmosphere rapidly decreases with increasing altitude, and M_A^2 rapidly decreases with altitude such that $M_A^2 \ll 1$ in the lower corona, where bulk convective motions are only a few kilometers per second (Table 6.1). This indicates that dynamic pressure is much less important than magnetic forces in controlling the flow of gas in the transition region and lower corona ($r \approx 1.04R_\odot$). In this region, we also have $\beta \approx 1$ and $U_{gravity}/C_A^2 \approx 1$, indicating that the pressure gradient and gravity forces are comparable. In fact, hydrostatic equilibrium is a reasonable approximation in the vertical direction since the sum of the pressure gradient and gravitational terms is much less than either one individually. In dimensionless form, the sum of these two terms is much less than unity, but the dimensionless order of the $\mathbf{J} \times \mathbf{B}$ term is ≈ 1, indicating that magnetic forces are generally dominant in controlling the dynamics. According to Table 6.1, in the region for which $r/R_\odot$ ranges from 1.04 up to about 3–10, all nonmagnetic terms in the single-fluid momentum equation are individually (and therefore together) less than the magnetic force term. This means that the field lines tend to channel the flow of plasma throughout all of the solar corona at least out to 3 solar radii, and thus the configuration of the open field lines determines the area function needed to understand the outflow of plasma.

Table 6.1 indicates that by a radial distance of about 10 solar radii, M_A^2 has increased to unity. In the solar wind near 1 AU (Table 5.2), $M_A^2 \approx 100$, indicating that dynamic pressure is again dominant, as it was in the photosphere. A surface can be defined in the "outer" corona (or "inner" solar wind) on which $M_A^2 \approx 1$. This surface is called *the source surface*, and as its name indicates the source of the interplanetary magnetic field (IMF) can be associated with a surface at a radial distance denoted R_s. Table 6.1 indicates that R_s is typically 3–10 $R_\odot$, although $2.5R_\odot$ is also commonly assumed. Field lines are assumed to be "open" to interplanetary space beyond R_s. Outside the source surface, the inertial force is greater than the $\mathbf{J} \times \mathbf{B}$ force, and the plasma is capable of "pushing" the magnetic field around, which has significant implications for the choice of method for determining the magnetic field. Outside the source surface, the field can still be assumed to be frozen into the plasma, and this makes it possible to use a kinematic dynamo approach (Section 5.1.5) in which a flow pattern is adopted and the field calculated solely from the magnetic induction equation. This will be carried out for the interplanetary field in Section 6.3.3. However, inside the source surface different methods must be applied because the magnetic field itself is important in determining the plasma flow. In general, this is a difficult problem and must be addressed by using numerical solutions of the full MHD equations, or even by using the multi-fluid conservation equations. However, a simple and surprisingly good approximation can be applied to this problem, as we will see in the next section.

6.3.2 Coronal magnetic fields – Potential field

The values in Table 6.1 indicate that the $\mathbf{J} \times \mathbf{B}$ term is the most important term in the momentum equation (6.63) for the solar corona below the source surface. Assuming then that all the other terms are very small throughout the corona inside the source surface (i.e., for $r < R_s$), the steady-state MHD momentum equation can be approximated as

$$\mathbf{J} \times \mathbf{B} \approx \mathbf{0}. \tag{6.64}$$

Recall that this type of momentum balance was discussed in Chapter 4. For a finite magnetic field, Equation (6.64) can be satisfied in two ways: (1) $\mathbf{J} = \mathbf{0}$ or (2) $\mathbf{J}$ is parallel to $\mathbf{B}$. The first possibility requires that no electrical currents exist in the region of interest, although external currents outside this region are possible; for the second possibility the magnetic field configuration must be in the form of force-free structures (see Equation 4.105 and Figure 4.11). Many coronal filaments are undoubtedly force free, but the first possibility ($\mathbf{J} = \mathbf{0}$) usually is an adequate approximation throughout most of the corona. However, this condition is violated in current sheets that are known to exist in the corona. A better approximation than the force-free or current-free ones, as represented by Equation (6.64), is to retain the pressure gradient and gravity terms in Equation (6.63) while still neglecting the inertial term (see Bagenal and Gibson, 1991). However, we will stick with the simpler, albeit less accurate, $\mathbf{J} = \mathbf{0}$ approximation.

Combining Ampère's law (minus the displacement current), $\nabla \times \mathbf{B} = \mu_0 \mathbf{J}$, plus the condition that $\mathbf{J} = \mathbf{0}$, we obtain

$$\nabla \times \mathbf{B} = \mathbf{0}. \tag{6.65}$$

By analogy with electrostatics and with the definition of a scalar electric potential for a conservative electric field (see Chapter 1 for a brief review), as a consequence of Equation (6.65) we can define a *magnetic scalar potential* Ψ_B such that the magnetic flux density is proportional to the gradient of this potential:

$$\mathbf{B} = -\mu_0 \nabla \psi_B. \tag{6.66}$$

Note that $\nabla \times \mathbf{B}$ is automatically zero with this form of the magnetic field since the curl of a gradient is identically zero. The electrostatic potential can be defined everywhere, but one must always be cautious with the magnetic scalar potential (unlike the more general magnetic vector potential $\mathbf{A}$), which can be defined *only* in those regions where the current density is zero. We can proceed further by using another of Maxwell's equations, $\nabla \cdot \mathbf{B} = 0$, together with Equation (6.66) to get

$$\nabla^2 \psi_B = 0. \tag{6.67}$$

Equation (6.67) is *Laplace's equation for the magnetic scalar potential.*

Laplace's equation is an elliptic partial differential equation, and its solution throughout some region defined by a closed boundary surface requires a boundary condition everywhere on the surface surrounding the region, either on ψ_B or on the gradient of ψ_B in a direction normal to the boundary surface, $\partial \psi_B / \partial s$. Figure 6.9 shows the region between the surface of the Sun and the source surface where $\mathbf{J} \approx \mathbf{0}$ and, hence, where Laplace's equation for ψ_B can be used. We need to define ψ_B or $\partial \psi_B / \partial s$ at the surface of the Sun (the inner boundary of our region of interest) and we can eliminate the need for a boundary condition at the source surface by assuming that $R_s \to \infty$ since $R_s \gg R_\odot$. We can use the photospheric magnetic field as a function of position on the surface (using the polar coordinates θ and ϕ) for our inner boundary condition, which is then

$$-\mu_0 \nabla \psi_B |_{r=R_\odot} = \mathbf{B}_{\text{photospheric}}(\theta, \phi). \tag{6.68}$$

Actually, we can simply use the radial (i.e., normal) component of Equation (6.68). Now we solve Equation (6.67) for several simple scenarios subject to the boundary condition (6.68). Once we have obtained a solution, $\psi_B(r, \theta, \phi)$, for $r > R_\odot$, we

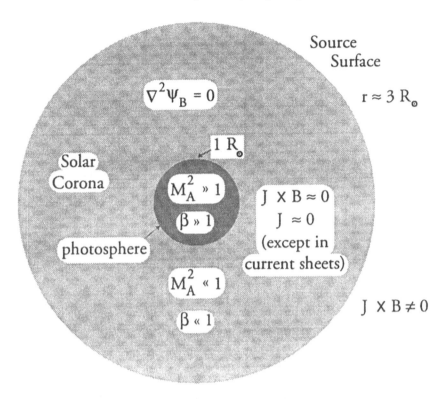

Figure 6.9. Schematic of the source surface and region of the corona where potential fields can be used.

can find $\mathbf{B} = -\mu_0 \nabla \psi_B|_{r=R_s}$ at the source surface and then use this field as a boundary condition for a determination of the magnetic field outside the source surface ($r > R_s$). A different method of finding the magnetic field has to be used outside the source surface since $\mathbf{J} \neq \mathbf{0}$ outside the source surface and Equation (6.67) is invalid.

Consider first a very simple photospheric field, one that is purely radial and uniform over the solar surface:

$$\mathbf{B}(r = R_\odot, \theta, \phi) = B_\odot \hat{\mathbf{r}}. \tag{6.69}$$

The magnetic scalar potential in this case is only a function of distance r (i.e., $\psi_B(r)$), and the appropriate Laplace's equation for a spherically symmetric geometry is

$$\frac{1}{r^2} \frac{\partial}{\partial r} \left(r^2 \frac{\partial \psi_B}{\partial r} \right) = 0. \tag{6.70}$$

After integrating once over r, we obtain

$$\frac{\partial \psi_B}{\partial r} = \frac{C}{r^2}, \tag{6.71}$$

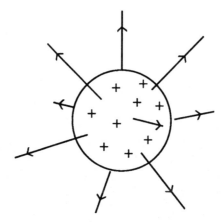

Figure 6.10. Monopole solution of Laplace's equation. This solution cannot be true globally because it would then violate one of Maxwell's equations.

where C is a constant of integration that must equal $-B_\odot R_\odot^2/\mu_0$, from the boundary condition (6.69). From Equation (6.66), we then have

$$\mathbf{B}(r, \theta, \phi) = \hat{\mathbf{r}} B_\odot / r^2. \tag{6.72}$$

This is a very simple result (Figure 6.10), which should have been obvious even without having to solve Laplace's equation, but there is one slight problem. If we take a closed surface integral of $\mathbf{B}$ for any surface surrounding the Sun, we find that it is equal to $B_\odot 4\pi R_\odot^2$; however, combining the Maxwell's equation (see review in Chapter 1) $\nabla \cdot \mathbf{B} = 0$, with Gauss's theorem indicates that any closed surface integral of $\mathbf{B}$ must be equal to zero. Clearly, our boundary condition, given by Equation (6.69), is incorrect! We know that the Sun is not full of magnetic monopoles.

The next simplest boundary condition, which does obey Maxwell's equation, is to retain uniform radial photospheric fields but to let them have opposite polarity in the two hemispheres (e.g., positive for $\theta > 90°$ (northern hemisphere) and negative for $\theta < 90°$ (southern hemisphere)). Laplace's equation can be solved independently for each hemisphere and the solution (6.72) still obtained, but now $B_\odot$ is positive for $\theta > 90°$ and is negative for $\theta < 90°$, as pictured in Figure 6.11. However, there remains a problem. Looking at the solution (Figure 6.11) it is clear that $\nabla \times \mathbf{B}$, and hence the current density $\mathbf{J}$, is indeed zero in each hemisphere, but a current sheet must exist in the equatorial plane between the two hemispheres (with the current flowing eastward). Hence, $\nabla \times \mathbf{B} \neq \mathbf{0}$ right at the equator and a magnetic scalar potential cannot be defined there. Actually, this is not a real problem because we can define two separate regions in the two hemispheres in which $\mathbf{J} = \mathbf{0}$. The $\mathbf{J} = \mathbf{0}$ condition is locally violated but not in the region where Laplace's equation was solved. Of course, the existence of a current sheet also has implications for the plasma dynamics.

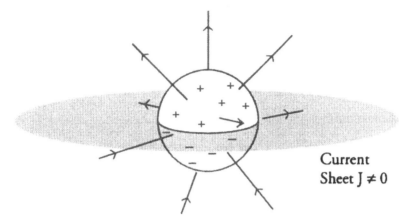

Figure 6.11. Separate monopole solutions ($\mathbf{J} = \mathbf{0}$) of opposite polarity for each hemisphere with a current sheet (with $\mathbf{J} \neq \mathbf{0}$) located in the equatorial plane.

Another simple solution of Equation (6.67) can be found for the case of a magnetic dipole. The boundary condition for Laplace's equation is not simple in this case because the photospheric field is nonradial and nonuniform. The average, large-scale magnetic field in the vicinity of the Sun does have a dipolelike appearance, although on small scales the observed solar field is clearly much more complex than a dipole field. The terrestrial dipole magnetic field was discussed in Chapter 3, and the same formulae apply here with suitable changes in scale size and field strength. In Problem 6.7, you are asked to derive the dipole field components, B_R and B_λ, given in Chapter 3 (Equation 3.67) from the following magnetic scalar potential for a solar dipole with magnetic moment $\mathbf{m}_\odot$:

$$\psi_B = \frac{\mathbf{m}_\odot \cdot \hat{\mathbf{r}}}{4\pi r^2} \quad \text{[units of A]}. \tag{6.73}$$

You are also asked in Problem 6.7 to demonstrate that expression (6.73) is indeed a solution of Laplace's equation. Note that we can use the polar angle θ in place of magnetic latitude λ; that is, $\theta = (\lambda - \pi/2)$ radians.

What if the magnetic field pattern on the Sun's surface does not correspond to a dipole field? A more general solution to Laplace's equation can be found by performing a *multipole expansion* of the scalar potential in *spherical harmonics* (e.g., Dennery and Krzywicki, 1967):

$$\psi_B(r, \theta, \phi) = \sum_{\ell=0}^{\infty} \sum_{m=-\ell}^{\ell} a_{\ell m} \left(\frac{R_\odot}{r}\right)^{\ell+1} Y_\ell^m(\theta, \phi), \tag{6.74}$$

where the Y_ℓ^m are the spherical harmonics.

Usually in such mathematical problems boundary conditions are imposed on ψ_B on some surface, but the observed quantity in this case is the photospheric magnetic field (and in particular the radial component); hence the relevant boundary condition is on the derivative of ψ_B normal (i.e., radial) to the Sun's surface. Taking the radial component of Equation (6.68) we find

$$B_r(r = R_\odot, \theta, \phi) = \hat{\mathbf{r}} \cdot \mathbf{B}_{\text{photospheric}}(\theta, \phi)$$

$$= -\mu_0 \hat{\mathbf{r}} \cdot \nabla \psi_B|_{r=R_\odot} = -\mu_0 \partial \psi_B / \partial r|_{r=R_\odot}. \quad (6.75)$$

We can then define the function

$$h(\theta, \phi) \equiv -B_r(r = R_\odot, \theta, \phi) R_\odot / \mu_0. \quad (6.76)$$

The coefficients in the spherical harmonic expansion (6.74) can be determined in terms of $h(\theta, \phi)$ as

$$a_{\ell m} = \frac{1}{\ell + 1} \int_0^{2\pi} d\phi \int_0^\pi \sin \theta \, d\theta \, \bar{Y}_\ell^m(\theta, \phi) h(\theta, \phi). \quad (6.77)$$

The complex conjugate of Y_ℓ^m appears in this equation. Note that $a_{\ell m}$ can, in general, be complex, so as to keep the scalar potential real.

The nature of the scalar potential given by the expansion (6.74) can be better understood by examining the first few spherical harmonics (Dennery and Krzywicki, 1967):

$$Y_0^0 = \frac{1}{2\sqrt{\pi}}$$

$$Y_1^0 = \frac{1}{2} \sqrt{\frac{2}{\pi}} \cos \theta$$

$$Y_1^{\pm 1} = -/+ \frac{1}{2} \sqrt{\frac{3}{2\pi}} \sin \theta e^{\pm i\phi}$$

$$Y_2^0 = 2 \cos^2 \theta - \sin^2 \theta$$

$$Y_2^{\pm 1} = \cos \theta \sin \theta e^{\pm i\phi}$$

$$Y_2^{\pm 2} = \sin^2 \theta e^{\pm 2i\phi}. \quad (6.78)$$

The spherical harmonics become increasingly complex in the polar and azimuthal angles with larger and larger values of the index ℓ. Putting the functions from Equation (6.78) into the first few terms of the expansion gives

$$\psi_B = a_{00} \frac{R_\odot}{r} \frac{1}{2\sqrt{\pi}} + \left(\frac{R_\odot}{r}\right)^2 \frac{1}{2} \sqrt{\frac{3}{2\pi}} (a_{1,-1} \sin \theta e^{-i\phi}$$

$$+ a_{1,1} \sin \theta e^{+i\phi} + a_{1,0} \cos \theta) + \cdots. \quad (6.79)$$

The first term of Equation (6.79) is inversely proportional to radial distance r and is the monopole solution. This is fine for electrostatics but not for magnetostatics, although a couple of pages back, we combined two separate "monopole" solutions in opposite hemispheres with opposite polarity. In general, the first acceptable terms are those in Equation (6.79) that are proportional to the inverse square of r. If the boundary condition at the Sun's surface is independent of azimuthal angle ϕ, and if we only retain the terms shown in Equation (6.79), then

$$a_{00} = a_{1,-1} = a_{1,1} = 0. \tag{6.80}$$

We are left with

$$\psi_B = \left(\frac{R_\odot}{r}\right)^2 \frac{1}{2}\sqrt{\frac{3}{2\pi}}(a_{1,0}\cos\theta). \tag{6.81}$$

This is obviously just the dipole magnetic potential given by Equation (6.73), with a suitable definition of the coefficient $a_{1,0}$! Recall from Chapter 3 that the magnetic field strength varies as $1/r^3$, which is consistent with the scalar potential varying as $1/r^2$.

The ℓth term in the expansion (6.74) is proportional to

$$\psi_B \approx \frac{1}{r^{\ell+1}}. \tag{6.82}$$

For the general case, all terms starting with the dipole term must be included (i.e., quadrupole, octupole, etc.). The r-dependence of (6.82) indicates that even if the magnetic field starts off being very complicated near the Sun's surface, the higher-order terms fall off faster with r than the lower-order terms, so that with increasing radial distance the field looks simpler and simpler.

Figure 6.12 is a somewhat oversimplified schematic of a typical coronal magnetic field pattern inside the source surface. The field is basically a potential field ($\mathbf{J} = \mathbf{0}$), although a current sheet with $\mathbf{J} \neq \mathbf{0}$ starts to form near the source surface over the active region. As expected from Equation (6.82), the field simplifies with increasing radial distance. Notice that the field polarity pattern ($+$ or $-$) is more complicated on the solar surface than on the source surface; some actual data illustrating this will be shown in the next section. This schematic could be a "cut" through any plane, but if it is the equatorial plane, then the change in polarity near the source surface is called a *sector boundary*. Regions of colder coronal plasma (and therefore less bright in the X-ray part of the spectrum) tend to be located on open magnetic field lines and are called *coronal holes*. The plasma on closed field lines (i.e., field lines linked with active regions) tends to be hotter and brighter (see Figure 1.2). Figure 6.13 shows a more realistic magnetic field pattern deduced from observations.

We will return to the issue of the large-scale structure of the coronal, and interplanetary, magnetic field in Section 6.4. In the next section, we will consider how the average interplanetary magnetic field behaves, given a specified field pattern on the source surface.

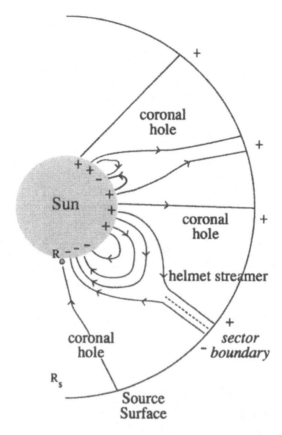

Figure 6.12. Schematic of a "typical" field pattern (not an actual solution) in the solar corona inside the source surface (located at approximately $3R_{\odot}$ for this example). The pluses and minuses on the surfaces denote field polarity. The dashed line near the source surface is a current sheet as well as a sector boundary that separates regions in interplanetary space of opposite polarity. The current sheet originates from a structure called a helmet streamer.

6.3.3 The interplanetary magnetic field

In the last section, we described how the magnetic field in the solar corona could be determined by solving Laplace's equation. But if we wish to find the magnetic field outside the source surface ($r > R_s$), then we must use a different method because the current density is not zero ($\mathbf{J} \neq \mathbf{0}$) and we can no longer define a scalar magnetic potential. Again, as discussed in Section 6.3.1, a full solution of the MHD equations, or multi-fluid equations, provides a more accurate picture of how the *interplanetary magnetic field (IMF)*, or solar wind field, varies. We will use a simpler method of describing the IMF.

Inside the source surface ($r < R_s$), we can see from Table 6.1 that $M_A^2 \ll 1$, and consequently magnetic forces (i.e., $\mathbf{J} \times \mathbf{B}$) dominate the dynamics, whereas outside the source surface, $M_A^2 \gg 1$, and the dynamic pressure of the solar wind flow is the dominant factor in the dynamics. Outside the source surface, the solar wind flows unimpeded by the magnetic field, and the magnetic field "goes along for the ride." In

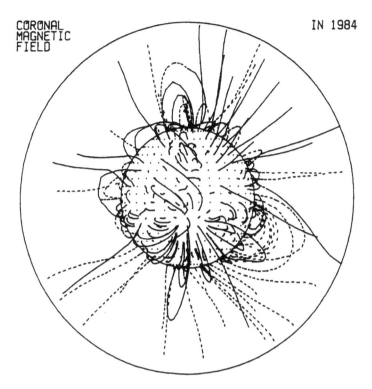

Figure 6.13. A three-dimensional structure of magnetic field lines in a corona in 1984. Solid lines are located at the observer's side of the sky plane, and dashed lines are beyond the sky plane. The inner and the outer circles are the photosphere and the so-called source surface of 2.5 solar radii, respectively. (From Hakamada et al., 1991.)

this case, the plasma does not flow along magnetic field lines but flows radially (or almost radially) outward from the Sun. Furthermore, the solar wind speed is almost constant outside the source surface, with most of its acceleration having taken place closer to the Sun. Hence, we can make the following approximation for $r > R_s$:

$$\mathbf{u}(r, \theta, \phi) = u_{sw}\hat{\mathbf{r}}, \qquad (6.83)$$

where u_{sw} is a constant solar wind speed.

The magnetic Reynolds number of the solar wind plasma is much greater than unity, and thus the magnetic field is frozen into the plasma and is carried outward by the solar wind flow. This process can be described using the magnetic induction equation introduced in Chapter 4 (Equation (4.76)):

$$\frac{\partial \mathbf{B}}{\partial t} = \nabla \times (\mathbf{u} \times \mathbf{B}). \qquad (6.84)$$

Equation (6.84) can be used to determine the large-scale structure of the interplanetary magnetic field. However, before proceeding we need to consider a further complication – the Sun rotates with a 27-day period ($T_{\odot\mathrm{rot}}$), and the "feet" of the magnetic field lines we wish to map out into interplanetary space are fixed onto the

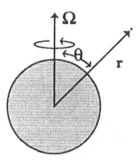

Figure 6.14. Geometry of rotating Sun. The angular velocity vector, Ω, points to the celestial north. The position vector is $\mathbf{r}$.

rotating Sun (or on the source surface). A frame of reference rotating with the Sun has a velocity $\mathbf{u}_\Omega$ relative to the solar system inertial frame given by the expression

$$\mathbf{u}_\Omega(r, \theta) = \Omega \times \mathbf{r} = \Omega r \sin \theta \hat{\phi}, \tag{6.85}$$

where Ω is the angular frequency vector of the Sun, which points to the celestial north and which has a magnitude $\Omega = 2\pi / T_{\odot\mathrm{rot}}$ (see Figure 6.14). The polar angle is θ, and $\hat{\phi}$ is the unit vector in the azimuthal direction. We wish to find solutions to the magnetic induction equation that are steady state *in the rotating frame of reference.*

The plasma flow velocity in the rotating frame of reference ($\mathbf{u}'$) is equal to

$$\mathbf{u}'(r, \theta, \phi) = \mathbf{u}(r, \theta, \phi) - \mathbf{u}_\Omega(r, \theta). \tag{6.86}$$

Combining Equations (6.84) and (6.86), we obtain

$$\frac{\partial \mathbf{B}}{\partial t} = \nabla \times [(\mathbf{u}' + \mathbf{u}_\Omega) \times \mathbf{B}] = \nabla \times (\mathbf{u}' \times \mathbf{B}) + \nabla \times (\mathbf{u}_\Omega \times \mathbf{B}). \tag{6.87}$$

After a judicious application of some vector calculus identities and Maxwell's equations (see Problem 6.10), we can write Equation (6.87) as

$$\left(\frac{\partial \mathbf{B}}{\partial t} \right)_R = \nabla \times (\mathbf{u}' \times \mathbf{B}), \tag{6.88}$$

where $(\partial \mathbf{B}/\partial t)_R$ denotes the time derivative in the rotating reference frame. A term, $\Omega \times \mathbf{B}$, also results from evaluating the right-hand side of Equation (6.87), but this term is canceled out when the time derivative term is converted to the rotating coordinate system. We could have designated the field in this reference frame as $\mathbf{B}'$, but this really is not necessary.

The total/convective derivative of some vector quantity $\mathbf{A}$ in a rotating reference frame, designated R, is related to the derivative in a fixed/inertial frame, designated F, by

$$\left(\frac{D\mathbf{A}}{Dt} \right)_F = \left(\frac{D\mathbf{A}}{Dt} \right)_R + \Omega \times \mathbf{A}. \tag{6.89}$$

We now write Equation (6.88) as

$$\left(\frac{\partial \mathbf{B}}{\partial t}\right)_R = \nabla \times (\mathbf{u}' \times \mathbf{B}) \quad \text{or} \quad \left(\frac{\partial \mathbf{B}}{\partial t}\right)_R = -\nabla \times \mathbf{E}', \qquad (6.90)$$

where we note that the electric field in the rotating frame of reference is just $\mathbf{E}' = -\mathbf{u}' \times \mathbf{B}$, given that we have assumed zero resistivity as well as neglecting other terms in the generalized Ohm's law (see Chapter 4).

We now look for steady-state solutions to Equation (6.90) (i.e., $(\partial \mathbf{B}/\partial t)_R = 0$):

$$\mathbf{0} = \nabla \times (\mathbf{u}' \times \mathbf{B}) \quad \text{or} \quad \mathbf{0} = -\nabla \times \mathbf{E}'. \qquad (6.91)$$

Because the curl of $\mathbf{E}'$ is zero, we can represent it as the gradient of a scalar electric potential Φ' (the prime denotes the rotating frame of reference): $\mathbf{E}' = -\nabla\Phi'$. The steady-state solution must obey the relation

$$\nabla\Phi' = \mathbf{u}' \times \mathbf{B}. \qquad (6.92)$$

We are not interested in solving the full MHD equations for $\mathbf{u}'$ (or $\mathbf{u}$); instead, we simply adopt Equation (6.83) and use the definition of the rotational velocity given by Equation (6.85). In this case, $\mathbf{u}'$ is completely specified. Actually, a small correction must be made to Equation (6.83). At the source surface the solar wind plasma must also corotate with the field as well as expand outward, although we can assume that at larger distances ($r \gg R_s$) this corotation is no longer enforced. We can write the solar wind velocity as

$$\mathbf{u} = u_{sw}\hat{\mathbf{r}} + \mathbf{u}_{\Omega S}, \qquad (6.93)$$

where $\mathbf{u}_{\Omega S} = \mathbf{u}_\Omega(r = R_S, \theta)$ and where we take $\mathbf{u}_\Omega(r, \theta)$ from Equation (6.85). Using the revised solar wind velocity (6.93) and Equation (6.85), we can specify the solar wind velocity vector in the rotating frame of reference, $\mathbf{u}'$:

$$\mathbf{u}'(r, \theta, \phi) = \mathbf{u}(r, \theta, \phi) - \mathbf{u}_\Omega(r, \theta) = u'_\phi \hat{\boldsymbol{\phi}} + u'_r \hat{\mathbf{r}}, \qquad (6.94)$$

with $u'_r = u_{sw}$ and $u'_\phi = \Omega \sin\theta (R_s - r)$. Figure 6.15 is a schematic showing a streamline in the rotating frame of reference.

We can see by taking the dot product of Equation (6.92) with the unit vectors in the direction of the flow and in the magnetic field direction, $\hat{\mathbf{u}}'$ and $\hat{\mathbf{b}}$, respectively, that flow streamlines in the rotating frame are equipotential lines and magnetic field lines are also equipotential lines. The magnetic field specified at the source surface, $\mathbf{B}(r = R_s, \theta, \phi) = \mathbf{B}_S(\theta, \phi)$, plus Equations (6.92)–(6.94), can be used to provide a boundary condition on the gradient of the potential at the source surface. Poisson's equation for the potential Φ' can be found by taking the divergence of Equation (6.92); unfortunately, the charge density that appears in this Poisson's equation depends on the magnetic field, which is not known (see Problem 6.9). However, the equipotential lines discussed above plus the boundary condition at

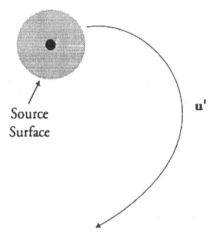

Source
Surface

$\mathbf{u}'$

Figure 6.15. Streamline of solar wind flow as seen from a frame of reference rotating with the Sun. The streamline is described by an Archimedean spiral. In the solar system inertial frame of reference, the solar wind flows outward (almost) radially.

the source surface can be used to map out the potential everywhere, after which the magnetic field can be found by using Equation (6.92). However, it is really easier in this case just to numerically solve the original magnetic induction Equation (6.84).

Let us consider a simpler situation; let us assume that the field at the source surface is radial (this is *almost* true in practice): $\mathbf{B}_S(\theta, \phi) = B_S(\theta, \phi)\hat{\mathbf{r}}$. The velocity vector $\mathbf{u}'$ is also radial at the source surface; hence $\mathbf{u}' \times \mathbf{B}_S$ is equal to zero and $\nabla \Phi' = 0$ on the source surface. The whole source surface is an equipotential surface in this case! Since streamlines are equipotentials also, the value of Φ' on the source surface maps everywhere out into space. Thus, $\nabla \Phi' = 0$ everywhere on and outside R_s. Equivalently, the electric field in the rotating reference frame is everywhere zero ($\mathbf{E}' = \mathbf{0}$). We see from Equation (6.92) that this has the consequence that $\mathbf{B}$ must be parallel to $\mathbf{u}'$ (and Equation (6.94) tells us what this velocity vector is):

$$\mathbf{0} = \mathbf{u}' \times \mathbf{B} \quad \text{or} \quad \mathbf{u}' \parallel \mathbf{B}. \tag{6.95}$$

The ratio of the components of $\mathbf{u}'$ and $\mathbf{B}$ must therefore be equal, so

$$\frac{B_\phi}{B_r} = \frac{u'_\phi}{u'_r} = \frac{\Omega \sin\theta (R_s - r)}{u_{sw}}, \tag{6.96}$$

where the components of $\mathbf{u}'$ were taken from Equation (6.94). Note that the magnetic field is the same in all reference frames for nonrelativistic situations.

The curve describing a magnetic field line can be found using Equation (6.96) and is an Archimedean spiral:

$$\frac{r \sin\theta \, d\phi}{dr} = \frac{B_\phi}{B_r} = -\frac{\Omega r \sin\theta}{u_{sw}} \quad \text{or} \quad \frac{d\phi}{dr} = -\frac{\Omega}{u_{sw}}. \tag{6.97}$$

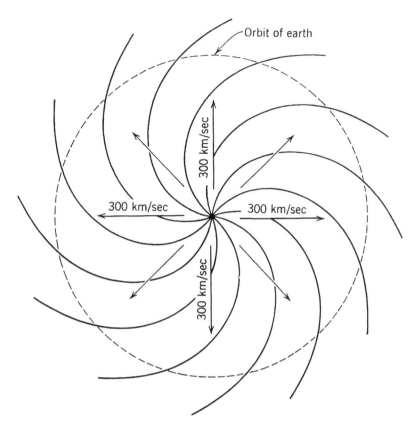

Figure 6.16. Parker spiral interplanetary magnetic field lines in the equatorial plane ($\theta = 90°$). (From Parker, 1963.) Copyright © (1963 John Wiley and Sons, Inc.). Reprinted by permission of John Wiley and Sons, Inc.

It was assumed in Equation (6.97) that $r \gg R_s$. The azimuthal angle ϕ can be found by integrating Equation (6.97) outward from the source surface to get

$$\phi(r) = \phi_s - \frac{\Omega}{u_{sw}}(r - R_s), \tag{6.98}$$

where ϕ_S is the azimuthal angle of the field line footprint on the source surface.

Figure 6.16 shows some field lines as described by Equation (6.98). Magnetic field measurements made by spacecraft traversing interplanetary space confirm that *on the average* the IMF is indeed described by an Archimedean spiral (often called the *Parker spiral* after E. Parker, who was the first to explain the average IMF structure). Physically, the spiral can be understood by picturing a parcel of plasma flowing radially outward, carrying a "piece" of field line with it. As the plasma flows outward in one direction, the end of the field line stays attached to the source surface, which rotates around with the Sun, so that the field line is no longer radially directed.

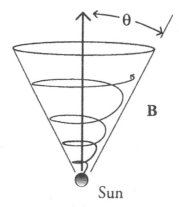

Figure 6.17. Sketch of IMF field line for higher heliospheric latitudes. The magnetic field line wraps around a cone whose surface has an angle θ with respect to the rotation axis. The magnetic field does not have a polar component.

The angle between the IMF and the radial direction χ (or *IMF spiral angle*) is given by

$$\tan \chi = \frac{(r - R_{\rm s})\Omega \sin \theta}{u_{\rm sw}}. \qquad (6.99)$$

The structure of the magnetic field off the equatorial plane (i.e., at higher heliospheric latitudes) is illustrated in Figure 6.17.

Example 6.4 (IMF at 1 AU in the equatorial plane) For solar wind speed $u_{\rm sw} = 400$ km/s, at 1 AU, and on the equator ($\theta = 90°$), the IMF spiral angle is equal to

$$\tan \chi \cong \frac{r\Omega}{u_{\rm sw}} \cong \frac{150 \times 10^6 \, {\rm km} \times 2.9 \times 10^{-6} \, {\rm s}^{-1}}{400 \, {\rm km/s}} \approx 1 \quad {\rm or} \quad \chi \approx 45°.$$

Considering this angle in another way, the solar wind moving at 400 km/s takes 100 hours (or 4 days) to travel 1 AU, in which time the Sun has rotated 4 days/27 days $\times 360° \approx 50°$, which is close to the 45° value. The solar rotation period is 27 days.

The direction of the field was given by Equation (6.96) but not its magnitude. The magnetic field can be written in terms of its radial component as

$$\mathbf{B} = B_r(r, \theta, \phi) \left(\hat{\mathbf{r}} - \frac{\Omega \sin \theta (r - R_{\rm s})}{u_{\rm sw}} \hat{\phi} \right). \qquad (6.100)$$

An expression for B_r can be found by demanding that Maxwell's equation, $\nabla \cdot \mathbf{B} = 0$, be satisfied in spherical coordinates:

$$0 = \frac{1}{r^2} \frac{\partial}{\partial r}(r^2 B_r) - \frac{\Omega(r - R_{\rm s})}{r u_{\rm sw}} \frac{\partial B_r}{\partial \phi}, \qquad (6.101)$$

where the divergence in spherical coordinates was used. In the vicinity of the source surface ($r \approx R_s$), the second term is zero and $r^2 B_r \approx constant$. To proceed beyond this, we assume that the magnetic field is uniform in magnitude and polarity over some patch or area of the source surface, as pictured in Figure 6.18. In this case, $\partial B_r / \partial \phi = 0$ over this area, as well as in the volume mapped out from this area by the curved field lines. The solution of Equation (6.101) then becomes

$$B_r(r, \theta, \phi) = B_{Sr}\left(\frac{R_S}{r}\right)^2. \tag{6.102}$$

Equations (6.96) and (6.102) specify the interplanetary magnetic field in terms of the magnetic field on the relevant region of the source surface $B_{Sr} = B_S$. The magnetic field for the volume of space associated with a given area of source surface is given by

$$\mathbf{B} = B_S\left(\frac{R_S}{r}\right)^2 \left(\hat{\mathbf{r}} - \frac{\Omega \sin\theta (r - R_s)}{u_{sw}}\hat{\phi}\right). \tag{6.103}$$

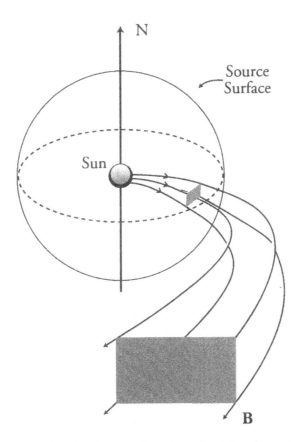

Figure 6.18. Schematic showing the mapping of a region of source surface with the same magnetic field out into the interplanetary medium.

The associated magnitude of the IMF can now be calculated from the field components:

$$|\mathbf{B}| = B_S \left(\frac{R_s}{r}\right)^2 \left(1 + \frac{\Omega^2(r - R_s)^2}{u_{sw}^2}\sin^2\theta\right)^{1/2}.\qquad(6.104)$$

In the limit of small radial distance or for high solar magnetic latitudes (i.e., near the poles, $\theta = 0°$ or $180°$), the magnetic field strength as given by Equation (6.104) decreases as $1/r^2$, whereas at very large values of r, $|\mathbf{B}|$ varies as $\sin\theta/r$.

Example 6.5 (IMF at 1 AU) Assume that near the source surface ($R_s \approx 10R_\odot$), the field strength is $B_S = 2{,}000$ nT (Table 5.2). Then Equation (6.104) indicates that in the polar region at 1 AU, the field magnitude is

$$|\mathbf{B}| = B_S \left(\frac{R_s}{r}\right)^2 = \frac{B_S}{21^2} \cong 5\,\mathrm{nT},\qquad(6.105)$$

where we used the fact that $r/R_\odot = 210$. However, the Earth's orbit is usually located near the solar equatorial plane, where $\theta \approx 90°$; Equation (6.104) then becomes

$$|\mathbf{B}| = B_S \left(\frac{R_s}{r}\right)^2 \left(1 + \frac{\Omega^2 r^2}{u_{sw}^2}\right)^{1/2} \cong 5\,\mathrm{nT}\left(1 + \frac{\Omega^2 r^2}{u_{sw}^2}\right)^{1/2} \cong 7\,\mathrm{nT},\qquad(6.106)$$

where we used the estimate of $\Omega r/u_{sw}$ from Example 6.4. Thus, we find that the interplanetary magnetic field strength near 1 AU is a few nT, which agrees with measurements made of this quantity.

In this section the source of the interplanetary magnetic field and its structure was discussed. Similarly, in Section 6.2 the average flow properties of the solar wind were explained using gas dynamics. We saw in Chapter 5 that the atmosphere of the Sun and its magnetic field possess a complex structure. The solar wind also has spatial and temporal structure on both small and large scales. The structure of the solar wind and IMF primarily relates to structure on the surface and in the atmosphere of the Sun. For example, as was discussed in this section, complex surface magnetic fields can be (approximately) mapped out to the source surface using potential theory. These fields were shown to become simpler with increasing radial distance. From the source surface the magnetic field (and any structure in this field present at that source surface) is carried out into interplanetary space by the solar wind.

6.4 Solar wind structure and composition

A mostly theoretical discussion was presented in Sections 6.1 through 6.3, but experimental data will be the main emphasis of Section 6.4. The observed solar wind exhibits a complex spatial and temporal structure on many different scales. Most of this structure, particularly on the larger scales, can be associated with structure on the Sun itself. In this section, both large-scale structure such as sector structure and solar wind streams and smaller scale (i.e., time scales less than a few hours or length scales much less than an AU) structure such as interplanetary shocks and waves will be discussed, although somewhat more emphasis will be given to the former topic.

We start in Section 6.4.1 with a short description of solar wind composition and we continue in Section 6.4.2 with the topic of the average radial structure of the solar wind (i.e., the dependence of solar wind parameters on heliospheric distance and on latitude). In Section 6.4.3 we will consider other large-scale structure (particularly sector structure and stream structure), and in Section 6.4.4, small-scale structure will be discussed. In Section 6.4.5 a somewhat tangential topic, but an important one, will be considered: solar wind electrons. For a more detailed treatment of any of these subjects, you should consult one or more of the references listed at the end of this chapter. Especially valuable, if you can find it in your library, is the proceedings of the first major conference devoted to the solar wind (known as *Solar Wind I*), which took place in 1972. Five other such conferences (with published proceedings) have taken place since then, and many theoretical and experimental advances have taken place since 1972, but for an introduction to the subject, and for large-scale aspects of the solar wind, the first volume is still the best. Other useful references on the solar wind are also listed at the end of this chapter.

6.4.1 Observed composition of the solar wind

The gas in the solar corona is collisional. For example, according to Table 5.2, the electron–electron collisional mean free path, λ_{coll}, is much less than a relevant macroscopic length scale (i.e., for the corona this is the solar radius). Near the base of the corona at $r = 1.01 R_\odot$, $\lambda_{coll} \approx 10\,\text{km}$, or $10^{-4} R_\odot$. And at $r \approx 3 R_\odot$, $\lambda_{coll} \approx 0.2 R_\odot$, which is still significantly smaller than $R_\odot$. The ionization state, or composition, of the corona is thus determined by collisional processes, and because the coronal temperature is about one million degrees, the coronal gas is fully ionized, with heavier species existing in highly ionized charge states such as O^{+6}, Si^{+10}, and Fe^{+10}. This composition is evidenced by the EUV and soft X-ray line emission from the corona, as discussed in Chapter 5. As the solar wind flows outward from the Sun, the density rapidly drops as r^{-2} and the collisional mean free path becomes much larger than it was in the corona. Taking typical solar wind parameters for 1 AU, we can estimate that the proton–proton collisional mean free path is roughly 5–10 AU (see Problem 6.13), which is much greater than 1 AU, indicating that the solar wind plasma well outside the corona has become collisionless. However, the type of collision revelant to composition is electron impact ionization of ion species. The mean free path for ionizing collisions is less than or of the order of the radial distance only within a few solar radii of the Sun. At 1 AU, this mean free path is about 10^3 AU (Problem 6.13). In the corona the ionizations are balanced by electron–ion recombination, another collisional process. Hence the composition of the solar wind is determined primarily by the composition of the solar corona, and as the gas flows out from the Sun, the charge state is "frozen in" at some distance not too far from the Sun, still in the outer corona. Hence, the solar wind plasma is fully ionized and heavy species are highly ionized.

Flux or density measurements of both light and heavy ion species have been made in the solar wind by plasma instruments on many spacecraft. For example, the measurements reported on by Bame (1972) are typical. Bame reports that the average He to H density ratio is

$$\langle n_{He}/n_H \rangle = 0.045, \tag{6.107}$$

although the H to He ratio has been measured to be as low as 0.0025 and as high as 0.25 (i.e., in solar wind associated with solar flares). Note that the hydrogen is entirely in the form of protons and the helium in the form of alpha particles. Summing over all charge states, the average oxygen to helium ratio is

$$\langle n_O/n_{He} \rangle = 0.01, \tag{6.108}$$

and the ratios of silicon to oxygen and iron to oxygen are

$$\langle n_{Si}/n_O \rangle = 0.21 \quad \text{and} \quad \langle n_{Fe}/n_O \rangle = 0.17. \tag{6.109}$$

Figure 6.19, taken from the Bame paper, shows measured counting rates as a function of E/Q (energy over charge). The E/Q of protons is just the proton energy (in eV or volts). The peak E/Q for protons is at 450 V and is associated with the bulk flow of the solar wind. Note that an energy of 450 eV corresponds to a solar wind speed of 300 km/s. The width of the proton peak reflects the random thermal velocity of the protons. The peak He flux is located at an E/Q twice that of protons, as one would expect for alpha particles. Also notice the presence of high-charge states of oxygen, silicon, and iron, reflecting the thermal conditions deep in the solar corona where the composition was "fixed."

6.4.2 Observed radial and latitudinal structure of the solar wind

Average values of solar wind density, velocity, and temperatures for 1 AU were listed in Table 5.2 and Figure 6.7. "Typical" values of the solar wind density and speed at 1 AU are about $5\,\text{cm}^{-3}$ and 400 km/s, respectively. A typical magnitude of the interplanetary magnetic field is 5 nT. Figures 6.20 and 6.21 show some solar wind density and speed measurements from the *VELA 3* and *HEOS 1* spacecraft for 1 AU. The average density from the sample reproduced in Figure 6.20 is $7.7\,\text{cm}^{-3}$, but notice that the density can be anywhere from $2\,\text{cm}^{-3}$ up to more than $20\,\text{cm}^{-3}$.

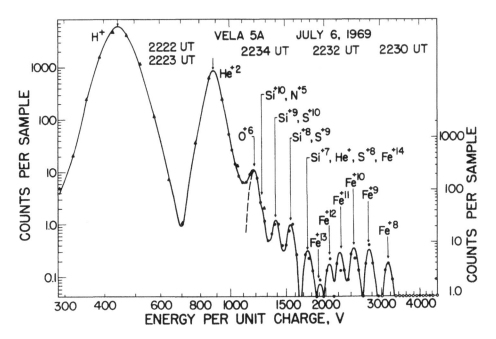

Figure 6.19. Measured composition of the solar wind. Data are from the solar wind analyzer and the heavy ion analyzer on the *VELA 5A* spacecraft. (From Figure 16 of Bame, 1972.)

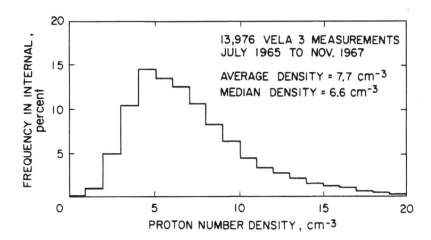

Figure 6.20. Solar wind densities measured by the *VELA 3* spacecraft at 1 AU. (From Wolfe, 1972.)

Similarly, the solar wind speed can vary greatly – from 200 km/s up to 700 km/s for the data shown in Figure 6.21. Both low- and high-speed "streams" exist in the solar wind.

Now let us consider how solar wind parameters vary with radial distance from the Sun. First, let us review what our theoretical expectations are, from earlier in this chapter. We expect that the density should vary inversely as r^2 (Equation 6.61) and

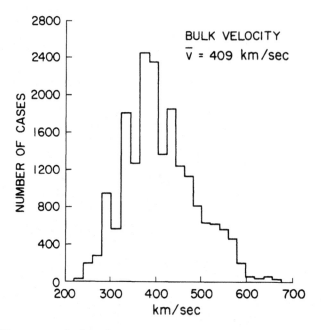

Figure 6.21. Histogram of solar wind speeds measured by the *HEOS 1* spacecraft at 1 AU. (From Wolfe, 1972.)

that the solar wind speed should be roughly independent of r, for distances from the Sun considerably beyond the critical point. Figure 6.22 shows proton number densities measured by the *Voyager 1* and *2* spacecraft (Belcher et al., 1993). The best fit has a radial dependence of $r^{-2.1}$, whereas for a constant solar wind speed one expects the density to vary as r^{-2}, but this is within the uncertainty of the data. Data compiled from the *Pioneer 10* and *11* missions by Smith and Barnes (and not reproduced here) indicate that u_{sw} is essentially constant between 1 and 20 AU, although many temporal variations were apparent in the data.

Now let us consider the solar wind temperature. The theoretical expectations for the dependence of the solar wind temperature on radial distance are not simple or clear. Equation (6.60) predicted that C_S^2 (proportional to the temperature) for a simple one-fluid plasma varies as

$$T(r) \approx C/r^{2(\gamma-1)}, \tag{6.110}$$

where γ is the polytropic index used in the Bernoulli's energy relation (6.12) and C is some constant. But as we have discussed earlier, what should we use for γ? The ideal gas value is 5/3, but this gives unreasonable values for the solar wind outflow speed and for the critical point of the flow (i.e., the transonic point). We earlier found that the critical radius and solar wind speed become reasonable for a value $\gamma \approx 1.15$. The justification we used for this was that the plasma is really composed of two fluids (electrons and ions) and that heat conduction can supply heat to the electron gas making the "effective" γ much less than the ideal gas

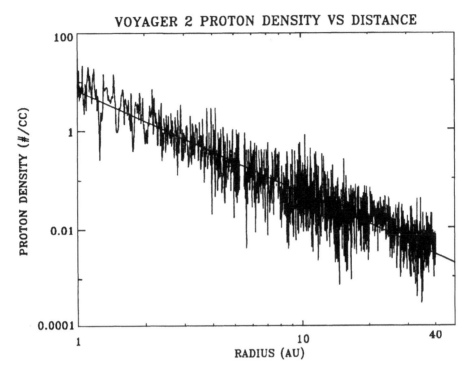

Figure 6.22. Solar wind proton number density versus radial distance from *Voyager 2* measurements. The solid line is a least-squared fit to the data and is proportional to $r^{-2.1 \pm 0.3}$. (From Belcher et al., 1993.)

value. Indeed, measured electron temperatures in the solar wind at 1 AU ($T_e \approx (1 - 3) \times 10^5$ K) are typically several times greater than measured proton temperatures ($T_p \approx (0.3\text{–}1) \times 10^5$ K), indicating that the two fluids are not entirely coupled thermally.

The variation of temperature with r predicted by Equation (6.110) using $\gamma \approx 1.15$ is $T \approx C/r^{0.3}$. Note that the solution of the heat conduction equation given by Equation (6.11) also happens to give a radial variation very close to this: $T \approx C/r^{2/7} = C/T^{0.286}$. Using $1/r^{0.3}$ to extrapolate a coronal temperature value of $T_c = 1.5 \times 10^6$ K at $r = 2R_\odot$ out to 1 AU, we find that $T \approx 0.28 \, T_c = 3 \times 10^5$ K, which is closer to the measured solar wind electron temperature (T_e) than to the solar wind proton temperature (T_p). Figure 6.23 shows proton temperatures measured by *Pioneers 10* and *11* as a function of distance. The temperature varies as $1/r^{0.57}$ – a somewhat faster rate of decrease with r than was predicted by Equation (6.110).

Next, we consider the measured radial variation of the interplanetary magnetic field. The magnetic field strength measured by the magnetometers on board the *Pioneer 10* and *11* spacecraft (Smith and Barnes, 1983) is shown in Figure 6.24 for radial distances between 1 and 11 AU. On the average, the measured field strength

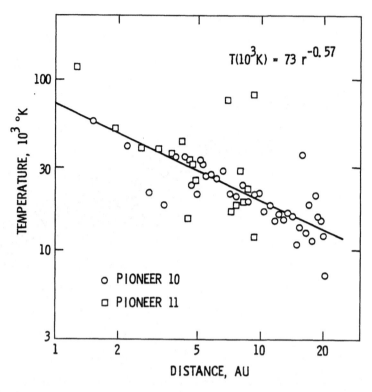

Figure 6.23. Solar wind proton temperature versus radial distance from *Pioneer 10* and *11* measurements. (From Smith and Barnes, 1983.)

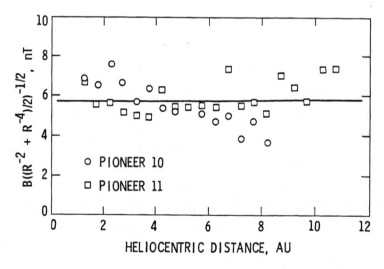

Figure 6.24. Magnetic field strength measured by *Pioneers 10* and *11* versus radial distance. The field strength has been divided by $[(R^{-2} + R^{-4})/2]^{1/2}$, where R is the heliocentric distance, r, in units of AU. (From Smith and Barnes, 1983.)

can be represented as a Parker spiral. The radial variation of the field strength for the ecliptic plane ($\theta = 90°$) and for $\Omega/u_{sw} \approx 1$ (AU)$^{-1}$ can be shown using Equation (4.103) to vary as $1/r$ at large heliocentric distances. The average IMF field strength is approximately 5.5 nT at 1 AU and decreases to 0.4 nT by $r = 10$ AU. Smith and Barnes also showed that the azimuthal component of the IMF varies, on average, as r^{-1}, in conformance with our theoretical expectations.

Next, let us consider the latitudinal variation of the solar wind. Until very recently all *in situ* measurements of the solar wind have been obtained in the vicinity of the ecliptic plane (i.e., mostly for low heliographic magnetic latitudes, depending on the orientation of the heliospheric current sheet) because of the difficulty of achieving trajectories out of this plane. The *Ulysses* spacecraft (launched by the European Space Agency on October 6, 1990) used a gravitational assist from Jupiter (encountered on February 6, 1992) to propel itself out of the ecliptic plane (Smith et al., 1995). From 1992 to 1995 this spacecraft traveled from low to high solar latitudes while moving from about 5 AU to 2 AU , it then went back to low latitude again by spring 1995. Figure 6.25 displays solar wind velocity data from this mission (Phillips et al., 1995). The typical solar wind speed increases from about 450 km/s at lower latitudes up to about 750 km/s at polar latitudes. An alternation with a period of a solar rotation period between low and high solar wind speed regions (called *streams*) is also evident in Figure 6.25 for lower latitudes but not at latitudes

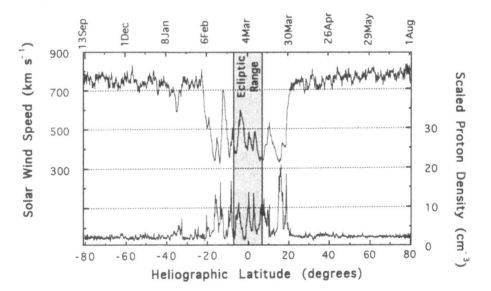

Figure 6.25. Solar wind data from the Los Alamos National Laboratory solar wind plansa experiment on board the NASA/ESA *Ulysses* spacecraft. Observations were assembled with assistance from the NASA Jet Propulsion Laboratory. The top curve is solar wind speed, the bottom curve is the proton density scaled to a 1 AU distance. The date of the observation is indicated at the top, and the solar latitude of the spacecraft is indicated at the bottom. (From Phillips et al., 1995.)

greater than about 45°. A few coronal mass ejection events are also present at low latitudes.

A few remote measurements of some solar wind properties at high latitudes have also been made closer to the Sun by using radio waves scattered from fluctuations in the interplanetary medium electron density. This is the method of *interplanetary scintillations* and it gives information on a line of sight region that is approximately 0.4–1 AU. For example, Coles and Rickett (1976) used this technique to show that the solar wind is slowest in the equatorial plane ($u_{sw} \approx 420$ km/s) and fastest in the polar regions ($u_{sw} \approx 560$ km/s). Recall from Section 6.3.2 that the polar regions tend to be associated with open field lines. We also found out in Chapter 5 that the polar regions are generally darker in the X-ray part of the spectrum and are called *polar holes*. These polar holes are apparent in the *Yohkoh* image of Figure 1.2. Other X-ray dark regions, called *coronal holes*, sometimes appear at lower latitudes on the Sun and are also associated with open field lines. We will see in the next section that these coronal hole regions are associated with faster than average solar wind speeds.

6.4.3 Sector and stream structure – The heliospheric current sheet

Suppose that the magnetic field on the source surface has regions of opposite polarity, due to the magnetic field structure on the photosphere; then as the field is carried away from the Sun by the solar wind this "sector structure" is also carried out into interplanetary space (see Figures 6.12 and 6.26). Structure in the IMF should derive from, in some modified fashion, the source surface structure, which in turn derives from the photospheric field structure. In addition to these variations in the IMF, the solar wind velocity and density also exhibit large-scale variations.

Let us first consider *sector structure* of the IMF. A schematic showing the solar wind in the equatorial plane with four sectors is given in Figure 6.26. The number of sectors in the solar wind varies from about two to four, typically. A given pattern (on the largest scales only) generally persists for several solar rotation periods, only slowly evolving; hence, as the Sun rotates with a 27-day period, the IMF has a 27-day time periodicity for a fixed observer (such as a spacecraft circling Earth). Looking at Figure 6.26, we see that an observer would see the polarity switching every 7 days. Sector structure is the largest scale IMF structure; many structures with smaller scales are also present.

A few pages ago the source field was described in terms of the photospheric field by means of potential theory. We saw that the magnetic field structure becomes "simpler" with increasing radial distance; in some sense the source surface field is an averaged surface field. Severny et al. (1970) compared measured IMF data for four solar rotation periods with simple "mean" disk surface fields (see Figure 6.27). That is, the complicated field measured over the solar disk at a given time is simply averaged, although before a comparison is made with the measured IMF at 1 AU a

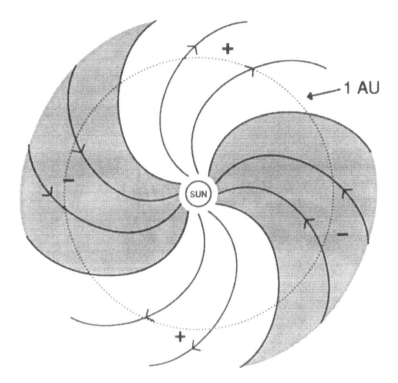

Figure 6.26. Schematic of sector structure of the solar wind. Four sectors are shown.

4.5-day lag was introduced to account for the solar wind transit time. Several points can be made about this figure:

(1) The gross measured solar wind sector structure is matched by the mean solar surface field structure. This means that there is a strong relation between the solar field and the IMF (not surprisingly) and also that the source field is an "average," in some sense, of the surface field.
(2) Several sectors are evident in the solar wind and in the mean solar field. In this particular case, there are four sectors – two with positive (+) polarity and two with negative (−) polarity.
(3) The IMF sector structure is more or less repeated in time with a 27-day period. As just discussed, this is due to solar rotation. The large-scale field does appear to slowly evolve on an even longer time scale.
(4) A good deal of smaller-scale structure (1–2 days) in the IMF also is apparent but is not predicted by the simple mean field model. This smaller-scale structure does *not* persist from solar rotation to solar rotation. Although not apparent on this figure there are IMF structures with periods as small as 2–3 hours, but these are not associated (one to one) with source surface features and are instead thought to be MHD waves (or MHD turbulence).

Consider now structure that has been observed in the solar wind density and speed. Solar wind speed and densities measured by instruments on board the *Mariner 2*

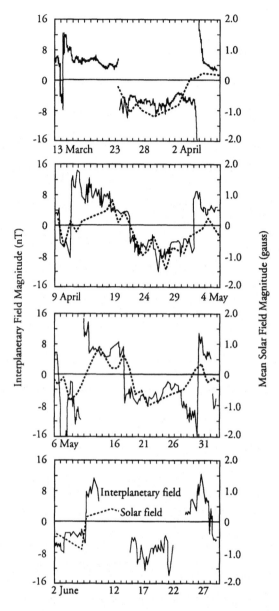

Figure 6.27. Observed "mean" solar field values (dashed lines) and 3-hour averages of measured interplanetary field at 1 AU are shown. The solar observations are "lagged" by 4.5 days to allow for the average time it takes the solar wind to traverse 1 AU. Each panel covers a solar rotation period. Adapted and redrawn from Severny et al. (1970). Also see Schatten (1972) in *Solar Wind 1*.

spacecraft as a function of time are shown in Figure 6.28. You should note the existence of regions of low ($u_{sw} \approx 350\,\mathrm{km/s}$) and high ($u_{sw} \approx 700$ km/s) speed regions, or streams, within each rotation period. These are called *low-speed streams* and *high-speed streams*. Also note the strong 27-day periodicity: The same streams reappear every 27 days, albeit with some modification. This is also evident in

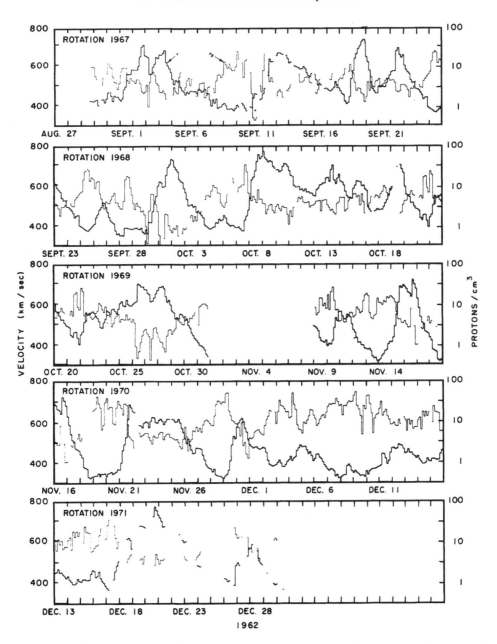

Figure 6.28. *Mariner 2* solar wind data are shown. Three-hour averages of solar wind speed (darker curve) and proton density (lighter curve) are shown for September through December 1962. Each panel covers a period of 27 days. (From Neugebauer and Synder, 1966.)

Figure 6.25 for low latitudes. This periodicity simply reflects the fact that the Sun rotates with this period, as does the source surface, such that if one region of the Sun produces high-speed streams, for example, it rotates around to face the Earth every 27 days. The proton temperature (not shown here) varies from about $T_p \approx 10^4$ K to 2×10^5 K, with the larger values being associated with the *leading edges* of high-speed streams. The solar wind density (Figure 6.28) and the IMF field strength

(Figure 6.27) are also larger on the leading edges of high-speed streams. The IMF polarity almost always changes in the region *between* high-speed streams (i.e., in the low-speed streams) rather than in the high-speed streams themselves. In other words, high-speed streams are almost always located entirely in one sector.

An image of the coronal X-ray brightness was reproduced in Figure 1.2. Regions of low X-ray intensity are called *coronal holes*. Large coronal holes are almost always present at the poles and are called *polar holes*. In other words, the equatorial region is generally brighter in the X-ray part of the spectrum than in the polar regions. The brightest X-ray regions are associated with active regions in the lower corona and the chromosphere, and these are associated with sunspot groups on the photosphere. These regions of especially hot coronal plasma are associated with *closed* magnetic field lines. Coronal and polar holes are associated with *open* magnetic field lines. Open field line regions are thought to be the source of the solar wind. The centers of the coronal holes are thought to be the source of high-speed streams, and the slower solar wind is thought to originate near the edges of coronal holes. Hence, in general, the solar wind speed is larger at high latitudes (associated with the polar holes) and smaller near the equator. This agrees with the observed solar wind latitudinal variation (Figure 6.25).

Let us consider a very simple two-region case similar to that shown in Figure 6.11. We can be somewhat more realistic by assuming a dipole field close to the Sun, but with a current sheet (like that in Figure 6.11) for radial distances near the source surface and beyond (see the helmet streamer in Figure 6.12). Furthermore, the solar magnetic pole is usually not aligned with the rotation axis; that is, *the heliospheric current sheet* is tilted with respect to the ecliptic plane, as pictured in Figure 6.29. The coronal brightness associated with such a tilted dipole plus current sheet field pattern is shown in Figure 6.30. The magnetic neutral line at large distances coincides with the magnetic equator, and it appears as a sinusoidal curve when plotted versus solar latitude and longitude. The coronal bright regions are associated with the closed (quasi-dipole) field lines near the magnetic equator, and polar holes, with opposite magnetic polarity, are present at high latitudes.

A more realistic (in fact observed) brightness pattern is shown in Figure 6.31. A tilted dipole again provides a good description but smaller-scale complexities are also apparent. The bright regions are located near the equator in a sinusoidal pattern. The poles are darker in X-ray emission and are associated with opposite magnetic polarity. The solar wind and IMF observed at Earth derive from conditions near the equatorial region (0° solar latitude; however, this can correspond to different solar magnetic latitudes). Hence, following along the 0° line in Figure 6.31 we find only two sectors – a negative one near 90° longitude and a positive one near 270° longitude, with sector boundaries near 0 and 180° longitude. We expect to see high-speed solar wind streams near the coronal dark regions and slower solar wind in between. Figure 6.31 also shows measured solar wind speed for the same time as the solar observations. Indeed, there are only two sectors and the high-speed streams are located in the middle of each sector.

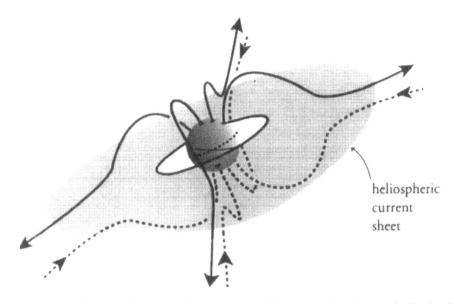

Figure 6.29. Schematic of simple solar magnetic field pattern – tilted magnetic dipole plus current sheet. The heliospheric current sheet is shown.

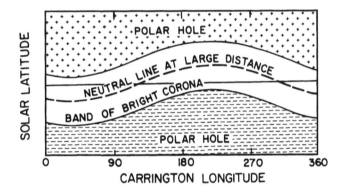

Figure 6.30. Coronal X-ray brightness pattern plotted versus solar latitude and longitude for the tilted magnetic dipole plus current sheet. The magnetic equator, or neutral sheet, appears as a sinusoidal curve in these coordinates. (From Hundhausen, 1977.)

6.4.4 Small-scale solar wind structure

Now we briefly consider solar wind structure for length scales smaller than sectors and streams. This smaller-scale structure was apparent in some of the observations already reviewed and includes structure associated with sector boundaries and stream–stream boundaries, interplanetary shocks, and MHD waves and turbulence.

Interplanetary shocks are small scale in the sense that their passage past a spacecraft only requires an hour or so, but they are associated with larger-scale phenomena in the interplanetary medium such as stream structure. The two known causes of interplanetary shocks are: (1) *stream–stream interactions* and (2) *coronal mass ejections* or *coronal transients*. We now consider each of these mechanisms in more detail.

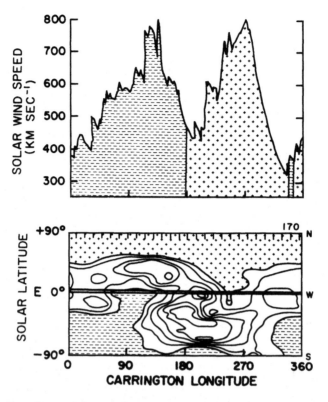

Figure 6.31. (Top) Measured solar wind speed near 1 AU for solar rotation 1616. (From Hundhausen, 1977.) (Bottom) Measured coronal X-ray brightness contours versus solar latitude and longitude for same period. Magnetic polarity is also indicated in the polar regions.

6.4.4.1 Stream–stream interactions

The plasma in a fast solar wind stream naturally tends to overtake slower streams. The interface between the two regions starts off being smooth, but as the interaction region moves away from the Sun the density gradients at the interface steepen (as discussed in Section 4.9) and a shock forms. This typically happens for radial distances greater than 1 AU or so. In fact, two separate shocks form in the interaction region: a forward shock and a reverse shock (see Figure 6.32). The forward shock propagates outward with respect to the ambient solar wind ahead of it (i.e., the *slow-speed stream*), and the reverse shock propagates inward with respect to the ambient solar wind ahead of it, which is at smaller r (i.e., the *high-speed stream*). The density in the interaction region between the two shocks is enhanced, or compressed. An ideal interaction region would appear to a distant observer to corotate with the Sun; hence, in a rotating frame of reference the interaction region appears to be stationary. Thus, the name *corotating interaction region (CIR)* is also given to this phenomenon. At very large heliocentric distances, in the outer solar system, CIRs appear almost as concentric circles centered at the Sun, owing to the large IMF spiral angle.

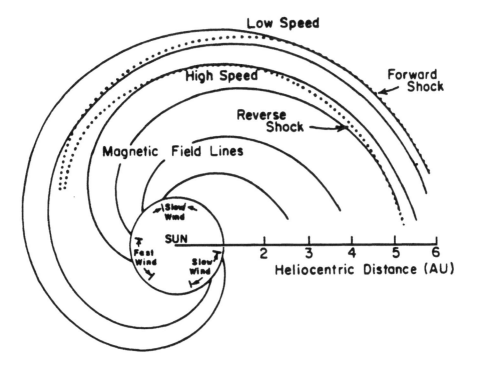

Figure 6.32. Schematic of a stream–stream interaction region (or CIR) with both forward and backward interplanetary shocks. The region between the two shocks has enhanced density. (From Fisk, 1982.) Copyright 1983 Kluwer Academic Publishers. With kind permission from Kluwer Academic Publishers.

Let us consider the passage of a typical interplanetary shock (e.g., a forward shock associated with a CIR). The preshock solar wind speed might be 390 km/s (the slow stream), and the postshock solar wind speed might be 470 km/s. The shock is moving at 500 km/s in the Sun's reference frame and is moving at a speed of 110 km/s with respect to the ambient medium it is propagating into. The magnetosonic speed (or MHD fast-mode wave speed) of the ambient medium is $C_{MS} \approx 50$ km/s, so that the magnetosonic Mach number is $M_{MS} \approx 2.2$. This Mach number is typical for an interplanetary shock. The gas flow speed relative to the shock is supersonic (or supermagnetosonic), as it must be for a shock to exist. As we will see in Chapter 7, the bow shocks in the solar wind upstream of the planets typically have Mach numbers in excess of 5 and are thus much stronger than interplanetary shocks. Nonetheless, particle acceleration can still take place at interplanetary shocks because of the large horizontal scale of the shocks and the fact that more than one shock at a time is present in the solar wind.

6.4.4.2 Coronal mass ejections

Sometimes large prominences or helmet streamers on the Sun become unstable and energy is converted into the bulk acceleration of a large chunk of coronal gas – that

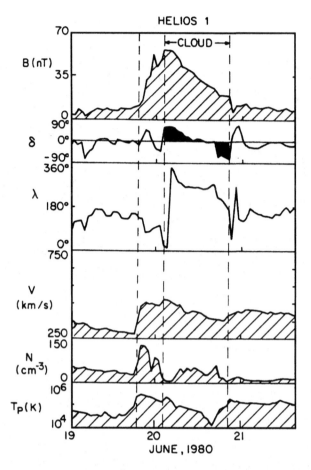

Figure 6.33. Data from *HELIOS 1* spacecraft for passage of a magnetic cloud and the interplanetary shock preceding it. The shock is located between the dashed lines near the 20 June mark. (Top panel) Magnetic field strength. (Middle panels) δ and λ: latitude and longitude of magnetic field in solar-ecliptic coordinates (see text). (Lower panels) The measured solar wind speed V, density N, and proton temperature T_p. (From Burlaga et al., 1982.)

is, a coronal mass ejection (CME). Also, it is now thought that CMEs usually occur without a solar flare, although the formation mechanism is not understood. Particle acceleration can also take place in the shock front ahead of the CME. Some of the CME-related interplanetary disturbance is caused by the pile-up of solar wind plasma and IMF ahead of the CME (i.e., in the CME *sheath*).

The interplanetary shock preceding the CME is evident in the data shown in Figure 6.33. The interplanetary shock in Figure 6.33 was actually observed in the spacecraft frame of reference, which is not too different from the Sun's frame of reference. The shock analysis performed in Chapter 4 (Section 4.9) was for the frame of reference in which the shock front is stationary. In the spacecraft frame the upstream (unshocked gas) flow speed is $u_1 = 270$ km/s (away from the Sun) and the downstream speed is $u_2 = 460$ km/s. The upstream plasma density (unshocked) is $n_1 \cong 30$ cm^{-3} and the downstream (shocked) value is $n_2 \cong 90$ cm^{-3}, giving a shock jump of $Z_s = n_2/n_1 \cong 3$. We can transform to the shock frame by requiring that the upstream and downstream flow speeds (u_1' and u_2', respectively) in this reference frame be such as to give the same shock jump condition as the densities require, $u_1'/u_2' = Z_s \cong 3$. The shock front speed that accomplishes this is $u_{\text{shock}} = 555$ km/s, in which case we have $u_1' = -285$ km/s and $u_2' = -95$ km/s (see Problem 6.12). The Rankine–Hugoniot relations relate the shock jump to the incident, or upstream, Mach number M_1. In this case, you can show (Problem 6.12) that $M_1 \cong 3$ for $\gamma = 5/3$. We should identify this as the magnetosonic Mach number since we are dealing with a magnetized plasma.

The magnetic structure in a CME is often complicated and is sometimes even helical, or force free, with $\mathbf{J} \times \mathbf{B} \approx \mathbf{0}$, as shown in the schematic (Figure 6.34)

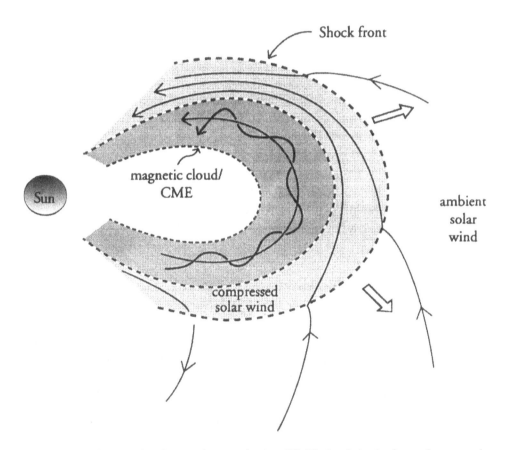

Figure 6.34. Schematic of coronal mass ejection (CME) that is in the form of a magnetic cloud. The size depicted here is about 1 AU. The coronal gas acts as a "piston" pushing out into the slower ambient solar wind. A shock wave moves out ahead of the disturbance. The arrows indicate the flow velocity and the thin solid lines represent interplanetary magnetic field lines. The magnetic structure is force free in this case (see Burlaga et al., 1991).

(cf. Burlaga et al., 1991). CMEs are now thought to be the prime cause of large *geomagnetic storms*. Particularly important characteristics of CMEs in this respect are the enhanced solar wind dynamic pressure associated with CMEs and also the change in IMF direction. The north–south IMF component (that is, the B_z component in *solar-ecliptic coordinates*) is most relevant. A change of B_z from northward to southward has a dramatic effect on the magnetosphere, as we will see in Chapter 8.

Note that in the solar-ecliptic coordinate system, the x axis points radially away from the Sun in the ecliptic plane, the z axis is normal to the ecliptic plane, and the y-axis completes the coordinate system and is also in the ecliptic plane.

6.4.4.3 Solar wind turbulence

Now we consider *solar wind turbulence*. Even the "small-scale" structures we have so far discussed involve "averages" over a few hours or so or have large-scale organization. For example, interplanetary shocks tend to be narrow but are organized on a large scale. The solar wind also contains structure, evident in spacecraft measurements, with time periods of the order of hours or less. Figure 6.35 is an example of solar wind density and magnetic field data that illustrates the existence of structure with spacecraft time scales of hours down to minutes. These structures are either quasi-static structures (such as tangential discontinuities) being convected out with the solar wind or they are MHD waves (cf. Barnes, 1983).

The MHD waves have been shown to be either (1) Alfvén waves (transverse magnetic perturbations with no associated compression in B or plasma density) propagating through the ambient solar wind at the Alfvén speed C_A or (2) magnetosonic (i.e., fast-mode) waves propagating at the magnetosonic speed $C_{MS} = [C_S^2 + C_A^2]^{1/2}$. It is evident from the small variation in the magnetic field strength as well as from the correlation between the plasma velocity and magnetic field directions that most of the waves in the solar wind (e.g., Figure 6.35) are Alfvén waves. Review the Section in Chapter 4 on MHD waves, and recall that $C_A \approx C_{MS} \approx 40$ km/s in the solar wind, which is much less than the typical solar wind speed $u_{sw} \approx 400$ km/s. Consequently, these waves are almost stationary in the solar wind frame and are convected past a spacecraft (or the Earth) at approximately the solar wind speed. Thus, a spatial length scale (or wavelength λ) is manifested in the measurements (such as those shown in Figure 6.35) as a time variation with a period $T \approx \lambda/u_{sw}$. Most of the waves are known to be Alfvén waves propagating away from the Sun in the solar wind frame, thus suggesting an origin in the solar corona (Goldstein et al., 1995). We can relate the frequency f of an Alfvén wave, as observed by a spacecraft, to the wavenumber of the wave as

$$2\pi f \approx k(u_{sw} \pm C_A) \approx ku_{sw}, \tag{6.111}$$

where we have neglected factors of order unity associated with the IMF and solar wind directions.

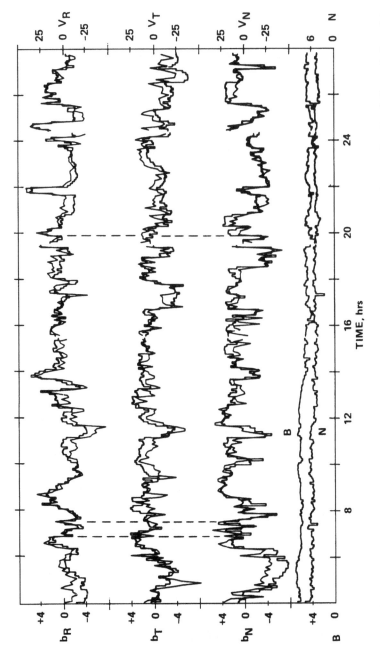

Figure 6.35. Magnetic field measurements showing "turbulent" structure on many time scales. Dark lines (left scale) trace three components of the magnetic field relative to the mean field in units of nT (R, T, and N refer to radial (x), tangential (y), and normal (z) directions in solar-ecliptic coordinates). The light lines (right scale) are the corresponding components of the measured plasma velocity relative to the mean flow (units of km/s). (From Belcher and Davis, 1971.)

As evident from the measurements, the waves are not monochromatic but are present at many different frequencies. We are really talking about MHD turbulence rather than coherent waves. Any "signal" that is a function of some variable such as time, t, can be Fourier decomposed. This was also discussed in Chapter 3 with respect to pitch-angle diffusion of particles by waves. Let us consider the magnetic field $\mathbf{B}(t)$ such as might be measured by a magnetometer on board a spacecraft in interplanetary space. First, we separate out the fluctuating (or wave) part of the field by subtracting the background IMF ($\mathbf{B}_0$):

$$\mathbf{B}_1(t) = \mathbf{B}(t) - \mathbf{B}_0(t). \tag{6.112}$$

The *correlation function* for the i and j components, $R_{ij}(\tau)$, can be computed by taking the following average over time:

$$R_{ij}(\tau) = \langle B_{1i}(t)B_{1j}(t+\tau)\rangle. \tag{6.113}$$

τ is the time lag.

Note that $R_{ii}(0) = \langle B_{1i}^2 \rangle$ is just the variance of the ith component of the field. The *power spectrum* (or *power spectral density*) is given by the Fourier transform of $R_{ij}(\tau)$:

$$P_{ij}(f) = \int_{-\infty}^{\infty} R_{ij}(\tau)e^{+i2\pi f\tau}\,d\tau. \tag{6.114}$$

Figure 6.36 displays the power spectrum computed for the north–south component of the IMF (i.e., $P_{zz}(f)$ versus f) as measured by *Mariner 4's* magnetometer. The units of power are $[\text{nT}]^2/[\text{s}]$. Note that the integral of $P_{zz}(f)$ over f is just the variance of the B_z component of the field – that is, this integral is equal to B_{rms}^2, where "rms" denotes root mean square. For the case of Figure 6.36, $B_{\text{rms}}^2 = 4.3\,\text{nT}^2$; that is, $B_{\text{rms}} = 2.1\,\text{nT}$ and $B_{\text{rms}}/B_0 \approx 0.4$, given that $B_0 = 5\,\text{nT}$. The total electromagnetic energy density of the turbulent waves is given by $B_{\text{rms}}^2/2\mu_0$. A break in the spectrum is evident at the *correlation wavenumber*,

$$k_c = 6 \times 10^{-12}\,\text{cm}^{-1}. \tag{6.115}$$

The *correlation length* L_c is given by $L_c = 1/k_c$. In this case, $L_c \approx 2 \times 10^{11}\,\text{cm} = .01\,\text{AU}$. For $k < k_c$, the spectrum is flat, and for $k \gg k_c$ the power obeys a power law, $P \approx k^{-5/3}$, which is a typical k-dependence for any turbulent phenomenon.

L_c is the characteristic length of the magnetic fluctuations. The magnetic field values at two locations separated by a distance much greater than L_c are uncorrelated, whereas the field values at two locations separated by a distance considerably less than L_c are correlated.

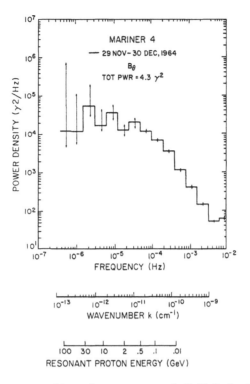

Figure 6.36. Power spectrum of interplanetary magnetic field B_z (basically the same as B_θ) computed from field measurements made by the *Mariner 4* spacecraft. The top abscissa is frequency. The wavenumber abscissa is discussed in the text, and the resonant proton energy is discussed in Section 6.6. Note that 1γ is equal to 1 nT. (From Jokipii and Coleman, 1968.) Also see Jokipii (1971).

6.4.5 Solar wind electrons

The electron and ion gases in the solar wind are not thermally coupled; that is, the solar wind is not really a single-fluid plasma, although for simplicity we treated it as such for much of this chapter. For example, as discussed earlier, typical temperatures in the solar wind are $T_e \approx 2 \times 10^5$ K for the electrons and $T_p \approx 5 \times 10^4$ K for the protons. In fact, the electron distribution function is not even fully Maxwellian. The measured solar wind electron distribution function shown in Figure 6.37 is typical. The electron gas can be divided into two populations: (1) a *core* Maxwellian with temperature $T_c \approx 10^5$ K (or $T_e \approx 10$ eV) and density $n_c \approx 3.1 \text{cm}^{-3}$ and (2) a hotter *halo* population (approximately Maxwellian) with temperature $T_h \approx 10^6$ K and density $n_h \approx 0.11 n_c$. The "break" between the core and halo distribution is located at an energy of about 70 eV. The distribution function is typically somewhat larger for electrons moving away from the Sun than for those electrons moving toward the Sun; this cannot be seen in the angle-averaged distribution shown in Figure 6.37. This anisotropy results in a net flux of heat, **Q**, away from the Sun. The net heat

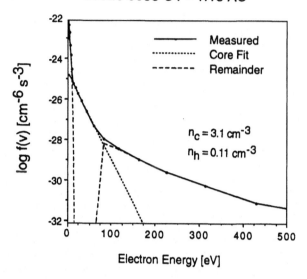

90323 0955 UT - 1.16 AU

Figure 6.37. Electron distribution function as a function of electron energy measured in the solar wind at 1.16 AU by a Los Alamos plasma electron instrument aboard the *Ulysses* spacecraft. The measured distribution function is shown as the solid line and has been averaged over all directions. The dotted line shows a Maxwellian fit for the core electrons. The dashed line labeled "Remainder" includes a hot halo population at high energies and spacecraft photoelectrons at very low energies. Notice that it takes approximately two Maxwellian populations to account for the data (excluding the non–solar wind low-energy spacecraft electrons) – a colder Maxwellian for the core distribution and a hotter Maxwellian for the halo. (From McComas et al., 1992.)

flux near 1 AU is typically about $Q \approx 8 \times 10^{-6}$ J/m^2/s, and 95–99% of this flux is carried by the halo population; the core electron distribution is almost isotropic. The core electrons have a thermal speed of about 2,000 km/s. This greatly exceeds the bulk flow speed of the solar wind, $u_{sw} \approx 400$ km/s.

The electron distribution function for each population is really a bi-Maxwellian; that is, different temperatures apply to the parallel and perpendicular directions with respect to the IMF (i.e., $T_\parallel$ and $T_\perp$, respectively). Again, the core electrons are almost isotropic with $T_\parallel / T_\perp \approx 1.04$, whereas for the halo population $T_\parallel / T_\perp \approx 1.2$ (Feldman et al., 1975).

Why are the core and halo populations so different ? This problem is not fully understood, but the degree to which each population is collisionless plays an important role. The electron–electron Coulomb momentum transfer collision frequency is given by (Chapter 2)

$$\nu_{ee} = \frac{54 n_e}{T_e^{3/2}},$$

(6.116)

where the electron density is specified in units of cm^{-3}. This collision frequency strongly depends on the electron temperature. The collision frequency for core electrons (with other core electrons) from Equation (6.116), using 1 AU parameters, is $\nu_{ee} \approx 3 \times 10^{-5}\,s^{-1}$. Given a thermal speed of $\nu_{e,\,th} \approx 2{,}000$ km/s for $T_e \approx 10^5$ K, then the mean free path is $\lambda_{mfp,core} \approx 0.4$ AU. Thus, the core electrons actually have a significant probability of undergoing a collision as they stream from the corona out to 1 AU. Since the core population is marginally collisional, a relatively isotropic Maxwellian distribution can be maintained. The core population thus "reflects," to a significant extent, "local" (1 AU) dynamical conditions. However, the collision frequency of halo electrons with core (or other halo) electrons is much less than the core electron collision frequency according to Equation (6.116), owing to the much higher halo temperature. The mean free path for halo electrons is $\lambda_{mfp,halo} \approx 30$ AU, which is obviously much greater than 1 AU. Clearly, the halo population is virtually collisionless near 1 AU. For comparison, the mean free path for proton–proton Coulomb collisions is roughly $\lambda_{mfp,pp} \approx 5$ AU (see Problem 6.13), indicating that the proton gas is also virtually collisionless near 1 AU.

The core electron gas is sensitive to local (1 AU) conditions, but the halo population reflects the properties of the solar corona where the density and temperature are much higher than at 1 AU. In Problem 6.13 you are asked to determine at what radial distance the halo population ceases to be collisional; the answer is $r \approx 3R_\odot$ – in the "heart" of the corona. Hence, the halo population represents conditions in the corona where $T_e \approx 10^6$ K$(\approx T_{halo})$. The halo population is the 1 AU remnant of the original coronal electron population. This also explains the outward heat flux associated with the halo population.

A second reason why the core electrons, which constitute the bulk of the total electron density, cannot freely stream along the interplanetary magnetic field at their thermal speed is that an ambipolar polarization electric field is set up in order to keep the plasma quasi-neutral (see Chapter 4). The protons flow out from the Sun at only $u_{sw} \approx 400$ km/s, and without an electric field the bulk of the electrons would "outrun" the protons at the relatively large electron thermal speed. The polarization potential between the corona and 1 AU is approximately 60 volts; note in Figure 6.37 that the break in the electron distribution, separating core and halo populations, is also at about 60–70 eV.

The solar wind electron distribution is altered from the usual solar wind distribution downstream of a shock (either a planetary bow shock or an interplanetary shock). As one would expect, the electron temperature increases across a shock, but not as much as would be expected from single-fluid MHD theory or as much as the proton temperature jumps. Furthermore, the electron distribution function departs even further from a strict Maxwellian form downstream of a shock. The halo population is not strongly affected by the shock, and the core electron distribution becomes "flat-topped" rather than Maxwellian.

6.5 The heliosphere and cosmic ray modulation

The region of space influenced by the Sun and solar wind is called *the heliosphere*. The heliosphere and solar system are in some sense synonymous. Outside the heliosphere lies the *very local interstellar medium (VLISM)*. The structure of the heliosphere is discussed in Section 6.5.1 and cosmic rays are discussed in Sections 6.5.2 and 6.5.3. The bulk of the heliospheric plasma (i.e., the solar wind) consists of flowing electrons, protons, and some heavier ions, with temperatures of the order of 10 eV or less. However, a nonthermal, very high energy "tail" exists for both the electron and ion distributions. These nonthermal particles are called *cosmic rays* and have very low densities, but they are very energetic. The energetic electrons and ions coming from outside the heliosphere are called *galactic cosmic rays*; those coming from the Sun are called *solar cosmic rays*. But whatever their origin, the solar wind and IMF act as a medium through which these particles must travel.

6.5.1 The heliosphere and the interstellar medium

Parker (1961) was the first to propose a realistic structure for the heliosphere, but much work has been done on this topic since then, and the review paper by Suess (1990) covers this material very nicely. However, almost all of this work has been theoretical because no spacecraft has yet traveled far enough from the Sun to reach the heliopause. *Voyagers 1* and *2* and *Pioneers 10* and *11* have all reached heliocentric distances of at least 40 AU. The schematic in Figure 6.38 illustrates what we currently *think* the overall structure of the heliosphere is like. The heliopause is a contact surface (see Chapter 4) that separates the external very local interstellar medium from plasma of solar origin. Table 6.2 summarizes what we know about the very local interstellar medium. The VLISM is that region of the interstellar medium located immediately outside the heliosphere (within a distance

Table 6.2. *Properties of the very local interstellar medium*[a]

Neutral component	
Density:	H 0.1 cm^{-3}(± 0.03 cm^{-3})
	He 0.01 cm^{-3}
Temperature (H and He):	10^4 K
Flow speed (relative to Sun):	23 km/s (± 3 km/s)
Ionized component	
Density (n_e):	<0.003 cm^{-3} ($n_e \approx 10^{-3}$ cm^{-3})
Temperature and flow speed:	probably the same as for the neutral component
Cosmic rays:	total pressure $= 1.3(\pm 0.2) \times 10^{-13}$ N/m^2
Magnetic field:	$|\mathbf{B}| \approx 0.3$ nT (uncertainty of a factor of 10!)

[a]Largely adopted from Suess (1990).

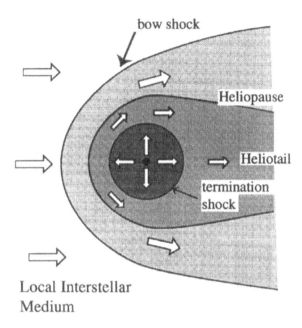

Figure 6.38. Schematic structure of the heliosphere. The heliopause separates solar wind plasma from plasma of interstellar origin and is thought to lie between 50 and 150 AU. It is unclear whether or not the bow shock exists, because it is not known whether the flow of the local interstellar gas is sub- or supermagnetosonic, but the termination shock almost certainly exists.

of $\approx 2{,}000$ AU or 0.01 pc). The local interstellar medium (LISM) is the region within about 100 pc from the solar system. The characteristics of the neutral component of the VLISM are known reasonably well, as is the cosmic ray component, but unfortunately this is not true of the ionized component and of the magnetic field. The VLISM consists of both plasma and neutral components and is moving with respect to the solar system at about 23 km/s.

The heliosphere acts as an obstacle to the flow of VLISM plasma, which must flow around the heliopause. Supposing that the plasma part of the VLISM is moving supersonically, then a bow shock should form in this flow. However, although the VLISM *might* be moving supersonically, there are indications that it is moving sub-Alfvénically and, most importantly, submagnetosonically, so that a bow shock might *not* exist outside the heliopause. Inside of the heliopause, the solar wind plasma flows radially outward from the Sun, as we know, carrying with it the interplanetary magnetic field. The IMF inside the heliopause is wound up as a spiral, with about 20 complete windings if the heliopause is located near 100 AU (see Problem 6.11). The heliopause acts as an obstacle to the solar wind, and because solar wind is supersonic, a standing shock wave must form in the solar wind. This shock is called the *solar wind termination shock* or just *termination shock*. In the region of subsonic flow between the termination shock and the heliopause, the solar wind flow is diverted and flows down *the heliotail*.

The location of the heliopause can be estimated by using momentum (or pressure) balance. Recall that at the beginning of this chapter we excluded the possibility of an interplanetary medium in hydrostatic balance because then the pressure even at very great distances from the Sun would greatly exceed the interstellar pressure, estimated as $p_{ism} \approx 10^{-13}$ N/m^2. It was for this reason that a dynamic outflowing solar wind was considered necessary. However, the solar wind dynamic pressure decreases approximately as r^{-2}, and at some large enough heliocentric distance it must equal p_{ism}. The heliopause should be located at the distance where

$$\rho_{sw} u_{sw}^2 = p_{ism}. \tag{6.117}$$

The thermal and magnetic pressure components of the solar wind were neglected in Equation (6.117).

We can use Equation (6.117) to determine the heliocentric distance of the heliopause (and termination shock). The total interstellar plasma pressure p_{ism} must include the dynamic pressure $p_{dyn} = \rho_{ism} u_{ism}^2$ and thermal pressure (electrons and ions) of the plasma, as well as the magnetic pressure $p_B = B^2/2\mu_0$ and the cosmic ray pressure p_{cr}:

$$p_{ism} = \rho_{ism} u_{ism}^2 + 2 n_e k_B T_e + p_B + p_{cr}, \tag{6.118}$$

where $\rho_{ism} = n_e \langle m \rangle$ is the mass density, $\langle m \rangle$ is the mean mass of the plasma, $\langle m \rangle \approx m_p$, and u_{ism} is the flow speed of the VLISM with respect to the solar system. We assumed in Equation (6.118) that the electron and ion temperatures were equal. Note that we did not include any pressure from the neutrals because the interstellar neutrals do not "see" the heliopause and flow unimpeded into the heliosphere. The interstellar H and He penetrate deeply into the solar system where the atomic hydrogen is observed using resonantly scattered solar Lyman alpha photons. Somewhere between the orbit of Neptune (30 AU) and Saturn (10 AU) the H atoms (and other interstellar neutrals) get photoionized by solar photons or undergo charge transfer reactions with solar wind protons and are converted to ions. These ions are "picked up" by the solar wind flow and carried outward. However, the neutrals cannot exert a force on the heliopause and are excluded from p_{ism}.

Parameters from Table 6.2 can be used to estimate that the ratio of magnetic to thermal to dynamic pressure is given by $p_B : p_{dyn} : p_{therm} \approx 300 : 3 : 1$, with a total pressure (still minus the cosmic ray pressure) of 4×10^{-14} N/m^2 (or 4×10^{-13} dynes/cm^2). Hence, magnetic pressure appears to be dominant, but the magnetic pressure is uncertain by about two orders of magnitude! Furthermore, the actual magnetic force ($\mathbf{J} \times \mathbf{B}$) on the upstream side of the heliosphere is likely to be as much as a factor of two greater than p_B owing to the curvature force associated with magnetic field line draping around the heliopause. This additional force increases the total pressure up to $\approx 10^{-13}$ N/m^2. The pressure ratios that were just estimated indicate that the plasma is submagnetosonic (Problem 6.15) and, thus, that a bow shock probably does not exist outside the heliopause.

What about the cosmic ray pressure in Equation (6.118)? The very energetic particles that constitute cosmic rays can at least to some extent penetrate through the heliopause. Galactic cosmic rays are detected at the Earth, but we have not yet observed the thermal interstellar plasma. The magnetic field in the outer heliosphere does affect at least the lower energy cosmic rays, and thus a gradient of cosmic ray pressure should exist and can "apply a force" to the heliopause. But is this pressure gradient sharp enough right at the heliopause itself to significantly affect its location, or is the gradient spread throughout a larger part of the heliosphere? Nonetheless, because $p_{cr} \approx 10^{-13}$ N/m^2 its contribution to p_{ism} can only be within a factor of two or so of our previous estimate of p_{ism}. Therefore, our earlier estimate of $p_{ism} \approx 10^{-13}$ N/m^2 is reasonable. Given this estimate plus the known properties of the solar wind we can estimate using Equation (6.117) that the heliopause should be located at a distance from the Sun of $r_{helio} \approx 100$ AU (Problem 6.16).

6.5.2 Particle transport theory

The theory of how energetic charged particles move along magnetic field lines in the presence of magnetic fluctuations is now discussed. Cosmic rays (CRs) constitute only a very small fraction of the total density of the plasma, but cosmic ray energies greatly exceed the thermal energy of the bulk plasma (i.e., of the solar wind). Cosmic rays have a negligible collision probability, and the Vlasov equation can be used to determine their distribution function. The CR distribution function is not Maxwellian. If the IMF had no small-scale irregularities, then single particle motion as described in Chapter 3 would accurately describe CR motion in the heliosphere. In particular, we would then expect that the first adiabatic invariant (the magnetic moment $\mu_m = mv_\perp^2/2B$) would hold. However, as we saw earlier in this chapter, the solar wind contains magnetic fluctuations (or waves), and the magnetic moment should not be exactly conserved for particle motion along a field line with small-scale magnetic fluctuations with scale sizes of the order of the gyroradius. The magnetic field can be written in terms of a background IMF ($\mathbf{B}_0$) and a wave field ($\mathbf{B}_1$): $\mathbf{B} = \mathbf{B}_0 + \mathbf{B}_1$. The particles, in this case, undergo wave–particle interactions. We discussed wave–particle interactions in Chapter 3 (Section 11), and although that section focused on radiation belt charged particles, it still provides a starting point for discussing cosmic ray transport in the heliosphere. Useful references for this topic are listed at the end of the Chapter (Jokipii, 1971; Fisk, 1982a,b,c); this chapter closely follows (although it highly simplifies) the treatment of this topic given in those references.

The *quasi-linear diffusion equation* for the evolution of the distribution function of a charged particle species in the presence of wave–particle interactions was given by Equations (3.93) and (3.97). We can write this diffusion equation in a form appropriate for studying CR transport along the IMF:

$$\frac{\partial f}{\partial t} + (u_{sw} \cos \chi + \mu v)\frac{\partial f}{\partial s} = \frac{\partial}{\partial \mu}\left(D_{\mu\mu}\frac{\partial f}{\partial \mu}\right), \tag{6.119}$$

where χ is the angle between the solar wind direction and the IMF, $\mu = \cos \alpha$ with α as the particle pitch angle, s is distance along the IMF, t is time, and $D_{\mu\mu}$ is the pitch-angle diffusion coefficient for wave–particle interactions. The distribution function of the particle species is given by $f(s, \mu, v, t)$, where $v = |\mathbf{v}|$ is the particle speed. The energy diffusion term, involving D_{vv}, has been omitted from Equation (6.119), for the sake of simplicity, as has the source term and an "adiabatic heating/cooling" term that accounts for the effects of spatial changes in u_{sw}. The advection term on the left-hand side now includes the solar wind motion since we are dealing with a moving plasma. The omitted energy diffusion term could be used to study particle acceleration by waves (and shocks – the so-called diffusive shock acceleration mechanism, which has been used to explain the acceleration of galactic cosmic rays at supernova shocks), but for simplicity we neglect this term. Furthermore, pitch-angle diffusion is a "faster" process than energy diffusion (see Section 3.11, Equations (3.98) and (3.99)) if the wave speed is much less than the particle speed. For CRs, the particle speeds are much greater than u_{sw}, and the relevant wave speed is the Alfvén speed C_A, which is much less than u_{sw}; that is, $v_{wave} = C_A \ll v$.

An expression for $D_{\mu\mu}$ appropriate for MHD waves (see Chapter 3 or the above references) is

$$D_{\mu\mu} = \frac{\pi \Omega^2}{2B_0^2} \frac{1 - \mu^2}{|\mu| v} P\left(k = \frac{\Omega}{|\mu| v}\right), \qquad (6.120)$$

where Ω is the cosmic ray gyrofrequency and $P(k)$ is the power spectral density of the MHD waves as a function of wavenumber k. We discussed the power spectrum for magnetic fluctuations in the IMF in Section 6.4.4. Notice that the diffusion coefficient depends on the power evaluated at a specific wavenumber: the resonant wavenumber $k = \Omega/|\mu| v$. The general wave–particle resonance condition was given by Equation (3.100) but here the ω term is much less than the other terms. When an MHD wave and particle are in resonance, the particle undergoes one gyration in the time it takes to traverse one wavelength during its motion along the IMF. Recall that the particle speed along $\mathbf{B}_0$ is given by $v_{\parallel} = \mu v$. The power for a turbulent wave spectrum, such as is found in the solar wind, is proportional to the wave energy density and therefore to $\langle B_1^2 \rangle = B_{rms}^2$. Hence, $D_{\mu\mu}$ is equal to $[\langle B_1^2 \rangle / B_0^2]\Omega$ multiplied by some dimensionless factor proportional to the relative power at the resonant wavenumber (see Equation (3.96)). For particles whose gyroradius is comparable to the turbulent correlation length $r_L \approx k_c^{-1}$, we have strong pitch-angle scattering and $D_{\mu\mu} \approx (\pi/2)\Omega(1 - \mu^2)[\langle B_1^2 \rangle / B_0^2]$, where k_c is the correlation wavenumber of the magnetic turbulence. The time it takes for significant pitch-angle scattering to take place is (again consult Section 3.11) $\tau_{\mu\mu} \approx 1/4D_{\mu\mu}$. For strong pitch-angle scattering we have $\tau_{\mu\mu} \approx \Omega^{-1}$ ($\approx$ one gyroperiod). However, very low energy CRs have speeds low enough such that $k \gg k_c$ and the rate of pitch-angle scattering is much less than the maximum value.

Particles can also be scattered across magnetic field lines by waves, as we saw in Chapter 3 for cross-L-shell diffusion. We can broaden the number of spatial variables for f beyond just s and μ to include coordinates perpendicular to the background magnetic field (i.e., x and y) so that $f = f(x, y, s, \mu, v, t)$. For simplicity, (x, y, s) can be represented at a position vector $\mathbf{s}$. We then must add a spatial diffusion term to Equation (6.119), $\partial/\partial x[D_{xx}\partial f/\partial x]$, and an identical term for the variable y. We can write the spatial diffusion coefficient D_{xx} as

$$D_{xx} = \frac{\langle \Delta x^2 \rangle}{\Delta t} \approx \frac{\frac{1}{2}r_L^2}{\tau_{\mu\mu}} \approx r_L^2 D_{\mu\mu}, \tag{6.121}$$

where the approximation has been made that during one average "wave–particle collision event," which takes a time $\tau_{\mu\mu}$, a particle's gyrocenter can shift by about a gyroradius r_L.

We need to understand the behavior of the cosmic ray distribution function over the scale of the heliosphere, but before we do that, we must understand the scattering properties of the heliosphere in a local region first. Let us suppose that the $f(\mathbf{s}, \mu, v, t)$ is *almost* isotropic because of rapid pitch-angle scattering. Closely following Jokipii (1971), we expand f about the variable μ to obtain

$$f(\mathbf{s}, \mu, t) \cong \frac{1}{2}[U(\mathbf{s}, t) + n_1\mu + \cdots], \tag{6.122}$$

where $U(\mathbf{s}, t)$ is an isotropic "density" (units of $\mathrm{cm}^{-3}\mathrm{eV}^{-1}$) and $n_1(\mathbf{s}, t)$ is a perturbation such that $|n_1| \ll U$. The variable v has been suppressed. Note that the integral of $f(\mathbf{s}, \mu, t)$ over μ from -1 to 1 yields U. The following spatial diffusion equation can be written for the cosmic ray density U:

$$\frac{\partial U}{\partial t} = K_\parallel \frac{\partial^2 U}{\partial z^2} + K_\perp \left[\frac{\partial^2 U}{\partial x^2} + \frac{\partial^2 U}{\partial y^2} \right], \tag{6.123}$$

where $K_\parallel$ and $K_\perp$ are spatial diffusion coefficients for diffusion parallel and perpendicular, respectively, to the background IMF for cosmic rays. For the heliosphere as a whole we require a different form of the spatial diffusion equation, but first we investigate the coefficients $K_\parallel$ and $K_\perp$.

The parallel diffusion coefficient can be found using Equations (6.119) and (6.121). The form of this coefficient given in Fisk (1983b) is

$$K_\parallel = \frac{v^2}{4} \left[\int_{-1}^{1} \frac{1 - \mu^2}{D_{\mu\mu}} d\mu \right]. \tag{6.124}$$

The coefficient $D_{\mu\mu}$ can be found from Equation (6.120) and the power spectrum shown in Figure 6.36; however, we must recall that parameters in Equation (6.120) such as B_0 and Ω do depend on the heliocentric distance r.

We can also derive an expression for $K_\perp$ by simply averaging $D_{\mu\mu}$ over μ:

$$K_\perp \cong \frac{1}{2}r_L^2 \langle D_{\mu\mu} \rangle = \frac{r_L^2}{4} \int_{-1}^{1} D_{\mu\mu} d\mu. \tag{6.125}$$

Now let us return and write an equation for the "global" behavior of U. Again following Jokipii (1971), we have

$$\frac{\partial U}{\partial t} + \nabla \cdot (\mathbf{u}_{sw} U + \mathbf{\Phi}_{CR}) = 0. \tag{6.126}$$

Here we have included the flux of CRs associated with the outflowing solar wind; the magnetic fluctuations responsible for the wave–particle scattering move with the solar wind. If the only significant gradients in U are in the radial direction, then the diffusive flux of cosmic rays is given by

$$\mathbf{\Phi}_{CR} = -K_{rr} \frac{\partial U}{\partial r} \hat{\mathbf{r}}, \tag{6.127}$$

where K_{rr} is the radial spatial diffusion coefficient. We have omitted from Equation (6.126) an important term associated with the "adiabatic cooling" of the cosmic rays, but the main characteristics of CR behavior can still be understood without this term. The coefficient K_{rr} depends on both the parallel and perpendicular diffusion coefficients, as follows:

$$K_{rr} = K_{\parallel} \cos \chi^2 + K_{\perp} \sin \chi^2. \tag{6.128}$$

The angle χ between the radial direction and the IMF was derived in Section 6.3 and is specified by $\tan \chi = (r\Omega/u_{sw})$. In the inner solar system, χ is about 45° or less and $K_{rr} \approx K_{\parallel}$, whereas in the outer heliosphere, χ is close to 90° and $K_{rr} \approx K_{\perp}$.

6.5.3 Galactic cosmic rays

Galactic cosmic ray protons with energies significantly greater than 1 GeV have little trouble making their way through the heliosphere to Earth with little diminution of flux. However, the CR flux at Earth is increasingly attenuated, or modulated, as the energy decreases below 1 GeV. In addition, *galactic cosmic ray modulation* has a solar cycle dependence with cosmic rays being attenuated more during solar cycle maximum and during periods of increased solar activity in general than during solar cycle minimum. Cosmic ray proton data are shown in Figure 6.39, and the solar cycle modulation is apparent at lower energies.

The particle transport theory described in Section 6.5.2 can be used to qualitatively explain CR modulation, and a simplified version is presented here. We first estimate that $K_{rr} \approx K_{\perp}$ and then use Equation (6.125) and a simple form of $D_{\mu\mu}$ to estimate that the spatial diffusion coefficient is

$$K_{rr} \approx K_{\perp} \approx 10^{17} \beta \ \mathrm{m^2/s} \quad \text{for } r = 1 \text{ AU}, \tag{6.129}$$

where $\beta = v/c$ and c is the speed of light. Because the power spectrum was explicitly used to derive Equation (6.129), it is not apparent that K_{rr} is proportional

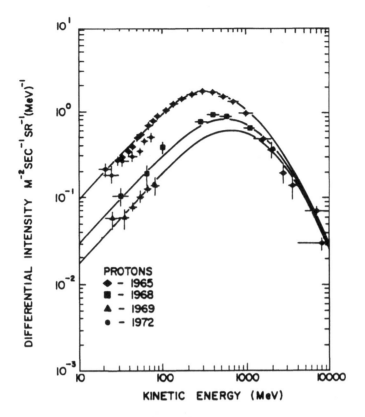

Figure 6.39. Flux of cosmic ray protons versus energy for different years. Solar activity was higher in 1972 than in 1965. (From Fisk, 1983.) Copyright 1983 Kluwer Academic Publishers. With kind permission from Kluwer Academic Publishers.

to the relative strength of the magnetic fluctuations, $\langle B_1^2 \rangle / B_0^2$. Here we used the power spectrum for 1 AU discussed in Section 6.4.4 (and Figure 6.36), for which $\langle B_1^2 \rangle / B_0^2 \approx 0.15$. However, the fluctuations tend to damp out at very large heliocentric distances, and rather than deal with this by explicitly making K_{rr} a function of r, K_{rr} is kept constant out to some distance D (the *cosmic ray modulation distance* – see Jokipii for details). This distance is assumed to be the effective outer boundary of the region in which modulation is possible. Also note that Equation (6.129) breaks down for CR energies much greater than several GeV, in that the gyroradius becomes comparable to D, such that K_{rr} is much greater than Equation (6.129) indicates.

Now we return to Equations (6.126) and (6.127), looking for a steady-state solution ($\partial U / \partial t = 0$); we have

$$\nabla \cdot (\boldsymbol{u}_{\mathrm{sw}} U + \boldsymbol{\Phi}_{\mathrm{CR}}) = \frac{1}{r^2} \frac{\partial}{\partial r} [r^2 (\boldsymbol{u}_{\mathrm{sw}} U + \Phi_{\mathrm{CR},r})] = 0, \qquad (6.130)$$

where $\Phi_{\mathrm{CR},r}$ is the radial component of the cosmic ray flux, and where we have used the spherical form of the divergence.

Integration of Equation (6.130) gives

$$r^2\left(u_{sw}U - K_{rr}\frac{\partial U}{\partial r}\right) = constant. \qquad (6.131)$$

We set the constant equal to zero since there are no significant sources or sinks of galactic cosmic rays inside the heliosphere, and we integrate the resulting equation from distance r out to the modulation boundary:

$$U(r) = U_\infty \exp\left(-\int_r^D \frac{u_{sw}}{K_{rr}}\,dr'\right) = U_\infty \exp\left(-\int_r^D \frac{dr'}{\lambda_{CR}}\right), \qquad (6.132)$$

where we have defined the CR modulation length scale as $\lambda_{CR} = K_{rr}/u_{sw}$. Using Equation (6.129) for very energetic cosmic rays ($\beta \approx 1$), we find $\lambda_{CR} \approx 2.5$ AU. The unmodulated cosmic ray density outside the heliosphere, and outside the modulation region, is denoted U_∞. Assuming that K_{rr}, and thus λ_{CR}, are constant for $r < D$, we can write this solution for $r \ll D$ as

$$U(r) \cong U_\infty \exp\left(-\frac{D}{\lambda_{CR}}\right). \qquad (6.133)$$

Setting D equal to 5 AU and for CR energies of roughly 1 GeV, Equation (6.133) yields $U(1\text{ AU}) \approx 0.1U_\infty$. However, if we assume that $K_\parallel$ varies as r, then Equation (6.132) gives $U/U_0 \approx (r/D)^{1/\lambda_{CR}(AU)} \approx 0.5$ for $r \approx 1$ AU. The actual modulation is roughly a factor of two at 1 GeV but is thought to take place much further out in the heliosphere than 5 AU, although Equation (6.132) is qualitatively correct. Equations (6.132) and (6.129) together predict that the CR modulation should be more severe at lower energies than at higher energies, because K_{rr} is proportional to v/c, and this is indeed borne out by the observations.

The above theoretical picture of cosmic ray modulation was grossly oversimplified in four ways:

1. Large-scale drifts (e.g., grad-B drifts) were omitted but have been shown to be important as well as diffusion. It has been demonstrated that an observed 22-year solar cycle in the flux of galactic cosmic rays can be explained by particle drift inward from the heliopause at low magnetic latitudes (i.e., near the heliospheric current sheet) and then drifts away from the Sun at higher magnetic latitude for 11 years and then, for the other 11 years, the large-scale particle drift is inward from the heliopause at high magnetic latitudes and then outward near the heliospheric current sheet (cf. Kota and Jokipii, 1982).
2. The simple diffusion model was implicitly assumed to apply to low heliospheric latitudes (i.e., near the heliospheric current sheet), where the Parker spiral is tightly wound. Cosmic ray transport at higher latitudes, where the field is less wound up, can also be important.
3. In the outer heliosphere corotating interaction regions (CIRs) tend to merge and become *merged interaction regions* (MIRs). As these MIRs propagate outward at the solar wind speed, they act as significant obstacles to the inward diffusion of galactic cosmic rays (cf. Burlaga et al., 1991).

4. *Anomalous cosmic rays* also contribute to the energetic particle spectrum between energies of about 100 MeV and 1 GeV. These particles start as relatively cold interstellar neutrals (H, He, O, etc.) that enter our solar system and are ionized by solar extreme ultraviolet radiation. Once ionized, the new ions are "picked up" by the solar wind (this process will be discussed in Chapter 7 with reference to cometary ions) and carried out to the termination shock, where they are accelerated up to cosmic ray energies.

6.5.4 Solar cosmic rays – Data and theory

In Chapter 5, we learned that solar flares can produce energetic particles as well as X-ray emission. These particles can also be further accelerated by interplanetary shocks. Typically, solar cosmic rays are mainly protons and have energies of tens of MeV. Figure 6.40 is a schematic of the typical time evolution of the flux of energetic protons such as would reach the Earth during a *solar particle event* (or *solar proton event*) that is associated with a solar flare. These energetic particles can enter the magnetosphere and then the atmosphere of the Earth near the geomagnetic poles. Large events can produce copious ionization in the mesosphere and upper stratosphere. Dissociation of molecular nitrogen also takes place, and the resulting "odd nitrogen" production results in significant, albeit temporary, ozone depletions.

Notice in Figure 6.40 that the arrival of the 20 MeV protons at the Earth is delayed, in comparison with the 500 MeV protons, by about 30 minutes. The complete rise time of the proton flux is an hour or so, but it takes several days for the proton fluxes to decay away, even though the original solar flare is over in a few hours. This time evolution of solar cosmic ray protons can be explained using the particle transport theory presented in Section 6.5.2 and, in particular, the spatial diffusion Equations (6.126) and (6.127). First, we must estimate the radial diffusion coefficient for wave–particle interactions.

The radial diffusion coefficient in the inner solar system is approximately (Equation 6.128) equal to the parallel spatial diffusion coefficient, $K_{rr} \approx K_{\parallel}$. We can derive the approximate expressions for $K_{\parallel}$ given by Jokipii (1971) by using Equation (6.124) plus a simple form of the pitch-angle diffusion coefficient $D_{\mu\mu}$:

$$K_{\parallel} \approx \begin{cases} 5 \times 10^{17} R_G^{1/2} \beta \text{ m}^2/\text{s} & \text{for } R_G < 2 \text{ GeV} \\ 1.5 \times 10^{17} R_G^2 \beta \text{ m}^2/\text{s} & \text{for } R_G > 2 \text{ GeV.} \end{cases} \tag{6.134}$$

The power spectrum for 1 AU shown in Figure 6.36 was used. R_G is the *proton rigidity* in units of GeV. The rigidity is given by $R_G = \gamma m v c$, with $m = 980$ MeV being the proton rest mass, v the particle speed, c the speed of light, and $\gamma = 1/(1-\beta^2)^{1/2}$. For a 50 MeV proton, we find (Problem 6.20) that $K_{\parallel} \approx 5 \times 10^{16}$ m^2/s. We know from earlier chapters that a spatial diffusion coefficient can be expressed in terms of the average particle velocity and the scattering mean free path as $K_{\parallel} = (1/3)v\lambda_{\text{mfp}}$. For 50 MeV protons, the wave–particle mean free path in the inner solar system is $\lambda_{\text{mfp}} \approx 0.01$ AU.

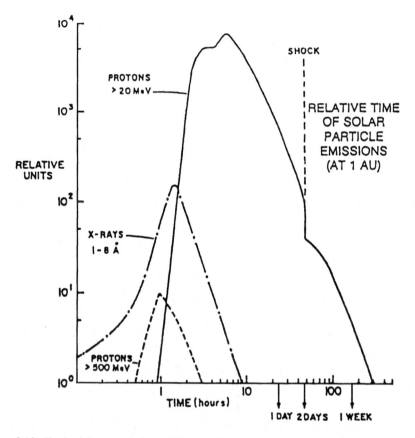

Figure 6.40. Typical time evolution of fluxes of various quantities for a solar particle event observed at Earth. The X rays, white light (not shown), and very energetic protons reach the Earth very quickly, such that the displayed time evolution is associated with the solar flare rather than with propagation in the interplanetary medium. In contrast, the time evolution of the lower energy protons is primarily determined by propagation effects. (From Smart and Shea, 1989.) Also see Gorney (1990).

Considerable effort has been devoted over the past couple of decades to solving the time-dependent diffusion Equation (6.126) both analytically and numerically, subject to realistic solar flare boundary conditions. Here, we merely illustrate that the time scales evident in Figure 6.40 are reasonable. Consider a typical solar proton energy of 50 MeV. The very first particles, which avoid any significant amount of wave–particle scattering, arrive at the Earth after a transit time of $\tau = (1\,\text{AU})/v \approx$ 30 minutes, where $v \approx 0.3\,c$. However, the bulk of the particles are subject to spatial diffusion and have to "random walk" their way out to the Earth from the Sun. In fact, some particles "overshoot" 1 AU and have to diffuse back to 1 AU from larger heliospheric distances. The time scale for spatial diffusion (as discussed in earlier chapters) is given by $\tau_{\text{diff}} \approx r^2/K_{rr}$, where the length scale of relevance is $r = 1$ AU, in which case $\tau_{\text{diff}} \approx (1.5 \times 10^{11}\,\text{m})^2/(5 \times 10^{16}\,\text{m}^2\,\text{s}^{-1}) \approx 5$ days. The time scale for decay of the proton spectrum in Figure 6.40 is indeed a few days. The details of the evolution of any particular event depend on details such as the location of the solar flare on the Sun, the IMF configuration, and the amount of MHD turbulence in the interplanetary medium at the time. A significant complication is that particle acceleration at interplanetary shocks also contributes to the particle spectrum at Earth.

Problems

6.1   Derive the hydrostatic pressure relation, (6.4), using the temperature profile given by Equation (6.3).

6.2   Derive Equation (6.9) for the coronal gas temperature as a function of radial distance, $T(r)$, starting with the steady-state heat conduction Equation (6.8).

6.3   Demonstrate that for steady-state, source-free conditions the continuity equation can be transformed to $\rho u A = constant$ along a streamline, where u is the gas speed and A is the cross-sectional area bounded by the streamlines.

6.4   Complete the description of simple steady-state flow in a tube of varying cross-sectional area by deriving Equations (6.36)–(6.39).

6.5   **Dusty gas outflow from a cometary nucleus** Gas flows radially outward from the surface of a sphere of radius r_0. The mass of the spherical cometary nucleus is small enough that the gravitational acceleration is negligible. Dust is carried off the surface by the gas, but the dust flow speed u_d is almost zero because each dust grain is so massive. The number density of dust grains, $n_d(r)$, as a function of radial distance r from the nucleus can be written in terms of the dust density n_{d0} at the surface:

$$n_d(r) = n_{d0}(r_0/r)^2.$$

The collision frequency of the gas molecules with the dust grains can be written as

$$\nu = n_d \sigma C_s$$

with gas–dust collision cross section $\sigma = constant$. C_s is the gas speed of sound.

The steady-state continuity, momentum, and energy (polytropic) equations for the gas outflow are

$$\frac{\partial}{\partial r}(\rho u r^2) = 0$$

$$\rho u \frac{\partial u}{\partial r} + \frac{\partial p}{\partial r} = -\rho \nu u.$$

ρ, p, and u are the gas mass density, pressure, and radial outflow speed, respectively. For an energy equation use the polytropic equation $p/\rho^\gamma = constant$.

Use these equations plus information given in this chapter to derive the following equation for the Mach number, $M = u/C_s$, as a function of r:

$$\frac{1}{M}\frac{dM}{dr} = \frac{\Lambda(r)}{1 - M^2}.$$

Find the function $\Lambda(r)$. Plot this function versus r for $n_{d0} = 10\,\text{cm}^{-3}$ and $r_0 = 10\,\text{km}$ and for $\sigma = 3 \times 10^{-7}\,\text{cm}^2$ (appropriate for 1 μm sized grains). Also assume that $\gamma = 5/3$ and that $M \approx 1$ (that is, the function you find will only be approximate). The critical point r_c is located where

$$\Lambda(r_c) = 0.$$

Find the critical point (or radial distance) r_c.

The mean free path of a gas molecule due to collisions with dust is given by $\lambda = 1/(n_d\sigma)$. What limits must exist on the mean free path (and therefore on the dust density) in the vicinity of the comet's surface, λ_0, such that the critical radius lies above the surface of the cometary nucleus ($r_c > r_0$)?

6.6   Demonstrate that Equations (6.48) and (6.54) for the critical radius are consistent. For $s = 3$ and $\gamma = 1.15$ derive expressions for the variation of pressure, density, and velocity versus radial distance for large values of the radial distance.

6.7   Demonstrate that the dipole magnetic field (as described in Chapter 3 – Equation 3.67) can be found with the scalar magnetic potential given by Equation (6.73). Also demonstrate that this scalar potential is a solution of Laplace's equation.

6.8   Suppose that the radial component of the magnetic field at the solar surface is given by $B_r(r = R_\odot, \theta, \phi) = -a\cos\theta + b\sin^2\theta\cos 2\phi$ with $b = a/3$, where a is a constant. Find the magnetic scalar potential and the magnetic field, $B(r, \theta, \phi)$ for $r > R_\odot$.

6.9   Use Equations (6.92)–(6.94) plus Maxwell's equations to demonstrate that a Poisson's equation can be written for the electrostatic potential in the rotating frame:

$$\nabla^2 \Phi' = -\frac{\rho_c}{\varepsilon_0}$$

with charge density

$$\rho_c = \frac{u_{sw}}{c}\frac{J_r}{c},$$

where c is the speed of light and J_r is the radial component of the current density.

6.10 Starting from Equation (6.87), derive Equation (6.88) for the time variation of the magnetic field in the rotating coordinate system, by using vector
calculus identities (which can be found at the beginning of this book or
in most undergraduate electromagnetism texts), Maxwell's equations, and
expression (6.85) for the rotational velocity. Note that you also have to put
the convective derivative of the magnetic field in terms of the rotating co-
ordinate system, which introduces a term $\mathbf{\Omega} \times \mathbf{B}$, where $\mathbf{\Omega}$ is the rotational
velocity vector.

6.11 The magnetic field lines for the average interplanetary magnetic field (IMF)
follow Archimedean spirals. Find the heliocentric distance r in Astronomical Units (AU), where a field line has wrapped itself around the Sun once.
Assume that the solar wind speed is $u_{sw} = 400$ km/s.

What is the angle χ between the radius vector and the field vector $\mathbf{B}$ at
this point?

What is the magnitude of the field, $|\mathbf{B}|$, at this point if the source surface
is located at $R_s = 10R_\odot$ and if the magnetic field at R_s is $B_s = 10^{-6}$ T?

Also determine the number of times the magnetic field has wound around
the Sun by a heliocentric distance of 100 AU.

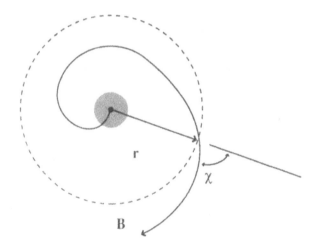

6.12 The passage of a coronal mass ejection (CME) and an associated interplanetary shock preceeding it was measured by the *HELIOS* spacecraft,
as shown in Figure 6.33. The shock jump calculated from the plasma
density is about $Z_s = 3$. From this value of Z_s and the data shown
in Figure 6.33, show that the shock speed is $u_{shock} \approx 555$ km/s. Next
use this speed to determine the upstream and downstream plasma speeds.
Are these speeds consistent with the shock jump Z_s found from the density data? Show that the upstream Mach number is $M \approx 3$, and also
determine the downstream Mach number and the pressure jump across
the shock. Strictly speaking, MHD versions of the shock jump relations

(or Rankine–Hugoniot relations) should be used, but the gas dynamic versions of these relations, as given in Chapter 4, are adequate in this case.

6.13 (a) Calculate the proton–proton collision mean free path, λ_{mfp}, in the solar wind near 1 AU. At what distance from the Sun does the proton gas become collisional (i.e., $\lambda_{mfp} \approx r$)?

(b) Repeat (a) for both the core and halo electron populations in the solar wind.

(c) The collisional rate coefficient (see Chapter 2) for the electron impact ionization of an ion species with ionization potential χ_k is approximately (Zirin, 1988):

$$k_{ioniz} \simeq 1.1 \times 10^{-8} n T^{1/2} \chi_k^{-2} e^{-5040\chi_k/T} \quad [cm^3\, s^{-1}],$$

where n is the principle quantum number of the species being ionized, T is the electron temperature in units of K, and χ_k is the ionization potential of this species and charge state (k) in units of eV. Estimate ionization collision times and mean free paths for the solar corona and the solar wind for a "typical" coronal ion species with $n = 2$ and $\chi_k = 100\,eV$.

6.14 In Section 6.3.3, magnetic field lines were assumed to be equipotentials. This cannot be strictly true because of the ambipolar electric field term in the generalized Ohm's law associated with the electron pressure gradient. Estimate the electric potential difference along an interplanetary field line between the solar corona and a 1 AU distance.

6.15 The very local interstellar medium (VLISM) is thought to be moving relative to the heliosphere at a speed of $u_{ism} \approx 20$ km/s. Using VLISM parameters given in Table 6.2, calculate the following quantities for the VLISM: C_S, C_A, M_s, M_A, and M_{ms}.

6.16 Estimate the heliocentric distance to the heliopause if the pressure of the interstellar plasma just outside (i.e., VLISM) is $p_{ism} \approx 10^{-13}$ N/m^2.

6.17 Determine the pitch-angle diffusion coefficient $D_{\mu\mu}$ and the associated time scale $\tau_{\mu\mu}$ for 10 MeV, 100 MeV, and 1 GeV protons for the solar wind near 1 AU using the power spectrum shown in Figure 6.36. Note in Figure 6.36 that the power spectral density $P(f)$ is in units of [nT]^2/[Hz], where f is frequency, whereas Equation (6.120) requires the power in terms of the wavenumber, $P(k)$, which has units of [nT]^2/[cm^{-1}]. Compare $\tau_{\mu\mu}$ with the solar wind transit time for 1 AU and also with a cosmic ray transit time for 1 AU.

6.18 Another form of the parallel diffusion coefficient was presented by Jokipii (1971):

$$K_{\parallel} = \frac{2v^2}{9} \left[\int_{-1}^{1} D_{\mu\mu}\, d\mu \right]^{-1}.$$

Derive this equation by taking the moments of Equation (6.119) and using Equation (6.120).

6.19   A crude estimate of the power spectral density for a turbulent spectrum for wavenumbers less than the correlation length (i.e., for $k < k_c$) is given by $P \approx \delta B^2 / k_c$, where $\delta B^2 = \langle B_1^2 \rangle$ and δB is the rms magnetic fluctutation level. Demonstrate that the following approximations are reasonable (neglecting singularities near $\mu \approx 0$, which disappear in a more accurate approach):

$$D_{\mu\mu} \approx \frac{\pi}{2} \left(\frac{\delta B}{B} \right)^2 \frac{\Omega^2}{v k_c} (1 - \mu^2)$$

$$K_\perp \approx \frac{2\pi}{15} \left(\frac{\delta B}{B} \right)^2 \frac{\beta c}{k_c}.$$

Equation (6.129) in the text follows from this last equation when specific values are used for the parameters.

6.20   Derive the second expression in Equation (6.134) for the spatial diffusion coefficient for cosmic rays, $K_\parallel$, using a simple form of $D_{\mu\mu}$ plus Equation (6.124), as well as the power spectrum of the magnetic field shown in Figure 6.36. Note that the gyrofrequency presented in Chapter 3 was nonrelativistic; here you will need the relativistic gyrofrequency $\Omega = q B_0 / \gamma m$, where $\gamma = [1 - v^2/c^2]^{-1/2}$. Also determine $K_\parallel$ for a 50 MeV proton.

Also show that the parallel spatial diffusion coefficient is approximately

$$K_\parallel \approx \frac{v^2}{3\pi} \left(\frac{\delta B}{B} \right)^{-2} \frac{\beta c k_c}{\Omega^2}.$$

From this one can estimate that $K_\parallel \approx (3 \, \mathrm{s}^{-1}) \beta r_L^2$ so that $K_\parallel \approx 10^{17} \, \mathrm{m}^2 \, \mathrm{s}^{-1}$ near 1 AU and $K_\parallel \approx 3 \times 10^{18} \, \mathrm{m}^2 \, \mathrm{s}^{-1}$ near 5 AU. Note that near 5 AU, the cosine of the typical IMF spiral angle can be shown to be $\cos^2 \chi \approx 0.04$ so that the contribution to K_{rr} is about $\cos^2 \chi K_\parallel \approx 10^{17} \, \mathrm{m}^2 \, \mathrm{s}^{-1}$, as it is at 1 AU.

Bibliography

Bagenal, F. and S. Gibson, Modeling the large-scale structure of the solar corona, *J. Geophys. Res.*, **96**, 17663, 1991.

Bame, S. J., Spacecraft observations of the solar wind composition, p. 535 in *Solar Wind* (known as Solar Wind I), ed. C. P. Sonett, P. J. Coleman Jr., and J. M. Wilcox, *NASA Publ. SP-308*, 1972.

Barnes, A., Hydromagnetic waves, turbulence, and collisionless processes in the interplanetary medium, p.155 in *Solar-Terrestrial Physics: Principles and Theoretical Foundation*, ed. R. L. Carovillano and J. M. Forbes, D. Reidel Publ. Co., Dordrecht, The Netherlands, 1983.

Belcher, J. W. and L. Davis Jr., Large amplitude Alfvén waves in the interplanetary medium, *J. Geophys. Res.*, **76**, 353, 1971.

Belcher, J. W., A. J. Lazarus, R. L. McNutt, Jr., and G. S. Gordon, Solar wind conditions in the outer heliosphere and the distance to the termination shock, *J. Geophys. Res.*, **98**, 2177, 1993.

Burlaga, L. F., L. W. Klein, N. R. Sheeley Jr., D. J. Michels, R. A. Howard, M. J. Koomen, R. Schwenn, and H. Rosenbauer, A magnetic cloud and a coronal mass ejection, *Geophys. Res. Lett.,* **9**, 1317, 1982.

Burlaga, L. F., F. B. McDonald, N. F. Ness, and A. J. Lazarus, Cosmic ray modulation: Voyager 2 observations, 1987–1988, *J. Geophys. Res.,* **96**, 3789, 1991.

Coles, W. A. and B. J. Rickett, IPS observations of the solar wind speed out of the ecliptic, *J. Geophys. Res.,* **81**, 4797, 1976.

Davis, L., Jr., The interplanetary magnetic field, p. 93 in *Solar Wind* (known as Solar Wind I), ed. C. P. Sonett, P. J. Coleman Jr., and J. M. Wilcox, *NASA Publ. SP-308,* 1972.

Dennery, P and A. Krzywicki, *Mathematics for Physicists,* Harper and Row Publ., New York, 1967.

Feldman, W. C., J. R. Asbridge, S. J. Bame, M. D. Montgomery, and S. P. Gary, Solar wind electrons, *J. Geophys. Res.,* **80**, 4181, 1975.

Fisk, L. A., Solar cosmic rays – their injection, acceleration, and propagation, p. 201 in *Solar-Terrestrial Physics: Principles and Theoretical Foundation,* ed. R. L. Carovillano and J. M. Forbes, D. Reidel Publ. Co., Dordrecht, The Netherlands, 1983a.

Fisk, L. A., Solar modulation of galactic cosmic rays, p. 217 in *Solar-Terrestrial Physics: Principles and Theoretical Foundation,* ed. R. L. Carovillano and J. M. Forbes, D. Reidel Publ. Co., Dordrecht, The Netherlands, 1983b.

Fisk, L. A., The acceleration of energetic particles in the solar wind, p. 231 in *Solar-Terrestrial Physics: Principles and Theoretical Foundation,* ed. R. L. Carovillano and J. M. Forbes, D. Reidel Publ. Co. (Kluwer Academic Publishers), Dordrecht, The Netherlands, 1983c.

Goldstein, M. L., D. A. Roberts, and W. H. Matthaeus, Magnetohydrodynamic turbulence in the solar wind, *Annu. Rev. Astron. Astrophys.,* **33**, 283, 1995.

Gorney, D. J., Solar cycle effects on the near-Earth space environment, *Rev. Geophys.,* **28**, 315, 1990.

Hakamada, K., M. Kojima, and T. Kakinoma, Solar wind speed and HeI (1083 nm) absorption line intensity, *J. Geophys. Res.,* **96**, 5397, 1991.

Hundhausen, A. J., *Coronal Holes and High Speed Streams,* Colorado Associated Univ. Press, Boulder, CO, 1977.

Jokipii, J. R., Propagation of cosmic rays in the solar wind, *Rev. Geophys.,* **9**, 27, 1971.

Jokipii, J. R. and P. J. Coleman Jr., Cosmic-ray diffusion tensor and its variation observed with Mariner 4, *J. Geophys. Res.,* **73**, 5495, 1968.

Kota, J. and Jokipii, J. R., Cosmic rays near the heliospheric current sheet, *Geophys. Res. Lett.,* **9**, 656, 1982.

Kundu, P. K., *Fluid Mechanics,* Academic Press, San Diego, 1990.

McComas, D. J., S. J. Bame, W. C. Feldman, J. T. Gosling, and J. L. Phillips, Solar wind halo electrons from 1–4 AU, *Geophys. Res. Lett.,* **19**, 1291, 1992.

Neugebauer, M. and C. W. Snyder, Mariner 2 observations of the solar wind, 1. Average properties, *J. Geophys. Res.,* **71**, 4469, 1966.

Parker, E. N., The stellar-wind regions, *Astrophys. J.,* **134**, 20, 1961.

Parker, E. N., *Interplanetary Dynamical Processes,* Interscience/Wiley, New York, 1963.

Phillips, J. L., et al., Ulysses solar wind plasma observations from pole to pole, *Geophys. Res. Lett.,* **22**, 3301, 1995.

Schatten, K. H., Large-scale properties of the interplanetary magnetic field, p. 65 in *Solar Wind* (known as Solar Wind I), ed. C. P. Sonett, P. J. Coleman Jr., and J. M. Wilcox, *NASA Publ. SP-308,* 1972.

Severny, A., J. M. Wilcox, P. H. Scherrer, and D. S. Colburn, Comparison of the mean photospheric magnetic field and the interplanetary magnetic field, *Solar Physics,* **15**, 3, 1970.

Smart, D. F. and M. A. Shea, Proton events during the past three solar cycles, *J. Spacecr. Rockets*, **26**, 403, 1989.

Smith, E. J. and A. Barnes, Spatial dependences in the distant solar wind: Pioneers 10 and 11, p. 521 in *NASA Conf. Publication NASA CP2280, Solar Wind 5*, 1983.

Smith, E. J., R. G. Marsden, and D. E. Page, Ulysses above the Sun's south pole: An introduction, *Science*, **268**, 1005, 1995.

Suess, S. T., The heliopause, *Rev. Geophys.*, **28**, 97, 1990.

Withbroe, G. L., The temperature structure, mass, and energy flow in the corona and inner solar wind, *Astrophys. J.*, **327**, 442, 1988.

Wolfe, J. H., The large-scale structure of the solar wind, p. 170 in *Solar Wind* (known as Solar Wind I), ed. C. P. Sonett, P. J. Coleman Jr., and J. M. Wilcox, *NASA Publ. SP-308*, 1972.

Zirin, H., *Astrophysics of the Sun,* Cambridge Univ. Press, Cambridge, UK, 1988.

7

The solar wind interaction with planets and other solar system bodies

We learned in the previous chapter that the solar wind is an almost collisionless plasma consisting mainly of protons and electrons flowing outward from the Sun supersonically and super-Alfvénically at several hundred kilometers per second. The interplanetary magnetic field is carried out into the solar system by the solar wind. Planets and other solar system bodies act as obstacles to the flow of the solar wind, but the nature of this interaction strongly depends on the characteristics of the planet. Chapter 7 deals with the solar wind flow around planets and other objects. A very brief introduction to this topic was given in Chapter 1. Further reading material on this topic can be found in the bibliography at the end of this chapter. Chapter 8 will deal with the internal dynamics of the terrestrial magnetosphere as well as with the magnetospheres of the outer planets.

7.1 Types of solar wind interaction

7.1.1 Nature of the obstacle

The manner in which the solar wind interacts with objects, or bodies, in the solar system depends, naturally, on the characteristics of that object. Relevant characteristics include its heliocentric distance (r), its size, whether or not it has an atmosphere and ionosphere, and the strength of its intrinsic magnetic field. Table 7.1 lists some relevant characteristics for all the planets and for other solar system bodies. For example, the density of the solar wind (and the solar wind dynamic pressure) on the average decreases inversely with the square of the heliocentric distance (i.e., $n_e \propto 1/r^2$), and obviously this strongly affects how the solar wind interacts with an object. The heliocentric distance of Jupiter is about 5 AU; hence the solar wind dynamic pressure at Jupiter is only, on average, about 4% what it is at the Earth. An intrinsic magnetic field can act as obstacle to the solar wind flow, and thus the strength of this field is another important characteristic. The magnetic dipole moment provides a good way of quantifying the intrinsic field strength. Planets with large magnetic dipole moments include the Earth, Jupiter, Saturn, Uranus, and

Table 7.1. *Characteristics of planets and other solar system bodies*

Planet	Mass $(10^{23}$ kg)	Equatorial radius (km)	Equatorial grav. acceleration (ms^{-2})	Average heliocentric distance (AU)
Mercury	3.33	2,439	3.8	0.46
Venus	48.7	6,050	8.6	0.72
Earth	59.8	6,378	9.88	$(149.6 \times 10^6$ km $= 1$ AU)
Mars	6.42	3,398	3.72	1.52
Jupiter	18,990	71,400	22.88	5.2
Saturn	5,686	60,330	9.05	9.51
Uranus	870	25,560	8.65	19.2
Neptune	1,030	24,765	11.0	30.1
Pluto	0.14	1180	0.66	39.4
Io	0.89	1816	1.8	5.78 R_J (Jupiter)
Titan	1.36	2575	1.37	20.3 R_s (Saturn)
Comets	$\approx 10^{-9}$	$\approx 3.-10.$	$\approx 10^{-3}$	—

Planet	Length of year	Period of rotation (days)	Magnetic dipole moment relative to Earth	Surface pressure of atm. (bars)
Mercury	58.6 days	58.6	3.8×10^{-4}	$\approx 10^{-14}$
Venus	224.7 days	-243	$<5 \times 10^{-5}$	80
Earth	365.3 days	1	1.0^c	1
Mars	687.0 days	1.03	$<2 \times 10^{-4}$	5×10^{-3}
Jupiter	11.86 yr.	0.41	1.9×10^4	0.3^b
Saturn	29.45 yr.	0.43	6.0×10^2	0.5^b
Uranus	84.01 yr.	0.72	4.9×10^1	0.3^b
Neptune	164.79 yr.	0.66	2.5×10^1	0.3^b
Pluto	248 yr.	6.39	$<7 \times 10^{-4}$	1.6×10^{-4}
Io	—	0.77^a	~ 0	
Titan	—	15.9^a	~ 0	1.5
Comets	—	≈ 1	~ 0	$\approx 10^{-7}$

^{a} Sidereal period.
^{b} Cloud tops.
^{c} Earth's magnetic dipole moment is 7.9×10^{15} T m^3.

Neptune; it is the magnetic field of these planets that acts as the obstacle to the solar wind rather than the planets themselves. Mercury only has a small dipole moment ($M_{\mathrm{merc}} = 1.5 \times 10^{12}$ Tm^3), yet even for this planet the magnetic field acts as the obstacle to the solar wind. In contrast, Venus and Mars are known to have very small intrinsic magnetic fields (but it is not known how small – only lower limits on the dipole moments are available). We do know that for Venus the ionosphere acts as the obstacle to the solar wind. In the next subsection, four general types of solar wind interaction are discussed.

7.1.2 Types of solar wind interaction with solar system bodies

A rather arbitrary categorization of the types of solar wind interaction in the solar system is given by: (1) Lunar type, (2) Earth type, (3) Venus type, and (4) Comet type. We discuss each of these next.

7.1.2.1 Lunar type

Earth's moon possesses neither a large-scale magnetic field nor any atmosphere to speak of. Consequently, the solar wind directly impacts the lunar surface and is absorbed. A bow shock does not form in the flow because no significant pressure perturbation is created when the incident flow is absorbed. However, as will be discussed in Section 7.4, a plasma wake does exist downstream of the Moon. Other examples of objects with this type of interaction include the asteroids, inactive cometary nuclei, and certain satellites of other planets such as Phobos (Mars).

7.1.2.2 Earth type

The outer core of the Earth is a rotating electrically conducting fluid and produces a dynamo magnetic field that is closely approximated by a dipole field in the region outside the Earth itself for radial distances within a few Earth radii. The solar wind plasma has great difficulty penetrating this large intrinsic magnetic field and is diverted around a large region containing this field. This region is called the *magnetosphere*. A bow shock exists in the solar wind flow around the magnetospheric obstacle. A full description of this type of solar wind interaction must be very extensive and includes such concepts as magnetic merging of the geomagnetic and interplanetary fields, the convection electric field, and geomagnetic storms. Other examples of objects with an Earth-type solar wind interaction include Mercury, Jupiter, Saturn, Uranus, and Neptune. Although the intrinsic magnetic fields of all these planets act as the main obstacle to the solar wind flow, significant differences exist in the internal magnetospheric dynamics of these various planets. The plasma flow in the magnetospheres of Mercury and Earth is primarily controlled by the solar wind electric field, whereas the plasma flow in the magnetospheres of the outer planets is mainly rotationally controlled (i.e., *Jupiter-type* magnetospheres). Bow shocks exist in the solar wind flow around all Earth-type obstacles.

7.1.2.3 Venus type

Venus rotates only very slowly (see Table 7.1), which might explain its lack of any significant intrinsic magnetic field. However, due to photoionization of neutrals by solar extreme ultraviolet radiation, Venus has a dense neutral atmosphere and also a significant ionosphere. The ionospheric plasma is a very good electrical conductor and acts as an obstacle to the solar wind flow with its imbedded interplanetary magnetic field due to what is basically a diamagnetic effect. A bow shock exists because the solar wind flow is supermagnetosonic and must become submagnetosonic in order to flow around the obstacle. Other possible examples of objects with a Venus-type solar wind interaction are Mars and Titan. However, we are *still* not sure whether or not Mars has a small intrinsic magnetic field that might also contribute to the obstacle. This lack of understanding of Mars is in spite of the several missions to this planet including the recent Russian *PHOBOS* mission. Both Venus and Mars have bow shocks. Titan has both a dense atmosphere and an ionosphere; Titan's orbit around Saturn at a distance of 10 Saturnian radii sometimes brings it within the solar wind and at other times within the magnetosphere of Saturn. Nonetheless, it is thought that the Titan interaction with either the solar wind or with magnetospheric plasma is Venus-like, albeit with significant differences.

7.1.2.4 Comet type

Cometary nuclei are chunks of ice and dust with diameters of typically only a few kilometers. The intrinsic magnetic fields of cometary nuclei are negligible. A cometary nucleus is an inactive, "bare" object throughout most of its orbital period, which is spent far from the Sun. In this case, its solar wind interaction is lunar type. But when a cometary nucleus approaches within a couple of AU of the Sun, its surface is heated by solar radiation so that the ice sublimates, producing water vapor (and other volatiles). The gravitational attraction of the nucleus is negligible, and thus the gas flows outward to great distances, forming an extensive atmosphere, or cometary coma, millions of kilometers in extent. The outflowing gas carries dust along with it, some of which is shaped by radiation pressure into a dust tail. Photoionization of the cometary neutrals produces both heavy and light ions that are assimilated into the solar wind flow, thus "mass-loading" it. A cometary ionosphere is also formed close to the nucleus. A weak bow shock exists around active comets. The interaction of the solar wind with Venus and the Titan interaction also have some "comet-like" aspects.

These various types of solar wind interaction will be explored in greater detail in the remainder of this chapter. Furthermore, a section on ionospheres will be included prior to the discussion on the solar wind interaction with the ionosphere of Venus.

7.2 Earth-type interaction – Location of the magnetopause

The intrinsic magnetic field acts as an obstacle to the solar wind and shields a volume of space from the direct access of the solar wind. This volume (or obstacle) is called the *magnetosphere*, and the boundary surface separating the magnetosphere from solar wind plasma is called the *magnetopause*. We will not be concerned with the internal dynamics of the magnetosphere in this chapter; the role of the magnetosphere as an obstacle to the solar wind is not seriously affected (at least to about the 95% level) by the details of its internal structure or dynamics. Of course, the converse is not true; the solar wind strongly affects the internal dynamics of the magnetosphere and ionosphere, as we will see in Chapter 8. In this section, we will see how an intrinsic magnetic field acts as an obstacle to the external flow. We will also derive an expression for the radial distance (from the Earth) to the dayside magnetopause. The magnetospheres of other planets with intrinsic magnetic fields will also be briefly discussed.

7.2.1 Location of the magnetopause – Earth

Why does an intrinsic magnetic field act as an obstacle? One answer is that in space plasmas with very large magnetic Reynolds numbers, the plasma and magnetic field are "frozen" together, and external plasma flow with its own magnetic field cannot easily interpenetrate another magnetized plasma region. Observationally, the terrestrial magnetopause is, on the average, located at a radial distance of $r_{mp} = 10\,R_E$ along the Earth–Sun line. This distance is not terribly sensitive to the solar wind conditions; typical variations over a few days of the solar wind dynamic pressure are a factor of 2 or 3, yet r_{mp} only varies by about 15%.

We can determine the position of the magnetopause by using some simple MHD principles presented in Chapter 4. The location of the magnetopause can be determined by invoking static pressure balance. The magnetospheric plasma ($r < r_{mp}$) is a low-β plasma; that is, magnetic pressure is the dominant pressure component. This plasma is also moving slowly enough so that we can neglect its dynamic pressure. The solar wind plasma, however, has a β of about 1 and is supersonic. Dynamic pressure dominates for the solar wind. But since the dayside magnetopause is not in direct contact with the supersonic solar wind, in order for it to be able to flow around the magnetosphere, a standing shock sets up (i.e., the *bow shock*, which we discuss further in Section 7.5), and the flow becomes subsonic.

The region between the magnetopause (i.e., the obstacle) and the bow shock, containing shocked solar wind, is called the *magnetosheath*. At the magnetopause itself, the normal component of the flow must be zero or at least small; the magnetopause as a whole must move if a pressure imbalance exists. In the magnetosheath just sunward of the subsolar magnetopause, the upstream solar wind dynamic pressure ($\rho_{sw} u_{sw}^2$) is almost entirely converted into thermal pressure: $p \approx \rho_{sw} u_{sw}^2$. Recall the discussion of dynamic pressure in Section 4.6.6. A more accurate three-dimensional treatment of the magnetosheath flow indicates that the thermal plasma pressure just outside the subsolar magnetopause is actually

$$p = 0.85\,\rho_{sw} u_{sw}^2, \tag{7.1}$$

where $\rho_{sw} \approx n_{sw} m_p$ is the mass density of the solar wind and u_{sw} is the solar wind flow speed upstream of the bow shock. The thermal pressure in the magnetosheath just outside the magnetopause is equal to $p = 0.85 \rho_{sw} u_{sw}^2 \cos^2 \phi$, where ϕ is the angle between the normal to the magnetopause surface and the solar wind direction. At the subsolar point, $\phi = 0°$ and the pressure is given by Equation (7.1). Much of the thermal pressure just outside the magnetopause is further converted into magnetic pressure due to compression of the IMF as the flow stagnates; however, the total pressure, $p + p_B$, is still given by Equation (7.1), and for the purposes of determining the location of the magnetopause, we do not lose anything by neglecting the IMF and simply using Equation (7.1).

The magnetopause separates a low-β plasma with magnetic pressure $p_B = B^2/2\mu_0$, where B is the magnetic field just inside, from a high-β plasma with thermal pressure given by Equation (7.1). At this point you should review Section 4.6.2 on static MHD pressure balance and also Section 4.9.1 on MHD discontinuities because the magnetopause boundary is a *tangential discontinuity*. The total pressure, $p + p_B$, must remain constant across the boundary; that is, the condition for static balance at the boundary is simply $p = p_B$, or

$$p(\text{magnetosheath}) = \frac{B^2(\text{magnetosphere})}{2\mu_0}. \tag{7.2}$$

Magnetic pressure in the magnetosphere balances thermal pressure in the magnetosheath (i.e., shocked solar wind). Precisely this situation was illustrated in Figure 4.9, which also indicates that such a boundary must be a current layer. The magnetic pressure gradient force is just another way of expressing the $\mathbf{J} \times \mathbf{B}$ force on a plasma, and the magnetic pressure at a sharp interface is given by $p_B \approx |\mathbf{K} \times \mathbf{B}|$, where the current density integrated over the width of the current layer is just

$$K = \int_{\Delta r} J \, dr \approx \frac{B}{\mu_0}, \tag{7.3}$$

where Δr is the width of the current layer and the units of K are [A/m].

A good physical picture of the force balance at the magnetopause was given by Ratcliffe (1972). Figure 7.1 illustrates how an influx of plasma (from the left) into a region of magnetic field naturally gives rise to a current layer whose width is approximately equal to the proton gyroradius, $\Delta r \approx r_L$. Magnetic reconnection is thought to be taking place on the subsolar magnetopause, as will be discussed in Chapter 8. The simple picture of the force balance given in Figure 4.9 or 7.1 is adequate for our present purpose of determining the distance to the magnetopause r_{mp}. The sketch of the magnetosphere in Figure 7.2 shows the location of the magnetopause, as well as the main pressure terms in each region (i.e., dynamic pressure upstream of the bow shock, thermal pressure in the subsolar magnetosheath, and magnetic pressure in the magnetosphere).

In order to use Equation (7.2) to find r_{mp} we have to know the magnetospheric magnetic field strength at the magnetopause. We start by simply assuming that the magnetic field is that of a magnetic dipole, as discussed in Section 3.8 and given

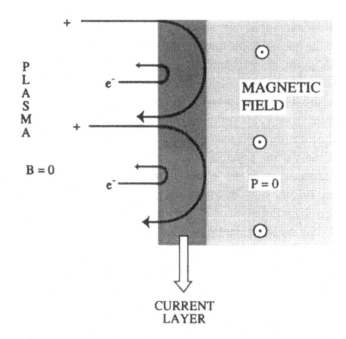

Figure 7.1. Schematic of current layer. A region of uniform magnetic field exists on the right side of the diagram. Protons and electrons impinge on the field region from the left and are deflected in opposite directions by the magnetic field (due to the Lorentz force), with the net effect that the particles are reflected from the layer. The direction of gyromotion is such that a region of current is set up with integrated current density $K \approx enu$, where n and u are the density and incident speed of the particles. The layer is about a gyroradius thick. The current density is such that the $\mathbf{K} \times \mathbf{B}$ force equals the incident dynamic pressure $\rho u^2 = nm_p u^2$, which also happens to equal the thermal pressure if this were a subsonic gas. Effects like charge separation and polarization electric field have been neglected here, but these are revisited briefly in Chapter 8. See Ratcliffe (1972) for further discussion of this scenario. The MHD approach is taken in the text rather than this "kinetic" approach, but they really are equivalent.

by Equation 3.67. The field strength as a function of radial distance is given by

$$B(r) = B_E \left(\frac{R_E}{r} \right)^3,$$ (7.4)

where R_E is the Earth's radius and $B_E = 3 \times 10^{-5}$ T is the field strength at the equator on the Earth's surface. Using Equation (7.4) for B in Equation (7.3) plus Equation (7.1) for the pressure p, the magnetopause force balance Equation (7.2) becomes

$$\frac{B_E^2}{2\mu_0} \left(\frac{R_E}{r} \right)^6 = \rho_{sw} u_{sw}^2.$$ (7.5)

Solving Equation (7.5) for $r = r_{mp}$, we find that the distance to the subsolar magnetopause is

$$\frac{r_{mp}}{R_E} = \left(\frac{B_E^2}{2\mu_0 \rho_{sw} u_{sw}^2} \right)^{1/6}$$ (7.6)

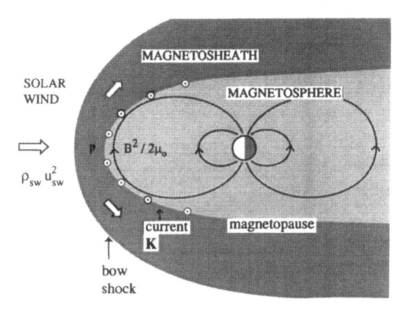

Figure 7.2. Schematic of the solar wind interaction with the terrestrial magnetosphere. The solar wind undergoes a shock at which dynamic pressure is converted into thermal pressure. The main pressure component in the upstream solar wind is dynamic pressure ($\rho_{sw}u_{sw}^2$), whereas thermal pressure (p) dominates the dynamics downstream of the bow shock. Inside the magnetosphere, magnetic pressure is most important ($B^2/2\mu_0$). The region of shocked solar wind plasma is called the magnetosheath. The magnetosheath plasma flows around the magnetopause.

This is the classic *Chapman–Ferraro* equation (see the bibliographic reference to the classic articles by these authors at the end of the chapter). The Chapman–Ferraro theory (i.e., Equation (7.6)) indicates that the magnetopause location is inversely proportional to the 1/6 power of the solar wind dynamic pressure. Naturally, a larger dynamic pressure pushes the magnetopause closer to the Earth. However, the 1/6 exponent in Equation (7.6) also tells us that even large variations of $\rho_{sw}u_{sw}^2$ produce only modest changes in r_{mp}. In Example 7.1, Equation (7.6) is used to show that $r_{mp} = 7.5\, R_E$ for typical solar wind conditions.

Example 7.1 (Location of the dayside terrestrial magnetopause) The solar wind at 1 AU typically has density $n_{sw} \approx 7\,\mathrm{cm}^{-3}$ and flow speed $u_{sw} \approx 400$ km/s, giving a dynamic pressure of

$$\rho_{sw}u_{sw}^2 \approx 2 \times 10^{-9}\,\mathrm{N/m^2}.$$

Substitution into Equation (7.6) then gives the following value for the subsolar radial distance from the Earth's center to the magnetopause: $r_{mp} \approx 7.5\, R_E$.

There is a slight problem with Example 7.1. The average distance to the magnetopause observed by spacecraft is more like $r_{mp} = 10\, R_E$ than 7.5 R_E. This is

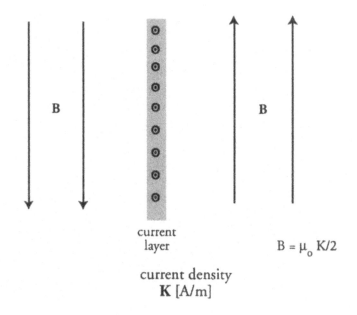

current
layer

$B = \mu_o K/2$

current density
K [A/m]

Figure 7.3. Schematic of the magnetic field produced by a current layer with current density K [units of A/m].

not a small error; in order for Equation (7.6) to predict the correct value of r_{mp}, we would need a magnetic field strength about a factor of two greater than the value we used!

How can this discrepancy be remedied? The geomagnetic dynamo field itself cannot be changed, but the magnetopause is not only a boundary separating the solar wind from the magnetosphere – it is also a current layer. And current layers produce magnetic fields of their own. Consider an infinite current sheet carrying current per unit length K [units of A/m], as shown in Figure 7.3. Ampère's law tells us that the magnetic field produced by this current configuration is uniform on each side of the sheet with opposite directions. The field strength is given by

$$B = \mu_0 K/2 \quad \text{(for region right of sheet)}$$

and

$$B = -\mu_0 K/2 \quad \text{(for region left of sheet).} \tag{7.7}$$

To find the total magnetic field this magnetic field from the current layer, call it $B_K = \mu_0 K/2$, must be added to the original dipole magnetic field B_{dipole}; thus,

$$B = B_{\text{dipole}} + B_K. \tag{7.8}$$

In the near vicinity of the dayside magnetopause (where we can approximate the magnetopause as a planar current sheet), B_K is given by Equation (7.7).

The magnetic field outside the magnetopause must be zero ($B = 0$) (we are still neglecting the IMF) and thus B_K (outside) must exactly cancel B_{dipole}. That is, we must have $B_K = -B_{\text{dipole}}$, giving $|K| = 2B_{\text{dipole}}/\mu_0$. Just inside the magnetopause, where the current layer appears planar, we then must have

$$B = B_{\text{dipole}} + B_K = B_{\text{dipole}} + \mu_0 K/2 = 2B_{\text{dipole}}. \tag{7.9}$$

The magnetic field strength just inside the magnetopause is twice the dipole value. You can also think of the magnetic field as being "compressed" as the solar wind "pushes" the magnetopause Earthward. Reformulating the force balance Equation (7.5) including the correct magnetic field and solving for r yields a new expression for the magnetopause distance:

$$\frac{r_{\text{mp}}}{R_{\text{E}}} = 2^{1/3} \left(\frac{B_{\text{E}}^2}{2\mu_0 \rho_{\text{sw}} u_{\text{sw}}^2} \right)^{1/6}. \tag{7.10}$$

Equation (7.10) is the same as (7.8), except for the extra $2^{1/3}$ factor. Revisiting Example 7.1, but using Equation (7.10), we find that the magnetopause distance for typical solar wind conditions is

$$r_{\text{mp}} = 2^{1/3} \, 7.5 \, R_{\text{E}} = 9.5 \, R_{\text{E}}. \tag{7.11}$$

An average subsolar magnetopause distance of 9.5 Earth radii is much more reasonable than a value of 7.5 R_{E}. On those rare occasions when $\rho_{\text{sw}} u_{\text{sw}}^2$ is ten times greater than the "typical" value, then Equation (7.10) indicates that

$$r_{\text{mp}} \approx [1/10]^{1/6} 9.5 \, R_{\text{E}} \approx 6.5 \, R_{\text{E}}. \tag{7.12}$$

This distance is approximately the distance of a geosynchronous satellite orbit.

Another method of estimating the effects of the magnetopause current layer is the method of image dipoles (see Ratcliffe again). For the region outside the magnetopause, the magnetic field caused by the magnetopause current layer can be considered to be from an "image" dipole located at the Earth's center with magnetic moment equal and opposite to the actual terrestrial dipole moment. Naturally, for the region exterior to the magnetosphere, the field contributions from the sum of the two dipoles exactly cancel, leaving zero external magnetic field. Similarly, for the region inside the magnetopause, the effects of the current layer can be simulated by placing an image magnetic dipole, with the same direction and magnitude as the actual terrestrial dipole, a distance of r_{mp} outside the magnetopause surface (see Figure 7.4). The total magnetic field just inside the magnetopause is clearly $2B_{\text{dipole}}$.

The method of images also enables us to estimate the magnetic perturbations produced by the current layer throughout the magnetosphere. For example, in Problem 7.1 you are asked to determine these magnetic perturbations at the surface of the Earth both for average solar wind conditions and for very high solar wind

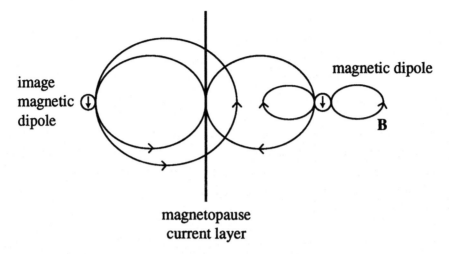

Figure 7.4. Shown is the "image" magnetic dipole appropriate for determining the magnetic field due to the magnetopause current layer in the magnetosphere region. Some field lines from each individual dipole are also shown but not the actual field lines. For the exterior region, the image dipole would be located coincident with the real dipole at the Earth's center but with the opposite orientation.

dynamic pressure conditions, such as might occur during the passage of a coronal mass ejection event that leads to a geomagnetic storm. You will, hopefully, find that for typical solar wind conditions, when $r_{mp} \approx 10\ R_E$, the magnetic perturbation at the Earth's surface is $\Delta B \approx +4$ nT, whereas for very extreme (i.e., once a solar cycle?) conditions where $r_{mp} \approx 5$–$6\ R_E$, $\Delta B \approx +25$ nT. Recall that in Chapter 3 we discussed the D_{st} index of geomagnetic activity. The D_{st} is basically the horizontal component of the magnetic field at low latitudes on the Earth's surface. We stated in Chapter 3 that during geomagnetic storms the ring current (in the inner magnetosphere) generates magnetic fields of about 100–200 nT that oppose the geomagnetic field. If you examine Figure 3.17, which shows a typical D_{st} evolution during a storm, you see that during the *initial phase* of the storm, the D_{st} actually increases by about 10–20 nT, before becoming strongly negative during the *main phase* of the storm. This increase of the surface field is just due to the magnetopause current layer. Geomagnetic storms will be discussed again toward the end of Chapter 8.

7.2.2 Location of the magnetopause – Mercury

Mercury has a small magnetic dipole with a dipole moment of $M_{mer} = 3 \times 10^{12}\ \text{Tm}^3$ (or $4 \times 10^{-4}\ M_E$). Almost all our information on the plasma environment of Mercury was obtained by instruments on board the *Mariner 10* spacecraft (cf. Russell, 1988). Mercury has an Earth-type solar wind interaction in which an intrinsic magnetic field acts as an obstacle to the solar wind. Both a magnetopause and a bow shock

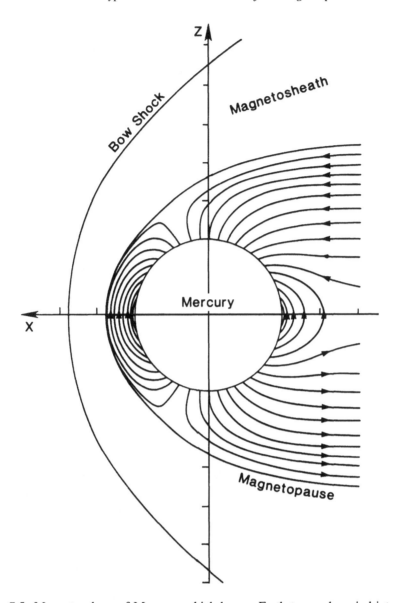

Figure 7.5. Magnetosphere of Mercury, which has an Earth-type solar wind interaction. Magnetic field lines are indicated as well as the magnetopause and bow shock. (From Jackson and Beard, 1977.) Also see Russell et al. (1988).

exist. Substorms analogous to those in the terrestrial magnetosphere appear to take place; this phenomenon will be briefly discussed near the end of Chapter 8. Mercury does not have a substantial atmosphere or ionosphere. Figure 7.5 shows a model Mercurian magnetic field based on *Mariner 10* magnetic field measurements. As discussed in Example 7.2, the location of its magnetopause can be determined using Chapman–Ferraro theory.

Example 7.2 (The magnetopause of mercury) The average solar wind dynamic pressure at Mercury is much greater than it is at Earth because the orbital radius of Mercury is only 0.39 AU. Scaling by the heliospheric distance squared and assuming average solar wind conditions ($\rho_{sw}u_{sw}^2 = 2 \times 10^{-9} \mathrm{Nm}^{-2}$ at 1 AU), we find at Mercury that $\rho_{sw}u_{sw}^2 = 1.3 \times 10^{-8} \mathrm{Nm}^{-2}$. The larger solar wind pressure and the smaller magnetic moment means that the magnetopause at Mercury should be located much closer to the planet than it is at Earth. Let us estimate the subsolar magnetopause distance r_{mp}. Equations (7.6) and (7.10) can be rewritten for Mercury, using B_{mer} for the surface field strength in place of B_E, and using the radius of Mercury R_{mer}, as well as the appropriate solar wind dynamic pressure. We find (see Problem 7.2) that the distance to the subsolar magnetopause at Mercury using the Mercury equivalents of Equations (7.6) and (7.10), respectively, is given by

$$r_{mp}/R_{mer} = 1.04 \quad \text{(without magnetopause currents)}$$

and

$$r_{mp}/R_{mer} = 1.3 \quad \text{(with magnetopause currents).} \tag{7.13}$$

Indeed, a magnetopause distance of $1.3 R_{mer}$ is consistent with the measurements made by *Mariner 10* (see Figure 7.5).

7.2.3 Location of the magnetopause – The outer planets

Most of what we have learned about the magnetospheres of the outer planets has come from measurements made by the *Pioneer 10* and *11* and the *Voyager 1* and *2* spacecraft. In particular, *Voyager 2* took "the grand tour," visiting all four of the gas giant planets. All the outer planets possess large intrinsic magnetic fields. The simple Chapman–Ferraro theory can be used to estimate the magnetopause locations at Saturn, Uranus, and Neptune, and you are asked to do precisely that in Problem 7.3 using the parameters given in Table 7.1 plus your knowledge of solar wind properties. The resulting subsolar magnetopause distances for Saturn, Uranus, and Neptune are approximately $r_{mp} = 19$, 22, and 21, respectively, relative to the respective planetary radii. These magnetopause distances are reasonably consistent with the observed values. The solar wind interaction with these three planets is Earth type, although the internal details of the magnetospheres of these planets differ from Earth and from each other in many respects. A very brief introduction to these magnetospheres will be given in Chapter 8; the reader is also referred to Bagenal (1985).

The intrinsic magnetic field at Jupiter is very large, but magnetospheric plasma pressure is also very important in the pressure balance at the magnetopause; hence the solar wind–Jupiter interaction is not entirely Earth type. Figure 7.6 depicts a simple schematic of the Jovian magnetosphere. Details of the internal dynamics

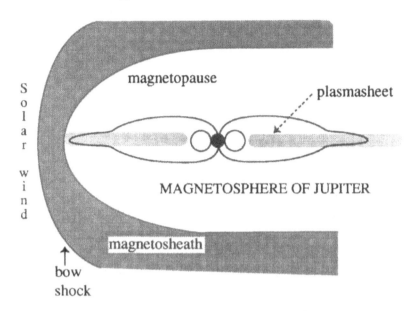

Figure 7.6. Schematic of the magnetosphere of Jupiter. Notice the extensive plasmasheet located near the magnetic equator, as indicated by the shading.

are left to Chapter 8, but rotation is very important in the Jovian magnetosphere. Jupiter rapidly rotates with a 10-hour period, and the magnetospheric plasma, which is largely supplied by the satellite Io, has a strong tendency to rotate also. The rapidly rotating magnetospheric plasma is forced outward by centrifugal forces, which has two effects on the outer Jovian magnetosphere. First, plasma pressure is much greater in the outer magnetosphere than one would otherwise expect. Most of this pressure is confined to the equatorial plane – this layer of enhanced plasma density and pressure is called the *plasmasheet*. The plasma beta inside the magnetosphere in the vicinity of the magnetopause was found by the *Voyager* and *Pioneer* missions to be approximately $\beta \approx 5$. Second, the outward centrifugal force on the plasma and effects due to plasma pressure anisotropies stretch the field lines. That is, currents flow in the plasmasheet and the field is no longer dipolar in the outer magnetosphere. This situation is similar to the formation of the heliospheric current sheet (recall Figure 6.29).

Let us consider the consequences of the extra plasma pressure and the stretched field lines for the solar wind interaction with the magnetosphere of Jupiter. In Problem 7.4 you found that the simple Chapman–Ferraro theory (Equation (6.10)), in which magnetic pressure balances solar wind dynamic pressure, gives a magnetopause distance of $r_{mp} \approx 38 \, R_J$. However, typical observed (from the spacecraft encounters) magnetopause stand-off distances are between about 45 and $100 R_J$. *Voyager 1* crossed the dayside magnetopause at a distance from Jupiter of about $47 R_J$, at which point the internal field strength was about 16 nT (Acuna and Ness,

1976), and *Voyager 2* crossed the dayside magnetopause at a distance of $62 R_J$. In order to push our theoretical magnetopause distance further out we need to take into account the plasma pressure and the nondipole (stretched) field configuration when balancing internal pressure with the solar wind pressure:

$$\rho_{sw} u_{sw}^2 = [p + p_B]_{\text{internal}} = (1 + \beta) p_{B,\,\text{internal}}$$
$$= (1 + \beta) \frac{(B_{\text{internal}})^2}{2\mu_0}, \tag{7.14}$$

where p and p_B refer to thermal and magnetic pressures, respectively, just inside the magnetopause. Let us assume that the plasma just inside the Jovian magnetopause has $\beta \approx 5$ (Vasyliunas, 1983) and that the internal field is again just the dipole field plus the field produced by the magnetopause current layer, $B_{\text{internal}} = 2 B_{\text{dipole}}$. We then find from Equation (7.14) (Problem 7.4) that $r_{mp} = 51 R_J$. Now, if we further assume that currents in the outer magnetosphere have enhanced the magnetic field strength to about $2 B_{\text{dipole}}$, then $B_{\text{internal}} \approx 4 B_{\text{dipole}}$, and we find from Equation (7.14) that $r_{mp} \approx 63 R_J$, which is a very reasonable value.

Clearly, in the case of Jupiter, to understand the solar wind interaction it is very important that the internal magnetospheric dynamics be taken into account. In Chapter 8, we will briefly revisit the magnetosphere of Jupiter and discuss why the thermal pressure is so enhanced.

7.3 Ionospheres

7.3.1 Introduction

An *ionosphere* is the partially ionized plasma region that coexists with the uppermost atmosphere of a planet. Ionospheric electron densities typically range from a few thousand to a few million electrons per cubic centimeter. An ionosphere is created by the ionization of the neutral atoms and molecules associated with the upper atmosphere of a planet. The ionization is mainly caused by either photoionization or energetic particle impact ionization. In this section we will consider questions such as: How are ionospheres formed? and What determines the plasma density as a function of altitude? We will also discuss the ionospheric electrical conductivity, a topic that we will also return to later in Chapter 8 when the subject of ionosphere–magnetosphere coupling will be taken up. Many books have been written about the terrestrial and planetary ionospheres, and in this one short section we will only be able to review a few essential features of ionospheric physics and chemistry. The reader is referred to the following references, listed at the back of this chapter: Rees (1989), Ratcliffe (1972), Banks and Kockarts (1973), Atreya (1986), and the chapters by Richmond, Schunk, and Cravens in the book *Solar–Terrestrial Theory* (1983).

The terrestrial ionosphere has been studied since the development of radio communication about a century ago, but most of our understanding of the physical and chemical processes operating in the ionosphere has been achieved in the space age by experiments carried on board rockets and satellites. Ionospheres have been detected on all of the planets in our solar system, except Pluto, which undoubtedly also has an ionosphere. Comets and some planetary satellites (e.g., Saturn's satellite Titan, Neptune's satellite Triton, and Jupiter's satellite Io) are also known to possess ionospheres. Both remote sensing techniques and *in situ* measurements have been used to study planetary ionospheres. For example, the *Mariner 5* and *10* spacecraft, as well as several Soviet spacecraft, observed the ionosphere of Venus remotely using the radio occultation technique. In the radio occultation technique, the upper atmospheric regions of a planet intersect the ray path of radio waves traveling to Earth from a spacecraft transmitter. The index of refraction for radio waves is altered by the ionospheric plasma and changes the phase of the signal in a measureable manner.

Instruments on board the *Pioneer Venus* spacecraft made *in situ* measurements, starting in 1978, in the ionosphere of Venus. For example, the ion mass spectrometer on board *Pioneer Venus Orbiter (PVO)* measured the ion composition. *PVO* was also equipped with instruments designed to study the neutral atmosphere, such as a neutral mass spectrometer. The *Mariner 4, 6, 7,* and *9* spacecraft measured the electron density in the Martian ionosphere using the radio occultation technique. In 1976, retarding potential analyzers on the *Viking 1* and *2* landers measured electron and ion densities in the Martian ionosphere on their way down to the surface. The *Viking* landers also carried neutral mass spectrometers. Ionospheres were observed at Jupiter and Saturn via radio occultation by the *Pioneer 10* and *11* spacecraft and the *Voyager 1* and *2* deep space probes. *Voyager 2* also detected ionospheres at Uranus and Neptune as well as on Neptune's satellite Triton and Saturn's satellite Titan.

7.3.2 *The neutral upper atmosphere*

The characteristics of a planet's ionosphere depend on the characteristics of the neutral upper atmosphere of that planet, such as the neutral composition and temperature. The major constituents of the terrestrial atmosphere are molecular nitrogen (80%) and molecular oxygen (19%); some carbon dioxide (.03%), argon (1%), and water vapor are also present. The atmosphere of the Earth is divided into four regions, according to the temperature structure: the *troposphere* (altitudes from the surface up to about 18 km), the *stratosphere* (18–50 km), the *mesosphere* (50–90 km), and the *thermosphere* (90 km and upwards). Collisions between molecules become infrequent in the very rarefied upper part of the thermosphere, and the atoms and molecules move in ballistic trajectories. This collisionless region is called the *exosphere*, and its lower boundary is called the *exobase*, which is located near 200 km. The ionosphere coexists with the thermosphere and the exosphere.

The temperature structure of the lower terrestrial atmosphere is determined by the absorption and scattering of radiation (mainly visible light coming in from the

Sun and infrared radiation leaving the atmosphere) and also by convective mixing in the troposphere. Visible electromagnetic radiation is largely able to penetrate the atmosphere and make it down to the Earth's surface. Ultraviolet radiation is absorbed by ozone in the stratosphere. Extreme ultraviolet radiation with wavelengths, less than about 100 nm is absorbed in the thermosphere. The temperature structure of the thermosphere is mainly determined by energy input into this region associated with the absorption of solar EUV radiation and, in the polar region, with the energy deposition from precipitating energetic auroral electrons. The energy, or heat, from this energy input is thermally conducted down to the base of the thermosphere (or *thermopause*) near 100 km, where the energy is lost via emission of infrared (IR) radiation from CO_2. The neutral temperature (T_n) near the thermopause is about 200 K, but T_n increases with altitude in the thermosphere to an *exospheric temperature* of $T_{ex} \approx 1{,}000$–$2{,}000$ K, depending on the level of solar activity, with larger values of T_{ex} occurring during solar cycle maximum.

The vertical pressure structures of planetary atmospheres are well described by hydrostatic balance, as discussed in Chapter 4 (Equations (4.92)–(4.95)) and in Chapter 5 (Equation (5.10)). The downward gravitational force on a fluid parcel is balanced by an upward pressure gradient force. Integration of the appropriate momentum equation gives the following hydrostatic balance equation for the atmospheric pressure as a function of altitude:

$$p(z) = p_0 \exp\left[-\int_0^z \frac{dz'}{H(z')}\right] \quad \text{with } H(z) = \frac{k_B T(z)}{\bar{m}g}. \tag{7.15}$$

This equation is just Equation (5.10) with altitude z replacing radial distance r. The reference pressure at the surface is p_0 ($= 1$ bar $= 10^5$ N/m^2 for Earth), and the atmospheric scale height is given by $H(z)$. The neutral temperature as a function of altitude is $T(z)$, $\bar{m}$ is the mean molecular mass, and g is the acceleration due to gravity. Note that Equation (7.15) applies to all planetary atmospheres and not just to the Earth's, although, obviously, the parameters appearing in the expression such as gravitational acceleration g differ from planet to planet (Table 7.1). The variation of the neutral number density versus altitude, $n_n(z)$, can be found from Equation (7.15) by using the equation of state, $p = n_n(z)k_B T(z)$. As was shown in Example 4.2, the neutral scale height near the surface of the Earth is about 9 km; the atmospheric pressure near the top of Mt. Everest is about a factor of e^{-1} less than the surface pressure.

Equation (7.15) is not accurate throughout most of the thermosphere. The atmospheric composition remains well mixed by atmospheric eddy motions for altitudes (z) below the *homopause* (located at $z_{hom} \approx 100$ km for the Earth, Venus, and Mars); however, above the homopause, atmospheric constituents separate out according to their molecular mass. The details of this *diffusive separation* process are outside the scope of this book, but it can be rather simply (if not entirely accurately) represented by using diffusion equations (see Rees, 1989). The net result of diffusive separation is that the partial pressure of each species p_j ($j = N_2, O_2, O, He$, etc.) is given by an equation identical to (7.15) but with the average mass $\overline{m}$ replaced by the individual species mass m_j (e.g., $m_{N_2} = 28$ amu, $m_O = 16$ amu, etc.):

$$p_j(z) = p_{j0} \exp\left[-\int_{z_0}^{z} \frac{dz'}{H_j(z')}\right] \quad \text{with } H_j(z) = \frac{k_B T(z)}{m_j g}, \qquad (7.16)$$

where H_j is the individual species scale height. The reference height is z_0, which is often conveniently chosen to be the homopause altitude z_{hom}. The scale height for terrestrial N_2 with $T \approx 1{,}500$ K is $H_{N_2} \approx 50$ km.

Another factor influencing the composition of upper atmospheres is the photodissociation of molecules either by solar UV and EUV radiation or by energetic auroral particles. For example, atomic oxygen is created in the upper atmosphere by the photodissociation of O_2 via the reaction

$$h\nu + O_2 \rightarrow O + O. \qquad (7.17)$$

The atomic oxygen produced at high altitudes in the thermosphere diffuses downward through the background atmosphere and recombines near an altitude of 100 km. Diffusive separation dictates that O becomes more abundant than N_2 at very high altitudes. Molecular nitrogen is also dissociated by solar EUV radiation to form atomic nitrogen; however, a large fraction of the atomic nitrogen chemically reacts with oxygen and is converted into nitric oxide (NO), which has a maximum density of about 10^7–$10^8 \mathrm{cm}^{-3}$ near an altitude of 110 km. NO is an important molecule for the ionosphere because it is the only common atmospheric species that can be photoionized by solar Lyman alpha photons ($\lambda = 121.6$ nm). Again, for

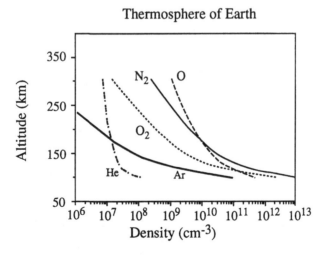

Figure 7.7. Neutral atmospheric densities for Earth with values taken from Table A1.2 of Rees (1989). These values are from the MSIS-1986 model atmosphere for 19 February 1979. The neutral temperatures at 100 km, 150 km, 200 km, and 300 km are 191 K, 779 K, 1093 K, and 1269 K, respectively

details of the upper atmosphere composition see Rees (1989). Figure 7.7 is a plot of densities versus altitude for several important neutral species in the terrestrial thermosphere for solar maximum conditions and moderate magnetic activity conditions. Notice that atomic oxygen gradually becomes more abundant than molecular nitrogen as altitude increases.

The major neutral constituent in the atmospheres of both Venus and Mars is carbon dioxide. The surface pressure at Venus is 90 bars and at Mars is only 6 mb. The surface temperature at Venus is about 600 K and at Mars is only about 250 K. However, we are more interested (for the topic of ionospheres) in the characteristics of the thermospheres of these two planets than in their lower atmospheres. The atmospheric density near 100 km is about the same at Earth, Venus, and Mars, although the structure at higher altitudes differs because the thermospheres of Venus and Mars are colder than the terrestrial thermosphere, due to the large cooling rate associated with the high abundance of the IR-active molecule CO_2. The exospheric temperatures of Venus and Mars are only about 280 K and 150 K, respectively. One consequence of these low temperatures is that the neutral density in the thermospheres of these two planets falls off more rapidly than at Earth; in Problem 7.5 you are asked to find the O and CO_2 scale heights for Venus and Mars. For the major neutral species, CO_2, we find that $H_n = 5$ km and 7 km for Venus and Mars, respectively. Figure 7.8 shows densities from an empirical model of the thermosphere of Venus based on data from the *Pioneer Venus Orbiter*, especially the neutral mass spectrometer. Notice that at higher altitudes, atomic oxygen becomes the most abundant

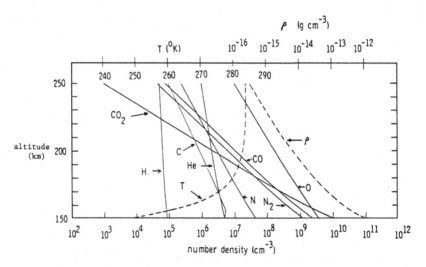

Figure 7.8. The thermosphere of Venus. The curves show the relationship between altitude and neutral number densities of various species. The mass density ρ and the neutral temperature are also shown with scales at the top of the plot. (From Keating et al., 1985.) Reprinted from *Advances in Space Research*, vol. 5, G. M. Keating et al., p. 117, in *The Venus International Reference Atmosphere*, eds. A. J. Kliore et al., copyright 1985, with kind permission from Elsevier Science ltd., The Boulevard, Langford Lane, Kidlington OX5 1GB, UK.

neutral, just as in the Earth's thermosphere. The O and CO in the atmosphere are produced by the photodissociation of CO_2 (analogous to reaction 7.17).

The outer planets (Jupiter, Saturn, Uranus, and Neptune) are gas giants and do not have solid surfaces, although they probably have solid cores. Their main atmospheric constituent is molecular hydrogen (H_2), although methane (CH_4) and other hydrocarbons are also present (see references listed at end of the chapter, especially Atreya). Atomic hydrogen is also present in the upper atmosphere due to the dissociation of H_2. Our information about the atmospheres of these planets is mainly from the *Voyager 1* and *2* (*V1* and *V2*) spacecraft although information for Jupiter is just now becoming available from the *Galileo* mission. Occultations by the upper atmospheres of these planets of ultraviolet light from the Sun and stars, and measured by the ultraviolet spectrometers on board *V1* and *V2*, were used to deduce neutral upper atmospheric densities and temperatures.

In spite of the large gravitational acceleration (Table 7.1), the neutral scale heights in the thermospheres of these planets are quite large (200 km and 400 km, respectively, for Jupiter and Saturn – see Problem 7.5), due to the low molecular mass of H_2 and to the high thermospheric temperatures. The exospheric temperatures are about $T_n \approx 1{,}100$ K for Jupiter and ≈ 900 K for Saturn (cf. Atreya, 1986). Model calculations using only solar EUV radiation to heat the upper atmospheres of these planets yield T_n values of only ≈ 150 K. It is thought that the energy deposited in the upper atmosphere by energetic auroral particles from the magnetosphere is very important in determining the thermal structure of the thermospheres of Jupiter and Saturn.

7.3.3 Formation of ionospheres

The ionospheric composition of a planet depends on the composition of the neutral thermosphere of that planet and also on ion–neutral chemistry. For example, a thermospheric neutral species, designated M, can be photoionized by solar EUV photons via

$$h\nu + M \rightarrow M^+ + e. \tag{7.18}$$

The photon energy $h\nu$ must exceed the ionization potential I_M of species M for the photoionization reaction (7.18) to proceed. For most atmospheric species, I_M lies in the range of 10 eV to 20 eV, which means that for photoionization to occur the photon wavelength must be less than ≈ 100 nm or the photon energy must exceed ≈ 10 eV. Note that h is Planck's constant and $\nu = c/\lambda$ is the photon frequency. The excess energy in reaction (7.18) goes to the photoelectron produced by the photoionization event,

$$E_{\text{photoelectron}} = (h\nu - I_M). \tag{7.19}$$

Photoelectrons are produced with energies ranging from zero up to hundreds of eV, although the average energy is about 15 eV.

Where in the atmosphere are the photoions, M^+, produced? Some knowledge of radiative transfer is needed to answer this question. The solar photon flux in the EUV wavelength range is produced in the solar corona and chromosphere and is dominated by the strong emission lines of highly excited ion species. Rees (1989) shows this flux as a function of wavelength. The solar chromosphere and corona were discussed in Chapter 5. The radiative transfer Equations (5.28) and (5.29), used in solar physics, are also relevant here although in slightly modified form in that flux (units of photons $cm^{-2} s^{-1}$) rather than intensity (units of photons $cm^{-2} s^{-1} sr^{-1}$) is employed. For a plane parallel atmosphere the photon flux at some altitude and at some wavelength depends on the photon flux incident at the top of the atmosphere ($F_{\lambda 0}$):

$$F_\lambda(z, \chi) = F_{\lambda 0} \exp(-\tau_\lambda), \qquad (7.20)$$

where the optical depth at wavelength λ and altitude z is given by

$$\tau_\lambda = \frac{1}{\cos \chi} \int_z^\infty n_n(z') \sigma_\lambda dz', \qquad (7.21)$$

where $n_n(z)$ is the neutral density as a function of altitude and σ_λ is the photoabsorption cross section at λ. χ is the solar zenith angle; that is, the angle between the Sun and the vertical direction. The integration is carried out from altitude z out to the top of the atmosphere. Actually, the optical depths for all individual neutral species in an atmosphere should be added together to obtain the total optical depth.

The photoabsorption cross section is almost equal to the photoionization cross section for ionizing photons, with values of the order of $\sigma_\lambda \approx 10^{-17} cm^2$. An estimate of the optical depth at an altitude z can be made by assuming that the atmospheric scale height in Equation (7.16) is independent of altitude (i.e., an isothermal atmosphere), in which case the integral in Equation (7.21) is easily evaluated, giving

$$\tau_\lambda \approx \frac{1}{\cos \chi} n_n(z) H_n \sigma_\lambda, \qquad (7.22)$$

where H_n is the scale height of the dominant neutral species. For Earth, where N_2 is the major species below 200 km with $H_n \approx 50$ km, and for an overhead Sun ($\chi = 0$), we have from Equation (7.22) that $\tau_\lambda \approx 5 \times 10^{-11} n_n (cm^{-3})$ for a typical EUV photon. The solar intensity is reduced from its unattenuated value at the top of the atmosphere by one e-folding, where $\tau_\lambda \approx 1$. This corresponds to a value of $n_n \approx 2 \times 10^{10} cm^{-3}$ or an altitude of $z \approx 150$ km according to the model atmosphere shown in Figure 7.7.

The production rate of ions of species M, $P_M(z)$, at an altitude z depends on the solar intensity at that altitude and on the density of species M at that altitude:

$$P_M(z) = \int_0^{\lambda_M} F_\lambda(z) n_n(z) \sigma_\lambda d\lambda \quad [\text{ions cm}^{-3} \text{ s}^{-1}], \qquad (7.23)$$

where the integration is over wavelength from zero up to the ionization threshold (λ_m). The EUV photon flux is given by Equation (7.20). At higher altitudes, where the optical depth is very small and F_λ equals $F_{\lambda 0}$, the ion production rate is proportional to the neutral density and, hence, decreases with z exponentially. At low altitudes, where the optical depth is larger, P_M decreases with decreasing z. $P_M(z)$ has a maximum at the altitude where the optical depth is unity. This type of production rate profile, or the electron density profiles (as we will see) that result from it, is called a *Chapman layer* profile (after the English scientist S. Chapman, who first developed the formalism in the 1930s).

In the optically thin region at high altitudes the solar intensity is independent of altitude, and Equation (7.23) can be written as some constant, R_M, multiplied by the density n_n; that is, $P_M(z) = R_M n_n(z)$. For ionization of atomic oxygen at 1 AU, $R_M \approx 10^{-6} \text{s}^{-1}$.

The major ion species produced in the terrestrial ionosphere are N_2^+, O_2^+, and O^+, which arise via the reactions

$$h\nu + N_2 \rightarrow N_2^+ + e,$$

$$h\nu + O_2 \rightarrow O_2^+ + e,$$

and

$$h\nu + O \rightarrow O^+ + e. \qquad (7.24)$$

In addition to photoionization by solar EUV photons, electron impact ionization is also important in the auroral zone at high latitudes. The peak ionization rates occur near 150 km for N_2^+ and O_2^+ and near 180 km for O^+. The peak ionization rate for N_2 is about $10^5 \text{ cm}^{-3} \text{ s}^{-1}$.

The major ions produced in the ionosphere of Venus and Mars are CO_2^+, CO^+, and O^+, from reactions similar to (7.24). Some of the O^+ ions in the lower ionosphere of Venus are produced from dissociative photoionization of CO_2. The results of a numerical calculation of ion production rates in the dayside ionosphere of Venus are shown in Figure 7.9. The altitude of peak ion production for Venus occurs at unit optical depth for EUV photons in a mainly CO_2 atmosphere. Given that the EUV photoabsorption cross section for CO_2 is about $2 \times 10^{-17} \text{cm}^2$ and using the model atmosphere shown in Figure 7.8 it can be demonstrated that peak ion production rate takes place near 140 km.

It is not enough to just produce ions. What happens to these ions? Continual production by itself would result in a completely ionized atmosphere. This is obviously ridiculous; the ions and electrons in the ionospheric plasma must recombine at

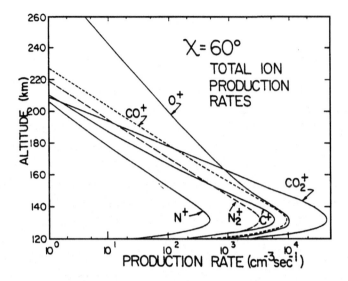

Figure 7.9. Ion production rate profiles in the dayside ionosphere of Venus for a solar zenith angle of 60 degrees at solar maximum conditions. (From Nagy, 1980.)

some point, and transport of plasma from one region to another can also take place. The behavior of the ionospheric plasma can be described using the conservation equations presented in Chapters 2 and 4.

7.3.4 Conservation equations for ionospheres

The fluid conservation equations discussed in Chapters 2 and 4 work very nicely for planetary ionospheres if the correct parameters and collision terms are employed. The single ion species continuity equation was given by Equation (4.4), but in many cases, one must work with the multi-species equations. The continuity equation for the density n_s of species s was given by Equation (2.40). In that equation, S_s represented the net source of particles of species s and is equal to the production rate of $s(P_s)$ minus the loss rate of $s(L_s)$:

$$S_s = P_s - L_s. \tag{7.25}$$

P_s must include both "primary" production from photoionization or electron impact ionization and production from chemistry. The loss L_s is mainly due to ion–neutral reactions and to electron–ion recombination. The example used in Chapter 2 was the ion–neutral reaction (2.44):

$$O^+ + N_2 \rightarrow NO^+ + N \qquad k_{26} \approx 10^{-12} \, \text{cm}^3 \, \text{s}^{-1}. \tag{7.26}$$

This reaction acts as a source of NO^+ ions and a chemical sink, or loss, of O^+ ions. Actually, the rate coefficients for different atomic states of O^+ are different, but we overlook this (see Rees).

The continuity equation (2.40) contains the bulk flow speed of species s, denoted $\mathbf{u}_s$. The momentum equation (2.61) for species s can be used to determine $\mathbf{u}_s$. The collision terms for this equation were given by Equation (2.64). Separate momentum equations for electrons and a single ion species were given in Chapter 4 (Equations (4.3) and (4.4)), and by combining these equations we obtained the single-fluid momentum equation (4.86). It was shown that for situations in which a large magnetic field acts to direct the bulk plasma flow, a diffusion approximation can be very useful; more specifically, the *ambipolar diffusion equation* for vertical plasma flow speed, u_z, was given by Equations (4.110) and (4.111). This section of Chapter 4 should be reviewed, because it is directly applicable to ionospheres! It is also possible to derive diffusion equations for the vertical flow speeds, u_{sz}, of individual ion species, starting from Equation (4.4) (Problem 7.7).

The combination of continuity equation and single-fluid momentum equation yields Equation (4.112). This equation is a second-order parabolic partial differential equation for the ion density ($n_i = n_e$) and can in general be solved numerically, although simple analytical approximations can also be made. For some conditions, the right-hand side of the equation, which contains the primary ion production and the chemical production and loss (i.e., the net ion production), is negligible in comparison with $d\phi_i/dz$ (or $d\phi_s/dz$), the vertical transport part. In this case, one has a source-free diffusion problem. For other conditions, both the $\partial n_s/\partial t$ and $\partial \phi_s/\partial z$ terms are much smaller than the net production S_s, and one is simply left with the equation $S_s \approx 0$; this is known as *photochemical equilibrium*. From Equation (7.25), the photochemical equilibrium equation for species s is:

$$P_s = L_s. \tag{7.27}$$

This equation states that at each altitude the "local" production rate of species s (including primary production and production from chemical reactions) must equal the "local" loss rate of this species. If photochemical equilibrium is valid for all ion species ($s = 1, \ldots, N$), then one must solve N algebraic equations (7.27) for N unknowns, n_s. These equations are in general coupled because the equation for one species contains in the chemical production or loss terms the densities of at least some of the other species. An auxiliary condition from quasi-neutrality must also hold: The sum of n_s for all ion species must equal n_e, the electron density.

When is it safe to apply photochemical equilibrium? The chemical lifetime of a species, denoted τ_{chem}, can be compared with the time, denoted τ_{trans}, it takes that species to be transported vertically a significant distance. If $\tau_{\text{chem}} \ll \tau_{\text{trans}}$, then we can safely neglect the nonchemical terms in the continuity equation and assume that $S_s = 0$. However, if $\tau_{\text{chem}} \gg \tau_{\text{trans}}$, then vertical transport of ions from one altitude to another is more important than local chemical processes. For an ion species whose loss is mainly from an ion–neutral chemical reaction, such as (7.26) for O^+, the chemical lifetime of that ion species is given by $\tau_{\text{chem}} \approx 1/kn_n$, where k is the relevant rate coefficient and n_n is the relevant neutral density. If the main chemical loss of species s is due to electron–ion recombination, then $\tau_{\text{chem}} \approx 1/\alpha n_e$, where α is the relevant electron–ion recombination rate coefficient.

Now we consider the vertical transport of ionospheric plasma. A particularly simple solution of the vertical diffusion equation is obtained if the ion flux is very small (i.e., $\phi_i \approx 0$), which is often true at high altitudes where τ_{trans} is very small. This approximation is called *diffusive equilibrium*, and for a single ion species plasma this is just the same as hydrostatic equilibrium, in which the pressure varies exponentially with a length scale given by the *plasma scale height H_p*. For an isothermal ionosphere, we find on rewriting Equations (4.97) and (4.98) that

$$n_e(z) = n_{e0} \exp\left(-\frac{z - z_0}{H_p}\right) \quad \text{with } H_p = \frac{k_B(T_e + T_i)}{m_i g}, \qquad (7.28)$$

where m_i is the mass of the major ion species and n_{e0} and z_0 are reference electron density and altitude, respectively. Equation (7.28) applies to the major ion species (and to the electron density) even if other ion species are present, as long as the densities of these species are considerably less than n_e. The diffusive equilibrium scale heights for "minor" ion species can be found from an equation similar to (7.28) (see part b of Problem 7.7). The electron temperature appears in Equation (7.28) because of the presence in the ion momentum equation of the ambipolar electric field, as considered in Equation (4.48). If we were to assume that $T_e = T_i$, then the plasma scale height can be written as $H_p = k_B T_i / \langle m \rangle g$, where for the special case of a single ion species the average mass (including electrons) is just $\langle m \rangle = m_i/2$. Thus, the plasma scale height formula differs from the neutral scale height formula (Equation 7.16) because the polarization electric field "ties" the ions and electrons together, necessitating the use of an "average" mass.

Suppose that the actual density profile in an ionosphere is not given by Equation (7.28), or its individual ion species equivalent, but is determined by other processes such as chemistry. Then, the diffusion equation (4.110) indicates that the plasma flows upward or downward with speed u_z depending on the actual altitude dependence of the density. Given a length scale H for the actual variation of the density, the order of magnitude of the plasma speed is given by

$$u_z \approx \sin^2 \theta_I D_a \frac{1}{H}. \qquad (7.29)$$

For a vertical magnetic field, $\theta_I = 90°$ so that $u_z \approx D_a/H$. D_a is the ambipolar diffusion coefficient given by Equation (4.111).

The time scale (τ_{trans}) for changes of density associated with the vertical transport part of the equation, $\partial \phi_i / \partial z$, can be estimated by dimensionally analyzing this equation or by noting that $\tau_{\text{trans}} \approx H/u_z$. Using Equation (7.29) for u_z, we then find that the lifetime "against vertical transport" is equal to

$$\tau_{\text{trans}} \approx \frac{H^2}{D_a} \frac{1}{\sin^2 \theta_I}. \qquad (7.30)$$

Notice the similarity of this diffusion time to the diffusion time scales for other types of diffusion problems such as pitch-angle diffusion and magnetic diffusion (Sections 4.4 and 5.1). The transport time decreases with increasing altitude because the ambipolar diffusion coefficient is inversely proportional to the neutral density via the ion–neutral momentum transfer collision frequency $\nu_{in} = k_{in}n_n$. In Problem 7.8 you are asked to plot τ_{chem} and τ_{trans} as a function of z for O^+ in the terrestrial ionosphere. Notice that τ_{chem} is proportional to n_n^{-1} and thus increases with altitude, whereas τ_{trans} is proportional to n_n and thus decreases with altitude. The transport time is quite large at lower altitudes where the ion–neutral collision frequency is high. Hence at low altitudes $\tau_{chem} \ll \tau_{trans}$ and photochemical equilibrium tends to be a good approximation, whereas at high altitudes $\tau_{trans} \ll \tau_{chem}$ and diffusive equilibrium is a good approximation.

The full set of fluid conservation equations not only includes the continuity and momentum equations but also the energy equations, which can be used to determine T_i and T_e. Energy equations were discussed both in Chapter 2 and in Chapter 4. A complete discussion of energy equations for ionospheres is really beyond the scope of this book, and the reader is referred to Banks and Kockarts (1973), Rees (1989), or other references listed at the end of this chapter. Nonetheless, a very short review of the energetics will now be provided.

Collision frequencies are high in the lower ionosphere, and collisional processes such as collisional heating and cooling are more important than transport of heat. In fact, at lower altitudes the local heating rate equals the local cooling rate. For electrons, cooling can occur via: (1) Coulomb collisions with colder ions, (2) elastic collisions with neutrals, and, most importantly, via (3) inelastic collisions with neutrals. For example, rotational and vibrational excitation of neutral molecules by electron impact are very important electron cooling mechanisms. In the Earth's ionosphere, electron excitation of the atomic oxygen fine structure lines is also an important electron cooling mechanism. The most important ion cooling mechanism is ion–neutral elastic collisions. In the lower ionosphere, the collision frequency is so large and the cooling so efficient that the plasma is well coupled to the neutral gas thermally and $T_e = T_i = T_n$. However, as z increases and the collision frequency becomes smaller, then differential heating of the electrons due to the addition of photoelectrons to the ionosphere or due to auroral electrons at higher latitudes on Earth (or due to the solar wind interaction at Venus) results in electron temperatures that exceed the ion and neutral temperatures. Electron–ion collisions are usually the main source of ion heat. The following relation almost always holds in planetary ionospheres: $T_e > T_i > T_n$. For example, in both the terrestrial and Venusian ionospheres $T_e \approx 5{,}000$ K at higher altitudes and $T_i \approx T_e/2 \approx 2{,}500$ K, whereas $T_n \approx 300$ K in the thermosphere of Venus and $T_n \approx 1{,}500$ K in the thermosphere of Earth.

Transport of heat must also be considered at higher altitudes. Sometimes, such as in the polar ionosphere of Earth, bulk transport of heat must be considered for the ion gas (this is important in the solar wind also), but usually the main heat transport mechanism in the ionosphere is vertical heat conduction, especially for the electrons. Electron heat conduction was also considered for the electron gas in the solar corona; see Equations (6.7)–(6.9). In Problem (7.9), you are asked to solve the electron heat conduction equation for the terrestrial ionosphere, making the simplifying assumptions that all cooling takes place at the "base" of the ionosphere and that all photoelectron heating is introduced above some reference altitude.

7.3.5 *Example of an ionosphere: The terrestrial ionosphere*

A plot of electron density in the terrestrial ionosphere was shown in Figure 4.8. We now use the information presented in Section 7.3 to explain some of the ionospheric regions indicated in this figure. The major ions produced in the D- and E-regions are N_2^+ and O_2^+, and the production rate profiles are described by a Chapman layer, as discussed earlier (Equation (7.23)). The peak ionization rate is about $10^5 \mathrm{cm}^{-3}\mathrm{s}^{-1}$ at 150 km. However, N_2^+ is only a minor ion, with a density orders of magnitude less than the electron density, because N_2^+ ions are rapidly removed from the ionosphere by ion–neutral reactions, namely

$$N_2^+ + O_2 \rightarrow O_2^+ + N_2 \qquad k_{31} = 5 \times 10^{-11} \, \mathrm{cm}^3 \, \mathrm{s}^{-1} \qquad (7.31)$$

and

$$N_2^+ + O \rightarrow NO^+ + N \qquad k_{32} = 1.4 \times 10^{-10} \, \mathrm{cm}^3 \, \mathrm{s}^{-1}. \qquad (7.32)$$

The N_2^+ ions are converted into O_2^+ and NO^+ ions, which are the major ions in the D- and E-regions of the ionosphere. NO^+ ions can also be produced from O_2^+ via the reaction

$$O_2^+ + NO \rightarrow NO^+ + O_2 \qquad k_{33} = 4.4 \times 10^{-10} \, \mathrm{cm}^3 \, \mathrm{s}^{-1}. \qquad (7.33)$$

Both O_2^+ and NO^+ undergo electron–ion dissociative recombination:

$$O_2^+ + e \rightarrow O + O \qquad \alpha_{34} = 1.9 \times 10^{-7}(300/T_e)^{1/2} \, \mathrm{cm}^3 \, \mathrm{s}^{-1} \qquad (7.34)$$

$$NO^+ + e \rightarrow N + O \qquad \alpha_{35} = 4.2 \times 10^{-7}(300/T_e)^{.85} \, \mathrm{cm}^3 \, \mathrm{s}^{-1}. \qquad (7.35)$$

For NO^+, this is the only loss process, and for this reason, NO^+ is known as a *terminal ion*. The rate coefficients for recombination depend on the electron temperature (see Rees for details and for references).

Ion–neutral reactions are quite fast in the E-region, and almost all O^+ or N_2^+ ions become either O_2^+ or NO^+. For simplicity, let us neglect O_2^+ and assume that NO^+ is the major ion; this is often a good assumption. The NO^+ density must then be equal to the electron density, in which case for photochemical equilibrium (Equation (7.27)) the total ion production rate, which is approximately $P_{N_2^+}$, neglecting the O_2^+ production, must equal the loss rate associated with reaction (7.35):

$$P_{N_2^+} = \alpha_{35} n_e^2. \tag{7.36}$$

Equation (7.36) can easily be solved for the electron density to give

$$n_e(z) = \left[P_{N_2^+}/\alpha_{35}\right]^{1/2}. \tag{7.37}$$

The electron density varies as the square root of the total ion production rate. If it were not for the presence of O^+ ions at higher altitudes, which we will discuss next, the electron density in the terrestrial ionosphere would have a maximum at the altitude where the production rate profile has its maximum. In any case, let us use Equation (7.37) to estimate the electron density at a particular altitude ($z = 150$ km); with $P_{N_2^+} \approx 10^5 \text{cm}^{-3} \text{s}^{-1}$ and given $\alpha_{35} \approx 10^{-6} \text{ cm}^3 \text{s}^{-1}$, we find that $n_e \approx 3 \times 10^5 \text{ cm}^{-3}$ (or $3 \times 10^{11} \text{ m}^{-3}$). Looking at Figure 4.8, we see that this is indeed the correct density for the daytime. At nighttime, the ion production from solar radiation obviously ceases, but there are other weak ion sources, and the ions produced during the day are only gradually removed chemically (see Problem 7.8). Furthermore, ions present at higher altitudes gradually diffuse downward into the E-region.

Why does the electron density continue to increase with altitude above the production maximum? The answer is that O^+ ions are produced at higher altitudes and O^+ ions cannot recombine dissociatively. The main chemical loss mechanism for O^+ is the relatively slow ion–neutral reaction (7.26). The optically thin production rate of O^+ ions is appropriate at higher altitudes: $P_{O^+} \approx 10^{-6} n_O$, where $n_O(z)$ is the neutral O density. In photochemical equilibrium, the production of O^+ is balanced by the loss via reaction (7.26); that is, the O^+ density is given by the relation $10^{-6} n_O = k_{26} n_{O^+} n_{N_2}$. Solving for the photochemical O^+ density, we find

$$n_{O^+} = [10^{-6}/k_{26}][n_O/n_{N_2}] \approx 10^6 [n_O/n_{N_2}] \quad [\text{cm}^{-3}]. \tag{7.38}$$

The O^+ density, which is equal to the electron density for z greater than ≈ 160 km, varies as the ratio of the neutral O density to the N_2 density and therefore increases with altitude because neutral atomic oxygen has a larger scale height than N_2. Equation (7.38) indicates that $n_{O^+} \approx 3 \times 10^5 \text{cm}^{-3}$ at 150 km for the neutral atmosphere shown in Figure 7.7, which is about equal to the NO^+ density earlier estimated for this altitude. However, at higher altitudes the NO^+ density continues to decrease with altitude, whereas the O^+ density increases, so that by an altitude of 250 km, n_{O^+} (and n_e) has reached a value of about 10^6 cm^{-3}. The region of the ionosphere near 200 km where O^+ is the major ion, but is still photochemically controlled, is designated the *F_1-region*.

The electron density cannot continue to increase indefinitely. The chemical lifetime of the O^+ ion also increases with altitude: $\tau_{\text{chem}} \approx 1/(k_{26} n_{N_2})$. As you will show in Problem 7.8, this lifetime exceeds the vertical transport time near an altitude of 250 km! Thus vertical transport of plasma (i.e., ambipolar diffusion) is more important than chemistry above about 250 km. The O^+ ions produced at higher altitudes diffuse down to altitudes below 250 km, where they then react with N_2. At altitudes where diffusion clearly dominates, diffusive equilibrium becomes a good approximation and the electron density is given by Equation (7.28). The electron density has a maximum roughly at the altitude where $\tau_{chem} \approx \tau_{trans}$. The ionospheric region where diffusion dominates is called the *F_2-region*. A plasma scale height of $H_p = 260$ km was calculated in Chapter 4 for the terrestrial *topside ionosphere* (i.e., the ionosphere above the peak).

In addition to vertical transport, the plasma in the terrestrial ionosphere can also be transported horizontally. In fact, the ions and electrons do not move at exactly the same speed, and electrical currents flow in the ionosphere. In the polar regions, this horizontal motion and the currents result from the application of electric fields that "map" down from the magnetosphere, as will be discussed in Chapter 8.

7.3.6 Example of an ionosphere: Venus

The lower ionospheres of Venus and Mars are well described by photochemical theory. The ionospheres of Venus and Mars are very similar in their composition and chemistry. The major neutral species is CO_2 and hence the major ion produced is CO_2^+ (Figure 7.9), although at higher altitudes O^+ production becomes important. Both CO_2^+ and O^+ are converted to O_2^+ by the ion–neutral reactions

$$CO_2^+ + O \rightarrow O_2^+ + CO \qquad k_{39} = 1.64 \times 10^{-10} \, cm^3 \, s^{-1} \qquad (7.39)$$

and

$$O^+ + CO_2 \rightarrow O_2^+ + CO \qquad k_{40} = 9.4 \times 10^{-10} \, cm^3 \, s^{-1}. \qquad (7.40)$$

Both of these reactions proceed rapidly in the lower ionosphere, and O_2^+ is the major ion species, although the atmosphere of Venus has less than a part per million of neutral O_2. The O_2^+ ions are removed via the dissociative recombination reaction (7.34). The photochemical density expression (7.37) is also applicable to the lower ionospheres of Venus and Mars, but with α_{34} rather than α_{35}, and with $P_{CO_2^+}$ rather than $P_{N_2^+}$. In Problem 7.14, you are asked to find the CO_2^+ and O_2^+ densities. The results of a numerical calculation including both chemistry and vertical transport are shown in Figure 7.10. The electron density at a given altitude is the sum of the ion densities at that altitude. The peak of the electron density profile is located where the production rate reaches a maximum.

What about O^+? At Earth, the O^+ density increases with altitude and O^+ is the major ion at high altitudes. At Venus, the O^+ density also increases with altitude and O^+ is the major ion. In fact, an expression analogous to Equation (7.38) can

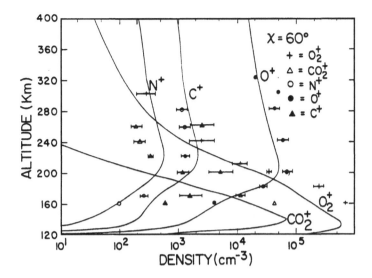

Figure 7.10. Ion density versus altitude for the dayside ionosphere of Venus. Adopted from a theoretical calculation for an unmagnetized ionosphere. Theoretical densities (solid lines) and densities measured by the ion mass spectrometer on board the *Pioneer Venus Orbiter* (OIMS). (From Nagy, 1980.)

be derived, but with the loss of O^+ coming from reaction (7.40):

$$n_{O^+} \approx [10^{-6}/k_{40}]\,[n_O/n_{CO_2}] \approx 10^3[n_O/n_{CO_2}] \quad [cm^{-3}]. \tag{7.41}$$

A big difference between Earth and Venus is that the rate coefficient of reaction (7.40) is much larger than the coefficient for reaction (7.26), with the consequence that the O^+ density at Venus is more than a factor of ten less than at Earth. Consequently, the maximum O^+ density at Venus, which also occurs at an altitude where $\tau_{chem} \approx \tau_{trans}$, is more than a factor of ten less than the maximum O^+ density at Earth. The electron density maximum at Venus exists not in the F-region but further down in the E-region. The electron density profile is a diffusive equilibrium type profile above the peak for those periods of time when the ionosphere remains unmagnetized and vertical ambipolar diffusion is able to operate. As we shall see in Section 7.4, the ionosphere of Venus occasionally becomes magnetized due to the solar wind interaction, in which case vertical plasma transport is strongly influenced by the $\mathbf{J} \times \mathbf{B}$ term in the momentum equation.

7.3.7 Other examples of planetary ionospheres

We now quickly review the ion composition and chemistry of the ionospheres of some other planets and bodies in the solar system. The major neutral species in the thermospheres of the outer planets is molecular hydrogen. Some atomic hydrogen is also present due to photodissociation and auroral electron impact dissociation of H_2. The neutral H_2 and H are ionized by solar EUV photons and by electron impact

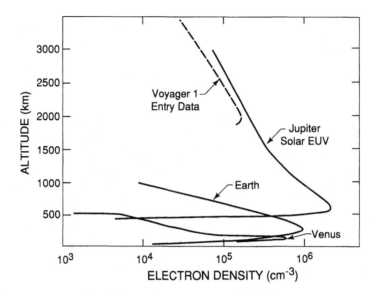

Figure 7.11. Comparison of ionospheric electron density profiles for Earth, Venus, and Jupiter. The dashed line is the profile measured by the *Voyager 1* spacecraft using the radio occultation technique. (From Cravens, 1983.) Copyright 1983 Kluwer Academic Publishers. With kind permission from Kluwer Academic Publishers.

(see Atreya, 1986, for details):

$$hv + H_2 \rightarrow \begin{cases} H_2^+ + e & \text{(90\% probability)} \\ H^+ + H + e & \text{(10\% probability)} \end{cases} \tag{7.42}$$

$$hv + H \rightarrow H^+ + e. \tag{7.43}$$

Although the largest ion production rate is for H_2^+, just as it is N_2^+ for Earth and CO_2^+ for Venus, H_2^+ is *not* the major ion species. H_2^+ is chemically removed from the ionosphere by the fast ion–neutral reaction

$$H_2^+ + H_2 \rightarrow H_3^+ + H. \tag{7.44}$$

The H_3^+ ions then recombine dissociatively via the fast reaction

$$H_3^+ + e \rightarrow H_2 + H \qquad \alpha_{45} \approx 10^{-7} \text{cm}^3 \text{ s}^{-1}. \tag{7.45}$$

In contrast, in standard models of the ionospheres of the outer planets (Figure 7.11) the only chemical loss process for H^+ is the following very slow *radiative recombination* reaction, in which a photon is emitted:

$$H^+ + e \rightarrow H + hv \qquad \alpha_{46} \approx 10^{-11} \text{cm}^3 \text{ s}^{-1}. \tag{7.46}$$

The electron density can be found by using an expression like Equation (7.37) but with α_{46} and the H^+ production rate in place of α_{35} and $P_{N_2^+}$. Other loss mechanisms for H^+ have also been invoked in the scientific literature; for a more complete treatment of the ionospheres of the outer planets, see Atreya (1986).

The largest satellite of Saturn, Titan, possesses a neutral atmosphere with a surface pressure ($p \approx 1.4$ bars) that is greater than the atmospheric pressure at Earth's surface. The major neutral species at Titan is N_2, although methane and other hydrocarbons are also present, with densities that are quite high in the thermosphere. Some hydrogen cyanide (HCN) is also present. Solar EUV photons and energetic electrons from the magnetosphere of Saturn can ionize the thermospheric neutrals. The major neutral produced is N_2^+ (reaction (7.24)), but the N_2^+ ions thus produced quickly react with methane and produce CH_3^+ ions, which in turn react with methane via the reactions

$$N_2^+ + CH_4 \rightarrow CH_3^+ + N_2 + H, \tag{7.47}$$

$$CH_3^+ + CH_4 \rightarrow C_2H_5^+ + H_2, \tag{7.48}$$

and

$$C_2H_5^+ + HCN \rightarrow H_2CN^+ + C_2H_4. \tag{7.49}$$

The $C_2H_5^+$ ions react with HCN and produce what is thought to be the major ion species in the ionosphere of Titan, H_2CN^+, which is the terminal ion species that recombines dissociatively. This suggested ion composition will be tested by measurements that will be made by the ion neutral mass spectrometer on board NASA's *Cassini* spacecraft, which is scheduled for launch in October 1997.

Now let us consider the neutral atmosphere and ionosphere of comets. Comets can either be periodic or make only a single appearance in the inner solar system. Our knowledge of comets has increased dramatically from spacecraft encounters with comet Giacobini–Zinner in 1985 and comet Halley in 1986. In particular, the *Giotto* spacecraft, built and launched by the European Space Agency, approached within 600 km of the nucleus of comet Halley, taking television pictures and measuring a variety of neutral and plasma parameters. The *nucleus* of a typical comet is only a few kilometers across and is mostly made of dust and water ice plus some other frozen volatiles. The mass ratio of dust to ice is approximately unity. When the nucleus is far from the Sun it remains inactive, but when it approaches within 2 AU or so, its surface heats up and water vapor sublimates from the ice, thereby producing a cometary atmosphere. The cometary atmosphere, called the *coma*, behaves very differently than other planetary atmospheres because the gravitational attraction is negligible for such a small object as a cometary nucleus. As a consequence, hydrostatic balance does not apply and the cometary atmosphere flows outward from the nucleus with a neutral outflow speed (u_n) of about 1 km/s. The neutral gas as it flows off the surface of the nucleus carries dust particles with it, so that a *dust coma* is also formed. The average dust grain is about $1 \mu m$ in diameter. Solar radiation exerts a small, but relentless, force on the small dust grains and they are "blown" away from the Sun, thus resulting in the formation of the *cometary dust tail*.

The neutral and gas outflow from a cometary nucleus can be studied using the hydrodynamical equations for neutrals and for dust. The momentum equations must include dust–gas friction, or drag, terms, but not gravity. The flow starts out from the surface subsonically with a neutral density of about 10^{12} cm^{-3} for a comet like Halley. The flow then goes through a transonic point (or critical point) at a radial distance not too far from the nucleus. This flow problem is analogous to the Laval nozzle problem and/or to the solar wind outflow problem. The effective area function has a constriction due to the gas–dust drag, whereas in the solar wind the "nozzle" is due to gravity. In Problem 6.5, you were asked to set up the cometary "dusty" hydrodynamics problem and to determine the critical distance of the flow for somewhat idealized conditions.

The flow speed becomes roughly independent of distance ($u_n \approx 1$ km/s) by a radial distance from the center of the nucleus of 10–20 km or so. The continuity equation for the neutral density can then be solved in a straightforward manner if one assumes spherical symmetry (Problem 7.15):

$$n_n(z) = \frac{Q}{4\pi u_n r^2}. \tag{7.50}$$

Q is the total gas production rate (molecules per second) of the comet, and r is the radial distance. Q is strongly dependent on the individual comet and on the heliocentric distance. Note that this functional dependence for the neutral density is very different from the exponential form we found for other planetary atmospheres, but recall that the solar wind electron density also varied as r^{-2} (Chapter 6). For comet Halley near 1 AU, $Q \approx 10^{30}$ s^{-1}.

The solar wind interacts very strongly with the neutral gas surrounding an active comet, but the plasma environment within a few thousand kilometers of a comet like Halley is largely shielded from the solar wind and an unperturbed ionosphere exists. The water vapor in the cometary atmosphere is photoionized by solar EUV radiation through the reaction

$$h\nu + H_2O \rightarrow H_2O^+ + e. \tag{7.51}$$

The ionization rate per neutral for H_2O is approximately $R_{H_2O} \approx 10^{-6}$ s^{-1} at 1 AU. However, the major ion in the cometary ionosphere is not H_2O^+ because the ion–neutral reaction

$$H_2O^+ + H_2O \rightarrow H_3O^+ + OH \qquad k_{52} = 2.1 \times 10^{-9} \text{ cm}^3\text{s}^{-1} \tag{7.52}$$

proceeds rapidly.

The major ion species is thus H_3O^+, which is removed via dissociative recombination (rate coefficient $\alpha \approx 10^{-6}$ cm^3 s^{-1}). It is easy to show (Problem 7.16) that the photochemical H_2O^+ density is independent of r and that the H_3O^+ density (and the electron density) varies as r^{-1}:

$$n_e(r) = \left(\frac{R_{H_2O}}{\alpha} \right)^{1/2} n_n(r)^{1/2} = \left(\frac{Q R_{H_2O}}{4\pi u_n \alpha} \right)^{1/2} \frac{1}{r}. \tag{7.53}$$

This expression provides a good description of the electron density variation at comet Halley out to about 10^4 km, as has been confirmed by *Giotto* measurements. The interaction of the solar wind with the atmosphere and ionosphere of a comet will be discussed in Section 7.5.

7.3.8 Electrical conductivity in a strongly magnetized ionosphere (Earth type)

The form of the generalized Ohm's law given by Equation (4.62) expresses the electric field $\mathbf{E}$ in terms of various parameters including the current density $\mathbf{J}$; however, it is sometimes more convenient to invert this expression and find $\mathbf{J}$ in terms of $\mathbf{E}$. For instance, this is true of the ionospheres of planets with strong intrinsic magnetic fields. In this section, we will first invert Equation (4.62) but will find that this is not adequate, and so we will then return to the multi-fluid momentum Equation (4.4) and derive another form of Ohm's law. These new forms of Ohm's law are a basic part of ionospheric electrodynamics, and they will also be useful in Chapter 8 when we will consider the coupling of the magnetosphere to the ionosphere.

We start with a simplified form of Equation (4.62). We neglect the electron inertial terms and we also neglect the ambipolar electric field term, which contains the gradient of the electron partial pressure. Although this is a poor assumption for the vertical direction, it is a good one for the horizontal direction because ionospheres are horizontally stratified with very small horizontal density and pressure gradients. And as we shall see in Chapter 8, we are particularly interested in horizontal electrical currents. We also work, for now, with the electric field ($\mathbf{E}'$) in the frame of reference of the plasma center of mass (which moves at velocity $\mathbf{u}$):

$$\mathbf{E}' = \mathbf{E} + \mathbf{u} \times \mathbf{B}. \tag{7.54}$$

The magnetic field is assumed to be specified. Even if electrical currents exist, the magnetic fields produced by these currents are much less than the intrinsic field for planets like Earth or Jupiter.

Now we write down what remains of the generalized Ohm's law:

$$\mathbf{E}' = \frac{1}{n_e e} \mathbf{J} \times \mathbf{B} + \eta_e \mathbf{J}, \tag{7.55}$$

where the first term on the right-hand side is the Hall term and the second term on the right-hand side is the Ohmic term. The motional electric field has been subsumed

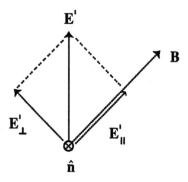

Figure 7.12. Geometry for ionospheric electrical conductivity.

into $\mathbf{E}'$. The resistivity was given by Equation (4.63) as $\eta = \eta_e = m_e \nu_{et}/(n_e e^2)$, where the total electron momentum transfer collision frequency, ν_{et}, includes both electron–ion and electron–neutral collisions. The subscript "e" has been added to denote resistivity just due to electron collisions. Now we can invert Equation (7.55) and find $\mathbf{J}$.

Consider the geometry shown in Figure 7.12. The electric field (in the frame of reference of the plasma) can be expressed in terms of its parallel and perpendicular components, $\mathbf{E}'_{\parallel}$ and $\mathbf{E}'_{\perp}$, respectively, as $\mathbf{E}' = \mathbf{E}'_{\parallel} + \mathbf{E}'_{\perp} = E'_{\parallel}\hat{\mathbf{b}} + E'_{\perp}\hat{\mathbf{l}}$, where $\hat{\mathbf{b}} = \mathbf{B}/|\mathbf{B}|$ is the unit vector in the magnetic field direction and $\hat{\mathbf{l}}$ is the unit vector along $\mathbf{E}'_{\perp}$. The unit vector in the $\mathbf{E} \times \mathbf{B}$ direction is denoted $\hat{\mathbf{n}}$. We can express the current density $\mathbf{J}$ in the coordinate system just defined (i.e., components in the parallel, perpendicular, and $\hat{\mathbf{n}}$ directions) as

$$\mathbf{J} = J_{\parallel}\hat{\mathbf{b}} + J_{\perp}\hat{\mathbf{l}} + J_{\mathrm{H}}\hat{\mathbf{n}}. \tag{7.56}$$

$J_{\parallel}$, $J_{\perp}$, and J_{H} are called the *parallel*, the *Pederson*, and the *Hall current densities*. The Pederson current is in the direction of the perpendicular component of the applied electric field and the Hall current is in the $\mathbf{E} \times \mathbf{B}$ direction.

We substitute Equation (7.56) into Equation (7.55), and then by writing the cross products as $\hat{\mathbf{l}} \times \hat{\mathbf{b}} = \hat{\mathbf{n}}$ and $\hat{\mathbf{n}} \times \hat{\mathbf{b}} = -\hat{\mathbf{l}}$, we find that Ohm's law has become

$$\mathbf{E}' = \eta_e J_{\parallel}\hat{\mathbf{b}} + \eta_e J_{\perp}\hat{\mathbf{l}} - \frac{B J_{\mathrm{H}}}{n_e e}\hat{\mathbf{l}} + \eta_e J_{\mathrm{H}}\hat{\mathbf{n}} + \frac{B J_{\perp}}{n_e e}\hat{\mathbf{n}}. \tag{7.57}$$

$\mathbf{E}'$ itself can be expressed in terms of its components. The parallel component of Equation (7.57) is

$$E'_{\parallel} = \eta_e J_{\parallel} \quad \text{or} \quad J_{\parallel} = \sigma_{0e} E'_{\parallel}, \tag{7.58}$$

where the *parallel electrical conductivity* (units of S/m – Siemans per meter) is just the reciprocal of the resistivity ($\sigma_{0e} = 1/\eta_e$).

After removing the parallel part of Equation (7.57), we are left with the $\hat{n}$ and $\hat{\perp}$ components. The following set of equations then must hold:

$$E'_\perp = -\frac{B}{n_e e} J_H + \eta_e J_\perp$$

$$0 = \eta_e J_H + \frac{B}{n_e e} J_\perp.$$

(7.59)

This pair of linear equations can be solved for the Hall and Pederson current densities. You will do this in Problem 7.17, using the fact that from the definition of η_e, one can recognize that

$$\frac{B}{n_e e} \frac{1}{\eta_e} = \frac{eB}{m_e} \frac{1}{v_{et}} = \frac{\Omega_e}{v_{et}}.$$

(7.60)

The electron gyrofrequency Ω_e is positive here. The solution of the equation set (7.59) for J_H and $J_\perp$ is

$$J_H = \sigma_H E'_\perp \quad \text{and} \quad J_\perp = \sigma_\perp E'_\perp,$$

(7.61)

where the *Hall* (σ_H) and *Pederson* ($\sigma_\perp$) *electrical conductivities* are given by

$$\sigma_\perp = \sigma_{0e} \frac{v_{et}^2}{v_{et}^2 + \Omega_e^2}$$

(7.62)

and

$$\sigma_H = \sigma_{0e} \frac{v_{et} \Omega_e}{v_{et}^2 + \Omega_e^2},$$

(7.63)

where Ω_e is taken to be positive. We can also write the current density as

$$\mathbf{J} = \sigma_{0e} \mathbf{E}_\parallel + \sigma_\perp \mathbf{E}'_\perp + \sigma_H \mathbf{E}'_\perp \times \hat{\mathbf{b}}.$$

(7.64)

Note that units of $\mathbf{E}'$ are V/m, units of $\mathbf{J}$ are A/m^2, and units of conductivity are S/m (or ohm^{-1}m^{-1} or Ω^{-1}m^{-1}).

As shown in Problem 7.17, the parallel conductivity is much larger than either the Hall or Pederson conductivities ($\sigma_{0e} \gg \sigma_H$, $\sigma_{0e} \gg \sigma_\perp$, and $\sigma_H \gg \sigma_\perp$) because the electron gyrofrequency is very much larger than the electron collision frequency.

A problem exists with this form of the GOL. The electric field is in the frame of reference of the bulk plasma flow (Equation (7.54)), so that if the electric field applied to the ionosphere, $\mathbf{E}$, is specified, one still needs to know the plasma flow speed $\mathbf{u}$ by some other means, such as from a full solution of the fluid equations. It would really be convenient if one could immediately find the current density from the applied electric field alone. In general, this is not possible; however, for ionospheres with strong magnetic fields one can achieve this without sacrificing too much accuracy. Consider a plasma composed of several ion species plus electrons (species $s = e$ or ions). The mass and charge for each species is given by m_s and

q_s, respectively (e.g., m_e and $q_e = -e$ for electrons). We find the current density by starting with its definition in terms of the density of species s, n_s, the charge q_s, and the bulk flow velocity of species s, $\mathbf{u}_s$:

$$\mathbf{J} = \sum_s n_s q_s \mathbf{u}_s. \tag{7.65}$$

The summation is over all plasma species, including electrons.

We can assume that we know the densities n_s from measurements or from theory (such as photochemical equilibrium theories in the E-region), and we therefore need to know the velocities $\mathbf{u}_s$ in terms of the applied electric field in order to use Equation (7.65) to find $\mathbf{J}$. We start with the momentum equation for species s, Equation (2.61), as discussed in Chapter 2, but we must greatly simplify this equation. First, we *only* consider collisions between species s and neutrals (i.e., only friction terms with ν_{sn}), and we neglect electron–ion and ion–ion collisions. This is a reasonable assumption for the ions but not for the electrons. We also neglect the inertial terms and the pressure gradient terms. The justification for the latter simplification is that *the horizontal component of the current is the most important*, and pressure gradients are small in this direction due to stratification (i.e., a typical vertical scale is 100 km and a typical horizontal scale is several thousand kilometers). The approximate momentum equation we are left with for species s is

$$\mathbf{0} = q_s \{ \mathbf{E}' + (\mathbf{u}_s - \mathbf{u}_n) \times \mathbf{B} \} - m_s \nu_{sn} (\mathbf{u}_s - \mathbf{u}_n). \tag{7.66}$$

Note that $\mathbf{E}'$ in this equation is not the same as the $\mathbf{E}'$ used earlier in this section. We now use $\mathbf{E}'$ to denote the electric field in the frame of reference of the neutral gas that has flow (or wind) velocity $\mathbf{u}_n$:

$$\mathbf{E}' = \mathbf{E} + \mathbf{u}_n \times \mathbf{B}. \tag{7.67}$$

Very often $\mathbf{u}_n$ is small and can be neglected, but, in any case, we can assume that we know what it is and do not generally have to solve for it.

Now we can solve the set of equations (three for each species) given by Equation (7.66) for $\mathbf{u}_s$. We do this in almost the same manner as we inverted the GOL, earlier in this section, and we obtain an equation identical to (7.64), although the expressions for the conductivities differ from those derived earlier. Equation (7.66) can be inverted to find the components of $\mathbf{u}_s$ along the $\mathbf{b}$, $\hat{\mathbf{n}}$, and $\hat{\perp}$ directions (see Problem 7.18). The resulting expressions for these components of $\mathbf{u}_s$ plus the definition of $\mathbf{J}$ given by Equation (7.65) can be used to rewrite the GOL in the form

$$\mathbf{J} = \sigma_0 \mathbf{E}'_{\parallel} + \sigma_{\perp} \mathbf{E}'_{\perp} + \sigma_{\mathrm{H}} \mathbf{E}'_{\perp} \times \hat{\mathbf{b}}. \tag{7.68}$$

This equation is identical to Equation (7.64) except that σ_0 has replaced σ_{0e}. Note that $\mathbf{E}'_{\parallel} = \mathbf{E}_{\parallel}$. Also, the expressions for the conductivities are different. For the parallel conductivity we find (again see Problem 7.17)

$$\sigma_0 = \sum_s \sigma_{0s} \quad \text{with } \sigma_{0s} = \frac{n_s q_s^2}{m_s \nu_{sn}}, \tag{7.69}$$

where ν_{sn} is the collision frequency between species s and the neutrals. The sum is over all plasma species. You can see that because $m_e \ll m_i$ (where i denotes any ion species), the contribution to the parallel conductivity from ions is negligible. Furthermore, since m_e is in the denominator the parallel conductivity is large. Actually, it is better to use the expression derived earlier for σ_{0e}, which is the same as Equation (7.69), except that the total electron collision frequency ($\nu_{et} = \nu_{ei} + \nu_{en}$, where ν_{ei} is the electron–ion collision frequency) is present instead of just the electron–neutral collision frequency.

We now continue and find new expressions for the Hall and Pederson conductivities. The Pederson conductivity becomes

$$\sigma_\perp = \sum_s \sigma_{0s} \frac{\nu_{sn}^2}{\nu_{sn}^2 + \Omega_s^2}. \tag{7.70}$$

Note that σ_{0s} was given by Equation (7.69) and only includes collisions of species s with neutrals. The Hall conductivity is

$$\sigma_H = -\sum_s \sigma_{0s} \frac{\nu_{sn} \Omega_s}{\nu_{sn}^2 + \Omega_s^2}, \tag{7.71}$$

where Ω_s has the sign of q_s. The electron contribution to σ_H is positive because Ω_s is negative.

In the above conductivity expressions, the contribution of a species s to the conductivities $\sigma_\perp$ and σ_H depends on the ratio of the collision frequency to the gyrofrequency (i.e., on ν_{sn}/Ω_s). At E-region heights, this ratio is extremely small for electrons ($\nu_{en}/\Omega_e \approx 1$ at an altitude of ≈ 70 km, and $\nu_{en}/\Omega_e \approx 10^{-3}$ at $z \approx 120$ km), and the electron contribution to $\sigma_\perp$ is very small. For ions, this ratio is of the order of unity due to the larger ion mass and smaller ion gyrofrequency. Hence, only ion species make significant contributions to $\sigma_\perp$ in the E-region. For the Hall conductivity, the electron contribution is positive and exceeds the negative ion contribution so that overall σ_H is positive. The collision frequency to gyrofrequency ratio for a typical E-region ion species (O_2^+ or NO^+) is $\nu_{in}/\Omega_i \approx 1$ at an altitude of 130 km, which is in the E-region. For altitudes lower than ≈ 130 km, the collision frequency rapidly becomes larger ($\nu_{in}/\Omega_i \gg 1$), and the ion contribution to $\sigma_\perp$ becomes equal to the ion parallel conductivity. For altitudes above about 130 km, $\nu_{in}/\Omega_i \ll 1$, and both $\sigma_\perp$ and σ_H become very small. Hence, in the F-region ionosphere and in the magnetosphere, the parallel conductivity is very large but the Hall and Pederson conductivities are almost zero. One implication of this "anisotropy" in the electrical conductivity is that magnetic field lines are almost equipotentials for altitudes above the E-region. Actually, in the magnetosphere and solar wind,

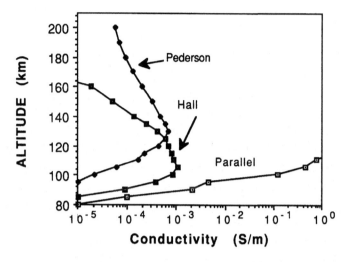

Figure 7.13. Conductivities versus height for the mid-latitude terrestrial ionosphere. Adapted from Richmond (1983).

terms left out of the simple ionospheric GOL start to be important again, and so we must be careful about categorically stating that field lines are equipotentials. Field lines are *not* always equipotentials in the magnetosphere, although this is often a good assumption.

Ionospheric electrical conductivities as a function of altitude are shown in Figure 7.13 for the terrestrial ionosphere. Both the Hall and Pederson conductivities are largest in the E-region, with values of the order of 10^{-3} S/m. Electrical currents can be generated in the ionosphere, according to the GOL, by applying an electric field $\mathbf{E}'$. For example (as we will see in Chapter 8), an electric field generated in the magnetosphere, $\mathbf{E}$, can be "mapped" down to the E-region ionosphere, where it drives both Pederson and Hall currents. This generally happens at higher latitudes, where the magnetic field lines reach further out into the magnetosphere. However, currents can also be driven at lower latitudes, even if $\mathbf{E}=\mathbf{0}$, due to the action of neutral winds. The electric field required by the GOL, Equation (7.68), was given by Equation (7.67). Even if $\mathbf{E}=\mathbf{0}$, we still have the motional electric field, $\mathbf{E}' = \mathbf{u}_n \times \mathbf{B}$, associated with neutral winds, which can drive ionospheric electrical currents. Wind-driven currents are called *atmospheric dynamos*. The neutral winds result from a number of causes, including solar and auroral heating of the upper atmosphere and atmospheric tides. The typical E-region neutral wind speed at low latitudes resulting from solar heating is of the order of 100 m/s, but it has a complex morphology, the description of which is outside the scope of this book (see the Richmond and Rees references). The dynamo currents from these winds are concentrated in the E-region owing to the nature of the Hall and Pederson conductivities.

When magnetosphere–ionosphere coupling is considered in Chapter 8, the quantities that will be needed are the height-integrated Hall and Pederson current densities, $\mathbf{K}_H$ and $\mathbf{K}_\perp$, respectively, which can simply be found by integrating Equation (7.68):

$$\mathbf{K}_\perp = \int_{bottom}^{top} \sigma_\perp \mathbf{E}' \, dz \qquad (7.72)$$

$$\mathbf{K}_H = \int_{bottom}^{top} \sigma_H \mathbf{E}'_\perp \times \hat{\mathbf{b}} \, dz, \qquad (7.73)$$

where the integration over altitude is from the bottom to the top of the ionosphere. Most of the contribution to these integrals comes from the E-region. The perpendicular electric field is almost independent of altitude, and it is a good approximation to remove it from the integral, in which case we have $\mathbf{K}_\perp = \Sigma_\perp \mathbf{E}'_\perp$ and $\mathbf{K}_H = \Sigma_H(\mathbf{E}'_\perp \times \hat{\mathbf{b}})$. The height-integrated conductivities are given by

$$\Sigma_\perp = \int_{bottom}^{top} \sigma_\perp \, dz \qquad (7.74)$$

and

$$\Sigma_H = \int_{bottom}^{top} \sigma_H \, dz, \qquad (7.75)$$

where the integration over altitude again runs from the bottom to the top of the ionosphere.

7.4 The solar wind interaction with nonmagnetic planets

7.4.1 Introduction

We now return to our discussion of how the solar wind interacts with solar system bodies. If a planet or body does not possess a strong intrinsic magnetic field, then the solar wind "directly" interacts with either the planet's neutral atmosphere or its ionosphere, as discussed in Section 7.1. In the next section we discuss the solar wind interaction with the ionosphere of Venus. The solar wind interaction with Mars and Venus are thought to be similar in many respects, but only Venus will be considered in Section 7.4.2. The cometary interaction and the lunar interaction will be discussed in Sections 7.4.3 and 7.4.4, respectively.

7.4.2 The solar wind interaction with Venus

This is a very extensive topic due to the large database that exists from the many spacecraft missions to Venus, especially the Soviet *Venera* missions and the U.S. *Mariner* and *Pioneer Venus* missions, plus the large amount of theoretical work associated with those missions. For a more complete coverage of this topic you are

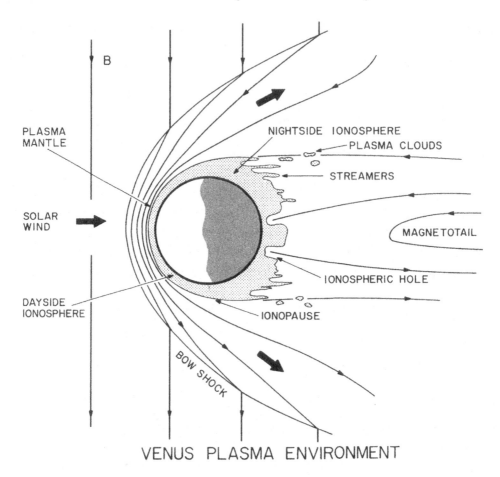

VENUS PLASMA ENVIRONMENT

Figure 7.14. Schematic of the plasma environment of Venus. The bow shock, ionopause, and magnetotail are shown, along with interplanetary magnetic field lines. The plasma mantle is the region just outside the ionopause in which plasma of both ionospheric and solar wind origin is present. The magnetic barrier coexists with the plasma mantle. (From Cravens, 1989.)

referred to books and articles cited at the end of this chapter, including the *Venus* (Hunten et al., 1983) and the *Venus Aeronomy* (Russell, 1991) books.

A schematic of the plasma environment of Venus is shown in Figure 7.14. The solar wind approaches Venus from the left-hand side of this schematic, carrying with it the IMF. The electrically conducting ionosphere acts as obstacle to the solar wind flow. An interface, called *the ionopause*, exists between the solar wind and ionospheric plasmas. The ionopause marks the upper boundary of the cold ionospheric plasma. A *bow shock* is present in the flow, and the newly subsonic (or *magnetosheath*) solar wind plasma continues to slow down as it approaches the ionopause. The flow stagnates just upstream of the subsolar ionopause, and dynamic pressure is converted into thermal pressure (see Equation (4.114)). Furthermore, the frozen-in magnetic field builds up in a region called the *magnetic barrier*, which is located just outside the ionopause. The *plasma mantle* region coexists with the magnetic barrier. In other words, as the flow slows down magnetic field lines "pile up" outside the ionosphere. The magnetic field buildup is enhanced by the loss of some hot plasma from the barrier, permitting magnetic pressure to "substitute" for thermal pressure in the momentum balance. See Equation (4.114), which does not apply exactly because it is one dimensional but which still is qualitatively correct.

The schematic also indicates the presence of a *magnetotail* and several types of ionospheric structures on the nightside. The magnetotail is an "induced" magnetotail, caused by the draping of interplanetary field lines. Measurements by the *Pioneer Venus Orbiter (PVO)* magnetometer indicate that the orientation of the field lines in the tail simply reflects the IMF direction.

Observations made by *PVO* plasma instruments and the magnetometer (see the *Venus Aeronomy* book) indicate that the ionosphere can exist in two states: (1) *unmagnetized* or (2) *magnetized*. Figure 4.10 shows some *PVO* data illustrating both states of the ionosphere. The ionosphere during Orbit 186 of *PVO* was clearly unmagnetized, although small-scale magnetic structure ($\Delta z \approx 10$ km) was clearly present. During Orbits 176 and 177, and especially the latter, large-scale magnetic fields were present throughout the dayside ionosphere. Notice the large magnetic fields at the highest altitudes in each panel of Figure 7.14; this magnetized region is the magnetic barrier. All the large-scale fields in these figures, both in the ionosphere and in the barrier, were primarily in the horizontal direction.

Let us consider some of the observed characteristics of each of the two ionospheric states. When the ionosphere is in its unmagnetized state (as it is most of the time), it does not have any significant large-scale magnetic fields, although small-scale structures such as *magnetic flux ropes* are usually present. In this case, the ionopause is located at high altitudes (i.e., $z > 300$ km) and is rather narrow ($\Delta z \approx 25$ km). A substantial nightside ionosphere is also present in this case. The unmagnetized state seems to occur whenever the solar wind dynamic pressure is substantially less than the maximum ionospheric thermal pressure. However, in those cases when the solar wind dynamic pressure is comparable to, or greater than, the maximum ionospheric thermal pressure, then the ionosphere is observed to be in a magnetized state. In this case, large-scale magnetic fields are induced throughout the dayside ionosphere, and a characteristic magnetic layer is present between 150 and 170 km. For the magnetized state, the ionopause is located at low altitudes ($z < 300$ km, but always above 200 km) and is broad ($\Delta z \approx 80$ km). The nightside ionosphere is weak or absent.

Now we look at the ionopause in a more quantitative manner. In fact, this topic has already been addressed in Chapter 4 (Section 4.6.2). The ionopause for the unmagnetized state of the Venus ionosphere can be treated as a tangential discontinuity at which magnetic pressure on the magnetic barrier side balances ionospheric thermal pressure (that is, $p = B^2/2\mu_0$). This ionopause pressure balance relation can be written as

$$n_e k_B (T_e + T_i) = B^2/2\mu_0, \tag{7.76}$$

where n_e, T_e, and T_i are the ionospheric electron density, electron temperature, and ion temperature just below the ionopause and B is the magnetic field strength at the bottom of the magnetic barrier. Typical values of the electron density and of B were shown in Figure 4.10; and $T_e \approx 5,000$ K and $T_i \approx 2,000$ K are typical temperature values, which are roughly independent of time and altitude at high altitudes.

PVO measurements have further shown that the peak magnetic pressure in the magnetic barrier is approximately equal to the solar wind dynamic pressure upstream of the shock:

$$B^2/2\mu_0 = 0.85\rho_{sw}u_{sw}^2 \cos^2 \chi, \tag{7.77}$$

where χ is the solar zenith angle ($\chi = 0$ at the subsolar point). Hence, as the solar wind dynamic pressure increases, B increases and so a higher electron density is needed to achieve the static balance prescribed by Equation (7.76). For this to happen the ionopause must move to a lower altitude because the topside electron density decreases exponentially with z (hydrostatic equilibrium or diffusive equilibrium – this has been discussed both in Chapter 4 and in Section 7.3). Similarly, if solar wind conditions change such that its dynamic pressure decreases, then the ionopause must move to a higher altitude to achieve pressure balance. This situation has indeed been confirmed by *PVO* data (e.g., Figure 4.10) and was also addressed by Problem 4.10. Now if the ionopause is pushed low enough, then the ionospheric thermal pressure is no longer sufficient to balance the solar wind pressure. In this case, we have a magnetized ionosphere, in which magnetic flux is "pushed" into the ionosphere and ion–neutral friction begins to be relevant to the momentum balance.

A general description of tangential discontinuities (TDs) was given in Section 4.6.2. All TDs, including the ionopause, are current layers. Figure 4.9 illustrates the electrical current configuration found in a TD. The small-scale structures in the ionosphere of Venus in its unmagnetized state are called *magnetic flux ropes*, and they were also discussed in Chapter 4. Recall that magnetic flux ropes in the Venus ionosphere are approximately force free and have magnetic field configurations consistent with Equation (4.104) ($\mathbf{J} \times \mathbf{B} = \mathbf{0}$) or Equation (4.105) ($\nabla \times \mathbf{B} = \alpha\mathbf{B}$). A schematic of a magnetic flux rope was shown in Figure 4.11.

Let us consider in more detail the magnetic field profile for a magnetized Venus ionosphere. The presence of a layer of enhanced magnetic field strength is evident near 170 km for this ionospheric state. Why? Magnetic flux is "pushed" from the magnetic barrier into the ionosphere of Venus when the solar wind dynamic pressure is large enough, and the plasma flow is downward throughout the ionosphere for this scenario. A calculated downward plasma flow speed is shown in Figure 7.15

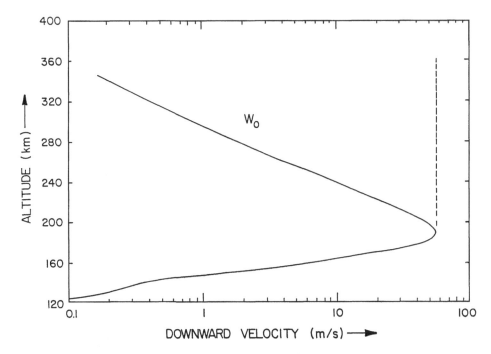

Figure 7.15. Vertical plasma velocity (solid line) for the magnetized state of the ionosphere of Venus. The plasma flow is directed downward. The dashed line is not relevant here. (From Cravens et al., 1984).

for the dayside ionosphere of Venus. Notice that the peak downward speed occurs near 200 km.

The MHD material in Chapter 4 is directly applicable to the problem of the iono-spheric magnetic field at Venus. In particular, the magnetic induction, or diffusion–convection, equation is applicable. Equation (4.77) is a one-dimensional form of this equation. The magnetic diffusion coefficient D_B is equal to the resistivity di-vided by the magnetic permeability of free space (i.e., $D_B = \eta/\mu_0$) and is therefore proportional to the total electron momentum transfer collision frequency, which in-creases with decreasing altitude. In the ionosphere of Venus, D_B is small at higher altitudes and is large in the lower ionosphere.

Now let us reconsider the magnetic diffusion–convection Equation (4.77) again. This equation can be solved numerically (and has been), but let us just examine it qualitatively. We start by estimating the magnetic Reynolds number R_m for the velocity profile shown in Figure 7.15. In Problem 7.20, you will demonstrate that the magnetic Reynolds number is equal to unity at $z \approx 140$ km and is much greater than this at higher altitudes. Consequently, the magnetic field is frozen into the plasma flow above 140 km. In a time-dependent situation, the magnetic field strength B should tend to decrease with time for $z > 200$ km where the flow is accelerating downward, and B should tend to increase for $z < 200$ km where the flow is decelerating.

Equation (4.78) was given as the steady-state solution for the one-dimensional frozen-in field problem; that is, $u(z)B(z) = constant$. This simple equation predicts that B has a minimum at 200 km where the downward plasma flow speed has a maximum, as was actually observed for Venus (Figure 4.10). However, this simple steady-state solution also predicts that B continues to increase with decreasing z. However, the magnetic field is *not* frozen into the flow below about 140 km; magnetic diffusion can then remove magnetic flux and reduce the magnetic field strength. In other words, the electrical currents responsible for the magnetic field, including the 170-km layer, Ohmically dissipate in the lower ionosphere where the collision frequency is high. In order to understand the ionospheric magnetic field more quantitatively, numerical solutions of the full MHD equations are required and more than one spatial dimension should be considered.

7.4.3 The solar wind interaction with comets

Our understanding of the cometary plasma environment has increased dramatically as a consequence of the many spacecraft missions to comet Halley and to comets Giacobini–Zinner and Grigg–Skjellerup. For those seriously interested in the subject of comets I would recommend, as a starting point, the two-volume book entitled *Comets in the Post-Halley Era* (Newburn et al., 1991). The solar wind interaction with comets, unlike the solar wind interaction with Venus, is primarily an interaction with the extensive cometary neutral atmosphere rather than with the cometary ionosphere. Figure 7.16 is a schematic of the cometary plasma environment. For an active comet like Halley, the bow shock is located at a subsolar distance from the nucleus of $\approx 3 \times 10^5$ km, which is very far from the nucleus. For comparison, the Venus subsolar bow shock is located at a distance of only $\approx 2{,}000$ km from the surface of Venus. Moreover, the *cometary bow shock* is weak (Mach number $M \approx 2$), whereas other planets have stronger shocks with $M \approx 5$–8. The plasma flow near the comet remains collisionless down to a radial distance of about 10^5 km (commonly called the *cometopause*), at which location charge transfer collisions with cometary neutrals start to become important. The cometary ionosphere can be defined as that region containing cold cometary plasma relatively undisturbed by the solar wind, and it is located within distances of $\approx 10^4$ km from the nucleus. Inside a distance of $\approx 5{,}000$ km for comet Halley, a field-free region exists called the *diamagnetic cavity*. The plasma flow inside the cavity is coupled to the neutrals by ion–neutral collisions and is directed outward. The boundary of this cavity is sometimes called the *contact surface* (CS in short) or sometimes just the *cavity boundary surface*. A *magnetic barrier* with enhanced magnetic field strength exists outside the cavity. The plasma is slow moving and almost stagnant within this magnetic barrier. A *cometary tail* is also indicated in Figure 7.16.

The central phenomenon in the solar wind interaction with comets is the addition of cometary ions to the solar wind flow over a very large volume of space.

COMETARY PLASMA ENVIRONMENT

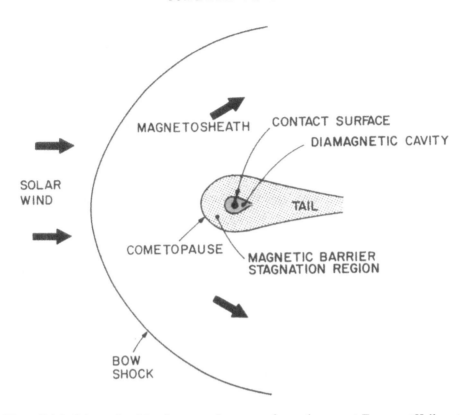

Figure 7.16. Schematic of the plasma environment of an active comet. For comet Halley at a heliocentric distance of 1 AU, the subsolar distance to the bow shock was about 0.3×10^6 km, whereas for comet Giacobini–Zinner at 1 AU it was about 5×10^4 km. (From Cravens, 1989)

The atmosphere, usually called the *coma*, of an active comet extends millions of kilometers. Cometary ions are mainly created "within" the solar wind flow by the photoionization of cometary neutral molecules and atoms by solar EUV photons via the reactions

$$hv + H_2O \rightarrow H_2O^+ + e \tag{7.78}$$

and

$$hv + O \rightarrow O^+ + e. \tag{7.79}$$

Reaction (7.79) is generally more important than (7.78) for the solar wind interaction because far from the nucleus (cometocentric distance $r > 10^5$ km), most of the water molecules have been photodissociated by solar radiation, leaving a neutral coma made of atomic oxygen and atomic hydrogen. However, in the cometary ionosphere ($r < 10^4$ km for comet Halley), reaction (7.78) is important and was already discussed in Section 7.3.

Naturally, the solar wind flow is affected by the addition of mass, particularly since the newly created cometary ions are heavy (i.e., O^+) in comparison with the solar wind protons. The mass addition results in mass-loading of the solar wind, which slows down the solar wind flow, in order for momentum to be conserved. As the flow slows down, magnetic field lines pile up ahead of the comet forming a magnetic barrier. The magnetic field drapes around an inner region surrounding the cometary nucleus because the ends of the field lines remain out in the more rapidly flowing solar wind far from the comet. Cometary plasma is "trapped" by the draped magnetic lines in the middle of the tail. A dust tail also is present for a different reason, as mentioned in Section 7.3.

The cometary ions created by reactions (7.78) and (7.79) are initially moving with the neutral outflow speed of about 1 km/s, and 1 km/s is close to zero in comparison with a typical solar wind speed of 400 km/s. The motional electric field and interplanetary magnetic field act together to "pick up" the newly created ion. The electric field in the cometary frame of reference (i.e., the rest frame) is given by the generalized Ohm's law, although only the motional electric field term is important here:

$$\mathbf{E} = -\mathbf{u}_{sw} \times \mathbf{B}. \tag{7.80}$$

The solar wind velocity vector is $\mathbf{u}_{sw}$ and $\mathbf{B}$ is the interplanetary magnetic field. The principles of single particle motion discussed in Chapter 3 can be used to explain this pick-up process. Example 3.1 describes the type of ion motion relevant to cometary ion pick up, and this example should be reread. The ion in this example is initially at rest. Figure 7.17 illustrates this pick-up process.

If the IMF happens to be perpendicular to the solar wind direction, then the $\mathbf{E} \times \mathbf{B}$ drift velocity is exactly equal to $\mathbf{u}_{sw}$, and cometary ions move with the solar wind. The gyration speed in this case is equal to u_{sw} also. The pitch angle of the cometary ions in the solar wind reference frame (SWRF), in this case, is equal to $90°$ (that is,

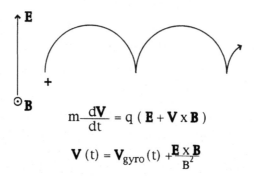

Figure 7.17. Trajectory of an ion in solar wind $\mathbf{E}$ and $\mathbf{B}$ fields. The ion is initially at rest. The ion velocity can be broken down into a gyration part, v_{gyro}, and an $\mathbf{E} \times \mathbf{B}$ drift part. The ion follows a cycloidal trajectory.

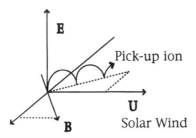

Figure 7.18. Trajectory of cometary pick-up ion in the solar wind for an IMF at some angle, θ_{vB}, with respect to the solar wind velocity vector. The $\mathbf{E} \times \mathbf{B}$ drift is not in the solar wind direction.

$v_\parallel = 0$). However, if the IMF is not orthogonal to $\mathbf{u}_{sw}$ (that is, $\theta_{vB} \neq 0$, where θ_{vB} is the angle between $\mathbf{B}$ and $\mathbf{u}_{sw}$) then the $\mathbf{E} \times \mathbf{B}$ drift is not in the solar wind direction but at an angle of $(90° - \theta_{vB})$ with respect to it. In this case, the ions are no longer moving entirely with the solar wind, and in the SWRF they have a parallel velocity of $v_\parallel = v_{\parallel 0} = u_{sw} \cos \theta_{vB}$ and a perpendicular velocity of $v_\perp = v_{\perp 0} = u_{sw} \sin \theta_{vB}$. The initial ion pitch angle in the SWRF is then $\alpha_0 = \tan^{-1}[v_\perp / v_\parallel] = \theta_{vB}$. Figure 7.18 illustrates this situation in the comet frame of reference.

Obviously, more than one cometary ion is created in the solar wind. If we assume that all the ions picked up by the solar wind merely act as "single particles," then the cometary ion distribution function is a delta function in $v_\perp$ and $v_\parallel$ given by

$$f(v_\perp, v_\parallel; \mathbf{x}) = [n_i(\mathbf{x})/(2\pi v_{\perp 0})]\delta(v_\parallel - v_{\parallel 0})\delta(v_\perp - v_{\perp 0}), \qquad (7.81)$$

where $\mathbf{x}$ is the position vector. The total density of cometary ions at location $\mathbf{x}$ is given by n_i and is determined by the ionization rate and solar wind flow parameters. Alternatively we could use the variables v and $\mu = \cos \alpha$. Figure 7.19 (see caption) is a schematic of velocity space; the velocity of newly created cometary ions is indicated. The ion distribution function given by (7.81) is called a *ring-beam distribution*. The "beam" term is appropriate because the ions are streaming parallel to the magnetic field with respect to the solar wind, and the "ring" refers to the gyration motion.

The single particle picture of cometary ion pickup, as just presented, is oversimplified. The presence of cometary ions at a single pitch angle generates a microscopic plasma instability of the type alluded to in Section 4.6.4. An analysis using the Vlasov Equation (2.11) plus Maxwell's equations can be used to show that a ring-beam distribution with a sufficiently large n_i generates ultra-low frequency (ULF) electromagnetic waves. These waves are essentially Alfvén waves and propagate along the IMF at the Alfvén speed C_A. The wave dispersion relation is given by $\omega = kC_A$. The peak wave production takes place for waves at a resonant wavenumber for the parallel direction of

$$k_{res} = \frac{\Omega_i}{v_{\parallel 0}}, \qquad (7.82)$$

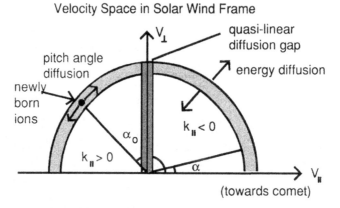

Figure 7.19. Velocity space in the solar wind reference frame. Velocities parallel and perpendicular to the IMF are used. The pitch angle is α and the pitch angle of newly created cometary ions is α_0, as defined in the text. Newly born cometary ions are created at the point $(v_{\perp 0}, v_{\parallel 0})$ on this diagram. The other aspects of this schematic are discussed in the text.

where Ω_i is the cometary ion gyrofrequency, found using the unperturbed IMF. These ULF waves have been observed by magnetometer experiments on cometary missions such as the *ICE* mission to comet Giacobini–Zinner (Tsurutani and Smith, 1986).

The motion of a cometary ion in the presence of ULF waves is such that the first adiabatic invariant of single particle motion (i.e., the conservation of the magnetic moment $\mu_m = mv_\perp^2/2B$) is no longer strictly obeyed. Without these ULF waves the pitch angle of a cometary ion remains at its pickup value ($\alpha = \alpha_0$), but the extra part of the Lorentz force associated with these waves, which have wavelengths of the order of the gyroradius, results in pitch-angle scattering. A description of this wave–particle interaction phenomenon can be found by transforming the Vlasov equation into a *quasi-linear diffusion equation*. You will recall from Sections 3.11 and 6.5.2 that the pitch-angle scattering coefficient $D_{\mu\mu}$ in the quasi-linear diffusion equation is proportional to the wave power at the resonant wavenumber.

The solar wind has its own indigenous magnetic fluctuations (Figure 6.36), and the cometary pickup process adds ULF waves near the resonant wavenumber as given by Equation (7.82). Within a distance from the cometary nucleus of a million kilometers, the wave power associated with the pickup process is greater than the solar wind wave power. The power spectrum qualitatively looks like the solar wind spectrum shown in Figure 6.36 but with a correlation wavenumber about equal to the resonant wavenumber as given by Equation (7.82). That is, we have $k_{cor} \approx k_{res}$. The total power increases as the distance to the comet decreases, and near the bow shock the magnetic fluctuations are extremely large with $(B_1/B_0)^2 \approx 1$.

The pitch-angle scattering caused by the ULF waves gradually isotropizes the cometary ion distribution function. Near the comet, the cometary ion distribution is sufficiently isotropized that it can described as a shell distribution,

$$f(v, \mathbf{x}) = [n_i(\mathbf{x})/4\pi u_{\mathrm{sw}}^2]\delta(v - u_{\mathrm{sw}}), \tag{7.83}$$

where v is the magnitude of the ion velocity. Note that the energy of the cometary ions in the SWRF is approximately $(1/2)m_i u_{\mathrm{sw}}^2 \approx 15$ keV, which is greatly in excess of the typical 1 keV solar wind proton kinetic energy. Cometary pickup ions are quite "hot."

As we can see from Equation (3.95), energy diffusion as well as pitch-angle diffusion can take place. The stochastic acceleration of charged particles via magnetic fluctuations is called *second-order Fermi acceleration*. Problem 3.7 addressed the physical mechanism underlying this process. The efficiency of energy diffusion is much smaller by a factor of about $(u_{\mathrm{sw}}/C_A)^2 \approx 100$ than that for pitch-angle diffusion. Nonetheless, significant ion acceleration takes place upstream of a comet and the initially narrow cometary ion shell distribution gradually broadens out. Ions can also be accelerated by encountering the cometary bow shock in two ways: (1) by *shock drift acceleration*, in which the particle drifts in the direction of the motional electric field, and (2) by *first-order Fermi acceleration*, in which a particle reflects off the shock and moves back upstream where it again reflects off a magnetic fluctuation. The proper treatment of these topics is beyond the scope of this book, and so we move on and consider the dynamical, or fluid, consequences of the cometary ion pickup by the solar wind.

The dynamics of plasma flow is not strongly dependent on the details of the plasma distribution functions but depends on the moments of these distribution functions including density and pressure. It is sufficient for our purposes to use the single-fluid conservation equations presented in Chapter 4 but with the inclusion of mass-loading terms associated with the addition of cometary ion pickup. We stick to a simple one-dimensional analysis. The spatial variable is the radial distance from the comet along the Sun–comet axis (r). This is an accurate approach outside the cometary bow shock. We start with the one-dimensional version of the continuity Equation (4.57). The mass density is given by $\rho(r) = n_p(r)m_p + n_i(r)m_i$, where n_p and n_i are the solar wind proton and cometary ion densities, respectively. The proton and cometary ion masses are $m_p = 1$ amu and $m_i \approx 16$, respectively. The net source in Equation (4.57) is the cometary ion production rate: $S_i = R_i n_n(x)$, where $R_i \approx 10^{-6}\mathrm{s}^{-1}$ is the ionization frequency for reaction (7.79), as discussed in Section 6.3. The cometary neutral density was given by Equation (7.50). Equation (4.57) now becomes

$$-\frac{d}{dr}(\rho u) = m_i R_i n_n(r)$$

$$= \frac{m_i R_i Q}{4\pi u_n r^2}. \tag{7.84}$$

The flow speed in the solar wind direction is denoted $u(r)$. Equation (7.84) is easily integrated to give the following expression for the total mass flux:

$$\rho u = \rho_\infty u_\infty + \frac{m_i R_i Q}{4\pi u_n} \frac{1}{r}$$

$$= \rho_\infty u_\infty + \frac{\Gamma}{r} \quad \text{with } \Gamma \equiv \frac{m_i R_i Q}{4\pi u_n}. \tag{7.85}$$

The quantity $\rho_\infty u_\infty$ is the mass flux as $r \to \infty$. Note that $\rho_\infty = n_{sw}m_p$ and $u_\infty = u_{sw}$. Equation (7.85) states that due to mass addition the total mass flux (protons and cometary ions) increases with decreasing distance from the nucleus. This expression is one dimensional and is valid upstream of the shock. Downstream of the bow shock two-dimensional flow effects become important and a different approach is required, as discussed in articles in the book *Comets in the Post-Halley Era* (Newburn et al., 1991). Equation (7.85) does remain qualitatively valid.

In order to deal with the flow speed $u(r)$, we use an appropriate one-dimensional, single-fluid momentum equation such as Equation (4.114). This equation is valid even with mass addition, as long as $u_n \ll u$. Using radial distance r instead of the variable x, we find

$$-\frac{d}{dr}\left(\rho u^2 + p + \frac{B^2}{2\mu_0}\right) = 0. \tag{7.86}$$

The Alfvénic Mach number in the solar wind is very large, so that we can neglect the magnetic pressure term in (7.86). The pressure p includes contributions from cometary ions as well as from solar wind electrons and protons, and p can become quite large near the comet due to the large energies (≈ 15 keV) of the cometary pickup ions. The integral form of Equation (7.86) can be written as

$$\rho u^2 + p = \rho_\infty u_\infty^2, \tag{7.87}$$

where the thermal pressure far upstream of the comet has been neglected on the right-hand side.

We can already see from Equation (7.87) why mass-loading slows down the solar wind. The mass flux, ρu, increases near the comet according to the continuity Equation (7.85), yet the pressure must increase because very hot ions are being added to the flow. Hence, to keep the left-hand side of (7.87) constant as the radial distance r decreases, the flow speed $u(r)$ must decrease. Intuitively, it is easy to understand that in order to conserve momentum the fluid must slow down when mass is being dumped into the flow. To represent this mass-loading effect more quantitatively, we need to specify the thermal pressure p. The thermal pressure of the solar wind protons and electrons is almost constant. We need to deal with the cometary ion pressure, and the distribution function of these ions is shell-like. The pressure associated with a distribution function was given by Equation (2.31) for an isotropic distribution. For the pressure of cometary ions (species i) this equation is

$$p_i = \frac{1}{3} m_i \int_0^\infty v^2 f(v) 4\pi v^2 \, dv. \tag{7.88}$$

The particle speed in the SWRF is denoted v. The distribution function of cometary ions, assuming rapid isotropization by waves, is a shell distribution specified by Equation (7.83). However, close to the comet, the distribution function is no longer an infinitesimally thin shell but is a thick shell. The slowing down of the mass-loaded solar wind means that ions are picked up at a speed less than u_{sw} yet ions added to the flow further upstream still have $u \approx u_{sw}$. Nonetheless, in Example 7.3, we make the approximation that the distribution function is simply given by Equation (7.83).

Example 7.3 (Simple mass-loaded solar wind solution) We wish to determine the flow speed $u(r)$ by solving the continuity and momentum equations. The pressure can be determined directly from the distribution function in lieu of an energy equation. To find a simple form of the cometary ion pressure let us assume that Equation (7.83) is valid throughout the region upstream of the cometary shock but with u substituted for u_{sw}. Equation (7.88) then gives the following expression for the pressure:

$$p_i = \frac{1}{3} m_i n_i(r) u^2. \tag{7.89}$$

The cometary ion mass flux, $m_i n_i(r) u(r)$, can be found using the definition $\rho(r) = n_p(r) m_p + n_i(r) m_i$ plus Equation (7.87):

$$n_i(r) m_i u(r) = \rho u - n_p(r) m_p u(r)$$

$$= \rho u - n_{sw} m_p u_{sw} = \rho u - \rho_\infty u_\infty = \Gamma / r, \tag{7.90}$$

where we used $n_p(r) m_p u(r) = \rho_\infty u_\infty$, which is the same as $n_p(r) u(r) = n_{sw} u_{sw}$, as can be demonstrated using the proton continuity equation. Equations (7.89) and (7.90) together give

$$p_i = \frac{1}{3} u \frac{\Gamma}{r}. \tag{7.91}$$

The cometary ion pressure expression contains both $u(r)$ and an explicit dependence on $1/r$.

The momentum Equation (7.87) can be transformed using Equations (7.89) and (7.91) to

$$\rho_\infty u_\infty^2 = u \left(\frac{\Gamma}{r} + \rho_\infty u_\infty \right) + \frac{1}{3} u \frac{\Gamma}{r}$$

$$= u \left(\frac{4\Gamma}{3r} + \rho_\infty u_\infty \right), \tag{7.92}$$

where the proton and electron pressure have been neglected. The solution of Equation (7.92) for the flow speed relative to the upstream solar wind speed is

$$\frac{u(r)}{u_\infty} = \frac{1}{1 + \dfrac{r_{\text{ml}}}{r}} \quad \text{with } r_{\text{ml}} \equiv \frac{4}{3}\frac{\Gamma}{\rho_\infty u_\infty} = \frac{4}{3}\frac{m_i R_i Q}{4\pi u_n \rho_\infty u_\infty}. \tag{7.93}$$

Equation (7.93) indicates that the flow speed decreases with decreasing r. The flow speed drops by a factor of two from u_{sw} at a distance of $r = r_{\text{ml}}$ according to Equation (7.93). For an active comet like Halley with a gas production rate at 1 AU of $Q = 10^{30}\,\text{s}^{-1}$, this distance is $r_{\text{ml}} \approx 500{,}000$ km (see Problem 7.22). Equation (7.93) predicts that $u \to 0$ as $r \to 0$, as is actually observed in spite of the fact that the equations used to find this solution are only approximate and were not appropriate for the three-dimensional flow downstream of the cometary bow shock.

In the cometary ion pressure expression used in Example 7.3 we assumed that all cometary ions had speed $v = u$; however, some ions are present at higher speeds. Hence the pressure was somewhat underestimated and the mass-loading was also underestimated. A more accurate solution, described by Galeev in *Comets in the Post-Halley Era*, for the region upstream of the bow shock is

$$\frac{u(r)}{u_\infty} = \frac{5}{8}\left(\frac{\rho_\infty u_\infty}{\rho u}\right)\left[1 + \sqrt{1 - \frac{16}{25}\frac{\rho u}{\rho_\infty u_\infty}}\right], \tag{7.94}$$

where the total mass flux, ρu, as a function of r is still given by (7.85). The rates of decrease of $u(r)$ predicted by Equations (7.93) and (7.94) are almost identical far upstream of the comet, but closer to the comet, the flow speed predicted by Equation (7.94) falls off significantly faster than the speed predicted by the simpler equation. In fact, there is a limit to the validity of Equation (7.94); keeping the quantity in the square root positive leads to a maximum or *critical mass flux* $(\rho u)_{\text{crit}} = 25/16$. Associated with this critical mass flux is a minimum flow speed $u = 0.4\,u_\infty$. The flow at this critical point is sonic; that is, $M = 1$. The radial distance at which this critical mass flux occurs depends on how active the comet is and can be determined using Equation (7.85). For comet Halley, this distance is

$$r_{\text{crit}} = (4/3)r_{\text{ml}} \approx 500{,}000 \text{ km}. \tag{7.95}$$

In practice, Equation (7.95) is not entirely accurate for a comet as active as Halley because photoionization of cometary neutrals results in an exponential attenuation of the neutral density, with a scale of about one million kilometers, in addition to the variation given by Equation (7.50).

The subsonic flow downstream of the critical point cannot be accurately described by a simple one-dimensional theory. Two- and three-dimensional numerical MHD or gasdynamic calculations, as well as actual spacecraft observations, have demonstrated that the cometary flow never reaches the critical point. A shock forms upstream of the critical point at a location where the Mach number is approximately 2. The mass flux at this $M = 2$ location is $\rho u \approx 1.3$, and the flow speed is $u = 0.65 \, u_\infty$. The flow speed and other flow parameters downstream of this shock can be found using the Rankine–Hugoniot (or shock jump) relations discussed in Chapter 4. From the shock jump relations for a $M = 2$ shock, one finds that the downstream flow speed is $u = .29 \, u_\infty$. The flow downstream of the shock along the Sun–comet axis continues to decelerate, as qualitatively indicated by Equation (7.93), and at some point the flow virtually stagnates. Both the cometary ion density and the thermal pressure build up as this happens, and the magnetic field strength also builds up. That is, frozen-in field lines pile up as flow slows down since $uB \approx constant$.

Figure 7.20 shows some results from a three-dimensional numerical MHD model of comet Halley. Notice the region located about 10^5 km upstream of the nucleus, located at the origin, with dense plasma and a large magnetic field. The field lines drape around the front of the comet and form a tail. The plasma density in the tail is relatively high. Momentum balance in the direction orthogonal to the tail balance must be satisfied. In the neutral sheet of the tail, the magnetic field strength is very small ($B \approx 0$). The magnetic pressure associated with the tail lobes located outside the tail must approximately balance the plasma pressure in the tail itself (sometimes called the *plasmasheet*):

$$[B^2/2\mu_0]_{\text{lobes}} \cong p_{\text{plasmasheet}}. \tag{7.96}$$

Essentially the same momentum balance perpendicular to the tail axis also operates in the Venus and terrestrial magnetotails, although the field in the latter case is largely an intrinsic field rather than an interplanetary field. The plasma concentrated in the plasmasheet is observable telescopically, and such observations of the *plasma* (or *ion*) *tails* of comets have been made for many decades (Brandt, 1982). In particular, cometary CO^+ ions have been observed using the "comet-tail" molecular band emission. Cometary ion tails appear as straight and narrow features, whereas the dust tail is both spatially and spectroscopically much broader (see Figure 7.21). Also evident in Figure 7.21 is a "disconnection" of the plasma tail of comet Halley, where the tail appears to be interrupted and to change direction. Such *disconnection events* or *DEs*, as they are called, are not uncommon, and they seem to initiate near the head of a comet and then propagate down the tail (Brandt, 1982). Statistical studies of DEs indicate that they are associated with the passage of the heliospheric current sheet (this is also a sector boundary or crossing as discussed in Chapter 6) past the comet (Yi et al., 1994). One promising theory suggests that DEs result from reconnection of the magnetic field ahead of the comet as the heliospheric current sheet is convected toward the comet and compressed as part of the solar wind–comet interaction (cf. Yi et al., 1994).

The solar wind–comet interaction as discussed so far has been collisionless.

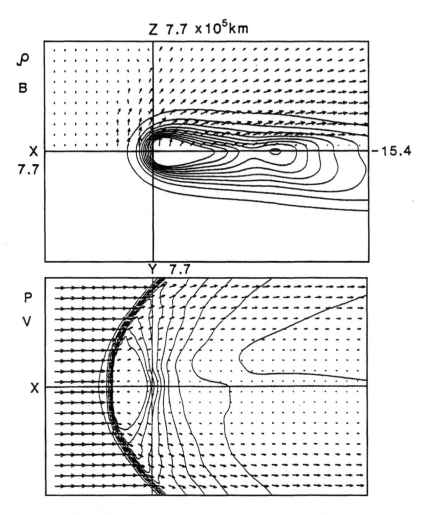

Figure 7.20. Mass density ρ, contours, and magnetic field vectors from a global one-fluid, three-dimensional MHD model are shown in the top panel. Velocity vectors and pressure contours are displayed in the lower panel. The Sun is to the left. The bow shock, plasma tail, and stagnation/magnetic barrier region are all evident. (From Ogino, et al., 1988.)

However, at a radial distance of about 10^5 km (for a Halley-type comet) the collision time of an ion becomes comparable to its transport time and one must incorporate collisions into the description of the solar wind interaction. Charge transfer collisions between the hot cometary ions and solar wind protons with cold cometary neutrals first become important near *the cometopause* (shown in the schematic Figure 7.16). For example, a fast solar wind proton can collide with a cold H_2O molecule and strip off an electron, leaving a fast neutral H atom and a cold H_2O^+ ion:

$$H_{fast}^+ + H_2O \rightarrow H_{fast} + H_2O^+. \tag{7.97}$$

In this manner, the original solar wind proton population and the energetic/hot

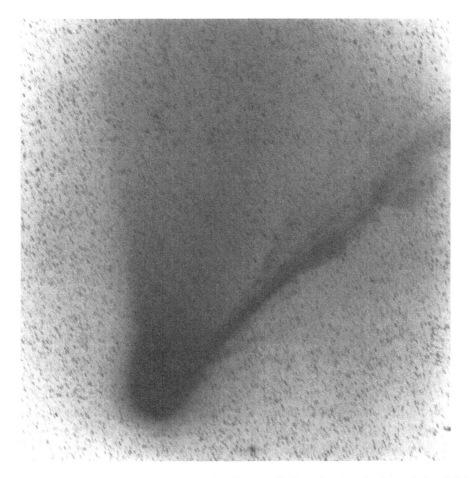

Figure 7.21. April 18, 1986, Photograph of comet Halley showing the ion tail (straight) and the dust tail (broad and fan-shaped). A disconnection event in the ion tail is evident at the upper right. Photograph by F. Miller, University of Michigan/CTIO: LargeScale Phenomena Network-International Halley Watch. The author also thanks Jack Brandt for help in obtaining this photograph.

cometary pickup ion population is lost and is replaced by cold cometary plasma. At distances much closer ($\approx$30,000 km for Halley) to the nucleus than the cometopause, the plasma is relatively cold (temperatures $\approx$1 eV) and slow ($u \ll u_{\mathrm{sw}}$), and one can call this region the *cometary ionosphere*, as discussed in Section 7.3.

In the magnetic barrier located near $r \approx 20,000$ km, the magnetic pressure is approximately equal to the unperturbed solar wind dynamic pressure. A region that is free of magnetic field, *the diamagnetic cavity*, was observed by the *Giotto* magnetometer to exist inside the magnetic barrier (Figure 7.22). The boundary of this cavity is located at a radial distance of $r = 4,500$ km. The magnetic field lines outside the cavity are parallel to this boundary. At Venus, the ionopause boundary separating the field-free ionosphere from the magnetic barrier is controlled by the balance between thermal and magnetic pressure. The momentum balance at the

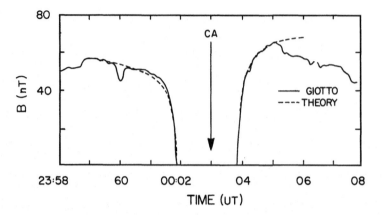

Figure 7.22. Magnetic field versus time near closest approach of the *Giotto* spacecraft to the nucleus of comet Halley. Shown are *Giotto* magnetometer data in the vicinity of the diamagnetic cavity plus results of a simple MHD model. The time of closest approach to the nucleus is marked CA. The spacecraft speed was 68 km/s and hence the diameter of the cavity is ≈ 9000 km. (From Cravens, 1989.)

cometary cavity surface (or contact surface CS) is *not* the same as at Venus, because the cometary thermal pressure is not high enough to balance the magnetic pressure and because the ion density does not decrease across the boundary as it does at the Venus ionopause. The chemical lifetime of the cometary H_3O^+ ions (Section 7.3) is sufficiently short that the plasma density remains unaltered by the dynamics, with n_e proportional to $1/r$, except for a very narrow ($\Delta z \approx 30$ km) region of enhanced density located right at the CS.

The magnetic pressure gradient force is directed inward in the region just outside the CS, and if thermal pressure cannot counteract this force, then what does? Ion–neutral friction provides the necessary outward force. In the field-free region inside the CS, the cometary plasma moves with the neutrals, which are flowing outward from the nucleus at a speed $u_n = 1$ km/s. Ion–neutral collisions tie the neutral and plasma fluids together in this region, and no other force is available to counteract this. However, in the magnetic barrier the ions are subject to the Lorentz force and have a flow speed $u \approx 0$ (i.e., stagnation). The neutrals do not feel the Lorentz force and continue merrily on their way, flowing past the stagnated ions with a relative speed of u_n. The outflowing neutrals exert a force on the ion gas due to *ion–neutral collisions* (i.e., friction). The outward ion–neutral friction force on the plasma balances the inward magnetic pressure gradient force (or $\mathbf{J} \times \mathbf{B}$ force). Starting from the single-fluid MHD momentum equation (4.80) and throwing out all other terms, we have

$$0 \cong \mathbf{J} \times \mathbf{B} - \rho v_{in}(\mathbf{u} - \mathbf{u}_n). \tag{7.98}$$

This equation can be further simplified by just working with the radial component, in which case the radial part of $\mathbf{J} \times \mathbf{B}$ is $-d/dr(B^2/2\mu_0)$. We can further simplify the momentum balance by assuming that $u = 0$ and by using Equations (7.50) and (7.53) for the neutral and electron densities, respectively. In this case, the only unknown in the momentum equation is the magnetic field strength, $B(r)$, which can be found by integration of our one-dimensional version of Equation (7.98). In Problem 7.23, you are asked to do precisely that, and you should find that

$$B(r) = \begin{cases} B_0 \left(1 - \dfrac{r_{cs}^2}{r^2}\right)^{1/2} & r > r_{cs}, \\ 0 & r < r_{cs}. \end{cases} \tag{7.99}$$

B_0 is the field strength as $r \to \infty$; actually, B_0 can be identified as the field strength in the magnetic barrier, in which case it can be estimated from the solar wind dynamic pressure. r_{cs} is the radial distance to the diamagnetic cavity boundary and you are also asked to find this in the problem. Not surprisingly, r_{cs} is larger for comets with larger gas production rates. The dashed curve in Figure 7.22, marked "theory," is from Equation (7.99).

7.4.4 Lunar interaction

The Moon lacks an atmosphere, or ionosphere, and does not possess a significant intrinsic magnetic field, and the lunar rock itself is a poor electrical conductor. As a consequence of these properties the moon is not able to deflect the solar wind around it; instead, the solar wind runs right into the Moon and is absorbed. A bow shock does not form ahead of the Moon as it did for other bodies, which were able to deflect the solar wind in one manner or another. This lack of a bow shock is discussed in more detail in the next section. Both Russell (1985) and Bagenal (1985) review the lunar interaction.

The solar wind plasma is stopped at the lunar surface, but the interplanetary magnetic field lines are easily able to pass through the Moon. Consequently, the field lines do not pile up and do not create an obstacle to the solar wind plasma (unlike Venus). The transport time for the frozen-in IMF to pass by the Moon in the flanks is approximately equal to the lunar radius divided by the solar wind speed: $\tau_{trans} = R_{moon}/u_{sw} \approx 10\,\text{s}$. The IMF that reaches the front of the Moon cannot be carried through it by the plasma, which is absorbed at the surface (creating a *plasma wake*). The magnetic field must diffuse through the Moon with a time constant, as discussed in Section 4.4, of $\tau_B \approx (R_{moon})^2/D_B = \mu_0 (R_{moon})^2/\eta_{lunar} \approx 10\,\text{s}$. The magnetic diffusion coefficient was expressed in terms of the resistivity of the lunar rock (η_{lunar}) and a value of $10^6\,(\text{S/m})^{-1}$ (for granite) was used for our crude estimate. The ratio of the transport to magnetic diffusion times yields a magnetic Reynolds number for this situation of $R_m \approx 1$. This low magnetic Reynolds number does not allow any significant amount of magnetic flux to pile up in front of the Moon, and the magnetic field passes right through into the wake.

The lunar plasma wake exhibits the structure illustrated schematically in Figures 7.23 and 7.24. A perturbation forms in the magnetic field just downstream

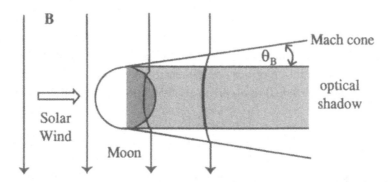

Figure 7.23. Schematic drawing of interplanetary magnetic field lines in the vicinity of the Moon. The magnetic disturbance propagates away from the shadow at the Alfvén speed, such that the edge of the disturbed field forms an angle of θ_B with respect to the solar wind direction.

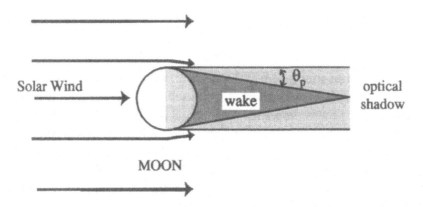

Figure 7.24. Schematic of the lunar plasma wake. The solar wind is absorbed on the ramside of the Moon but tends to flow back into the wake region at roughly the sound speed. The angle of the edge of this inflow is θ_p.

of the Moon due to differences in the transport and diffusion times of the field lines and also because some magnetic flux is carried into the wake by the inflow of plasma from the flanks. The field lines tend to bend inward into the wake (not visible from the perspective of Figure 7.23). This perturbation propagates along the IMF at the Alfvén speed C_A, but the magnetic field line is simultaneously "moving" downstream at the solar wind speed. Hence, the loci of disturbance (the *Mach cone*) has the following angle, θ_B, with respect to the solar wind direction:

$$\theta_B = \tan^{-1}\left(\frac{C_A}{u_{sw}}\right) = \tan^{-1}\left(\frac{1}{M_A}\right) \approx 6°, \qquad (7.100)$$

where the Alfvénic Mach number of the solar wind is roughly $M_A \approx 10$.

   The plasma itself flows inward to fill the wake with an average speed equal to the thermal (or sound) speed C_S, such that the edge of the "evacuated" region forms

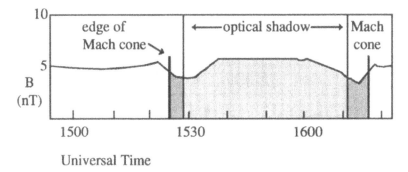

Universal Time

Figure 7.25. Simplified version of magnetic field strength data from a June 15, 1968, passage of *Explorer 35* through the wake region of the Moon (within a half-lunar radius of the surface). The abscissa is spacecraft time, which is proportional to the distance along the spacecraft orbit. Adapted from the review paper by Russell (1985).

the following angle:

$$\theta_p = \tan^{-1}\left(\frac{c_S}{u_{sw}}\right) = \tan^{-1}\left(\frac{1}{M_S}\right) \approx 6°, \qquad (7.101)$$

where the sonic Mach number of the solar wind is approximately 10. This "edge" is actually relatively diffuse because the actual particle speeds have a "thermal" distribution about the average speed.

Figure 7.25 shows the magnetic field strength measured by *Explorer 35* as it passed through the lunar wake at a distance from the surface of only 10^3 km or so. The region of slightly enhanced field in the wake is apparent in this figure as are the regions with slightly reduced field strength located right near the edge of the wake (due to flux removal by the inflow).

7.5 Comparative bow shocks

7.5.1 Introduction – Why bow shocks?

Shocks were discussed in Chapter 4 (Section 4.9) where a very brief discussion of why shocks exist in space was presented. Earlier in the present chapter, we indicated that bow shocks were known to exist in the solar wind flow around all the planets that have been visited by spacecraft (i.e., Mercury, Venus, Earth, Mars, Jupiter, Saturn, Uranus, and Neptune, as well as comets). Bow shocks exist because these planets acted as obstacles to the solar wind flow, in one way or the other, and for a supersonic solar wind to flow around an obstacle it must first become subsonic (see Figure 4.22).

Now let us consider in more detail why a supersonic flow around an obstacle requires a shock by considering the example of air flow around a steel ball. Suppose that the air is flowing at a supersonic speed. Then let us postulate that a steel ball mysteriously appears at rest in this flow at time $t = 0$. What happens to the gas that is

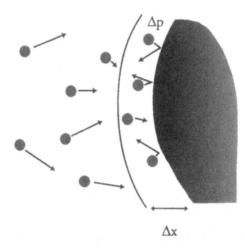

Figure 7.26. Schematic of air molecules flowing into a reflecting surface on a microscopic scale a short time after the surface "appears." A pressure enhancement, Δp, is created and begins to move away from the surface. The width of this layer, Δx, increases with time until a classic shock front develops.

moving toward this ball? The air is "unaware" of the ball at $t = 0$, but if we make the reasonable assumption that on a microscopic scale air molecules bounce, or reflect, off the steel surface, then after a very short time a very narrow layer develops near the surface, in which air molecules have significant motions in all directions (see Figure 7.26). New incoming molecules then collide with the reflected molecules on a length scale of the collision mean free path. The air in this thin boundary layer located just outside the surface has zero net (or bulk) speed and has enhanced density and pressure due to the presence of the reflected molecules. The boundary surface separating the region with enhanced pressure from the unperturbed upstream flow is an "embryo" shock wave, or shock front, and it moves outward from the surface as more and more upstream molecules plow into this region. In effect, the gas reflection off the obstacle surface causes a pressure buildup. The "embryo" shock front quickly becomes a classic shock wave with a thickness of the order of a few mean free paths. The region between the shock and the surface soon attains a macroscopic size. At this point ordinary hydrodynamics is applicable and the Rankine–Hugoniot relations are valid for the shock wave moving outward. The gas flow behind the shock is subsonic and is able to adjust to the presence of the steel ball. Eventually, the shock front moves far enough out such that all the impinging flow is able to move around the ball and then the front becomes stationary, always remaining in the same configuration ahead of the ball (as pictured schematically in Figure 4.22).

The scenario just described also applies to the solar wind flow around the planets in some sense, although the solar wind is a collisionless plasma, and the relevant microscopic length scale is *not* the collisional mean free path, as will be discussed in Section 7.5.3. But even though the microphysics of a collisionless shock differs from that of an ordinary gas shock, the *macroscopic* properties of planetary bow shocks are essentially the same as those discussed in Section 4.9, including the validity of the Rankine–Hugoniot relations.

However, why is a bow shock *not* observed to exist in the solar wind flow at the Moon? The surface of the Moon absorbs the solar wind plasma (as well as magnetic flux) and no particle reflection takes place. There is thus no way for a pressure buildup to occur outside the surface, and it was the pressure disturbance (i.e., a compression) that developed into the shock wave in the above example.

7.5.2 Comparative bow shocks

The sizes of the bow shocks around the various planets are very different because the obstacles have very different sizes. For example, the obstacle in the terrestrial case is the magnetopause, which is located at about 10 Earth radii, whereas the Venus obstacle is the ionosphere itself and has a radius that is only slightly greater than 1 Venus radius. However, even though the various bow shocks have quite

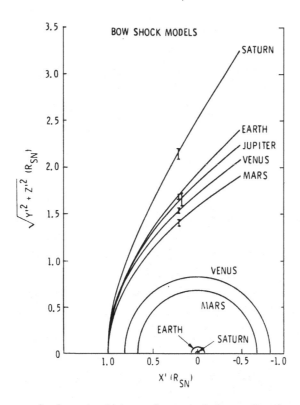

Figure 7.27. Comparative bow shock shapes for several planets, based on measured shock positions. The shocks have been scaled in size such that the subsolar distances to the shocks are the same. (From Russell, 1985.)

different sizes, the shapes are quite similar, as illustrated in Figure 7.27, which was taken from Russell (1985) (see the bibliography at the end of the chapter). The size scales have been adjusted in this figure to give the same subsolar shock distance. The obstacle sizes are comparable in the figure although obviously not the planet sizes. The bow shock shapes relative to the obstacle are clearly similar, but there are some differences. The flare angle of the terrestrial bow shock near the flanks is somewhat greater than the flare angles for the Venus and Mars shocks because the "magnetospheric obstacle" is somewhat "blunter" (i.e., less streamlined) than the Venus or Mars "ionospheric obstacles."

The jump conditions for planetary bow shocks are just the Rankine–Hugoniot relations presented in Section 4.9. Strictly speaking, a magnetized plasma is not an ordinary gas and one should use the MHD jump conditions, rather than the simpler gas dynamic conditions, which take into account the angle of the magnetic field with respect to the shock normal direction, θ_{Bn}, as well as the plasma β. Indeed, the nature of the shock jump and shock structure does depend on these, and other, plasma parameters, but nonetheless the Rankine–Hugoniot relations for ordinary gas dynamics provide a surprisingly good approximation! This is particularly true if the magnetosonic Mach number of the upstream flow, M_{ms}, is used in evaluating the jump conditions rather than just the sonic Mach number. Typically, in the solar wind, $M_{ms} \approx 6$–8 (Section 4.8.3), in which case both the density and magnetic field strength (for θ_{Bn} near $90°$ perpendicular shock) jump by a factor of ≈ 3.6 in going from upstream to downstream. The flow speed drops across the shock from a typical value of ≈ 400 km/s to a value of ≈ 100 km/s. And the plasma temperature increases from about 10^5 K up to a value of about 2×10^6 K. The magnetosheath plasma present downstream of the shock is slower, hotter, and denser than the unshocked solar wind plasma.

The above discussion (and Mach number values) apply to the subsolar part of the shock. The shock wave downstream of the shock "nose" is weaker than it is at the subsolar point because the relevant upstream Mach number refers to the flow speed normal to the shock surface, and the incident, normal Mach number is smaller as the angle between the upstream flow and the shock normal becomes larger. The nature of the shock (*quasi-parallel* or *quasi-perpendicular*) is also different for different parts of the shock front, as illustrated in Figure 7.28. The "quasi" just means "almost;" quasi-parallel (or q-$\parallel$) shocks are present if θ_{Bn} is between 0 and $45°$, whereas quasi-perpendicular (or q-$\perp$) shocks are present if θ_{Bn} is between $45°$ and $90°$.

7.5.3 Collisionless shocks – Microscopic structure

The difficulty with understanding shocks in space plasmas is that these plasmas are generally collisionless. The shocks are called *collisionless shocks*. The dissipation in ordinary gas dynamic shocks, which causes the specific entropy to

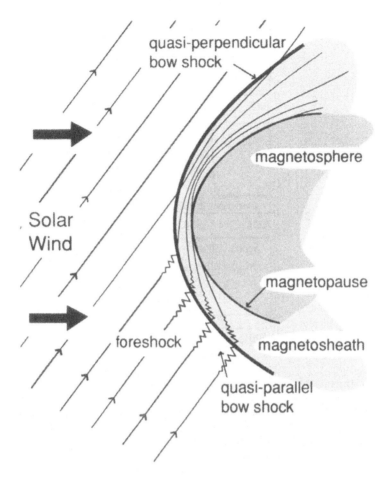

Figure 7.28. Schematic of bow shock shape and types of MHD shocks that are relevant to different parts of the terrestrial bow shock for a typical IMF direction. One part of the shock is quasi-parallel ($\theta_{Bn} \approx 0°$), and another part of the shock is quasi-perpendicular ($\theta_{Bn} \approx 90°$). The MHD jump conditions are somewhat different for the two types of shocks, and the microscopic structures are quite different.

increase, is caused by collisions, and these shocks have widths of the order of the collisional mean free path. However, the collision mean free path for the solar wind is approximately 1 Astronomical Unit, yet observationally the shock fronts of interplanetary shocks and planetary bow shocks are known to have widths of only 10^2–10^3 km. Figure 7.29 shows some *ISEE-1* and *-2* spacecraft magnetometer data for quasi-perpendicular terrestrial bow shock crossings. The jumps in the magnetic field strength across the shocks in Figure 7.29 are obvious. The plasma density also increases as the shock is crossed, as we know from Chapter 4, but is not shown here. The substructure of a collisionless shock contains several parts: (1) the *foreshock* region, in which increased plasma wave activity is often observed well upstream of the shock, (2) the *foot* region, in which the magnetic field just starts to increase,

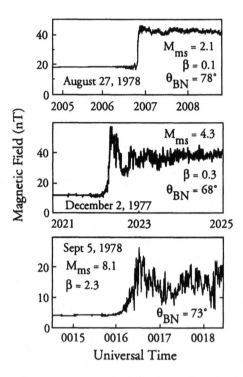

Figure 7.29. Structure of a quasi-perpendicular collisionless shock. Shown are magnetic field measurements of the terrestrial bow shock from the *ISEE-1* and 2 magnetometers. The spacecraft moved through the shock front. The relative velocity with respect to the shock was roughly 10 km/s and the shock width (or thickness) varied from about 50 km for the laminar shock up to about 200 km for the more turbulent higher Mach number shock. The different regions of the shock are discussed in the text. Adapted from Russell (1985).

(3) the *shock ramp*, in which most of the field strength increase takes place, and (4) the *shock overshoot* region. Quasi-perpendicular shocks are generally smooth or *laminar* for low values of M_{ms}. Higher Mach number quasi-perpendicular shocks are somewhat broader and contain more field fluctuations or *turbulence*. The *shock overshoot* also increases with increasing values of both the incident Mach number and the plasma beta.

The required dissipation for collisionless shocks must come from collisionless plasma processes, such as microscopic instabilities that can lead to the growth of electric and magnetic field perturbations. The type of instability present in a particular shock strongly depends on parameters such as the Mach number, the plasma beta, and the shock normal angle θ_{Bn}. A large difference exists between quasi-parallel and quasi-perpendicular shocks. For quasi-perpendicular shocks, ion gyration and reflection, and instabilities associated with counterstreaming ions, provide the shock dissipation. These processes operate on length scales of the order of a proton gyroradius. The shock foot width is approximately u_{sw}/Ω_p, where Ω_p is the proton gyrofrequency, and the width of the shock ramp is only about c/ω_{pi}, where c is the speed of light and ω_{pi} is the proton plasma frequency. Quasi-parallel shocks require a larger degree of plasma turbulence than quasi-perpendicular shocks in order for the required dissipation to be generated, and these shocks are much wider than quasi-perpendicular shocks.

7.5.4 Magnetosheath plasma flow

The solar wind plasma downstream of the bow shock of the Earth, or any planet, is subsonic in the general vicinity of the Sun–planet axis. The flow near the flanks, or sides, of the obstacle is actually supersonic, although the flow speed component normal to the shock surface is still supersonic upstream and subsonic downstream of the shock (or there would not be a shock). The region downstream of the shock but exterior to the obstacle, be it a magnetosphere or an ionosphere, is called the *magnetosheath*. The flow streamlines in the magnetosheath are almost entirely diverted around the obstacle; actually a very small fraction of the streamlines intersect (or are "absorbed" by) the obstacle surface.

Two- and three-dimensional numerical models of the flow around the terrestrial magnetopause have been carried out. Both magnetohydrodynamic and hydrodynamic (i.e., no magnetic field) models have been devised; the latter type of model gives very similar results to the first type of model except in the region just outside the obstacle. For global models, the coupled fluid equations (i.e., continuity, momentum, and energy equations) as presented in Chapter 4 are solved using appropriate numerical methods. Figure 7.30 shows numerical results from a single-fluid

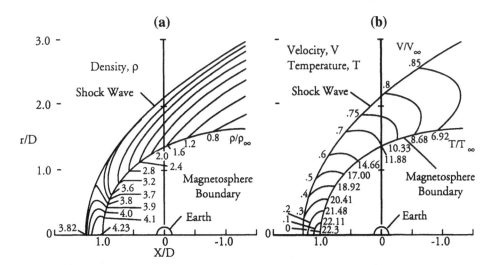

Figure 7.30. Results of a hydrodynamic solar wind flow model for the Earth for an upstream solar wind flow with a Mach number $M = 8$. Adapted from Spreiter et al. (1966) (also see Haerendel and Paschmann, 1982). (a) Density contours (relative to the upstream solar wind value) are shown. (b) Flow speed contours (relative to the upstream solar wind value) are shown. The same contours also apply to the temperature, although with different values.

hydrodynamical model of the solar wind flow around the terrestrial magnetosphere. The variables are shown relative to their values upstream in the unshocked solar wind (denoted with subscript ∞). Note that the proton density jumps across the bow shock, with the largest jump of a factor of 3.82 occurring at the subsolar point. The density downstream of the shock is roughly constant in the subsolar region but decreases in the flanks. The flow along the Sun–Earth axis, also called the *stagnation line*, is directed toward the Earth, but the flow everywhere else is largely tangential to the magnetopause. The flow speed is lowest near the subsolar part of the magnetosheath (i.e., *near the stagnation point*) but is larger near the flanks.

A sonic Mach number (M) can also be associated with each velocity contour in Figure 7.30 (see Problem 7.24). The Mach numbers near the subsolar point are less than unity; that is, the flow is subsonic. However, beyond what is called the *sonic line*, the magnetosheath flow is supersonic. In Figure 7.30 the sonic line approximately (within a few percent) corresponds to the $v/v_\infty = 0.5$ contour (Problem 7.24). Fluid flow downstream of a shock is supposed to be subsonic; yet the magnetosheath flow in the flanks is clearly supersonic. What happened? Actually, the component of the flow velocity along the shock normal direction *does remain subsonic* everywhere downstream of the bow shock, but the angle between the flow and this shock normal direction becomes rather large for the region downstream of the sonic line. In addition, the relevant upstream Mach number for applying the Rankine–Hugoniot relations must be calculated using the component of the upstream velocity along the shock normal direction; consequently, this Mach number is less than the "nominal" $M_\infty = 8$ for the part of the magnetosheath outside of the subsolar region.

Problems

7.1   Estimate the magnetic field perturbation ΔB at the surface of the Earth due to the electrical currents flowing on the dayside magnetopause for: (a) normal solar wind conditions when the subsolar magnetopause is located at a radial distance of 10 R_E, (b) during a geomagnetic storm when the magnetopause is located at 5.5 R_E.
        Hint: You can use the method of images.

7.2   Determine the distance to the subsolar magnetopause of Mercury using the Chapman–Ferraro theory, with and without taking into account the electrical current flowing in the magnetopause.

7.3   Determine the distances to the subsolar magnetopauses of Saturn, Uranus, and Neptune, with and without taking into account the electrical current flowing in the magnetopause.

7.4   Determine the subsolar magnetopause distance for Jupiter using Chapman–Ferraro theory. Now take into account the internal plasma pressure of the magnetosphere and redetermine this distance. First, assume that the dipole field is a good approximation and derive an expression for the distance r_{mp}. Second, take into account the internal plasmasheet current (i.e., field stretching) by assuming that $B_{internal} \approx 5 B_{dipole}$.

7.5 Determine the neutral scale height for the thermospheres of Venus, Earth, Mars, and Jupiter using the data given in Table 7.1 plus the exospheric temperatures discussed in the text.

7.6 Determine the altitude of peak ion production for Venus if the EUV photoabsorption cross section for CO_2 is $\approx 2 \times 10^{-17}$ cm^2. Use the model atmosphere shown in Figure 7.8

7.7 **Ion diffusion in an ionosphere**

(a) Derive the following diffusion equation for the flow speed parallel to the magnetic field, $u_{s\parallel}$, of an ion species s, starting from the momentum Equation (4.4) for species s:

$$
u_{s\parallel} = \frac{1}{\nu_{st}} (\nu_{si} u_{si} + \nu_{sn} u_{sn} + \nu_{se} u_{se})
$$
$$
- D_s \left(\frac{1}{n_s} \frac{\partial n_s}{\partial z} + \frac{1}{T_s} \frac{\partial T_s}{\partial z} + \frac{m_s g}{k_B T_s} - \frac{e E_\parallel}{k_B T_s} \right)
$$
$$
= \frac{1}{\nu_{st}} (\nu_{si} u_{si} + \nu_{sn} u_{sn} + \nu_{se} u_{se})
$$
$$
- D_s \left[\frac{1}{n_s} \frac{\partial n_s}{\partial z} + \frac{1}{T_s} \frac{\partial T_s}{\partial z} + \frac{m_s g}{k_B T_s} + \frac{T_e}{T_s} \left(\frac{1}{n_e} \frac{\partial n_e}{\partial z} + \frac{1}{T_e} \frac{\partial T_e}{\partial z} \right) \right],
$$

where $\nu_{st} = \nu_{si} + \nu_{sn} + \nu_{se}$ and $D_s = k_B T_s / (m_s \nu_{st})$.

Equation (4.4) was actually written for a single ion species $s = i$, but almost the same equation applies to an individual ion species in a multi-ion species plasma. Assume that species s can collide with electrons, neutrals, and another ion species denoted by index "i." You will need the electron momentum Equation (4.3) (or a generalized Ohm's law) to find the electric field. Neglect the convective derivative terms (i.e., inertial terms) as well as the mass-loading terms and terms of the order of m_e/m_i. For further simplicity assume that the magnetic field is in the z direction.

(b) Setting $u_{s\parallel} = 0$ in the diffusion equations that you have just derived, find the density scale height H_s for species s. Assume an isothermal ionosphere for which the electron and ion temperatures are denoted T_e and T_i, respectively. Also assume that the flow speeds $u_{i\parallel}$, $u_{e\parallel}$, and $u_{n\parallel}$ are all zero. The electron density as a function of altitude $n_e(z)$ is assumed to be known and has a scale height H_e.

7.8 Use Equations (7.29) and (7.30) to find the lifetime against vertical transport of O^+ ions as a function of altitude for the ionosphere of Earth using the neutral atmosphere shown in Figure 7.7. Plot this transport time and also the chemical lifetime of O^+ as a function of altitude. Note the altitude at which these times are equal. This is approximately the altitude of the

F_2 peak. Also estimate the vertical ion flow speed near the ionospheric peak.

7.9 **Electron heat conduction in the terrestrial ionosphere** In Problem (2.14) you found the electron temperature profile by solving the heat conduction equation. A heat flux near the "top" of the ionosphere was adopted. For the present problem estimate the total electron heat flux by integrating a heat production rate over altitude. A very crude electron heat production rate can be estimated by taking the total ion production rate and multiplying it by 1 eV (i.e., a typical 10 eV photoelectron energy multiplied by a 10% efficiency that takes into account that most of the photoelectron energy does not go into the thermal electron gas but is lost via inelastic collisions with neutrals). But below about 300 km, the electron energetics are also not conduction dominated; neglect the heat production below this altitude. Note that the atmosphere is optically thin at high altitudes and that the ion production is just the ionization frequency ($\approx 10^{-6}\,\mathrm{s}^{-1}$) times the neutral density. Apply your calculated electron heat flux to a boundary altitude of 700 km and use the results of Problem 2.14 to find the temperature profile $T_e(z)$. Assume that the bottom of the conduction region is at 300 km where the neutral (and electron temperature) is about 1,000 K.

7.10 **Equation for ion temperature in polar ionosphere** The ion energetics are collision dominated throughout most ionospheres, except at the highest altitudes. The ion temperature can then be solved by setting the collision terms by themselves equal to zero and solving the resulting energy equation. The main source of heat for the ions is ordinarily electron–ion collisions, but if the ions and neutrals have a large enough relative velocity then frictional heating (sometimes called *Joule heating*) can take place. Recall from Chapter 2 that the right-hand side of the energy equation (in temperature form) that contains the collision terms can be written as

$$collision\ terms = \left(\frac{\delta E_s}{\delta t}\right)_{\mathrm{coll}} - \mathbf{u}_s \cdot \left(\frac{\delta \mathbf{M}_s}{\delta t}\right)_{\mathrm{coll}} + \left(\frac{1}{2}m_s u_s^2 - \frac{3}{2}k_B T_s\right)S_s.$$

For an ion gas ($s = i$) in which ion–neutral collisions are the main process and for which the net production (S_s) can be neglected, this becomes

$$collision\ terms = \left(\frac{\delta E_s}{\delta t}\right)_{\mathrm{coll}} - \mathbf{u}_s \cdot \left(\frac{\delta \mathbf{M}_s}{\delta t}\right)_{\mathrm{coll}}$$
$$= \mu_{in}\nu_{in}n_i[(3k_B/m_i)(T_n - T_i) + |\mathbf{u}_i - \mathbf{u}_n|^2],$$

where μ_{in} is the ion–neutral reduced mass, n_i is the ion density, and ν_{in} is the ion–neutral momentum transfer collision frequency (note: $\nu_{in} = k_{in}n_n \approx 10^{-9}\,\mathrm{cm}^3\,\mathrm{s}^{-1}\,n_n$, where n_n is the neutral density in units of cm^{-3}).
Set these collision terms equal to zero and derive an expression for the ion temperature T_i. Now you can estimate the ion temperature in the polar

ionosphere where the magnetospheric convection electric field results in plasma flow speeds as high as 1 km/s. Determine the ion temperature for this plasma flow speed if $T_n = 1,000$ K (as it is in the terrestrial thermosphere) and if the neutrals are stationary.

7.11 Find the electron density in the E-region (in particular, at an altitude 150 km) as a function of time on the nightside of the Earth, by solving a time-dependent continuity equation. Neglect vertical transport, and assume that the ion production completely shuts off at dusk (which is not quite true).

7.12 Find a photochemical expression for the density of the minor ion N_2^+ in the daytime E-region ionosphere of Earth. Quantitatively estimate this density at $z = 150$ km.

7.13 Estimate the chemical lifetime of O^+ in the ionosphere of Venus as a function of altitude. Also estimate the vertical transport time (i.e., the ambipolar diffusion time) as a function of altitude. At what altitude are these two time scales equal?

7.14 Find photochemical expressions for the O_2^+ and CO_2^+ densities as functions of altitude in the ionosphere of Venus. In finding the O_2^+ density assume that $n_e \approx n_{O_2^+}$. Plot your results as a function of altitude from 130 km up to 180 km. Use the neutral atmosphere and the production rate profiles shown in the text.

7.15 Derive the expression (7.50) for the neutral density in a cometary coma by integrating the source-free continuity equation for a constant outflow speed u_n.

7.16 Find photochemical equilibrium expressions for the H_2O^+ density and for the electron density in a cometary ionosphere unperturbed by the solar wind.

7.17 **Ionospheric electrical conductivity expressions**

(a) Derive the generalized Ohm's law, Equations (7.61)–(7.64), starting with earlier equations.

(b) Estimate values of the parallel, Hall, and Pederson conductivities at altitudes of 120 km, 160 km, and 200 km in the terrestrial ionosphere using the terrestrial neutral atmosphere and ionosphere shown earlier in this chapter and using collision frequency expressions given in Chapter 2. Note that the parallel conductivity that you just determined is reasonable, but that the Hall and Pederson conductivities are not accurate for reasons discussed in the text.

7.18 Starting with the simplified momentum Equation (7.66) for species s, find the components of the flow velocity $\mathbf{u}_s$ along the $\mathbf{b}$, $\hat{\mathbf{n}}$, and $\hat{\perp}$ directions. Then use these results and Equation (7.65) for $\mathbf{J}$ to obtain Equations (7.68)–(7.71).

7.19 **Pressure balance at Venus's ionopause** How well does static pressure balance describe the Venus ionopause for Orbit 186 in Figure 4.9? Use $T_e = 5,000$ K and $T_i = 2,000$ K and calculate both magnetic and thermal pressures. Give a percent discrepancy for the pressure balance. Repeat this for Orbit 177.

7.20 **Low-altitude magnetic layer in the magnetized Venus ionosphere**

(a) Calculate the magnetic diffusion coefficient at $z = 120$ km, 140 km, and 200 km in the ionosphere of Venus using the neutral atmosphere and ionosphere data shown in Figures 7.8 and 4.10 and using approximate collision frequency information discussed in Chapter 2. You will have to extrapolate the neutral densities from Figure 7.8 to lower altitudes in some manner. Assume that $T_e = 5,000$ K and $T_i = 2,000$ K.

(b) Estimate the magnetic Reynolds number at $z = 120$ km, 140 km, and 200 km using the plasma speed $u(z)$ shown in Figure 7.15 plus a suitable length scale.

(c) Using "upper" boundary values at $z = 300$ km for the vertical flow speed as shown in Figure 7.15 and assuming that $B = 100$ nT at 200 km, determine and plot the steady-state solution to the magnetic–diffusion equation for z between 130 and 300 km.

7.21 As ULF waves in the SWRF convected past the *Giotto* spacecraft (which encountered comet Halley in March 1986) at approximately the solar wind speed, they were observed by the magnetometer on board. Sketch the power versus frequency (in Hz) that should have been seen by the *Giotto* magnetometer just upstream of the bow shock where $B_1^2 \approx 0.3\, B_0^2$ if the shape of the power versus wavenumber is assumed to be the same as the solar wind power spectrum (discussed near the end of Chapter 6). Note that you need to use the correct correlation wavenumber and you will need to renormalize the power spectrum. What is the correlation frequency for this power spectrum? Determine an expression for the pitch-angle scattering coefficient $D_{\mu\mu}$, and estimate a pitch-angle scattering time.

7.22 For a comet with a gas production rate of $Q = 10^{30}\,\mathrm{s}^{-1}$ and for typical unperturbed solar wind parameters at 1 AU, determine and plot the one-dimensional flow speed for the mass-loaded flow using Equation (7.93). At what radial distance from the nucleus is the solar wind speed half the value it is far upstream?

7.23 Use the one-dimensional version of the momentum equation (7.98) to determine the magnetic field profile described by Equation (7.99). Use the approximations discussed in the text. Also derive an expression for the radial location of the cavity boundary, r_{cs}, and evaluate r_{cs} for comet Halley.

7.24 Determine the sonic Mach number for each flow speed contour (these are also temperature contours but with different labels) shown in Figure 7.30. What contour most closely corresponds to the sonic line?

7.25 Apply the Rankine–Hugoniot relations to the terrestrial bow shock both at the subsolar point and in the flanks ($x = 0$) in Figure 7.30. Do your results agree with the data displayed in Figure 7.30? What is the upstream sonic

Mach number with respect to the shock normal angle if $M_\infty = 8$? What is the Mach number just downstream of shock relative to the shock normal angle?

Bibliography

Acuna, M. H. and N. F. Ness, The main magnetic field of Jupiter, *J. Geophys. Res.*, **815**, 2917, 1976.

Atreya, S. K., *Atmospheres and Ionospheres of the Outer Planets and Their Satellites*, Physics and Chemistry in Space, vol. 15, Springer-Verlag, Berlin, 1986.

Bagenal, F., Planetary magnetospheres, p. 224 in *Solar System Magnetic Fields*, ed. E. R. Priest, D. Reidel Publ. Co., Dordrecht, The Netherlands, 1985.

Banks, P. M. and G. Kockarts, *Aeronomy*, Academic Press, New York, 1973.

Brandt, J. C., Observations and dynamics of plasma tails, pp. 519–537 in *Comets*, Univ. Arizona Press, Tucson, 1982.

Chapman, S. and V. C. A. Ferraro, A new theory of magnetic storms, 1. The initial phase, *J. Geophys. Res.*, **36**, 77, 1931a.

Chapman, S. and V. C. A. Ferraro, A new theory of magnetic storms, 2. The initial phase (continued), *J. Geophys. Res.*, **36**, 171, 1931b.

Cravens, T. E., T. I. Gombosi, J. Kozyra, A. F. Nagy, L. H. Brace, and W. C. Knudsen, Model calculations of the dayside ionosphere of Venus: Energetics, *J. Geophys. Res.*, **85**, 7778, 1980.

Cravens, T. E., Comparative ionospheres, p. 805 in *Solar-Terrestrial Physics: Principles and Theoretical Foundations*, ed. R. L. Carovillano and J. M. Forbes, D. Reidel Publ. Co., Dordrecht, The Netherlands, 1983.

Cravens, T. E., H. Shinagawa, and A. F. Nagy, The evolution of large-scale magnetic fields in the ionosphere of Venus, *Geophys. Res. Lett.*, **11**, 267, 1984.

Cravens, T. E., Solar wind interaction with non-magnetic planets, p. 353 in *Solar System Plasma Physics*, ed. J. H. Waite Jr., J. L. Burch, and R. L. Moore, Geophys. Monograph 54, American Geophysical Union, Washington, DC, 1989.

Haerendal, G. and G. Paschmann, Interaction of the solar wind with the dayside magnetosphere, chap. 2 in *Magnetospheric Plasma Physics*, ed. A. Nishida, Center for Academic Publ. Japan, D. Reidel Publ. Co., Dordrecht, The Netherlands, 1982.

Hunten, D. M., L. Colin, T. M. Donahue, V. I. Moroz, eds., *Venus*, Univ. of Arizona Press, Tucson, 1983.

Jackson, D. J. and D. B. Beard, The magnetic field of Mercury, *J. Geophys. Res.*, **82**, 2828, 1977.

Keating, G. M., et al., Models of Venus neutral upper atmosphere: Structure and composition, *Adv. Space Res.*, **5**, 117, 1985.

Nagy, A. F., T. E. Cravens, S. G. Smith, H. A. Taylor Jr., and H. C. Brinton, Model calculations of the dayside ionosphere of Venus: Ionic composition, *J. Geophys. Res.*, **85**, 7795, 1980.

Newburn, R. L. Jr., M. Neugebauer, and J. Rahe, eds., *Comets in the Post-Halley Era*, Kluwer Academic Publ., Dordrecht, The Netherlands, 1991.

Ogino T., R. J. Walker, and M. Ashour-Abdalla, A three-dimensional MHD simulation of the interaction of the solar wind with comet Halley, *J. Geophys. Res.*, **93**, 9568, 1988.

Ratcliffe, J. A., *An Introduction to the Ionosphere and Magnetosphere*, Cambridge Univ. Press, Cambridge, UK, 1972.

Rees, M. H., *Physics and Chemistry of the Upper Atmosphere*, Cambridge Atmospheric and Space Science Series, Cambridge Univ. Press, Cambridge, UK, 1989.

Richmond, A. D., Thermospheric dynamics and electrodynamics, p. 523 in
Solar-Terrestrial Physics: Principles and Theoretical Foundations, ed. R. L.
Carovillano and J. M. Forbes, D. Reidel Publ. Co., Dordrecht, The Netherlands, 1983.

Russell, C. T., Planetary bow shocks, p. 109 in *Collisionless Shocks in the Heliosphere:
Reviews of Current Research*, Geophysical Monograph 35, American Geophysical
Union, Washington, DC, 1985.

Russell, C. T., D. N. Baker, and J. A. Slavin, The magnetosphere of Mercury, p. 514 in
Mercury, ed. F. Vilas, C. R. Chapman, and M. S. Shapley, Univ. Arizona Press,
Tucson, 1988.

Russell, C. T., ed., *Venus Aeronomy*, Kluwer Academic Publishers, Dordrecht, The
Netherlands, 1991.

Schunk, R. W., The terrestrial ionosphere, p. 609 in *Solar-Terrestrial Physics: Principles
and Theoretical Foundations*, ed. R. L. Carovillano and J. M. Forbes, D. Reidel Publ.
Co., Dordrecht, The Netherlands, 1983.

Spreiter, J. R., A. L. Summers, and A. Y. Alksne, Hydromagnetic flow around the
magnetosphere, *Planet. Space Sci.*, **14**, 223, 1966.

Tsurutani, B. T. and E. J. Smith, Strong hydrodynamic turbulence associated with comet
Giacobini-Zinner, *Geophys. Res. Lett.*, **13**, 259, 1986.

Vasyliunas, V. M., Comparative magnetospheres, p. 479 in *Solar-Terrestrial Physics:
Principles and Theoretical Foundations*, ed. R. L. Carovillano and J. M. Forbes,
D. Reidel Publ. Co., Dordrecht, The Netherlands, 1983.

Yi, Y., F. M. Caputo, and J. C. Brandt, Disconnection events (DEs) and sector boundaries:
The evidence from comet Halley 1985–1986, *Planet. Space Sci.*, **42**, 705, 1994.

8

The magnetosphere

The intrinsic magnetic field of the Earth acts as an obstacle to the solar wind and shields a volume of space, called *the magnetosphere*, from direct access of the solar wind. In Chapter 7, we considered the role of the magnetosphere as an obstacle to the solar wind and were mainly concerned with the region "external" to the *magnetopause*. The details of the internal dynamics of the magnetosphere do not seriously affect, at least to about the 95% level, the external solar wind plasma flow, but the solar wind does strongly affect the internal dynamics of the magnetosphere and ionosphere, as we will see in this chapter. This chapter will strongly emphasize macroscopic or fluid aspects of magnetospheric physics rather than the microscopic physics operating in the magnetosphere. Some aspects of the inner magnetosphere (i.e., the *ring current* and *radiation belts*) were already considered in Chapter 3.

The terrestrial magnetosphere has been extensively studied over the past 35 years with dozens of Earth orbiting satellites. The *International Sun Earth Explorer (ISEE)*, *Dynamics Explorer (DE)*, and *AMPTE* missions have been especially important, and in the near future we can expect useful information from recently launched spacecraft such as *Geotail* and *Polar*. The volume of observational and theoretical literature that exists, mainly in the *Journal of Geophysics Research–Space Physics*, has become immense. Much has been learned about how the magnetosphere works, although many key processes remain poorly understood. The magnetospheres of Mercury, Jupiter, Saturn, Uranus, and Neptune have also been studied with data from the *Mariner*, *Pioneer*, and *Voyager* missions.

Section 8.1 introduces the morphology of the magnetosphere as well as some terminology. The sections that follow treat the magnetospheric dynamics and electric current systems more quantitatively. In particular, the concept of magnetic reconnection presented at the end of Chapter 4 is used to determine the convection electric field inside the magnetosphere. The magnetospheres of the outer planets are discussed in Section 8.7. The reader is also referred to the references listed in the bibliography at the end of the chapter, including Haerendel and Paschmann (1982), Hill (1983), Vasyliunas (1983), Parks (1991), Ratcliffe (1972), Kivelson and Russell (1995), Elphinstone et al. (1996), Smith and Lockwood (1996), and Bagenal (1985).

8.1 Introduction to the morphology of the magnetosphere

Figure 7.2 showed a simple schematic of the magnetosphere, the magnetosheath, and the bow shock. As discussed in Chapter 7, the boundary surface separating the magnetosphere from solar wind plasma is called the magnetopause. The shocked solar wind plasma flows around the magnetosphere outside the magnetopause in the magnetosheath.

A schematic illustrating the main internal regions of the terrestrial magnetosphere is given in Figure 8.1. The magnetopause is also indicated in Figure 8.1. The magnetopause is not infinitesimally thin, but has a thickness of several ion gyroradii across which the plasma characteristics change from those characteristic of the magnetosheath to those characteristic of the magnetosphere. This *magnetopause boundary layer* is also called the *plasma mantle* at some locations (i.e., for the downstream, or magnetotail, and high latitude portions of the magnetopause), the *tail boundary layer* in the tail region at low latitudes, and the *low latitude boundary* layer in the subsolar part of the magnetopause. It is thought that some magnetosheath plasma can diffuse inward across this boundary layer. In the *closed model* of the magnetosphere, originally postulated by Axford and Hines (1961), the viscous stress associated with the diffusion of the external flowing magnetosheath plasma across the boundary layer is responsible for the convection (or motion) of the magnetospheric plasma. The boundary layer/plasma mantle is less important

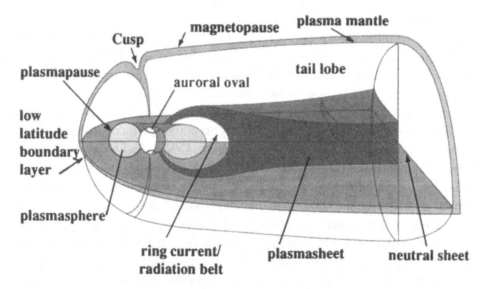

Figure 8.1. Schematic of the terrestrial magnetosphere. The different regions shown in this diagram are discussed in the text.

in the *open model* of the magnetosphere, in which some magnetic field lines (i.e., "open" field lines) reach from within the magnetosphere out to the interplanetary medium. The solar wind motional electric field directly "maps" into the magnetosphere along these open magnetic field lines. We will address this topic again later in this chapter.

The *cusp* (or *cleft*) is a region in which the boundary layer extends deep into the magnetosphere in a funnel-like fashion along open magnetic field lines. Magnetosheath plasma has its most direct access to the magnetosphere in the cusp regions (north and south).

The *plasmasphere* is a doughnut-shaped region located near the Earth with radial distances r less than a few R_E in the equatorial plane, and it contains relatively dense ($n_e > 10^2$ cm^{-3}) and cold ($T_e \approx T_i \approx 1$ eV) plasma of ionospheric origin. The magnetic field in this low latitude region is accurately described by a dipole field. The plasma in this region *corotates* with the Earth, as will be discussed in Section 8.4. The plasmasphere is an extension from lower altitudes of the topside F_2-region ionosphere, which serves as a source of plasma for the plasmasphere. The outer boundary of the plasmasphere is relatively sharp and is called the *plasmapause*. The electron densities outside the plasmapause are much lower than plasmaspheric electron densities. The plasmapause is typically located near an L-shell, $L_{pp} \approx 4$ (that is, in the equatorial plane $r_{pp} \approx 4R_E$), but L_{pp} strongly depends on the level of geomagnetic activity. The plasmapause moves inward to low L values during magnetically active times and is located at larger distances during quiet times.

The *plasmasheet* is a region containing hot plasma with densities $n_e \approx 0.1$–1 cm^{-3} and with an electron temperature of $T_e \approx 1$ keV. The typical ion energy or temperature is $T_i \approx 5$ keV. This region is slab shaped and is located in the equatorial plane, stretching tens of R_E down the Earth's *magnetotail*. The *neutral sheet* exists in the middle of the plasmasheet where the stretched magnetic field changes from sunward to antisunward orientation and has its minimum normal component. A *cross-tail current* flows from the dawnside to the duskside within the plasmasheet and is responsible for the stretched magnetotail geometry. The *tail lobes*, north and south, are located outside the plasmasheet and are characterized by large magnetic fields and very low plasma densities ($n_e \ll 0.1$ cm^{-3}). These field lines in the open model of the magnetosphere connect to the IMF. The magnetic pressure in the tail lobes balances the total pressure (mostly thermal pressure) in the plasmasheet. The boundary between the lobes and plasmasheet is called the *plasmasheet boundary layer* (PSBL). Figure 8.2 is a schematic of the cross section of the magnetotail facing toward the Sun. Notice that the PSBL flares up and down as it approaches the magnetopause. The cross-tail current flows across the plasmasheet and merges into currents flowing along the magnetopause in the plasma mantle.

The *ring current* region is located in the inner magnetosphere in the general vicinity of the plasmapause, but it extends both inwards and outwards to the plasmasheet. The ring current plasma is very hot, with proton energies of tens of keV, and it often

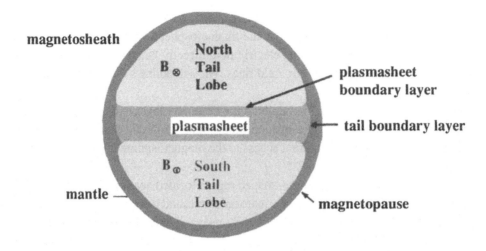

VIEW FROM MAGNETOTAIL FACING THE SUN

Figure 8.2. Cross section of the magnetosphere downstream (tailward) of the Earth.

coexists with colder plasmaspheric plasma populations. The *trapped radiation belts* are the higher energy extension of the ring current and contain particles with MeV energies. The radiation belts (also known as the *van Allen belts*) are located near $L \approx 3$. The cold plasma in the magnetosphere, including the plasmasphere, moves primarily with the $\mathbf{E} \times \mathbf{B}$ drift velocity. However, the more energetic ring current and radiation belt particles are subject to magnetic gradient and curvature drifts. These drifts plus a magnetization current produce a westward electrical current (i.e., the ring current). The total current is about one million amperes, but this is very dependent on the level of geomagnetic activity. At high levels of geomagnetic activity, the ring current energy density is high as is the total ring current, whereas during quiet times, both the energy density and the current are much lower. The radiation belts are much more stable than the ring current and exhibit little temporal variation. The ring current and radiation belt were discussed in Chapter 3.

Different parts of the magnetosphere can be mapped along magnetic field lines. This mapping is easy to do in the inner magnetosphere where the magnetic field is dipolar but is more difficult to carry out in the outer magnetosphere, especially in the magnetotail, where electrical currents distort and stretch the magnetic field. The inner magnetosphere (i.e., low L-shell values) maps to the low latitude ionosphere, and the outer magnetosphere and solar wind map down to the polar ionosphere. The polar plot of the terrestrial ionosphere in Figure 8.3 illustrates key regions. The *magnetic north pole* and circles of constant *magnetic latitude* are shown. As discussed in Chapter 3 the geographic and magnetic poles are offset by about 15°. Figure 8.3 also shows *local solar time* (LST) around the "polar dial" with noon at the top. At equinox, the terminator would be a straight line running across the plot from dawn to dusk.

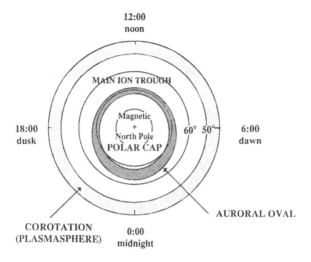

Figure 8.3. Polar plot of the ionosphere in magnetic coordinates.

In the currently accepted open model of the magnetosphere, the magnetic field lines in the *polar cap* are "open" and connect with the IMF in the solar wind and magnetosheath via the *cusp* and *tail lobe* regions of the magnetosphere. The cusp itself connects to the region near the noon edge of the polar cap near the *auroral oval*. The auroral oval separates regions of open and closed magnetic field lines. The *nightside* part of the auroral oval connects to the plasmasheet, and the dayside part, including the cusp, connects to the *dayside magnetopause* (or low latitude boundary layer). The auroral oval is typically located near 65–70° magnetic latitude, but during magnetically active time periods the oval expands to lower latitudes and during quiet times the oval contracts and is located at higher latitudes.

As the name implies, the auroral oval is where the *aurora borealis* (in the North) or the *aurora australis* (in the South) occurs. Energetic particles, especially electrons, precipitate from the magnetosphere into the upper atmosphere and produce the visible emissions that are observed as the "aurora." This *energetic particle precipitation* also causes significant ionization, heating, and dissociation in the thermosphere. We discussed some aspects of this in Chapter 7. The nature of the particle precipitation (i.e., the energy spectrum and time history of the incoming particles) differs for different local times around the oval. Especially energetic electron events known as *inverted-V* events occur during magnetic active times on the nightward part of the oval and are part of *auroral substorms*. *Field-aligned currents*, also called *Birkeland currents*, tend to be concentrated in the auroral oval. These currents are important for the dynamical coupling of the magnetosphere to the ionosphere, and they will be discussed in some detail later in this chapter.

The plasmasphere typically maps down to latitudes less than about 50°, as shown in Figure 8.3. The topside ionospheric electron and ion densities are, on average, lower for latitudes between the low altitude extension of the plasmapause and the auroral oval. This region of relatively low ionospheric density is called the *main ion trough* or sometimes just *plasma trough*; there is also an ionospheric *light ion trough* located elsewhere. Polar cap ionospheric densities are also rather low in comparison with those at lower latitudes for a couple of reasons: (1) There is less photoionization due to larger solar zenith angles, and (2) there exists outward escape of plasma along open, or very extended, magnetic field lines. Ionospheric densities in the auroral oval can be quite high. The plasma at low latitudes corotates with the Earth on the average, although some small local departures from corotation exist, whereas the plasma at higher latitudes largely moves in response to the *convection electric field* that can be mapped down to the ionosphere from the magnetosphere.

8.2 The open model of the magnetosphere

8.2.1 Open and closed magnetosphere models

Dungey (1961) was the first to propose that the Earth's magnetic field could connect to the interplanetary magnetic field via the magnetic reconnection process (this is the *open model of the magnetosphere*). The solar wind motional electric field can map down these "open" field lines into the magnetosphere, thus generating a *convection electric field* directed across the magnetosphere from the dawnside to the duskside. The plasma responds to this electric field largely by $\mathbf{E} \times \mathbf{B}$ drifting; that is, the magnetospheric plasma *convects* due to this electric field. The magnitude of the convection electric field depends on the efficiency of the reconnection process, which in turn strongly depends on the orientation of the IMF. Reconnection is most efficient for a southward-directed IMF (i.e., $B_z < 0$).

A model competing with the open model for many years was the *viscous interaction model* (or *closed magnetosphere model*) proposed by Axford and Hines (1961). In this model, diffusion of particles across the magnetopause boundary layer induces a potential drop across the magnetosphere, thus generating a plasma convection pattern similar to that of the open model. Solar wind/magnetosheath momentum is transferred by viscous stresses across a viscous boundary layer to the magnetospheric plasma located inside. The closed model has several difficulties: (1) The plasma is essentially collisionless so that an effective viscosity due to wave–particle interactions needs to be invoked in order to provide sufficient viscous momentum transfer across the magnetopause; (2) the model has no dependence on the IMF orientation, whereas observations show that magnetospheric convection, and geomagnetic activity in general, is very sensitive to this orientation; and (3) particle measurements, especially in the cusp region, provide evidence for an open field line configuration. The evidence is very strongly in favor of the open model; however, nature is complex, and a closed model description does appear to be more applicable during extended time periods of low geomagnetic activity, in which the IMF has been persistently northward.

Figure 8.4 shows the main features of the open model schematically, although it is highly oversimplified. The IMF is "convected" with the solar wind flow toward

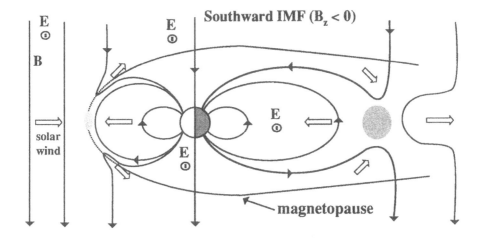

OPEN MODEL OF THE MAGNETOSPHERE

Figure 8.4. Schematic of the open model of the magnetosphere for southward interplanetary magnetic field (IMF). IMF field lines first "reconnect," or merge, with geomagnetic field lines at the dayside magnetopause. The field lines then convect antisunward, moving with the plasma, and eventually reconnect again far down the tail. The geometry in this schematic is idealized – the real magnetotail is much longer than is shown here. The "diffusion" regions, where the reconnection takes place, are indicated by shading. The motional electric field is everywhere directed out of the page, that is, from dawn to dusk.

the Earth. Plasmas with oppositely directed field lines encounter each other at the subsolar magnetopause. The magnetic field lines reconnect, or merge, in a "diffusion region" located on the dayside magnetopause. The reconnected field lines then convect tailward over the polar caps. The F-region ionospheric plasma in the polar cap also participates in this *antisunward flow* since it is magnetically connected with the solar wind flow. Magnetic flux piles up tailward of the Earth where field lines again merge in a diffusion region located in the plasmasheet. In the equatorial plane Earthward of the reconnection region, there is *sunward flow* due to the $\mathbf{E} \times \mathbf{B}$ drift associated with the convection electric field. This sunward, or *return flow*, carries magnetic field lines away from the tail reconnection region back into the subsolar reconnection region.

We will consider both the reconnection and the magnetospheric convection processes in more detail later. Although the open magnetosphere model is now the accepted model, many of the fundamental details of how it operates are not fully understood, and what is understood is difficult to describe in an introductory manner. In this chapter I try to capture the flavor of how the magnetosphere operates without getting bogged down in the "microscopic" details. The application of the reconnection process to the dayside magnetopause will be treated in a simplified, or even oversimplified, manner in Sections 8.2.2 and 8.2.3. Observations of the magnetopause will also be discussed in Section 8.2.6.

8.2.2 Reconnection at the dayside magnetopause and the convection electric field (southward IMF)

We now consider in more detail the reconnection process at the dayside magnetopause; reconnection in the magnetotail will be considered in Section 8.6. The starting point for this discussion is the Chapter 4 material on magnetic reconnection, which you should now review. In particular, recall the reconnection geometry as depicted in Figure 4.26, which looks very much like the dayside magnetopause reconnection region of Figure 8.4. Plasma flows into the *diffusion region* from both sides. The magnetic fields "carried" by this flow are in opposite directions on opposing sides of the magnetopause, and after "reconnection" or "merging" in the diffusion region the magnetic field is carried by the exit flow in a direction tangential to the current sheet. Our objective in this section is to estimate the electric field within the magnetosphere that results from the dayside reconnection process. To do this we need to figure out how to apply the picture of reconnection illustrated in Figure 4.26 to the magnetospheric geometry and also how to estimate the microscopic parameters that appear in the reconnection theory presented in Section 4.11. We will emphasize simple steady-state magnetic reconnection and only briefly discuss more complicated scenarios.

All solar wind streamlines would be diverted around (and external to) the magnetopause if reconnection did *not* take place (i.e., for a *closed* magnetosphere). However, reconnection at the dayside magnetopause allows some streamlines to penetrate this boundary and also allows the solar wind motional electric field access to the magnetosphere, albeit with reduced strength. Of critical importance to magnetospheric physics is the magnitude of the magnetospheric electric field, and in the open model this depends on the efficiency of the reconnection process. The electric potential is virtually constant along magnetic field lines in collisionless plasmas such as the solar wind and the magnetospheric plasma. Recall that this property of collisionless plasma was also used in Chapter 6 to deduce the large-scale properties of the interplanetary field. (There are exceptions to this, such as the parallel electric fields present on some auroral field lines). Hence, if we can keep track of the magnetic field lines in the reconnection scenario, then we can determine the electric potential and electric field in the magnetosphere starting from the solar wind electric and magnetic fields. We denote the electric and magnetic fields in the solar wind $\mathbf{E}_0$ and $\mathbf{B}_0$, respectively. Recall from Chapter 7 that

$$\mathbf{E}_0 = -\mathbf{u}_{\text{sw}} \times \mathbf{B}_0. \tag{8.1}$$

For purely southward IMF the magnitude is $E_0 = u_{\text{sw}} B_0$. The electric field inside the magnetosphere is denoted $\mathbf{E}_{\text{m}}$ and is approximately constant throughout the magnetosphere. This field is directed from dawn to dusk.

Consider the schematics of idealized magnetic reconnection at the dayside magnetopause for purely southward IMF shown in Figures 8.5, 8.6, and 8.7. The solar

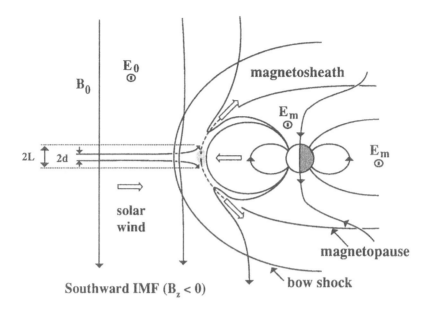

Figure 8.5. Schematic of the magnetosphere in the plane defined by the magnetic poles and the Sun–Earth axis. This diagram is for southward IMF. The vertical extent of the diffusion region is $2L$, and $2d$ is the separation of the streamlines in the solar wind that connect to this diffusion region. $\mathbf{E}_0$ and $\mathbf{E}_m$ are the electric fields in the solar wind and magnetosphere, respectively. The magnetosphere is assumed to have a diameter of between about $2r_{mp}$ to $3r_{mp}$ at the flanks.

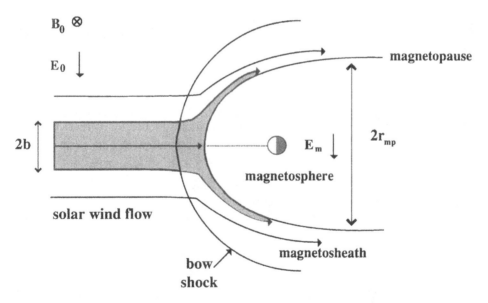

Figure 8.6. Schematic of the magnetosphere in the magnetic equatorial plane for southward IMF. Several streamlines are shown. The shaded region with width $2b$ in the solar wind and width of 2 to $3r_{mp}$ at the magnetopause encompasses all the equatorial streamlines that enter the magnetopause. Streamlines outside this region are completely diverted around the magnetosphere.

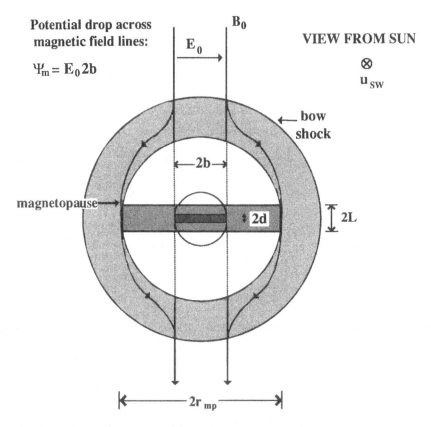

Potential drop across magnetic field lines:

$$\Psi_m = E_0\,2b$$

VIEW FROM SUN

Figure 8.7. Schematic of cross section of the magnetosphere in a plane perpendicular to the Sun–Earth axis and including the Earth. The view is from the Sun with the solar wind flow into the page. The outer shaded region between the magnetopause and the bow shock is the magnetosheath. The last (i.e., the outermost) magnetic field lines to intersect the magnetopause and reconnect are shown. The rest of the diagram is explained in the text.

wind flow diverges in order to flow around the magnetosphere after it passes the bow shock, but in the open magnetosphere model some streamlines that are close enough to the Sun–Earth axis intersect the magnetopause (i.e., the diffusion region), and the field lines carried by the flow on these streamlines reconnect (or merge) with geomagnetic field lines. The diffusion region in a simple steady-state model for southward IMF is beltlike with vertical extent $2L$ and stretches around the dayside magnetopause. The cross section of this region projected onto a plane normal to the solar wind direction is shown in Figure 8.7 as the outer shaded rectangle. The inner shaded rectangle in Figure 8.7 with height $2d$ and width $2b$ is the "footprint" of the diffusion region in the upstream unshocked solar wind. The magnetic field lines that start out in the solar wind as straight lines separated by a distance $2b$ are stretched out to width of about $2r_{mp}$ to $3r_{mp}$ by the time they intersect the magnetopause in the dawn–dusk plane. The distance to the subsolar magnetopause is denoted r_{mp}, whereas we are estimating that the distance to the magnetopause at the flanks is about $1.5r_{mp}$.

The potential difference between the two field lines pictured in Figure 8.7 is simply

$$\Psi_m = E_0 2b. \tag{8.2}$$

The potential difference across the magnetosphere is also Ψ_m because field lines are equipotentials. However, the electric field strength in the magnetosphere is less than it is in the solar wind because of the greater separation of the field lines; that is, Ψ_m is also equal to $E_m 3 r_{mp}$. Equating $E_m 3 r_{mp}$ to $E_0 2b$, we find that the ratio of the magnetospheric electric field to the solar wind electric field, denoted ζ, is equal to

$$\zeta = \frac{E_m}{E_0} = \frac{2}{3} \frac{b}{r_{mp}}. \tag{8.3}$$

The ratio ζ is also called the *efficiency of reconnection* and can be determined by finding the length b.

We proceed by using one-dimensional steady gas dynamics as presented in Chapter 6. You should review Section 6.2. The flow is pictured in Figure 8.5 and the spatial variable is distance along the Sun–Earth axis. Admittedly, the flow depicted in Figure 8.5 diverges drastically near the magnetopause and "one-dimensional" is hardly a good description, but some understanding can be obtained relatively easily in this way. Let x be the distance along the Sun–Earth axis. $A(x)$ is the cross-sectional area of a flow tube and $u(x)$ is the "one-dimensional" flow speed. The area of interest in the solar wind is enclosed in a circle of radius b with $A = \pi b^2$ (this circle is indicated in Figure 8.7). As the magnetopause is approached, $A(x)$ increases rapidly, reaching a value at the magnetopause of roughly $A \approx \pi (1.5 r_{mp})^2 \approx 2\pi r_{mp}^2$ (the surface area of a hemisphere).

The flow speed outside the bow shock is just the solar wind speed u_{sw}, and typically $u_{sw} \approx 400$ km/s. The solar wind flow undergoes a shock with incident Mach number $M \approx 6$ (see Chapter 7) such that just downstream of the subsolar bow shock the flow speed is u_2 as determined by the Rankine–Hugoniot relations (see Section 4.9). These relations yield the shock jump $Z_s = u_{sw}/u_2 = \rho_2/\rho_{sw}$ in terms of the upstream Mach number. For a magnetic field direction perpendicular to the flow it is also true that $B_2/B_0 = Z_s$, where B_2 is the postshock field strength. For a strong shock, $Z_s \approx 4$ and $u_2 \approx 100$ km/s.

The steady-state continuity equation for the flow in the subsolar magnetosheath is

$$\nabla \cdot (\rho \mathbf{u}) = 0, \tag{8.4}$$

where $\rho = n_p m_p$ and n_p is the proton density. This equation can also be written in terms of $A(x)$ as in Equation (6.17) (i.e., $\rho u A = constant$). If gravity is neglected then the expression for the variation of the Mach number derived in Section 6.2.1 can be combined with the continuity equation to obtain the following expression for the density variation $d\rho$:

$$\frac{d\rho}{\rho} = -\frac{dA}{A}M^2. \tag{8.5}$$

Expression (8.5) tells us that for subsonic flow ($M^2 \ll 1$), then $|d\rho/\rho| \ll |dA/A|$. That is, we have *incompressible flow*:

$$d\rho \approx 0 \quad \text{or} \quad \rho \approx constant. \tag{8.6}$$

For incompressible flow, from Equations (8.4) and (8.6) we have $\nabla \cdot (\mathbf{u}) = 0$, which can also be written in the form

$$u(x)A(x) = constant. \tag{8.7}$$

Note that the magnetic field strength in the subsolar (and subsonic) magnetosheath is also roughly constant, just as the proton density is; that is, $B \approx constant$ between the bow shock and the subsolar magnetopause (see Problem 8.2). Magnetosheath flow was also discussed briefly in Chapter 7. Figure 7.30 showed proton densities from the three-dimensional numerical hydrodynamical model of Spreiter et al. (1966). n_p is expected to be constant for incompressible flow, and the numerical model indicates that this is true within a factor of two or so over most of the dayside magnetosheath, particularly in the subsolar region.

For the flow tube depicted in Figures 8.5 and 8.7, u and A start off just downstream of the bow shock as u_2 and πb^2, respectively, and end up as u_0 and $2\pi r_{mp}^2$, respectively, at the magnetopause. In this case, Equation (8.7) becomes

$$u_0 2\pi r_{mp}^2 = u_2 \pi b^2 \tag{8.8}$$

or

$$\frac{b^2}{r_{mp}^2} = \frac{2u_0}{u_2}. \tag{8.9}$$

Given that the efficiency is defined by Equation (8.3), we use Equation (8.9) to write

$$\zeta \cong \frac{2}{3}\sqrt{\frac{2u_0}{u_2}} \cong \sqrt{\frac{u_0}{u_2}}. \tag{8.10}$$

We know what u_2 is and if we can just find the speed u_0 of the flow entering the diffusion region on the dayside magnetopause, then we have found the efficiency ζ and also, then, the magnetospheric electric field $E_m = \zeta E_0$. For a closed magnetosphere, no streamlines intersect the magnetopause and $u_0 = 0$, so that $\zeta = 0$ (at least from the reconnection process). But for an open model, flow carrying magnetic field enters the diffusion region with a finite speed. One complication is that u_0 is most appropriate for the subsolar magnetopause and is clearly not going to be the same over the whole equatorial magnetopause, but nonetheless we will loosely apply this speed to the whole diffusion region.

The problem of finding the efficiency has come down to the problem of finding the reconnection rate, which in Chapter 4 was expressed as the speed u_0 of the flow into the reconnection region. We now apply some of the results of Section 4.10 to magnetic reconnection at the dayside magnetopause. The inflow speed can be expressed in terms of the Alfvénic Mach number (M_{A0}) and the Alfvén speed (C_{A0}) of the inflowing plasma:

$$u_0 = M_{A0}C_{A0}. \tag{8.11}$$

We will obtain M_{A0} from reconnection theory, and from Chapter 4 we have $C_{A0}^2 = B^2/[\mu_0 n_p m_p]$. Note that both n_p and B are approximately constant in the subsolar magnetosheath (see Problems 8.1 and 8.2) and are thus equal to their postshock values. The postshock values of density and magnetic field strength are simply expressed in terms of the upstream values and the shock jump: $B = Z_s B_0$ and $n_p = Z_s n_{sw}$. The shock jump can be expressed in terms of the incident Mach number of the flow (Equation (4.184)), and for the solar wind $Z_s \approx 3$ or 4. The Alfvén speed downstream of the bow shock can be written as

$$C_{A0}^2 = Z_s C_{Asw}^2, \tag{8.12}$$

where C_{Asw} is the Alfvén speed in the unshocked solar wind. Using Equations (8.11) and (8.12), Equation (8.10) for the efficiency can be expressed as

$$\zeta \cong \sqrt{\frac{u_0}{u_2}} = \sqrt{\frac{M_{A0}C_{A0}}{u_2}} = Z_s^{3/4}\sqrt{\frac{M_{A0}}{M_{Asw}}}. \tag{8.13}$$

Equation (8.13) indicates that the efficiency is related to the ratio of the upstream solar wind Alfvénic Mach number, which is typically $M_{Asw} \approx 8$ (see Example 4.4), to the Alfvénic Mach number in the flow entering the diffusion region.

Now we utilize the Petschek/Sonnerup theory of steady-state reconnection as reviewed at the end of Chapter 8. Recall Equations (4.209)–(4.212). The maximum M_{A0} predicted by this theory is

$$M_{A0max} \approx \frac{\pi}{8}\frac{1}{\ln R_m^{os}}, \tag{8.14}$$

where R_m^{os} is the magnetic Reynolds number of the system, which can be expressed in terms of the incoming Alfvén speed, the system length L_{syst}, and the effective resistivity for the diffusion region. We have

$$R_m^{os} \approx \frac{\mu_0 C_{A0} L_{syst}}{\eta_{eff}}. \tag{8.15}$$

The collisional resistivity is negligible for a collisionless plasma (by definition) and gives an infinite magnetic Reynolds number, as defined by Equation (8.15).

In this case, looking at Equation (8.14), $M_{A0\,max} \approx 0$ and the reconnection rate approaches zero; however, the observational evidence demonstrates that the open magnetosphere model is appropriate, with a healthy rate of reconnection at the dayside magnetopause. The implication is that an effective collisionless electrical resistivity, greatly in excess of the collisional resistivity, must be present in the reconnection diffusion region. Plasma instabilities are known to be present in current layers, such as the magnetopause, that have sufficiently high current densities. Effective (or anomalous) resistivity results from the electric and magnetic field perturbations associated with the instabilities in these current layers. As discussed in Chapter 4, a reasonable value of the thickness of the diffusion region is the gyroradius ($a \approx r_L$, Equation (4.213)), in which case the resistivity given by Equation (4.214) is

$$\eta_{eff} = r_L \mu_0 u_0. \tag{8.16}$$

The issue of the thicknesses of the diffusion region and the magnetopause will be brought up again later.

The magnetic Reynolds number of the incident flow into the diffusion region now becomes

$$R_m^{os} \approx \frac{\mu_0 C_{A0} L_{syst}}{r_L \mu_0 u_0} = \frac{L_{syst}}{r_L} \frac{1}{M_{A0}} \cong \frac{r_{mp}}{r_L} \frac{1}{M_{A0}}, \tag{8.17}$$

where we also assumed that the system size is approximately the distance to the magnetopause r_{mp}. Taking $M_{A0} = M_{A0\,max}$ in Equation (8.17) and combining with Equation (8.14) we obtain an algebraic equation for $M_{A0\,max}$ that can be solved by employing an iterative procedure (see Problem 8.3). In the vicinity of the magnetopause, the gyroradius has the value $r_L \approx 100\,km$. Using this value of r_L and $r_{mp} \approx 10 R_E$, the Alfvénic Mach number for the reconnecting flow is

$$M_{A0max} \approx 0.04. \tag{8.18}$$

Now we can substitute this Alfvénic Mach number into Equation (8.13) to find that the efficiency of reconnection for the terrestrial magnetosphere is

$$\zeta \approx 0.2, \tag{8.19}$$

where $M_{Asw} = 8$ was used and $Z_s = 3$. Equation (8.19) indicates that the reconnection efficiency is about 20%, which is quite large. The values of b, d, and L can be found from the efficiency: $b = 0.2 r_{mp} \approx 2 R_E$, $d \approx 0.1 R_E$, and $L \approx 1 R_E$ (see Problem 8.4). Recall that L is the "height" of the diffusion region. Using these values we can see that about $4\,db/2\pi r_{mp}^2$ or 0.1% of the incident plasma actually intersects the diffusion region.

The electric potential drop, or difference, across the magnetosphere can now be estimated from the solar wind motional electric field, $E_0 = u_{sw} B_0 \approx (4 \times 10^5 \text{ m/s}) \times (5 \times 10^{-9} \text{ T}) \approx 2 \text{ mV/m}$, and from the efficiency:

$$\Psi_m = \zeta E_0 3 r_{mp} = \zeta u_{sw} B_0 3 r_{mp} \approx \zeta 3 \times 10^5 \text{ V} \approx 60 \text{ kV}. \qquad (8.20)$$

We have just estimated that the cross-tail potential is about 60 kV when the IMF is directed southward. This estimate is consistent with the observations made in the ionosphere and in the polar cap ionosphere during geomagnetically active time periods.

We did better than we had any right to expect considering the assumptions that went into the derivation of this result and considering that all the complicated microphysics was "swept under the rug" into η_{eff} (see Haerendel and Paschmann, 1982, and Cowley, 1985). Reality is more complicated; for example, there is evidence that transient reconnection rather than, or in addition to, steady-state reconnection is taking place. Also, the IMF is almost never exactly southward; we discuss a few of the implications of these complexities in the next section.

8.2.3 Energy input into the magnetosphere

The reconnection geometry on the dayside magnetopause is more complex when the IMF has a direction other than southward. We have been considering a rather simplified picture, but you should recognize that this topic is on the cutting edge of space physics and many alternative models of dayside magnetic reconnection are currently being debated (see Haerendel and Paschmann, 1982; Paschmann, 1979). It is known that the magnetospheric (or cross-tail) electric field is much smaller, and other geomagnetic phenomena such as the ring current and the aurora are much less active during time periods when the IMF does *not* have a significant southward component. This has been demonstrated by many authors, and a variety of empirical expressions have been developed that relate various geomagnetic parameters, such as the D_{st} index, to solar wind conditions. For example, Perrault and Akasofu (1978) (also see Haerendel and Paschmann, 1982, again) introduced the following expression for the *total solar wind power input into the magnetosphere* :

$$P_{sw-m} = u_{sw} B_0^2 (\sin(\theta/2))^4 l_0^2 \quad [\text{ergs/s}], \qquad (8.21)$$

where B_0 is the magnitude of the IMF, l_0 is a length parameter set equal to $7 R_E$, and θ is the angle of the IMF with respect to the z axis (z points northward). An angle $\theta = 0°$ corresponds to northward IMF and, in this case, the solar wind power input into the magnetosphere is zero according to Equation (8.21). An angle $\theta = 180°$ corresponds to a purely southward IMF, in which case P_{sw-m} has a maximum value that can be denoted $P_{sw-m,0} = u_{sw} B_0^2 l_0^2$. See Equation (8.21).

Equation (8.21) can be rewritten in SI units and in such a way as to bring out "the physics" (also see Volland, 1984):

$$P_{sw-m} = u_{sw} \frac{B_0^2}{2\mu_0} (\sin(\theta/2))^4 (2\pi r_{mp}^2) \quad \text{[watts]}. \qquad (8.22)$$

The power is clearly a product of the flux of magnetic energy (or electromagnetic energy) into the (approximate) surface area of the dayside magnetopause. The flux of magnetic energy is just the solar wind speed times the solar wind magnetic energy density $w_B = B_0^2/2\mu_0$. In Problem 8.5, you are asked to demonstrate that the *Poynting vector* can be used to derive this result. See the appendix for a review of electromagnetic theory. Let us also separate out the dependence of IMF angle θ by writing $P_{sw-m} = P_{sw-m,0}[\sin(\theta/2)]^4$. $P_{sw-m,0}$ is just the power for purely southward IMF.

The magnetic energy density comprises only a small fraction of the total solar wind energy density, which is dominated by bulk kinetic energy. What is the maximum possible energy input into the magnetosphere from the solar wind? The *total* energy flux of the solar wind into the cross-sectional area of the magnetosphere seems to be a reasonable answer and is equal to

$$P_{max/sw} = A_m F_{sw/ener}, \qquad (8.23)$$

where

$$A_m = \pi(1.5r_{mp})^2$$

is the cross-sectional area of the dayside magnetosphere. $F_{sw/ener}$ is the total energy flux of the solar wind and can be approximated by

$$F_{sw/ener} = (1/2)(\rho_{sw}u_{sw}^2)u_{sw}, \qquad (8.24)$$

since for the solar wind flow the kinetic energy of the bulk flow is much greater than the magnetic or thermal energy. Recall from Chapter 4 or 6 that $M_{Asw}^2 \gg 1$ and $M_{s,sw}^2 \gg 1$.

The ratio of $P_{sw-m,0}$ to $P_{max/sw}$ can be simply written as (Problem 8.5)

$$\frac{P_{sw-m,0}}{P_{max/sw}} \approx \frac{1}{M_{Asw}^2}, \qquad (8.25)$$

where a typical Alfvénic Mach number in the solar wind is $M_{Asw} \approx 7$.

Example 8.1 (Typical values for power inputs into the magnetosphere) Plugging in typical values for the solar wind parameters into Equations (8.22) and (8.23), we find that

1. $P_{max/sw} \approx 10^{13}$ W, although values ranging from 0.5 to 5×10^{13} W are possible, with the larger value appropriate for geomagnetically active conditions; and

2. $P_{sw-m,0} \approx 2 \times 10^{11}$ W.

The ratio $P_{sw-m,0}/P_{max/sw}$ is about 0.02 for typical solar wind parameters.

P_{sw-m} is the power dissipated in the near-Earth (within $\approx 1r_{mp}$) magnetosphere and in the polar ionosphere. The example demonstrated that this constitutes only a small fraction of the total solar wind energy incident on the magnetosphere, even for southward IMF when $P_{sw-m} = P_{sw-m,0}$. If one includes the distant tail region together with the near-Earth region (i.e., at least within a distance from Earth of $10 r_{mp}$ or so), then the fraction of the incident solar wind kinetic energy (KE) converted into magnetic energy (ME) in the vicinity of the magnetopause becomes much larger – as much as 50%. This larger power can be designated P_{tail} and is typically $P_{tail} \approx (1-5) \times 10^{12}$ W. Using Equations (8.3), (8.10), and (8.23), it can be shown that P_{tail} is roughly the solar wind KE input onto a cross-sectional area of πb^2, where the meaning of b is apparent in Figure 8.6; this can also be written as

$$P_{tail} \approx \zeta^2 P_{max/sw} \approx 10^{12} \text{ W} \tag{8.26}$$

for moderately active conditions. What is the fate of the energy that is input into the magnetosphere? This question is really the subject of the remainder of this chapter, but for the moment a quick answer will suffice. The form of the energy changes from place to place. For example, at the bow shock some of the solar wind KE is converted into electromagnetic energy (mostly magnetic energy or ME), and at the dayside magnetopause ME is converted back into KE again via the reconnection process. Further down the tail, KE is converted into ME in the vicinity of the magnetopause, and the resulting magnetic flux is convected into the tail lobes. This tail ME is again converted into KE via *magnetotail reconnection*, primarily during events called *magnetic substorms*.

The power dissipated in the magnetotail can be crudely estimated by taking the product of the total magnetotail current (in the plasmasheet – see the schematic Figure 8.1) and the cross-tail potential: $P_{dissip/tail} \approx I_{tail} \Psi_m \approx (B_{tail}/\mu_0) L_{tail} \Psi_m$. The cross-tail electric current in this expression was estimated using Ampère's law. L_{tail} is the length of the tail region that we wish to consider. For the near-Earth region, but outside the inner magnetosphere, $L_{tail} \approx r_{mp}$ and $B_{tail} \approx 20$ nT, so that $I_{tail} \approx 1$ MA. For moderately active conditions, $\Psi_m \approx 60$ kV and $P_{dissip/tail} \approx 1$ MA $\times$ 60 kV $\approx 10^{11}$ W. This is about the same as P_{sw-mag}. For a larger region (e.g., $L_{tail} \approx 10 r_{mp}$), naturally both I_{tail} and $P_{dissip/tail}$ are larger ($\approx 10^{12}$ W).

Some of the power input at the dayside magnetopause or in the magnetotail is "connected" to the ionosphere via magnetic field lines and by *field-aligned electrical currents*, as we will discuss in Sections 8.3 and 8.4. As a consequence, electrical currents flow within the polar cap ionosphere, and the power dissipated can be estimated by multiplying the current and the potential drop: $P_{ionos} \approx I_{ionos} \Psi_m \approx 10^{11}$ W. We again find 10^{11} W because $I_{ionos} \approx 1$ MA and because the cross-tail potential is also mapped across the polar ionosphere (Figure 8.3). The power dissipated in this way largely ends up as heat in the upper atmosphere – the ionosphere acts as a resistor and portions of the magnetosphere or magnetopause act as electrical generators.

Much of the field-aligned current flowing into the ionosphere is in the form of energetic particles, especially electrons, that deposit much of their energy via inelastic collisions with the atmospheric neutrals. Ionization is one important consequence of this *energetic particle precipitation* as is nonionizing excitation of atoms and molecules that leads to visible emissions. These emissions are observed as the *aurora borealis* in the northern hemisphere and as the *aurora australis* in the southern hemisphere, or simply the *aurora*. The auroral power is also of the order of 10^{11} W, as is demonstrated in Example 8.2.

What about the power associated with the ring current in the inner magnetosphere? First, recall from Chapter 3 that the total electrical current in the ring current (RC) is roughly 1 MA for moderately active conditions and that the total energy content of the ring current plasma is roughly $E_{RC} \approx 10^{15}$ J (Problem 8.6). This energy "originally" comes from the plasmasheet region and is "injected" into the ring current region during substorm events. Considering that it takes about 1–2 hours to build up the RC during an event, the average power during this period is again about 10^{11} W. During the subsequent recovery phase of the substorm, this energy is lost via charge transfer and other ring current loss processes.

Example 8.2 (Auroral power) Depending on the geomagnetic activity level, a typical auroral electron flux into the auroral oval is about 10^9 particles/cm^2/s and the average electron energy is about 5 keV. The corresponding energy flux is about 5×10^{12} eV/cm^2/s. For an auroral oval about 5° wide at a magnetic latitude of $\approx 70°$, multiplying surface area by the energy flux we find a total power of $P_{aurora} \approx P_{ionos} \approx 10^{11}$ W.

The values of the power discussed above were for moderately active conditions, which necessitates an IMF with a significant southward component. The degree of geomagnetic activity depends strongly on IMF angle according to Equations (8.21) and (8.22). The reconnection process must somehow be sensitive to this angle.

8.2.4 Reconnection at the dayside magnetopause – Nonsouthward IMF

According to Equation (8.21), the efficiency of reconnection at the dayside magnetopause and the energy input into the magnetosphere strongly depend on the IMF direction. The dependence of the potential drop Ψ_m on IMF angle θ goes roughly as $\sin^n(\theta/2)$, where $n \approx 1$–2. Figure 8.8 shows a schematic of some newly reconnected field lines for an IMF that is almost southward but has a small B_y (i.e., east–west) component. Whereas for a purely southward IMF, as considered in Section 8.2.2, the x-line and the magnetic equator at the magnetopause are the

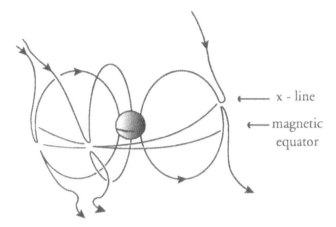

x - line

magnetic equator

Figure 8.8. Three newly reconnected magnetic field lines for a case in which the IMF is mainly southward but has a small east–west component. The merging line, or x-line, is tilted with respect to the magnetic equator.

same, in the present case the x-line is tilted with respect to the equatorial plane. The more nonsouthward the IMF, the greater the angle of tilt with respect to the equatorial plane; however, the x-line still has a configuration similar to that of the purely southward IMF case.

Many other models, or views, of dayside reconnection exist. For example, Crooker (1979) presents a picture in which most of the reconnection takes place in the vicinity of the cusps rather than near the magnetic equator for nonsouthward IMF. In this picture, the reconnection occurs at those locations on the magnetopause where the IMF and magnetospheric field lines are oriented almost at 180° with respect to each other. And Kan (1991) presented a model in which multiple x-lines are present rather than just the single tilted x-line that appears in Figure 8.8. Furthermore, in many models the reconnection does not proceed in a steady-state manner but in a sporadic time-varying manner; some observational evidence for this is discussed in the next section. Nonetheless, some appreciation for how nonsouthward IMF might lead to a reduced power into the magnetosphere (i.e., to a reduced efficiency ζ) can be obtained by considering the simple picture shown in Figure 8.8.

A schematic of the magnetosphere for nonsouthward IMF reconnection is given in Figure 8.9, which shows a cross-sectional view from the Sun. The "last" reconnecting field lines, which reconnect on the flanks of the magnetosphere, are shown. Interplanetary field lines that are further away from the magnetopause (or "outside" the two field lines shown in the figure) never reconnect, but field lines located "within" the two "last" field lines do reconnect on a portion of the magnetopause closer to the Sun. The field lines are electric equipotentials, and the potential drop across the magnetosphere is still roughly in the dawn-to-dusk (or east–west) direction, although the solar wind electric field $\mathbf{E}_0$ is no longer in a purely east–west direction. The distance between these last two reconnecting field lines, b, will still

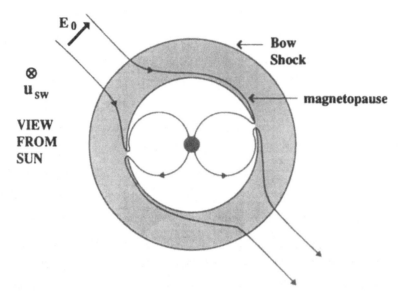

Figure 8.9. Cross-sectional view of the magnetosphere from the Sun for an IMF with a significant B_y component. The "last" two reconnecting magnetic field lines are shown. Note that these field lines also have a significant component in the x direction (into the page) in the magnetosheath, due to their being draped about the magnetopause.

determine the reconnection efficiency ζ. The question then is why should the value of b be smaller for this case than for the purely southward IMF case, everything else being equal. As discussed in the last section (Equation (8.9)), the value of b varies as the square root of the flow speed into the magnetopause, u_0, and somehow this speed must be lower for nonsouthward than for purely southward IMF. We will not treat this quantitatively here but will discuss why this is plausible.

Note that the interplanetary field lines actually "drape" around the magnetopause with a significant sunward/antisunward component due to the gradient of the flow speed across the magnetosheath. That is, the field lines are largely oriented tangentially to the magnetopause surface, except very close to the diffusion region. The magnetopause is a current layer that can be treated as a *tangential discontinuity*, except in the middle of the diffusion region, which only occupies a very small fraction of the magnetopause surface area (in fact, the diffusion region has never been unambiguously observed). Tangential discontinuities were discussed in Chapter 4, Section 4.9. A particular type of tangential discontinuity is the *rotational discontinuity* in which the magnetic field lines rotate but the field magnitude stays constant. The magnetopause can be approximated as a rotational discontinuity except near the diffusion region. The magnetopause is also obviously a *current layer*, and the current density **J** must be such as to produce the required rotation of the magnetic field (using Ampère's law). The observed structure and thickness of the magnetopause will be discussed in the next section.

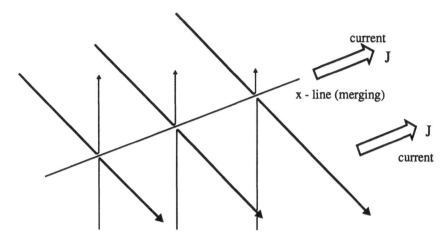

Figure 8.10. Expanded view of a portion of the dayside magnetopause as seen from the Sun. The heavy lines are the interplanetary field lines draped around the magnetopause and the light lines are the magnetospheric field lines right at the magnetopause. The field lines shown have just reconnected, or merged, along the tilted x-line shown. The magnetopause is a tangential discontinuity across which the magnetic field rotates.

Figure 8.10 shows an expanded view of a portion of the dayside magnetopause with a tilted x-line. The current density required by Ampère's law to explain the field rotation is in the direction of the x-line. The magnetic field on both sides of the magnetopause has a component along the x-line, denoted B_x, as shown in Figure 8.11. The current is needed only to "generate" the other components (denoted $B_{\perp i}$ or $B_{\perp o}$, for inside or outside the magnetopause, respectively). The motional electric field in the solar wind is orthogonal to $\mathbf{B}_{outside}$ (which is the same as $\mathbf{B}_0$), but the component of $\mathbf{E}$ that "drives" the reconnection is along the x-line. When the IMF is purely southward, then $B_x = 0$ and the dot product $\mathbf{E} \cdot \mathbf{J}$ is a maximum, but when the IMF has another direction then $B_x \neq 0$ and both $\mathbf{J}$ and $\mathbf{E}$ are directed along the x-line and are smaller; $\mathbf{E} \cdot \mathbf{J}$ is thus smaller.

Both Equations (4.195) and (4.203) had the magnetic energy dissipation rate being proportional to a volume integral of the "electromagnetic source function" $\mathbf{E} \cdot \mathbf{J}$ (also see the appendix). However, in the subsequent analysis, reconnection of oppositely directed field lines was assumed. Figures 8.8–8.11 indicate that the source $\mathbf{E} \cdot \mathbf{J}$ should be smaller for a tilted x-line. For example, in the extreme case of no change in magnetic field direction across the boundary (the purely northward IMF case), then both $\mathbf{J}$ and $\mathbf{E}$ are zero and *no* magnetic dissipation or reconnection takes place. In this case, the magnetosphere is *closed* and the rather small solar wind input into the magnetosphere that does occur must result from turbulent diffusion of energy and momentum across the magnetopause (this is the *closed magnetosphere model*). It is plausible that a repeat of the Chapter 4 analysis for the general case would result in a reconnection rate, or inflow speed u_0, that depends on the angle

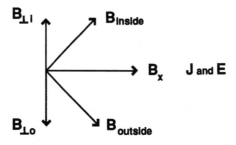

Figure 8.11. The detailed field geometry associated with Figure 8.10 is shown. B_x is the component of the magnetic field common to both $\mathbf{B}_{\text{inside}}$ and $\mathbf{B}_{\text{outside}}$ (the magnetic fields just inside and outside the magnetopause, respectively). B_x is the component in the direction of the x-line (or magnetic merging line). The remaining components of $\mathbf{B}_{\text{inside}}$ and $\mathbf{B}_{\text{outside}}$ are designated $B_{\perp i}$ or $B_{\perp o}$, respectively. The magnetopause current density, $\mathbf{J}$, is directed along the x-line. The component of the electric field along the x-line is also shown.

of the reconnecting field lines such that u_0 becomes smaller, as expected, when the field lines are not oppositely directed.

Another factor that can reduce the reconnection rate for nonsouthward IMF is a possible decrease of the effective resistivity that would be associated with a reduced width of the diffusion region, an increased value of the magnetic Reynolds number R_m^{os}, and reduced values of $M_{A0\,\text{max}}$ and u_0. It is plausible that the effective resistivity is smaller for nonsouthward than for southward IMF because the magnetopause current density is lower, as we have just seen, and it is likely that the microscopic plasma instabilities responsible for the resistivity are related to this current density (cf. Haerendel and Paschmann, 1982).

8.2.5 Transient reconnection and flux transfer events (FTEs)

A simple steady-state picture of reconnection on the dayside magnetopause with a single x-line was just presented in the previous section. As mentioned earlier, more sophisticated scenarios, both spatially and temporally, have been proposed (Kan, 1991). For example, in Figure 8.12 a cross-sectional snapshot of a portion of the magnetopause is presented in which two x-lines exist, at least at a particular time. Notice that a separate "island" of plasma and field exists between the two x-lines. Because an east–west component of the magnetic field is also usually present, the actual magnetic field configuration in an island is helical or ropelike.

There is evidence that dayside magnetic reconnection could be transient and that islands such as those in Figure 8.12 "convect" up or down, depending on where one is located on the magnetopause. Such an island would appear to move tailward, leaving a single x-line in the subsolar region, and then a new island might appear. Plasma and magnetic field events have been observed in the vicinity of the magnetopause that have been interpreted as signatures of these transient reconnection events; these events are called *flux transfer events* (FTEs). Figure 8.13 is a schematic of an

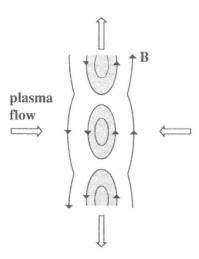

Figure 8.12. Magnetic field lines are shown for multiple x-line reconnection on the dayside magnetopause. The magnetosheath is on the left and the magnetosphere on the right side of this schematic. The magnetopause layer current is directed out of the page.

FTE as deduced from magnetic field measurements (see Kan, 1991, or Haerendel and Paschmann, 1982). Note that other FTE models or configurations are also consistent with the available data. Measurements will be discussed in the next section.

8.2.6　Observations of the magnetopause

The *ISEE 1* and *2* spacecraft have made *in situ* plasma and field measurements of the magnetopause and the adjacent boundary layer (NASA's *ISEE* mission had three spacecraft present throughout the magnetosphere in the late 1970s and the 1980s). A review of these observations can be found in Russell (1981) and in Haerendal and Paschmann (1982). An important example of a magnetopause crossing by the *ISEE* spacecraft that has been used as a proof of dayside reconnection was given by Paschmann et al. (1979) (cf. Russell, 1981, and Haerendal and Paschmann, 1982) and is reproduced in Figure 8.14. This magnetopause encounter took place on the dayside near 25° latitude and near noon local time. The z component of the IMF was negative, or southward, as can be judged by the field in the magnetosheath. The magnetopause (mp) is the layer in which the field changes direction – that is, the mp is a rotational discontinuity. The proton density n_p is low in the magnetosphere and higher in the magnetosheath; a gradual transition in n_p takes place across the *boundary layer* (or *low latitude boundary layer* – LLBL). The magnetopause is a dynamic region and moves in and out with speeds of roughly $v_{\rm mp} \approx 10$–20 km/s, which makes the magnetopause thickness $\Delta r_{\rm mp}$ difficult to determine using a single spacecraft. Fortunately, both *ISEE 1* and *2* were available and data from them could be used to determine $v_{\rm mp}$ and also $\Delta r_{\rm mp}$.

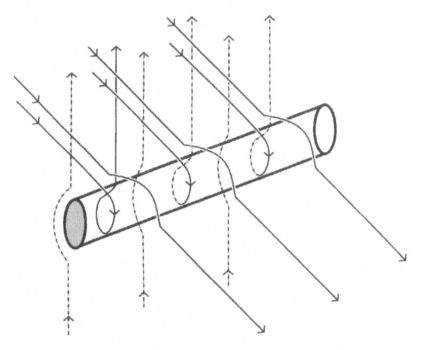

Figure 8.13. Schematic of flux transfer event (FTE) on the dayside magnetopause that is generally consistent with spacecraft observations. The tube represents an "island" of plasma that was recently reconnected near the bottom of the diagram and has subsequently moved upward/northward (and to the left) in response to the $\mathbf{J} \times \mathbf{B}$ force associated with the bend of the magnetic field lines. Not yet reconnected field lines inside and outside of the magnetopause are distorted by the passage of the plasma island; this is called an FTE. Another such tube of plasma (not shown below the diagram) exists and moves southward.

For the case shown in Figure 8.14, $\Delta r_{mp} \approx 500 \, \text{km}$, and the thickness of the adjacent boundary layer is comparable. The evidence for reconnection taking place nearby is contained in the rapid plasma flow that is present in the LLBL and mp regions, which is also where the magnetic pressure is smallest. The data are consistent with a diffusion region (where reconnection is taking place) located somewhere southward of the spacecraft, in which case the spacecraft is traversing the outflow region! Recall that the outflow speed ($V_P \approx 300–400 \, \text{km/s}$ in Figure 8.14) is approximately equal to the Alfvén speed according to steady-state reconnection theory, and this is indeed the case. In Figures 8.4 and 8.5, the *ISEE* spacecraft would be moving from inside the mp to the outside just northward of the x-line.

Russell (1981) points out that not all mp crossings show large outflow speeds, which seems to indicate that reconnection on the dayside magnetopause is often transient. In fact, the observed magnetic field in the general vicinity of the magnetopause exhibits transient events in which the component of the field normal to the mp surface undergoes localized positive to negative or negative to positive "bipolar" excursions. These events were interpreted as the nearby passages of magnetic flux

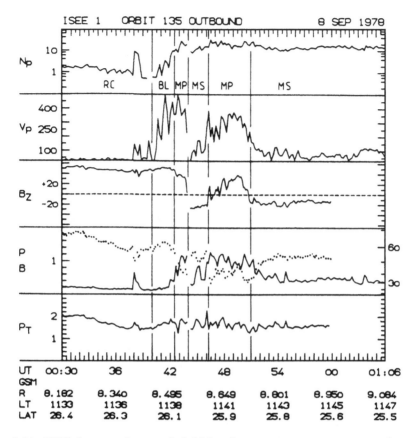

Figure 8.14. *ISEE* plasma and magnetic field data for a magnetopause passage on September 8, 1978. The various panels show, from the top, proton density (N_P), proton speed (V_P), z component of the magnetic field (B_z), plasma and magnetic pressures (solid and dotted lines, respectively), and total pressure (P_T). The abscissa shows several parameters: UT (universal time in hours and minutes), R (radial distance in R_E), LT (local time in hours and minutes), and Lat (latitude). The spacecraft started out in the magnetosphere (i.e., in the dayside ring current region, denoted RC), traversed the low latitude boundary layer (denoted BL) and then the magnetopause layer (denoted MP), before passing into the magnetosheath (denoted MS). The magnetopause evidently moved out temporarily, since the magnetopause layer was encountered a second time. Magnetopause crossings are evident from sign changes in B_Z. Reprinted with permission from *Nature* (Paschmann et al., *Nature*, **282**, 243, 1979). Copyright (1979) MacMillan Magazines Limited.

tubes associated with transient reconnection events (i.e., the FTE events previously discussed).

Russell (1981) also compiled measured values of Δr_{mp} from many magnetopause crossings and plotted them versus IMF angle (actually, the angle between the magnetosheath and magnetospheric fields); Figure 8.15 shows this plot. A clear difference is apparent for northward (angle less than 90° or so) and southward IMF. For southward IMF, we have a narrower magnetopause layer with $\Delta r_{mp} \approx 400\,\text{km} \approx 3$ proton gyroradii (r_{Lp}), whereas for northward IMF, we have a thicker mp with

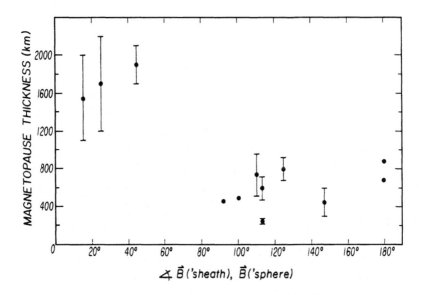

Figure 8.15. Measured magnetopause thickness, Δr_{mp}, versus the angle between the magnetic field in the magnetosheath and magnetosphere. (From Russell, 1981.) Reprinted from *Advances in Space Research*, Vol. 1, C. T. Russell, p. 67, copyright 1981, with kind permission from Elsevier Science Ltd., The Boulevard, Langford Lane, Kidlington OX5 1GB, UK.

$\Delta r_{\mathrm{mp}} \approx 1,600\,\text{km} \approx 12\,r_{\mathrm{Lp}}$. The low latitude boundary layer, just inside the mp on the dayside, is of comparable thickness to the mp layer. However, downstream in the tail, the boundary layer both at low latitudes (the LLBL) and at higher latitudes (the plasma mantle) can have thicknesses of the order of an R_{E} or more.

What do we expect Δr_{mp} to be? According to the simple schematic (Figure 7.1) of a thin current layer, Δr_{mp} is equal to $r_{\mathrm{Lp}}/2$, half the proton gyroradius. In this very simple picture, the region containing the magnetic field on the right does not contain any plasma and the ions streaming from the left are reflected due to the Lorentz force. The gyrating ions create a layer of current. But what about the electrons? If we were to assume that electrons with the same velocity as the ions stream into the layer in the same manner, then they would gyrate in the opposite sense from the protons and also create a downward current. However, we have overlooked a complication. The electron gyroradius is much less than the proton gyroradius, and the protons penetrate much deeper into the magnetic region than do the electrons. This results in charge separation with a narrow layer of negative charge located at the interface and a somewhat broader ($r_{\mathrm{Lp}}/2$) region of positive charge stretching farther in. The polarization electric field associated with this charge separation, if it existed, would be extremely large and directed outward, preventing the protons from actually penetrating too far into the magnetic layer. What then would the current layer thickness be?

The layer thickness, δ_{layer}, for a plasma on the left in which the electron and ion temperatures (or their kinetic energies for a non-Maxwellian plasma) are equal, is just the electron inertial length (cf. Haerendel and Paschmann, 1982) defined by

$$\delta_{\text{layer}} \approx c/\omega_{\text{pe}}, \tag{8.27}$$

where c is the speed of light and ω_{pe} is the electron plasma frequency of the plasma to the left of the boundary. For magnetosheath-type parameters, we find from Equation (8.27) that $\delta_{\text{layer}} \approx 1$ km, which is much less than the proton gyroradius ($r_{\text{Lp}} \approx 100\text{–}200$ km). Equation (8.27) can be derived by asking what electric potential is needed to repulse the incident protons and also by using a simple generalized Ohm's law, $\mathbf{E} = -\mathbf{u}_e \times \mathbf{B}$, where $\mathbf{u}_e$ is the bulk flow velocity of the electrons (see Problem 8.8). Note that the electrons can still be treated as a fluid, because the Debye length is still much less than δ_{layer} and the departure from quasi-neutrality is still small ($\delta n_e = |n_e - n_p| \ll n_e$). The electrical current in this picture is carried almost entirely by electrons moving upward at about the thermal speed.

The observed magnetopause thickness is *not* ≈ 1 km but is several proton gyroradii! That is, $\Delta r_{\text{mp}} \gg \delta_{\text{layer}}$ and the simple picture shown in Figure 7.1, which neglected the polarization electric field, appears to be closer to reality than the more "inclusive" theory. Why? The magnetized region is not a vacuum. Plasma is indeed present inside the magnetosphere, albeit with a density that is lower than the density in the magnetosheath. This plasma near the magnetopause partially "short circuits" the polarization electric field. In considering the generalized Ohm's law, Equation (4.62), we must also include the $-\nabla p_e$ term, which is able to oppose the motional electric field term. Without a large electric field, the ions are free to stream into the magnetic layer by about the distance of r_{Lp}, and these ions carry most of the current. Actually, the observed Δr_{mp} is several r_{Lp}, which suggests that diffusive processes must also play some role. Irregularities in the electric and magnetic fields near the mp can disrupt very simple gyration motion, creating a broader boundary across which the plasma diffuses. The diffusive-type picture is even more important for northward IMF than for southward IMF according to Figure 8.15.

In Section 8.2, we have considered how an electric field is generated in the magnetosphere via the magnetic reconnection process. Next we consider what happens *within* the magnetosphere due to the imposed magnetospheric electric field $\mathbf{E}_m$ that results from the dayside magnetic reconnection process.

8.3 MHD: Electrical current description

What is the internal response of the magnetosphere to the imposed magnetospheric electric field $\mathbf{E}_m$ resulting from the dayside magnetic reconnection process? Note that this field is usually called the *convection electric field*, which already tells us what this field does in the magnetosphere. We start out in Section 8.3 by revisiting MHD theory and considering it from the perspective of electrical currents. Before applying this MHD theory to magnetospheric convection in Section 8.4, we look at the stress balance for a simple scenario of two slabs of plasma linked by magnetic field lines.

8.3.1 MHD equations revisited

You should review Section 4.3.2 on the single-fluid MHD equations and also review Section 4.5. The MHD equations consist of a continuity equation, a momentum equation, an energy equation, and the magnetic induction equation. Here we are mainly concerned with the momentum equation, which handles the force balance in a plasma. The single-fluid momentum equation (4.80) is

$$\rho\frac{\partial\mathbf{u}}{\partial t} + \rho\mathbf{u}\cdot\nabla\mathbf{u} = -\nabla p + \mathbf{J}\times\mathbf{B} + \rho\mathbf{g} - \rho\nu(\mathbf{u}-\mathbf{u}_n) - P_i m_i(\mathbf{u}-\mathbf{u}_n), \quad (8.28)$$

where the pressure p is the sum of the electron and ion partial pressures. The other quantities were defined in Chapter 4. The magnetic force (or stress) is taken care of by the $\mathbf{J}\times\mathbf{B}$ term; the current density $\mathbf{J}$ appears here explicitly.

Another form of the single-fluid MHD momentum equation (i.e., Equation (4.86)) was derived in Chapter 4 by using Ampère's law, minus the displacement current, to eliminate $\mathbf{J}$ from Equation (8.28):

$$\rho\frac{\partial\mathbf{u}}{\partial t} + \rho\mathbf{u}\cdot\nabla\mathbf{u} = -\nabla\left(p + \frac{B^2}{2\mu_0}\right) + \frac{1}{\mu_0}\mathbf{B}\cdot\nabla\mathbf{B} + \rho\mathbf{g} - \rho\nu(\mathbf{u}-\mathbf{u}_n) - P_i m_i(\mathbf{u}-\mathbf{u}_n).$$
$$(8.29)$$

The current density does not explicitly appear in this equation; instead, the magnetic stress term is described in terms of the gradient of the magnetic pressure gradient term and the magnetic curvature force term, as described in Chapter 4. The magnetic diffusion–convection equation (4.69) is usually used in conjunction with Equation (8.29) and does not explicitly contain the current density. We can call this the *"pure MHD"* approach.

If a large "external" magnetic field is present, that is, a field created by external electrical currents as opposed to electrical currents flowing within the plasma, then the pure MHD approach can be difficult to apply in practice. For example, in the inner magnetosphere of the Earth, the intrinsic dipole magnetic field generated by currents in the Earth's interior exceeds by more than three orders of magnitude the magnetic field generated by currents flowing locally within the plasma ($\Delta B \ll B_0$, where ΔB is the field from "local" currents). Derivatives of the magnetic field, such as those that appear in Equation (8.29) or in the magnetic diffusion–convection equation, are difficult to carry out accurately in this case because one is attempting to evaluate a small difference between two large numbers.

A better approach to MHD for situations in which there is a large external field is to explicitly retain the $\mathbf{J}\times\mathbf{B}$ term in the momentum equation and to focus on finding $\mathbf{J}$ rather than derivatives of $\mathbf{B}$. We can call this the *current approach to MHD*. We now need to use Equation (4.82), in place of the diffusion–convection equation, to keep track of the current:

$$\nabla \cdot \mathbf{J} = 0. \tag{8.30}$$

Equation (8.30) means that electrical current (I_S) flowing through a closed surface S must be zero in steady-state situations. This can be seen by applying Gauss's integral theorem to Equation (8.30) for the volume V contained within the closed surface S:

$$\int_V \nabla \cdot \mathbf{J}\, d^3\mathbf{x} = \oint_S \mathbf{J} \cdot d\mathbf{S} = I_s = 0. \tag{8.31}$$

Equation (8.31) applied to electric circuits is just Kirchoff's first law; that is, electrical circuits must be closed for steady-state situations.

The schematic in Figure 8.16 shows a scenario in which ionospheric plasma is linked to magnetospheric, or solar wind, plasma via magnetic field lines. Stress balance between the two regions generally requires a $\mathbf{J} \times \mathbf{B}$ force, or magnetic stress, in both of the plasmas. The current density can be split into a horizontal component, $\mathbf{J}_{hor}$, and a component parallel to the magnetic field, $\mathbf{J}_\parallel = J_\parallel \,\hat{\mathbf{b}}$, where $\hat{\mathbf{b}}$ is a unit vector in the magnetic field direction:

$$\mathbf{J} = \mathbf{J}_{hor} + \mathbf{J}_\parallel. \tag{8.32}$$

The magnetic stress term is just $\mathbf{J}_{hor} \times \mathbf{B}$. When a horizontal current is needed for the momentum balance, then *field-aligned currents* (i.e., $J_\parallel$ is the field-aligned current density) are required to "close" the resulting electrical circuit and thus maintain $\nabla \cdot \mathbf{J} = 0$. Field-aligned currents are also called *Birkeland currents*.

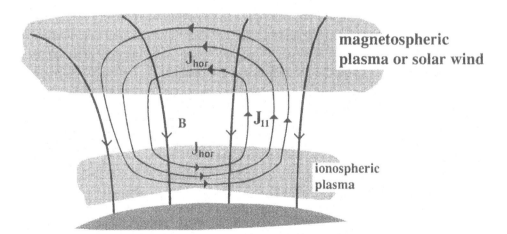

Figure 8.16. Schematic of an ionospheric plasma linked to magnetospheric plasma by field lines. Current density streamlines are indicated. The current component parallel to the magnetic field is denoted $J_\parallel$.

The field-aligned current density can be determined in terms of the horizontal current by using Equation (8.30):

$$\frac{\partial J_\parallel}{\partial s} = -\nabla \cdot \mathbf{J}_{\text{hor}}, \tag{8.33}$$

where s is distance along the magnetic field. This relation is especially useful for planetary ionospheres, which are slablike ionized layers surrounding planets. Integrating Equation (8.33) over the vertical extent of an ionosphere we obtain

$$J_{\parallel\,\text{top}} = -\frac{1}{\sin I}\nabla \cdot \mathbf{K}_{\text{hor}}, \tag{8.34}$$

where $J_{\parallel\,\text{top}}$ is the field-aligned current density out of the top of the ionosphere and I is the dip angle (the angle between the magnetic field and the surface). $\mathbf{K}_{\text{hor}}$ is the height-integrated horizontal current density given by

$$\mathbf{K}_{\text{hor}} = \int_{\text{bottom}}^{\text{top}} \mathbf{J}_{\text{hor}}\, dz, \tag{8.35}$$

where z is altitude and the integration is from the bottom to the top of the ionosphere.

Equations (8.30) through (8.35), plus the single-fluid momentum equation, are not sufficient by themselves to determine the plasma dynamics. A generalized Ohm's law (GOL) is also required; that is, we need a relation between the current density and the electric field. Recall that the ion and electron momentum equations were combined (Chapter 4) to obtain the single-fluid MHD momentum equation; this left one momentum equation "free" for other purposes. In Chapter 4, the electron momentum equation was used to derive a GOL, which was later used together with Faraday's law to derive the magnetic diffusion–convection equation. However, we could have used the ion momentum equation to find a different form of the GOL, and in Chapter 7, a version of the GOL was found that is appropriate for the highly magnetized ionospheric plasma. Expressions for the Pederson and Hall conductivities were given by Equations (7.70) and (7.71), and height-integrated Hall and Pederson conductivities (Σ_H and $\Sigma_\perp$, respectively) were defined by Equations (7.74) and (7.75). The height-integrated GOL for the ionosphere can be written as

$$\mathbf{K}_{\text{hor}} = \mathbf{K}_\perp + \mathbf{K}_H = \Sigma_\perp \mathbf{E}' + \Sigma_H \mathbf{E}' \times \hat{\mathbf{b}}. \tag{8.36}$$

Strictly speaking, we are interested in the horizontal components of the Pederson and Hall currents, but we make this an implicit assumption. $\mathbf{E}' = \mathbf{E} + \mathbf{u}_n \times \mathbf{B}$ is the electric field in the frame of reference of the neutral gas, where $\mathbf{u}_n$ is the neutral wind velocity.

If we know the electric field imposed on the ionosphere from the magnetosphere, then we can determine the field-aligned current out of the ionosphere (or into the magnetosphere) by using Equations (8.34) and (8.35) plus the height-integrated GOL, Equation (8.36). For situations with uniform conductivity, the contribution of the Hall current to the field-aligned current in Equation (8.34) is quite small (Problem 8.9) and the Pederson current makes the dominant contribution to the closing of the electrical circuit with the magnetosphere.

8.3.2 A simple scenario for MHD stress balance: A slab geometry

We now consider the simple scenario of two slabs of plasma linked by magnetic field lines; this approach is in the spirit of the article by Southwood (1985). Each of the slabs extends infinitely in the y direction and a distance L_x in the x direction. Each slab has a thickness H in the z direction. See Figure 8.17. The top slab could

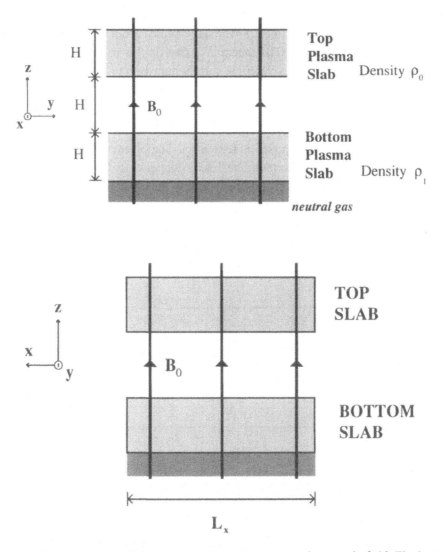

Figure 8.17. Two plasma slabs immersed in a strong external magnetic field. The bottom slab is immersed in a neutral gas. Two views are shown.

represent, loosely, the magnetospheric plasma and the bottom slab the ionosphere. The whole region is immersed in a strong external magnetic field in the z direction ($\mathbf{B}_0 = B_0\hat{\mathbf{z}}$). The plasma densities are ρ_0 and ρ_1 in the top and bottom slabs, respectively. We assume that the region between the two slabs consists of very low density plasma with infinite electrical conductivity. The top slab is presumed to be a collisionless plasma, but we suppose that a stationary neutral gas is present in the bottom slab so that ion–neutral collisions occur.

If both slabs are stationary, then the momentum coupling between them is trivial and no electrical currents are necessary ($\mathbf{J} = \mathbf{0}$ everywhere). But suppose that the plasma in the top slab is moving at some initial time, $t = 0$, with the velocity $\mathbf{u}_0 = u_0\hat{\mathbf{y}}$, and that the bottom slab is initially stationary: $\mathbf{u}_1 = \mathbf{0}$. The two regions are coupled by the magnetic field, and momentum is transferred between the two regions such that the plasma in the lower slab should also begin to move. We can describe this process using MHD theory. We first do this using the "pure" MHD approach, and next we use the electrical current approach to MHD.

8.3.3 MHD description of momentum coupling of two plasma slabs

The magnetic field in the top slab is frozen into the plasma ($\mathbf{E} = -\mathbf{u} \times \mathbf{B}$) and magnetic diffusion is not important. The field is not perfectly frozen into the bottom slab plasma, because this plasma is collisional, but the field lines nonetheless have a tendency to stick to the same plasma parcel. If we start with straight magnetic field lines connecting two stationary slabs and the top slab starts to move (but not the bottom – at least initially), then the field lines become bent as shown in Figure 8.18.

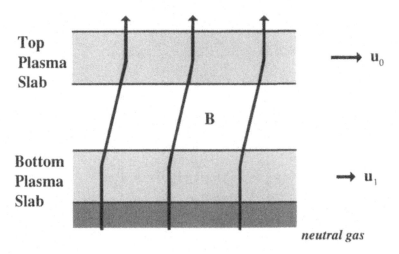

Figure 8.18. Plasma slabs in which the top slab moves faster relative to the bottom slab, bending the magnetic field lines.

The magnetic field in the middle region then has a y component:

$$\mathbf{B} = \mathbf{B}_0 + \mathbf{B}_1 \quad \text{with } \mathbf{B}_1 = B_y \hat{\mathbf{y}}. \tag{8.37}$$

The bending of the field lines is such that

$$\begin{aligned} \frac{\partial B_y}{\partial z} &< 0 \quad \text{top slab} \\ \frac{\partial B_y}{\partial z} &> 0 \quad \text{bottom slab.} \end{aligned} \tag{8.38}$$

The gradient of the magnetic field is approximately equal to $|\partial B_y/\partial z| \approx B_y/H$, where H is the spatial separation of the slabs.

Now let us consider the momentum balance. The single-fluid momentum equation for both slabs, neglecting gravity and mass-loading, can be written in the form

$$\rho \frac{D\mathbf{u}}{Dt} = -\nabla p + \mathbf{J} \times \mathbf{B} - \rho \nu \mathbf{u}. \tag{8.39}$$

The ion–neutral collision frequency is ν and the neutrals are assumed to be stationary. In the top slab, $\nu = 0$. Because we are for now taking a pure MHD approach, we rewrite $\mathbf{J} \times \mathbf{B}$, as in Equation (4.86), as

$$\mathbf{J} \times \mathbf{B} = -\nabla \left(\frac{B^2}{2\mu_0} \right) + \frac{1}{\mu_0} \mathbf{B} \cdot \nabla \mathbf{B}. \tag{8.40}$$

The magnitude of the magnetic field is greatest between the two slabs where $B_y \neq 0$, and a magnetic pressure gradient force exists upward on the bottom of the top slab and downward on the top of the bottom slab. We are interested in the horizontal momentum balance and merely suppose that the thermal pressure in both the top and bottom slabs is sufficiently large so that a static vertical pressure balance exists for all three regions. In the horizontal direction, the magnetic pressure gradient force is zero, and the magnetic curvature force in Equation (8.40) can be approximated as

$$\frac{1}{\mu_0} \mathbf{B} \cdot \nabla \mathbf{B} \approx \pm \frac{B_y B_0}{\mu_0 H} \hat{\mathbf{y}}, \tag{8.41}$$

where the $+$ sign is for the bottom slab and the $-$ sign for the top slab.

The magnetic curvature force seeks to *decelerate* the plasma in the top slab and *accelerate* the plasma in the bottom slab. Even if the bottom slab starts with $u_1 = 0$, after some elapsed time it will be moving at some speed in the $+y$ direction, as momentum is transferred via the curvature force from the top slab to the bottom slab. The horizontal component of the top slab momentum equation is now written as

$$\left[\rho_0 \frac{Du_0}{Dt} + \frac{\partial p}{\partial y} \right] = -\frac{B_y B_0}{\mu_0 H}. \tag{8.42}$$

If we suppose that $\partial p / \partial y = 0$, then it is easy to see that $Du_0 / Dt < 0$, and the plasma slows down due to the magnetic curvature force.

Now let us rewrite the horizontal component of the bottom slab momentum equation; letting $\partial p / \partial y = 0$, we have

$$\rho_1 \frac{Du_1}{Dt} = + \frac{B_y B_0}{\mu_0 H} - \rho_1 \nu u_1. \tag{8.43}$$

At $t = 0$, we have $u_1 = 0$ and $Du_1 / Dt = \partial u_1 / \partial t > 0$, and the friction force is zero. The lower slab plasma is accelerated by the magnetic curvature force, and after a short time the friction force opposes the acceleration. We now suppose that the curvature force comes into a steady-state balance with the friction force so that $Du_1 / Dt = 0$, in which case Equation (8.43) becomes

$$\rho_1 \nu u_1 = \frac{B_y B_0}{\mu_0 H} \quad \text{or} \quad u_1 = \frac{B_y B_0}{\rho_1 \nu \mu_0 H}. \tag{8.44}$$

The speed of the bottom plasma depends on B_y as well as on other parameters, but B_y depends on the behavior of both slabs and "couples" them together. B_y controls the force balance of the top slab also. The B_y in both Equations (8.42) and (8.44) is the same.

We consider two cases: (1) $\partial p / \partial y = 0$ in the top slab and (2) $Du_0 / Dt = 0$ in the top slab, at least after some adjustment period. For case (1), Equation (8.42) simplifies to

$$\rho_0 \frac{\partial u_0}{\partial t} = - \frac{B_y B_0}{\mu_0 H} = -\rho_1 \nu u_1, \tag{8.45}$$

where the second part comes from Equation (8.44). The top slab continually decelerates from its initial value due to the magnetic stress.

For case (2), a steady-state situation is set up in which a horizontal pressure gradient force counteracts the magnetic tension force and keeps the top slab moving at a constant speed u_0 in spite of the opposing magnetic curvature force; thus,

$$\frac{\partial p}{\partial y} = - \frac{B_y B_0}{\mu_0 H} = -\rho_1 \nu u_1. \tag{8.46}$$

Again, Equation (8.44) for the bottom slab was used to eliminate B_y. For case (1), the top slab deceleration is related to u_1, and for case (2) the bottom slab flow speed can be related to the horizontal pressure gradient in the top slab.

However, in both cases we are still missing a piece of information that allows us to completely specify all the parameters. What determines B_y? In the pure MHD approach, the magnetic field is calculated using the magnetic diffusion–convection equation, (4.69), but this can be tricky. Field lines must slip through the bottom resistive layer in order for a steady-state, or quasi-steady-state, magnetic field configuration to be achieved, and this requires that the magnetic convection term must balance the magnetic diffusion term: $\mathbf{0} \cong \nabla \times (\mathbf{u} \times \mathbf{B}) + \eta/\mu_0 \nabla^2 \mathbf{B}$. An approximate dimensional analysis of this expression yields

$$0 \approx \frac{u_0 B_0}{H} + \frac{\eta}{\mu_0} \frac{B_y}{H^2},$$

which gives

$$B_y \approx \frac{H \mu_0 u_0 B_0}{\eta}. \tag{8.47}$$

Note that Equation (8.47) can also be expressed as $B_y/B_0 \approx R_m$, where R_m is the magnetic Reynolds number for the bottom slab using a characteristic speed of u_0. We now have our expression for B_y. What should we use for the resistivity in the bottom layer? The resistivity associated with the Pederson conductivity as given by Equation (7.70) is appropriate for a strongly magnetized ionosphere, as discussed earlier, in which case we obtain

$$B_y \approx \left(H \mu_0 n_1 e \frac{\nu \Omega}{\nu^2 + \Omega^2} \right) u_0, \tag{8.48}$$

where $\Omega = e B_0/m$ is the ion gyrofrequency and n_1 is the plasma number density in the bottom slab. B_y is proportional to u_0, the flow speed of the upper slab.

We can now use expression (8.48) wherever B_y appears in either the upper or the lower slab momentum equations. Equation (8.44) for u_1 now becomes

$$u_1 = \left(\frac{\Omega^2}{\Omega^2 + \nu^2} \right) u_0. \tag{8.49}$$

The upper and lower slab flow speeds are now related. For the upper slab, and case (1), we can find a differential equation for the flow speed whose solution is (Problem 8.11)

$$u_0(t) = u_{00} \exp \left[-\frac{t}{\tau_{\text{slab}}} \right] \quad \text{with} \quad \tau_{\text{slab}} \approx \frac{1}{\nu} \frac{\rho_0}{\rho_1} \frac{\nu^2 + \Omega^2}{\Omega^2}, \tag{8.50}$$

where u_{00} is the upper slab flow speed at time $t = 0$. For a particular choice of parameters ($\Omega = \nu$ and $\rho_0 = \rho_1$) the time constant for the slowdown of the *upper* slab is just the collision time in the *lower* slab: $\tau_{\text{slab}} \approx 1/\nu$.

For case (2), Equation (8.46) gives us u_1 in terms of $\partial p/\partial y$, but to find u_0 we also need Equation (8.49), which was found using B_y from the diffusion–convection equation. Note that all these expressions contain both upper and lower slab parameters – the momentum balance of the two slabs is closely linked by the magnetic field. In particular, the B_y component of the magnetic field transfers momentum via the magnetic curvature force. However, for a "real" problem for which $|B_y| \ll |B_0|$, accurately determining B_y either experimentally or theoretically can be almost impossible. This leads us to try the electrical current approach to MHD, as discussed in the next section for our simple slab example.

8.3.4 *Momentum coupling of two plasma slabs:*
The electrical current approach

In this approach, we do not keep track of B_y but instead do our "bookkeeping" with the electrical current. In the momentum equation, (8.39), we do *not* make the substitution of Equation (8.40), as we did in the last section, but leave $\mathbf{J} \times \mathbf{B}$ in the equation. Of course, the current densities in the upper and lower slabs must be consistent with the magnetic field perturbations as prescribed by Ampère's law, but we need not explicitly concern ourselves with this. The $\mathbf{J} \times \mathbf{B}$ force in the upper slab acts to decelerate the flow, assuming again that the upper slab is initially moving with speed u_0 in the y direction, and the $\mathbf{J} \times \mathbf{B}$ force in the lower slab acts to accelerate that slab. The $\mathbf{J} \times \mathbf{B}$ force can be written as follows for the upper and lower slabs:

$$\text{top:} \quad \mathbf{J} \times \mathbf{B} \cong -J_{\perp 0} B_0 \hat{\mathbf{y}} \quad \text{(deceleration)}$$
$$\text{bottom:} \quad \mathbf{J} \times \mathbf{B} \cong +J_{\perp 1} B_0 \hat{\mathbf{y}} \quad \text{(acceleration).} \tag{8.51}$$

The upper and lower slab momentum equations now contain the Pederson current densities in the top and bottom slabs, $J_{\perp 0}$ and $J_{\perp 1}$, respectively:

$$\text{top:} \quad \left[\rho_0 \frac{Du_0}{Dt} + \frac{\partial p}{\partial y} \right] = -J_{\perp 0} B_0 \tag{8.52}$$

$$\text{bottom:} \quad \rho_1 \frac{Du_1}{Dt} = +J_{\perp 1} B_0 - \rho_1 \nu u_1. \tag{8.53}$$

Note that both $J_{\perp 0}$ and $J_{\perp 1}$ are positive quantities; the direction of $\mathbf{J}$ was explicitly taken care of with the signs in these equations. If we make the assumption, as in the last section, that after some period of acceleration the lower plasma slab moves at a constant speed u_1, with the $\mathbf{J} \times \mathbf{B}$ force balanced by the friction force, then Equation (8.53) simply becomes

$$u_1 = \frac{J_{\perp 1} B_0}{\rho_1 \nu}. \tag{8.54}$$

The flow speed in the lower slab depends on the current density as well as on the collision frequency. Similarly, the behavior of the upper slab depends on the current density in that medium.

How do we determine the current densities? We determine the current by applying $\nabla \cdot \mathbf{J} = 0$ plus the GOL to our simple slab geometry. Where does the electrical current in Figure 8.19 go, given that the slabs have finite width L_x? The answer is that currents must flow along the field lines between the two slabs, as illustrated in Figure 8.20. The current must follow a closed circuit! The existence of perpendicular currents like those shown in Figures 8.19 and 8.20 requires the existence of field-aligned currents. For our simple geometry we can easily be more

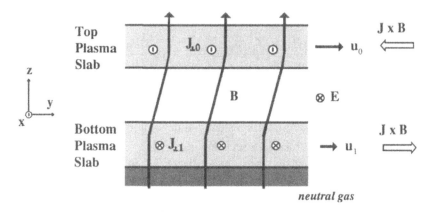

Figure 8.19. Plasma slabs with current densities indicated. The electric field must be the same throughout the whole region, including both slabs for a steady-state (or quasi-steady-state) scenario.

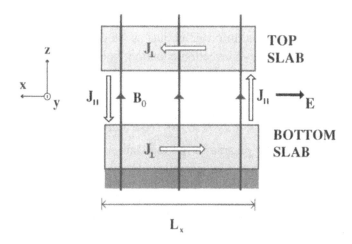

Figure 8.20. Side view of plasma slabs with the electrical current included.

specific, however. The total current per unit length in the y direction is just given by the height-integrated current densities $K_{\perp 0}$ and $K_{\perp 1}$ in the upper and lower slabs, respectively, and Equation (8.35) simply gives $K_{\perp 0} = J_{\perp 0}H$ and $K_{\perp 1} = J_{\perp 1}H$. To keep the total current per unit length the same "around the circuit," hence satisfying Equation (8.30), we must have for this simple example:

$$K_{\perp 0} = K_{\perp 1} \quad \text{or} \quad J_{\perp 0} = J_{\perp 1}. \tag{8.55}$$

The total field-aligned current per unit length, connecting the two slabs, is just $K_{\parallel} = K_{\perp 0} = K_{\perp 1}$. Thus, in Equations (8.52)–(8.54), we can use either $J_{\perp 0}$ or $J_{\perp 1}$. This allows us to use Equation (8.52) to find $J_{\perp 0}$ and then use it in Equation (8.54) to get

$$u_1 = \frac{J_{\perp 1} B_0}{\rho_1 v} = \frac{J_{\perp 0} B_0}{\rho_1 v} = -\frac{1}{\rho_1 v} \left[\rho_0 \frac{Du_0}{Dt} + \frac{\partial p}{\partial y} \right]. \tag{8.56}$$

If u_1 and $\partial p / \partial y$ happened to be known, then this equation could also be written as a differential equation for u_0; in either case, the top and bottom slabs are linked. Alternatively, if we are dealing with a steady-state situation for the upper slab then we have the second case of the last section.

We can again consider the two cases discussed in the last section, either $\partial p / \partial y = 0$ in the top slab or $Du_0 / Dt = 0$. Equation (8.56) then becomes

$$\begin{aligned} \rho_0 \frac{\partial u_0}{\partial t} &= -\rho_1 v u_1 \quad \text{(case 1)} \\ \frac{\partial p}{\partial y} &= -\rho_1 v u_1 \quad \text{(case 2)}. \end{aligned} \tag{8.57}$$

These equations are identical to Equations (8.45) and (8.46) in the last section, but we still need some additional information just as we did there. In that section, we used the magnetic diffusion–convection equation, but in the next section we will instead use Ohm's law and the electric field.

8.3.5 Role of the electric field

We have assumed that the plasma in the top slab is collisionless, in which case a reasonable approximation to the generalized Ohm's law is the motional electric field expression

$$\mathbf{E} = -\mathbf{u} \times \mathbf{B} \cong -\mathbf{u}_0 \times \mathbf{B}_0 \cong -u_0 B_0 \hat{\mathbf{x}}. \tag{8.58}$$

This is also a good approximation for the tenuous plasma that exists between the two slabs. Terms containing the current density $J_{\perp 0}$ are small, and it would be unproductive to obtain $J_{\perp 0}$ from the GOL. Equation (8.58) is consistent with the individual plasma particles in the upper slab undergoing $\mathbf{E} \times \mathbf{B}$ drift; that is, $u_0 = E / B_0$. Note that for a steady-state and one-dimensional situation, we can see from Faraday's law that the electric field must be equal to $E_x = -u_0 B_0$ everywhere, even in the lower slab. In other words, magnetic field lines are equipotentials, and the motional electric field is "mapped" from the upper slab down to the lower slab.

The appropriate GOL for the lower slab is the ionospheric one given by Equation (8.36) and discussed in Chapter 7. Here we assume that the neutral flow speed is zero ($\mathbf{u}_n = \mathbf{0}$) so that $\mathbf{E}' = \mathbf{E}$. We can apply Equation (8.36) to our simple example by using Equation (8.58) for the electric field. The Hall and Pederson currents become

$$\mathbf{K}_\mathrm{H} = \mathbf{E} \times \hat{\mathbf{b}} \Sigma_\mathrm{H} = u_0 B_0 \Sigma_\mathrm{H} \tag{8.59}$$

and

$$\mathbf{K}_\perp = \mathbf{E}\Sigma_\perp = -u_0 B_0 \Sigma_\perp, \tag{8.60}$$

where the height-integrated Hall and Pederson conductivities are simply

$$\Sigma_\perp \cong \sigma_\perp H = \frac{n_1 e^2 \nu H}{m(\nu^2 + \Omega^2)} \tag{8.61}$$

and

$$\Sigma_H \cong \sigma_H H = \frac{n_1 e^2 \Omega H}{m(\nu^2 + \Omega^2)}. \tag{8.62}$$

Equations (7.70) and (7.71) have been used for the "ionospheric" conductivities with only the ion terms retained in the summations over charged particle species (the electron contributions are very small). The collision frequency is just the ion neutral collision frequency ($\nu = \nu_{in}$).

The Hall current is in the y direction, which has infinite extension, and thus K_{H1} remains fixed and uniform. From Equation (8.34) then, the Hall current does *not* make a contribution to $J_\parallel$ because its spatial derivative is zero, nor for that matter does it make a contribution to the dynamics! In contrast, although the Pederson current is given by Equation (8.60) throughout most of the lower slab, the slab has a finite x extent and we must have $K_{\perp 1} = 0$ at the edges of the slab. Hence, a divergence of $K_{\perp 1}$ exists at the edges of the slab and $J_\parallel$ is produced according to Equation (8.34). In order to proceed, we note again that $K_{\perp 1} = J_{\perp 1} H$, and we use Equations (8.54), (8.60), and (8.61) to find $J_{\perp 1}$ so that we can write down $J_{\perp 1} B_0$, which appears in Equations (8.52)–(8.54), as

$$J_{\perp 1} B_0 = \rho_1 u_0 \nu \frac{\Omega^2}{(\nu^2 + \Omega^2)}. \tag{8.63}$$

Using this relation in Equation (8.54), we find that

$$u_1 = \frac{J_{\perp 1} B_0}{\rho_1 \nu} = u_0 \frac{\Omega^2}{(\nu^2 + \Omega^2)}. \tag{8.64}$$

Equation (8.64) is just Equation (8.49) again, but it was derived in a different way. From Equation (8.64) or (8.49) we see that the plasma flow speed in the ionospheric slab (the equivalent of the Earth's E-region) is less than the plasma speed in the upper slab, that is,

$$u_1 < u_0 = \frac{E}{B_0}. \tag{8.65}$$

Note that for our simple geometry, the $\mathbf{E} \times \mathbf{B}$ drift speed is the same everywhere and equal to u_0 (i.e., in both the upper and lower slabs). However, the plasma flow speed in the lower slab is less than the $\mathbf{E} \times \mathbf{B}$ drift speed. The plasma in Earth's F-region

$\mathbf{E} \times \mathbf{B}$ drifts, but the plasma in the E-region (also called the *dynamo region*) moves at speeds less than the $\mathbf{E} \times \mathbf{B}$ drift speed.

Two cases can be considered again, as in the last section, but the results are the same. For example, Equation (8.50) is again the solution for case (1), and for case (2) it is easy to show using either the pure MHD or the electric current MHD approaches that the flow speed in the upper slab must be equal to

$$u_0 = -\frac{(v^2 + \Omega^2)}{\Omega^2} \frac{1}{\rho_1 v} \frac{\partial p}{\partial y}. \tag{8.66}$$

Equation (8.66) also determines the electric field throughout the system with $|\mathbf{E}| = u_0 B_0$.

8.3.6 Energy balance and equivalent electric circuit

The electromagnetic (EM) energy relation (A.45) can be applied to the simple slab geometry of this section. The "local" net source of EM energy is equal to $-\mathbf{E} \cdot \mathbf{J}$. The current and electric field directions are opposing for the top slab ($-\mathbf{E} \cdot \mathbf{J} > 0$) and are in the same direction for the lower slab ($-\mathbf{E} \cdot \mathbf{J} < 0$); hence, the top slab is acting to produce EM energy (i.e., it is a *MHD generator*), whereas EM energy is being dissipated in the lower slab (i.e., it is a *resistor*). The total power generated per unit length ($P_{\text{EM},0}$ – top slab) or lost ($P_{\text{EM},1}$ – lower slab) is equal to the volume integral of $-\mathbf{E} \cdot \mathbf{J}$:

$$P_{\text{EM},0} = -P_{\text{EM},1} = \int_{\text{slab}} -\mathbf{E} \cdot \mathbf{J} \, dx \, dz \cong L_x H (u_0 B_0) J_{\perp 0}. \tag{8.67}$$

Another form of this power can be found by setting $J_{\perp 0} = J_{\perp 1}$ and using Ohm's law (in which only the Pederson term contributes) $J_{\perp 1} = \sigma_{\perp 1} E$.

If the power generated in the top slab is dissipated in the lower slab, then how, according to Equation (A.45), is the power transported from the top to the bottom slab for a steady-state situation? The Poynting vector, $\mathbf{S} = (\mathbf{E} \times \mathbf{B})/\mu_0$, which is a power density (W/m^2), takes care of EM energy transport, and in the region between the slabs the surface integral of $\mathbf{S}$ ought to be equal to $P_{\text{EM},0}$. For our case, the main component of $\mathbf{S}$ is in the y direction, the direction of the plasma flow, because $\mathbf{E}$ is in the negative x direction and $\mathbf{B}_0$ is in the z direction, but only the z component of $\mathbf{S}$ is relevant for finding the surface integral in the z direction. The EM power per unit length (W/m) transported between the top and bottom slab is equal to

$$\int_{\text{area}} \mathbf{S} \cdot \hat{\mathbf{z}} \, dx \, dy \cong S_z L_x = \frac{1}{\mu_0}[(-u_0 B_0 \hat{\mathbf{x}}) \times (B_0 \hat{\mathbf{z}} + B_y \hat{\mathbf{y}})] \cdot \hat{\mathbf{z}} L_x = \frac{u_0 B_0 B_y}{\mu_0} L_x, \tag{8.68}$$

where the integral over y is just over a unit length. The power per unit length obtained from either Equation (8.67) or Equation (8.68) are the same, as can be seen by recog-

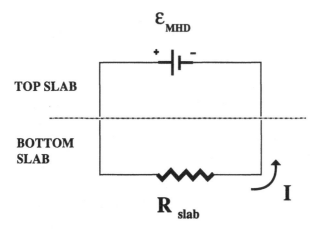

Figure 8.21. Equivalent electric circuit for the two-slab scenario.

nizing from Ampère's law that $J_{\perp 0} = B_y/(\mu_0 H)$. Electromagnetic energy is conserved! Where does the energy in the top slab come from? It comes from either the bulk kinetic energy (case 1) or from the thermal energy (case 2), but we would need to correctly apply the MHD energy equations (see Chapter 4) to the top slab to handle this. The EM energy dissipated in the lower slab (i.e. the "resistor") is lost as heat.

The electrical current approach to MHD for the two slabs can be represented using an "equivalent" electrical circuit, as illustrated in Figure 8.21. We consider just a piece of the slabs with length L_y in the y direction. The electromotive force (emf) generated by the top slab (the MHD generator) is denoted ε_{MHD} and is equal to

$$\varepsilon_{\text{MHD}} = u_0 B_0 L_x \quad \text{[V]}. \tag{8.69}$$

The electrical resistance of a length L_y of the lower slab is equal to

$$R_{\text{slab}} = \frac{L_x}{\Sigma_\perp L_y} \quad \text{[ohms]}. \tag{8.70}$$

A simple Ohm's law then gives the total electrical current flowing in the L_y segment:

$$I = \varepsilon_{\text{MHD}}/R_{\text{slab}} \quad \text{[A]}. \tag{8.71}$$

It is easily shown (Problem 8.12) that the current found from Equations (8.69)–(8.71) is consistent with Equation (8.60).

8.4 Magnetospheric convection

In the open model of the magnetosphere, magnetic reconnection allows part of the solar wind motional electric field to be mapped into the magnetosphere resulting in an electrical potential drop across the magnetosphere. This magnetospheric convection electric field drives magnetospheric convection and also is associated with electrical currents due to the MHD stress balance.

8.4.1 Magnetospheric plasma convection – Open field lines

In a "pure" MHD approach to the problem of magnetospheric flow, or convection, the electric field does not appear explicitly. However, the plasma flow velocity and the electric field are related via the generalized Ohm's law. For a low resistivity plasma we can use the ideal MHD version of the generalized Ohm's law: $\mathbf{E} = -\mathbf{u} \times \mathbf{B}$. The plasma flow velocity perpendicular to $\mathbf{B}$ can then be written as

$$\mathbf{u}_\perp = \frac{\mathbf{E} \times \mathbf{B}}{B^2}. \tag{8.72}$$

This also happens to be the $\mathbf{E} \times \mathbf{B}$ guiding center drift velocity for single particle motion in uniform electric and magnetic fields (Chapter 3). Note that the total drift motion of a particle should also include gradient and curvature drifts. However, the gradient and curvature drift speeds are proportional to the particle energy, and for low energy plasma $\mathbf{E} \times \mathbf{B}$ drift motion dominates. In the Earth's magnetosphere, Equation (8.72) is applicable to the "cold" (less than a few keV) thermal plasma, whereas gradient and curvature drifts are applicable to the more energetic ring current and radiation belt particles.

Equation (8.72) can be used to determine the convection velocity of the magnetospheric plasma, $\mathbf{u}_\perp$, if the electric field is known. For the Earth's magnetosphere, magnetic reconnection at the dayside magnetopause "opens" up the magnetosphere and allows the motional electric field to be mapped down the field lines, as discussed in Section 8.2. The resulting magnetospheric "convection" electric field was denoted $\mathbf{E}_m$ and is about 20% of the solar wind electric field during magnetically active times. Throughout most of the magnetosphere, including most of the tail, the electric field is about equal to $\mathbf{E}_m$, but assuming that magnetic field lines are equipotentials, the electric field is larger closer to Earth. In fact, the convection electric field maps all the way down to the ionosphere, where $\mathbf{E}$ is much greater than $\mathbf{E}_m$ because the field lines are closer together.

Let us first consider the magnetospheric plasma convection on the open magnetic field lines that reach from the polar cap ionosphere out through the magnetopause and into the solar wind. A glance at Figure 8.4 indicates that the plasma flow over the polar caps and in the parts of the magnetosphere near the magnetopause is antisunward.

Let us adopt a coordinate system in which the x axis points from the Earth to the Sun, the z axis points northward, and the y axis then must point from dawn to dusk in order to form a right-handed system. In this coordinate system the magnetospheric electric field is $\mathbf{E} = E\hat{\mathbf{y}}$. In the high latitude region of open field lines in the magnetosphere, the magnetic field points southward, $\mathbf{B} = -B\hat{\mathbf{z}}$, and the plasma convection is $\mathbf{u}_\perp = \mathbf{E} \times \mathbf{B}/B^2 \approx E/B(-\hat{\mathbf{x}})$. The flow is primarily in the negative x direction (that is, antisunward), although it is apparent from Figure 8.4 that the flow in the more distant magnetotail must also have a component toward the tail axis. The fields $\mathbf{E}$ and $\mathbf{B}$ are functions of the distance from the Earth. We can estimate the magnitude of the flow speed, $u_\perp = E/B$, by using a simple model.

The magnetic field near the Earth is approximately that of a dipole: $B = B_E(R_E/r)^3$ (see Chapter 3), where B_E and R_E are the surface field and the radius of the Earth, respectively. In electrostatics the electric field is just proportional to the gradient of the electric potential. We can thus estimate that $E \approx \Psi_m/l_m$, where Ψ_m is the potential difference of two field lines and the separation of the two field lines is denoted l_m. The potential drop across the magnetosphere is denoted Ψ_m, which is approximately 70 kV for moderately active times. Figure 8.7 shows how this potential is applied across the magnetosphere, and Figure 8.22 shows a cross section of the magnetosphere showing the "last two" field lines to open up on the flanks of the magnetosphere. The potential drop between these two field lines is Ψ_m. Near the magnetopause the field lines are separated by a distance of roughly $l_m \approx r_{mp}$, whereas near the Earth the separation is the width of the polar cap

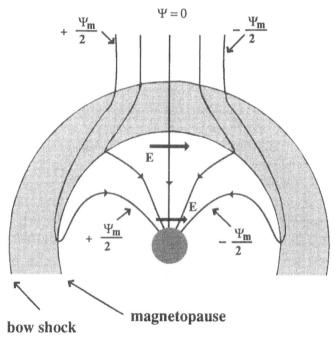

CROSS-SECTIONAL VIEW OF THE MAGNETOSPHERE FROM THE SUN

Figure 8.22. Cross-sectional schematic of the magnetosphere as viewed from the Sun. The last two reconnected magnetic field lines are shown.

($\approx$60 degrees latitude) so that $l_m \approx R_E$. Out in the solar wind $l_m = 2b$ according to Figure 8.7. We can estimate l_m at intermediate locations by again assuming that the field lines are dipolelike, although this is obviously not the case near the magnetopause. We keep the magnetic flux constant for a tube bounded on the extremes by our two field lines; that is, we assume $l_m^2 B = constant$ ($\approx B_E R_E{}^2$). In this case, we find (Problem 8.12) that

$$u_\perp \cong \frac{\Psi_m}{R_E B_E} \left(\frac{r}{R_E}\right)^{3/2}. \tag{8.73}$$

In the polar cap F-region ionosphere, where the electrical conductivity is high enough for field lines to be considered equipotentials and where $r = R_E$, the flow speed given by Equation (8.73) is just $\Psi_m/R_E B_E \approx 1$ km/s. Using the expression derived in Section 8.2 for Ψ_m in terms of the solar wind electric field and the reconnection efficiency ζ, Equation (8.73) can also be expressed in terms of the solar wind speed as (Problem 8.12)

$$\frac{u_\perp}{u_{sw}} \approx \frac{1}{3} \zeta \left(\frac{r}{r_{mp}}\right)^{3/2}, \tag{8.74}$$

where we have assumed that $r_{mp} = 10\,R_E$. For a typical value of $\zeta = 0.2$, we find that near the magnetopause the antisunward flow speed is $\approx u_{sw}\zeta/3 \approx 30$ km/s.

The region of antisunward flow in the F-region ionosphere is confined to high magnetic latitudes in both the northern and southern hemispheres where the field lines are open. This is really what we mean by the "polar cap" (see Figure 8.3). Figure 8.23 is a polar plot of the electrical potential and the flow streamlines in the northern hemisphere F-region. The flow is antisunward over the polar cap, and the other parts of the streamlines comprise the so-called *return flow* to be discussed later. Let us estimate the size of the polar cap. The polar cap is approximately bounded by the auroral oval, and thus we are also estimating the extent of the auroral oval. Let us do this on the sunward (or noon) side of the polar cap. For a rough estimate of the location of the poleward edge of the polar cap, we assume that the field is dipolar and map the dayside magnetopause back to the Earth. The effects of the magnetopause current layer on the magnetic field are neglected in doing this. Recall from Chapter 3 the definition of invariant latitude and, in particular, Equation (3.70). Identifying the L-value of the magnetopause as $L_{mp} \approx r_{mp}/R_E$, we find that the corresponding invariant latitude Λ_{mp} is given by

$$\Lambda_{mp} \cong \cos^{-1}\left(\sqrt{\frac{R_E}{r_{mp}}}\right). \tag{8.75}$$

For typical solar wind conditions, we saw in Chapter 7 that $r_{mp} \approx 10\,R_E$, so that from Equation (8.75) we find that $\Lambda_{mp} = 71.5°$; that is, the polar cap has an approximate radius of $20°$ magnetic latitude. For very high solar wind dynamic

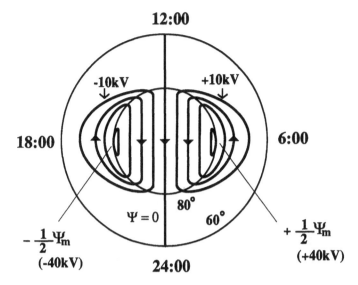

Figure 8.23. Polar plot of the high-latitude ionosphere. Equipotential lines are shown; these equipotential lines are also streamlines of ionospheric F-region plasma flow. The magnetic pole is in the center and magnetic latitude is shown. Local times are indicated with noon being at the top.

pressure conditions (as during a magnetic storm), $r_{mp} \approx 6R_E$ and $\Lambda_{mp} = 65°$; that is, the polar cap expands equatorward during magnetically active times.

8.4.2 Magnetospheric plasma convection – Closed field lines

It is known that the electric field $\mathbf{E}_m$ appears on closed field lines as well as on open magnetic field lines. The existence of this field is not entirely obvious because there are no "open" field lines along which the solar wind potential can be mapped. One can attribute the presence of this electric field on closed magnetic field lines to the reconnection process operating somewhere in the Earth's tail, as pictured schematically in Figure 8.4 for the Dungey model. For steady-state conditions ($\partial B/\partial t = 0$), it is apparent from Faraday's law that the cross-tail (and cross-magnetosphere) electric field must be uniform for a simple two-dimensional geometry as discussed in the last section of Chapter 4. From a pure MHD point of view, reconnection in the magnetotail creates flow that is Earthward in the equatorial plane on the Earthward side of the x-line. This is because of the MHD force balance that necessitates a flow velocity $\mathbf{u}$. The electric field can then be determined from the flow velocity using the generalized Ohm's law. From a "kinetic" point of view, one can picture a minute charge separation taking place in the magnetosphere with a layer of positive charge on the dawn magnetopause and a layer of negative charge on the dusk magnetopause; such a charge configuration leads to a more or less uniform magnetospheric electric field ($\mathbf{E}_m$) on both the open and closed magnetic field lines.

Magnetospheric convection in the equatorial plane observationally is not steady, even when the solar wind conditions are rather steady, but instead appears to be strongly enhanced during events that are called *magnetospheric substorms*. Magnetospheric substorms appear irregularly but typically occur every few hours and last 30 minutes or less. Hence, the electric field on closed field lines, as well as the associated sunward magnetospheric plasma convection, must also be temporally variable rather than steady state. Nonetheless, much can be learned by examining the time average (or steady) magnetospheric convection, as we will do next. Substorms are considered later (in Section 8.6).

The plasma flow velocity in the equatorial plane of the magnetosphere can be estimated using Equation (8.72) plus the simple cross-magnetosphere electric field $\mathbf{E}_m = E_m \hat{\mathbf{y}}$. This flow is clearly directed in the sunward direction because the magnetic field is directed northward, at least for the portion of the magnetosphere Earthward of any neutral line (or x-line) that might be present far downstream. That is, $\mathbf{u}_\perp = \mathbf{E} \times \mathbf{B}/B^2 \approx \hat{\mathbf{x}} E_m/B$. Near the Earth, the field is dipolelike so that we have $u_\perp = (E_m/B_E)(r/R_E)^3 \approx u_{sw}\zeta(B_0/B_E)(r/R_E)^3$ where B_0 is the IMF. The plasma flow speed in the equatorial plane strongly increases with increasing radial distance from the Earth. Near $r \approx 10R_E$, we expect that $u_\perp \approx 30$ km/s and near $r \approx 5R_E$ we only have $u_\perp \approx 4$ km/s. For radial distances less than about a few R_E in the equatorial plane (that is, for L-shells less than 5) the "observed" (or deduced) convection electric field is usually less than the E_m estimated above. This "shielding" of the inner magnetosphere from E_m, which will be briefly discussed later, appears to be less effective during magnetic substorms. Another limitation on our simple estimate of $u_\perp$ is that for radial distances greater than about $10R_E$ (or even less) the magnetic field is more tail-like than dipolar due to magnetospheric electrical currents such as the cross-tail current; therefore, the convection speed does not continue to increase as r^3 but levels off.

Equipotential surfaces in the equatorial plane of the magnetosphere are depicted in Figure 8.24 with zero potential being located in the midline. The inner magnetosphere shielding just referred to is neglected in this figure. The equipotential lines are also streamlines for the sunward flow. The flow carries plasma into the dayside magnetopause where it participates in the dayside magnetic reconnection/merging process.

The magnetospheric electric field in the equatorial plane maps along closed equipotential magnetic field lines into the ionosphere as illustrated in Figure 8.25 for the flank regions. The mapping of the electric potential onto the ionosphere generates the part of the pattern shown in Figure 8.23 that is equatorward of the polar cap. The ionospheric "return" flow associated with this region is still antisunward near noon and midnight local times but is sunward near the flanks (i.e., near dawn or dusk). The ionospheric flow for southward IMF conditions can be described as a *two-cell convection pattern*. The ionospheric flow speed can be estimated in the same way that we used for the open field line region; we again find that for mod-

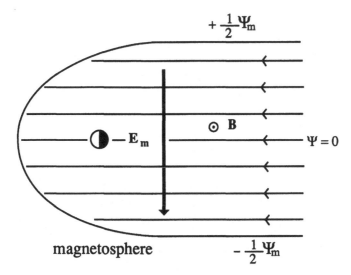

Figure 8.24. Equatorial plane cross section of the magnetosphere showing equipotential lines. These equipotential lines are also streamlines for low-energy plasma for which $\mathbf{E} \times \mathbf{B}$ drift motion is dominant.

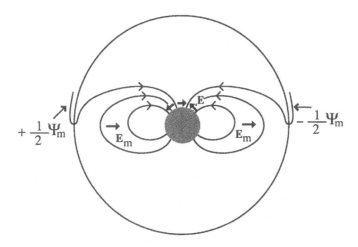

Figure 8.25. Cross-sectional schematic of the magnetosphere as viewed from the Sun showing magnetic field lines and electric field vectors.

erately active conditions $u_{\perp} \approx 1$ km/s. This is also apparent from Figure 8.23. For a steady situation, closed potential contours are needed, which indeed necessitates return flow of the plasma.

The electrostatic potential in the equatorial plane of the magnetosphere can be expressed as

$$\Psi = E_{\mathrm{m}} R_{\mathrm{E}} L \sin \phi, \tag{8.76}$$

where $\mathbf{E}_{\mathrm{m}}$ is assumed to point from dawn to dusk, L is the L-value, and ϕ is the azimuthal angle (or longitude) measured eastward from midnight (cf. Lyons and Williams, 1984; Volland, 1984). Equation (8.76) is consistent with the potential lines shown in Figure 8.24. An empirically based "shielded" potential can also be introduced for geomagnetically quiet times for $L < L_b$ (i.e., for the inner magnetosphere), where L_b is a characteristic L-shell roughly corresponding to the L-shell of the magnetopause ($L_b \approx 10$):

$$\Psi = E_m R_E L_b \left(\frac{L}{L_b}\right)^{q/2} \sin\phi, \tag{8.77}$$

where $q \approx 4$. Equation (8.76) still applies for $L > L_b$ or during geomagnetic storms even for $L < L_b$. Electrical currents flowing near the boundary between the plasma sheet and the inner magnetosphere create some charge separation that partially counteracts or *shields* E_m. See Volland (1984) for a more detailed treatment of this magnetospheric potential.

8.4.3 *Corotation electric field and the plasmasphere*

The neutral atmosphere by and large corotates with the solid Earth, although locally there are small departures from this corotation due to winds. The ionosphere also tends to corotate with the Earth because it is coupled to the neutral atmosphere by ion–neutral collisions and because ionospheric plasma is created, on the average, at the neutral atmosphere velocity when neutrals are ionized (see Chapter 7). An obvious exception to this ionospheric corotation is the polar cap two-cell convection associated with magnetospheric electric fields mapped into the ionosphere. The cold thermal plasma throughout the entire inner magnetosphere corotates, and not just the low altitude ionospheric plasma. This corotation is enforced by the *corotation electric field* created in the E-region ionosphere. The simple slab MHD picture presented in Section 8.3 can be used to understand the corotation electric field.

We start with the momentum equation in the ionosphere, which is essentially Equation (8.39) but with the neutral velocity $\mathbf{u}_n$ retained in the friction term as in Equation (8.28):

$$\rho\frac{D\mathbf{u}}{Dt} = -\nabla p + \mathbf{J} \times \mathbf{B} - \rho\nu(\mathbf{u} - \mathbf{u}_n), \tag{8.78}$$

where ν is the ion–neutral collision frequency, and the other terms were defined earlier. For steady-state conditions and for relatively low flow speeds, we can set the left-hand side of this equation equal to zero. Furthermore, we can neglect the pressure gradient term; it does contribute to some ionospheric "winds" with respect to the corotation but this is a minor effect. We can now convert Equation (8.78) to an equation analogous to Equation (8.54):

$$\mathbf{J}_{\perp\,ion} \times \mathbf{B} = \rho\nu(\mathbf{u} - \mathbf{u}_n), \tag{8.79}$$

where $\mathbf{J}_{\perp\,\text{ion}}$ is the Pederson current in the ionosphere. The electrical current in the ionosphere must "connect" with the current in the magnetosphere via field-aligned currents; however, if no significant stress exists on the plasma then this current is small ($\mathbf{J}_{\perp\,\text{ion}} \approx \mathbf{0}$). If this current is very small or zero, then from Equation (8.79) we see that the ionospheric plasma moves with the neutral gas:

$$\mathbf{J}_{\perp\,\text{ion}} = \mathbf{0} \Rightarrow \mathbf{u} = \mathbf{u}_n. \tag{8.80}$$

Conversely, if a significant Pederson current is present (due perhaps to the MHD force balance in the inner magnetosphere that is connected to the ionosphere via the dipole magnetic field lines), then the plasma does not move exactly with the neutrals ($\mathbf{u} \neq \mathbf{u}_n$).

The electric field and the current density are related by the generalized Ohm's law for the ionosphere given by Equation (7.68). Setting the Pederson current equal to zero in this equation, we find that $\mathbf{E}' = \mathbf{0}$, where we recall that $\mathbf{E}' = \mathbf{E} + \mathbf{u}_n \times \mathbf{B}$ (Equation (7.67)). Thus, if $\mathbf{J}_{\perp\,\text{ion}} = \mathbf{0}$, we have the following electric field:

$$\mathbf{E} = -\mathbf{u}_n \times \mathbf{B}. \tag{8.81}$$

Furthermore, if there is no neutral wind relative to the Earth's surface, and if the atmosphere is just corotating, then

$$\mathbf{u}_n = \mathbf{\Omega}_E \times \mathbf{r}, \tag{8.82}$$

where $\mathbf{\Omega}_E$ is the angular frequency of the Earth, which points north and has magnitude $\Omega_E = 2\pi/T_E$, and where the rotation period of the Earth is $T_E = 24$ hours. The radial position vector is denoted $\mathbf{r}$.

The *corotation electric field* is found by combining Equations (8.81) and (8.82) to give

$$\mathbf{E}_{\text{cor}} = -(\mathbf{\Omega}_E \times \mathbf{r}) \times \mathbf{B}. \tag{8.83}$$

If we follow along magnetic field lines above the ionosphere/atmosphere and into the inner magnetosphere, then the magnetospheric electrical current ($\mathbf{J}_{\perp\,\text{magn}}$) remains zero if $\mathbf{J}_{\perp\,\text{ion}} = \mathbf{0}$, and no field-aligned currents exist ($\mathbf{J}_{\parallel} = \mathbf{0}$). In this case, Equation (8.83) applies to the magnetosphere as well as to the ionosphere. The electric field "maps" along closed magnetic field lines and the field lines are equipotentials. Equation (8.72) plus Equation (8.83) can now be used to determine the plasma velocity in the inner magnetosphere (assuming a very low electrical resistivity):

$$\mathbf{u}_\perp = \frac{\mathbf{E}_{\text{cor}} \times \mathbf{B}}{B^2} = -\frac{((\mathbf{\Omega}_E \times \mathbf{r}) \times \mathbf{B}) \times \mathbf{B}}{B^2} = \mathbf{\Omega}_E \times \mathbf{r}. \tag{8.84}$$

Equation (8.84) tells us that *the plasma in the magnetosphere corotates with the Earth*. Note that this refers to cold plasma; more energetic charged particles (such

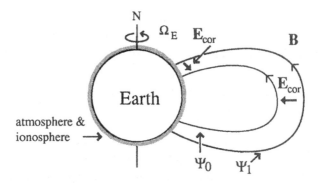

Figure 8.26. Mapping of corotation electric field from the ionosphere out into the magneto-sphere along two magnetic field lines whose electric potentials are Ψ_0 and Ψ_1. The plasma undergoes "solid body" rotation together with the planet and atmosphere in this case.

as in the ring current) also experience gradient and curvature drift as well as $\mathbf{E} \times \mathbf{B}$ drift.

The mapping of the corotation electric field is illustrated in Figure 8.26. The two dipole field lines shown are equipotentials with $\Psi_1 > \Psi_0$. The magnitude of the corotation electric field from Equation (8.83) is equal to

$$E_{\text{cor}} = |\mathbf{E}_{\text{cor}}| = \Omega_E \sin\theta \frac{B_E R_E^3}{r^2}. \tag{8.85}$$

The corotation electric field falls off as the square of the radial distance. The polar angle is denoted θ. The intersection of the equipotential surfaces with the equatorial plane are circles, and the plasma flow streamlines are parallel to these equipotential lines. The corotating plasma obviously moves counterclockwise when viewed from above the north pole. The equipotentials (and streamlines) in the Earth's ionosphere must obviously also be circles.

Figure 8.27 shows equipotential lines in the equatorial plane. The actual potential pattern can be approximated by simply adding together the magnetospheric and corotation electric fields (or potentials), as shown in Figure 8.27 and as discussed in Example 8.3. The pear-shaped region containing closed potential contours in Figure 8.27c is called the *plasmasphere*; its boundary, delineated by the last closed contour, is called the *plasmapause*. The plasmasphere is clearly more extended near dusk than it is near dawn. This *plasmaspheric evening bulge* results from the stagnation of the flow that occurs because the corotation electric field is equal and opposite to the magnetospheric electric field $\mathbf{E}_m$. This predicted evening bulge has been confirmed by density measurements in the inner magnetosphere. The thermal plasma in the plasmasphere corotates, whereas the plasma in the more distant equatorial plane, mostly consisting of the plasmasheet, convects toward the dayside magnetopause.

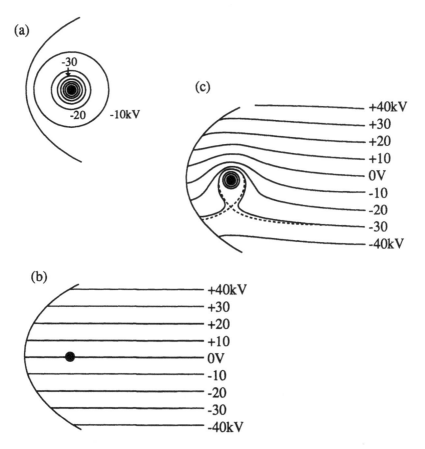

Figure 8.27. Equipotential lines in the equatorial plane for the following cases: (a) corotation electric field only, (b) only magnetospheric convection electric field, and (c) sum of both corotation and magnetospheric electric fields. The last closed potential contour marks the plasmapause boundary or outer edge of the plasmasphere. The solid line contours are equally spaced at 10 kV intervals. The dashed line potential value is −31 kV and is the "last" closed contour. The outermost solid closed contour has a potential value of −40 kV.

Example 8.3 (Electric potential in the equatorial plane) The electric potential in the equatorial plane associated with the corotation electric field given by Equation (8.85) can be expressed as (Problem 8.13)

$$\Psi_{\text{cor}} = -\frac{\Omega_E B_E R_E^2}{L} \cong \frac{-87\,\text{kV}}{L}, \tag{8.86}$$

where L is the L-shell value and where zero potential is located at infinity. The potential surfaces in the equatorial plane from Equation (8.86) are shown in Figure 8.27a. Similarly, the unshielded potential from Equation (8.76) can be expressed as $40\,\text{kV}(L/15)\sin\phi$ for $\Psi_m = 80\,\text{kV}$, and this is shown in Figure 8.27b.

The total electric potential in the inner magnetosphere (shown in Figure 8.27c) can be estimated by simply adding the corotation potential and the magnetospheric convection potential,

$$\Psi_{total} = \Psi_{cor} + \Psi_{mag} \cong \frac{-87\,kV}{L} + 40\,kV\left(\frac{L}{15}\right)\sin\phi. \qquad (8.87)$$

The dashed line in Figure 8.27c denotes the last closed equipotential contour and can be identified as the equatorial location of the plasmapause.

The L-value of the plasmapause at 18:00 local time can be determined by equating the corotation electric field to E_m (Problem 8.13):

$$L_{pp} \cong \sqrt{\frac{\Omega_E B_E R_E 3 r_{mp}}{\Psi_m}} \cong \sqrt{\frac{3,900\,kV}{\Psi_m(kV)}}. \qquad (8.88)$$

According to Equation (8.88), the L-shell of the evening plasmapause decreases as the level of magnetic activity, and hence the magnetospheric potential Ψ_m, increases. For the moderately high value of $\Psi_m = 80\,kV$ chosen here, $L_{pp} = 7$, as is evident in Figure 8.27. The plasmapause location at other local times can be found by using Equation (8.87) (Problem 8.13). The plasmapause is further from Earth at 18:00 LT than it is at other times.

The plasma density in the plasmasphere significantly exceeds the density found in the surrounding magnetosphere outside the plasmapause, as can be seen from the satellite data shown in Figure 8.28. The location of the plasmapause is quite clear for the cases shown; however, the radial density structure is often more complex, and a single "clean" plasmapause is not always evident, especially during magnetically quiet times immediately following an active period. Figure 8.28 clearly shows that as the magnetic activity level (as indicated by the K_p geomagnetic index) increases, the plasmapause moves closer to the Earth. The explanation can be found in Equation (8.88) given that the magnetospheric cross-tail potential, Ψ_m, and the K_p index are positively correlated. The structure of the plasmasphere in the direction normal to the magnetic equator was shown in the schematic of Figure 8.1 and is torus shaped.

The large-scale electric field determines how thermal plasma convects in the direction perpendicular to the magnetic field, but the plasma distribution *along* the magnetic field is determined by the parallel component of the momentum equation together with the continuity equation. For thermal, ionosphericlike plasma, the ambipolar diffusion equation provides a reasonable description of the parallel plasma flow (see Sections 4.6.5 and 7.3.4). For steady-state conditions, the density tends to follow a diffusive equilibrium profile, although the decrease of the gravitational acceleration with altitude must be taken into account. The source of cold thermal plasma is the ionosphere, and this plasma diffuses upward along the magnetic field lines into the plasmasphere.

Note that the main ion species in the plasmasphere is H^+ and not O^+. Earlier in this book the peak region of the F_2 ionospheric layer was discussed. The main ion

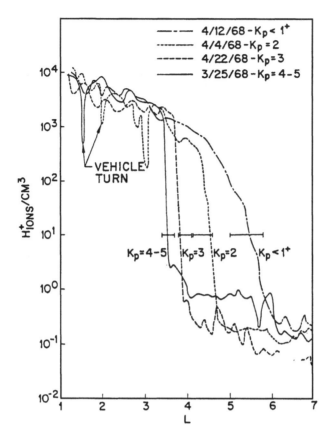

Figure 8.28. H$^+$ density versus L-shell (in the equatorial plane) as measured by the *OGO-5* spacecraft. The H$^+$ density is essentially equal to the electron density. The four ion density profiles are for days with different levels of magnetic activity as indicated by the K_p index (low K_p: low activity; high K_p: high activity). The plasmapause is located where the electron density falls off sharply. (From Chappell et al., 1970.)

species in this region is O$^+$. However, at higher altitudes the H density, associated with the extensive hydrogen geocorona, exceeds the neutral oxygen density and the following chemical reaction converts O$^+$ ions to H$^+$ ions:

$$O^+ + H \rightarrow O + H^+. \tag{8.89}$$

Furthermore, the plasma scale height for H$^+$ is greater than the scale height for O$^+$ due to the mass difference (recall Equation (7.28)), and this favors H$^+$ being the most abundant ion species at higher altitudes.

Ionospheric plasma flows up the magnetic field and supplies the inner magnetosphere with relatively cold plasma. This happens at all L-values, but the plasma density is higher in the plasmasphere than it is outside because the plasma inside is not lost due to horizontal transport. Horizontally, the plasma inside the plasmasphere just moves around and around the Earth due to corotation. However, plasma located outside the plasmapause never has a chance to build up in density because it is continually being convected away toward the magnetopause where it is lost to the magnetosheath/boundary layer flow. The convection speed near $L = 5$ is about 5 km/s; it thus takes the plasma about 5 hours to reach the magnetopause. The ionosphere cannot possibly supply plasma fast enough to replace this loss. The magnetospheric thermal plasma density outside the plasmapause never has a chance to reach diffusive equilibrium levels.

During magnetic storms, E_m (and Ψ_m) increases both because the solar wind electric field E_{sw} increases and because the reconnection efficiency ζ increases due to the IMF changing from northward to southward directed. However, the corotation electric field does not change. Previously closed streamlines (or potential contours) become open, and within a couple of hours the plasma on these field lines (which are no longer located within the plasmasphere) gets transported to the magnetopause and a new plasmapause forms at a lower L-value. Meanwhile, other events are taking place further away in the plasmasheet as well as at the magnetopause and in the ring current, as will be discussed later. After the storm has progressed to the recovery phase, when E_m has again decreased, the region of closed streamlines moves out again; however, a new plasmapause does not immediately form because it can take many hours for the magnetic flux tubes to *refill* due to the finite source of plasma from the ionosphere at lower altitudes. It can be shown from the continuity equation for a flux tube at L, neglecting losses (Problem 8.14), that the refilling time needed to go from an almost empty flux tube to one with density $\langle n_e \rangle$ (in units of cm^{-3}) is roughly given by the plasmaspheric flux tube content divided by the flux into the tube, that is,

$$\text{time} \approx 3L^4 R_E \langle n_e \rangle / \text{flux} \approx 2L^4 \langle n_e \rangle. \tag{8.90}$$

In Problem 8.14 you will show that for $L = 2, 3, 4$, and 5 the refilling times are roughly 1, 2, 3, and 5 days, respectively. The average electron density versus L in the plasmasphere can be taken from Figure 8.28.

The radial plasma density profiles during this refilling process generally have a more complex appearance than the profiles shown in Figure 8.28. Some density profiles measured during the process of refilling are shown in Figure 8.29. You should recall that in addition to this colder (few eV) thermal plasma the energetic ring current plasma also is present in the inner magnetosphere. The ring current was discussed in Chapter 3 and will be revisited later.

8.4.4 Plasma sources

The source of plasma for the plasmasphere was just stated to be the ionosphere, but what are the sources of plasma for the rest of the magnetosphere? As discussed by Moore in his 1991 review paper, "Origins of Magnetospheric Plasma" (see the bibliography at end of the chapter), the prevailing view of whether the solar wind or

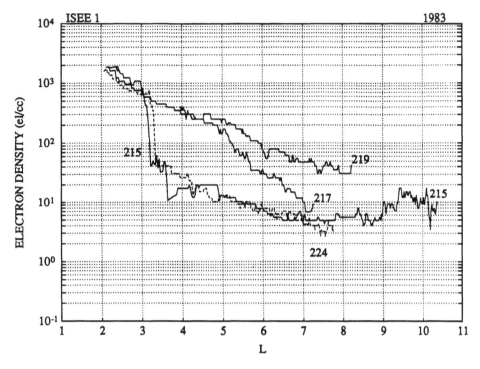

Figure 8.29. Electron density profiles versus L measured by the *ISEE-1* spacecraft. The profile for day 215 (of 1983) has a well-defined plasmapause at $L = 3$, but during the recovery period (days 217 and 219) the plasmasphere extends to higher L-values without a clear-cut plasmapause. A well-defined plasmapause near $L = 3$ again appears on day 224 following renewed magnetic activity. (From Carpenter and Anderson, 1992.)

the ionosphere is more important as a source of magnetospheric plasma has changed markedly over the past couple of decades. Twenty years ago it was thought that the solar wind, entering the magnetosphere through the plasma mantle/magnetopause boundary layer, was the dominant source of magnetospheric plasma outside the plasmasphere. However, the current view is that the ionosphere supplies most of the plasma out to a distance of at least 20 R_E. Obviously, all of the singly ionized O and He must come from the ionosphere, but even most of the protons are ionospheric H^+ ions rather than solar wind protons.

Moore defined a boundary, the *geopause*, inside of which the plasma is mainly, but not entirely, of ionospheric origin. Figure 8.30, taken from Moore (1991), is a schematic showing the location of the geopause. Clearly, the plasma mantle/magnetopause boundary layer (also see Figures 8.1 and 8.2) is populated by solar wind plasma or, to be more precise, magnetosheath plasma. This boundary layer is usually located adjacent to the magnetopause, although it actually penetrates deep into the magnetosphere in the *cusps* (or *clefts*), the funnel-shaped regions of open magnetic field lines that connect to the polar cap ionosphere near a local time of noon. Most solar wind plasma does not easily penetrate the cusp, even if

THE GEOPAUSE

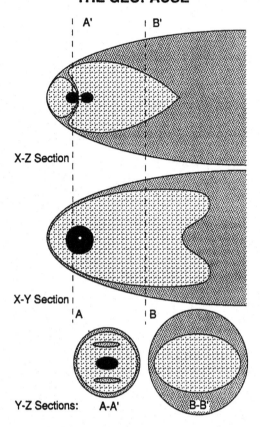

Figure 8.30. Schematic of the magnetosphere and geopause, which is the boundary between terrestrial and solar wind plasma dominance. The top panel is a view from the side. The middle panel is a view from over the north pole, and the bottom panel shows two cross-sectional cuts through the magnetosphere. The darkest shading near the Earth is the plasmasphere. The lighter shading denotes the geosphere and the darker shading beyond is the region dominated by solar wind plasma. (From Moore, 1991.)

the field lines are open, because the plasma is convected downstream more rapidly than it can move down the field lines. However, the more energetic magnetosheath electrons have sufficiently large parallel velocities that they can penetrate deeply into the cusps and even reach the ionosphere. One consequence of this is that these electrons *precipitate* into the polar cap region of the Earth and are responsible for part of the *diffuse aurora*.

The exact configuration of the magnetopause boundary layer further down the tail depends on the activity level and solar wind conditions, but it tends to broaden out due to plasma convection toward the neutral sheet (see Figure 8.4) and due to particle diffusion. The solar wind–dominated region (located outside the geopause) tends to track this boundary layer. Figure 8.30 shows that the geopause reaches the neutral sheet in the distant tail. Note that *some* ionospheric plasma can exist outside the geopause and some solar wind plasma can be present even inside this boundary. In particular, solar wind plasma having reached the neutral line region convects sunward supplying solar wind plasma to magnetospheric regions inside the nominal geopause boundary.

Now we consider how ionospheric plasma can populate the magnetosphere. As discussed in Chapter 7, plasma is created in the ionosphere by photoionization of atmospheric neutrals by solar EUV radiation or by electron impact ionization of neutrals by energetic auroral electrons. The plasma in the topside ionosphere can diffuse (that is, flow) up or down magnetic field lines depending on the distribution of plasma along the field line. In a completely filled plasmasphere the upgoing and downgoing plasma fluxes are balanced in general; in fact, a small net downward flux exists on the average because a net chemical loss of plasma is present at lower altitudes. However, during times of plasmaspheric refilling, the outer parts of the field lines are relatively empty and ionospheric plasma flows upward and gradually "refills" the flux tubes. As estimated in Problem 8.14, the ionospheric source can supply an ion flux as large as 10^8 to 19^9 cm^{-2} s^{-1}. At lower latitudes, this flux is mostly, but not entirely, in the form of H$^+$ ions, but at higher latitudes upward ionospheric fluxes of O$^+$ and He$^+$ ions are also present (cf. Moore, 1991; Chappell, 1988). For example, in the polar cap during active times, O$^+$ and H$^+$ fluxes of $\approx 10^7$–10^8 cm^{-2} s^{-1} have been observed by spacecraft. The total polar cap ionospheric source strength, integrated over both polar caps, for active times has been estimated by Chappell (1988) to be about 2×10^{25} ions/s for both species. The total ionospheric source strength from the auroral oval itself is expected to be about three times greater than this. The ionospheric plasma in the polar cap region, especially near the cusp/cleft, can be heated by plasma waves to temperatures of several electron volts during magnetically active times, whereas at other locations the ions are less energetic (≈ 1 eV or less).

Where does the upward flowing ionospheric plasma go? Clearly, at low latitudes this plasma populates the plasmasphere, but what about at higher latitudes? Figure 8.31, from Chappell (1988), shows what happens schematically; this picture and also the ionospheric source numbers above were largely derived from plasma measurements, and associated modeling efforts, from NASA's *Dynamic Explorer* mission in which two polar orbiting spacecraft, one of which was in a high apogee orbit and the other in a low orbit, were able to sample plasma on different parts of the same magnetic field line. The ionospheric plasma flowing upward out of the polar cap on open field lines is convected in an antisunward direction by the magnetospheric electric field, as discussed earlier in this chapter. Given a common ion temperature in the topside ionosphere of a few electron volts, the H$^+$ ions have a larger parallel velocity than the O$^+$ or He$^+$ ions. Hence, as shown in Figure 8.31, the H$^+$ ions move more quickly up the field into regions of faster antisunward

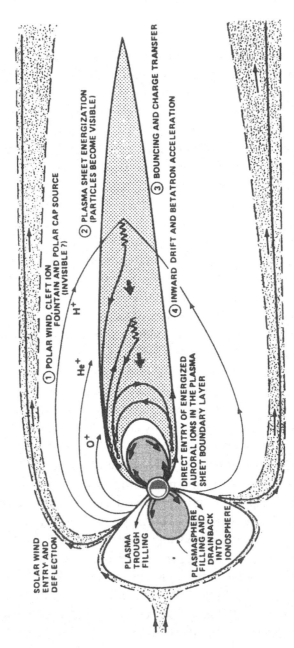

Figure 8.31. Schematic of how ionospheric plasma can populate different regions of the magnetosphere. Arrows show the flow direction of ions as they move from the ionosphere to the plasmasphere, plasma trough, plasma sheet, and tail lobe. In the plasma sheet, the ions flow back toward the Earth and are also energized. The plasmasphere is shown in dark shading and the plasma sheet in light shading. (From Chappell, 1988.)

Table 8.1. *Typical magnetospheric densities and temperatures*[a]

Region	Density $[\text{cm}^{-3}]$	Average Energy
Inner plasmasphere	10^4	1 eV
Outer plasmasphere	10^3	10 eV
Dayside plasma trough	5	—
Quiet plasmasheet	1	0.5 keV (electrons)
		3 keV (ions)
Active plasmasheet	0.3	1 keV (electrons)
		6 keV (ions)
Tail lobe	10^{-2}	100 eV (electrons)
		1 keV (ions)

[a] Some of this information is from Chappell (1988).

flow than do the heavier ion species. Hence, the lighter mass ions have more of a tendency than the heavier ions to be "injected" further down the tail. Eventually the ions convect toward the neutral sheet as a consequence of $\mathbf{E}_m \times \mathbf{B}$ drift in the tail lobes; this supplies the plasmasheet with ionospheric plasma. The convection in the plasmasheet is then toward the Sun.

Ions that reach the tail lobes and/or plasmasheet are no longer ionospheric in nature because the convection associated with E_m has energized these ions. Further energization can also occur in the sunward convection process as ions convect to regions with larger magnetic field strength because of betatron acceleration (i.e, conservation of the first adiabatic invariant). Some nonadiabatic energization can also take place in the neutral sheet due to chaotic particle motion and/or due to magnetic reconnection. In fact, during the intense convection events associated with magnetic storms some of the inward drifting ions are "converted" into energetic ring current particles.

What are the characteristics of magnetospheric plasma populations? The answer is extremely complex, and the electron and ion distribution functions are not always even Maxwellian. Table 8.1 provides a rough idea of typical densities and temperatures in various regions of the magnetosphere.

8.5 Magnetospheric currents and magnetosphere–ionosphere coupling

8.5.1 *Magnetospheric currents*

The magnetospheric and ionospheric plasma convection pattern discussed in Section 8.4 can be calculated using MHD theory. The force or stress balance associated with (or, actually, responsible for) the plasma convection necessitates field-aligned electrical currents, as was illustrated in Section 8.3 for a simple slab geometry. Different plasma regions, linked by magnetic field lines, have different magnetic stress terms, $\mathbf{J} \times \mathbf{B}$, as part of their overall force balance, and the different currents in the different regions are "reconciled" in the electrical current MHD description by means of parallel (field-aligned) currents. For example, in the terrestrial magnetosphere the plasma convection is associated with $\mathbf{J} \times \mathbf{B}$ forces and the current system throughout the magnetosphere must be consistent with this convection pattern. In particular, the solar wind and magnetospheric plasmas are connected to the ionosphere by magnetic field lines, and the requirements of the MHD force balance in different regions (i.e., ion–neutral friction in the ionosphere) results in field-aligned currents (or *Birkeland currents*) into or out of the ionosphere. Obviously, the energy balance in the magnetosphere/ionosphere system must also be consistent with both the convection pattern and electrical current pattern.

In this section, a simple steady-state picture of the magnetosphere current system will be presented, with an emphasis on the coupling of the magnetosphere to the ionosphere. However, the true magnetospheric plasma behavior is not steady but depends on solar wind conditions and is time dependent. Accordingly, in Section 8.6 we will modify our simple picture to account for time dependence; in particular, we will discuss magnetospheric substorms.

One of the most important magnetospheric current systems is that associated with the magnetopause. Recall the discussion of the size of the magnetosphere in Chapter 7. The magnetopause is a current layer (Figure 7.2), and from Equations (7.3) and (7.9) we can estimate from Ampère's law (Problem 7.1) that the surface current density is approximately $K_{\mathrm{mp}} \approx B/\mu_0 \approx 0.03$ A/m, which leads to a total dayside magnetopause current of approximately $I_{\mathrm{mp}} \approx r_{\mathrm{mp}} K_{\mathrm{mp}} \approx 2 \times 10^6$ A. The current must flow from dawn to dusk to be consistent with the direction of the magnetospheric field. Examining the magnetic field direction just inside the magnetopause from Figures 8.1 and 8.4 it is apparent that the current in the tailward part of the magnetopause must flow in a different direction than the dayside magnetopause current. Figure 8.32 illustrates the current pattern on the magnetopause. The dayside magnetopause current "closes" on the more tailward part of the magnetopause.

Figure 8.32 also shows the magnetopause current in the more distant tail closing in the neutral sheet region, which lies near the center of the plasmasheet. The cross-tail current in the near-Earth region (within $1\, r_{\mathrm{mp}}$ or so) is of the order of $I_{\mathrm{tail}} \approx 1\,\mathrm{MA}$; see Section 8.2.3. The cross-tail electrical current over a greater tail extent (i.e., $10\, r_{\mathrm{mp}}$ or so) is more like 10 MA.

Empirically, we know that approximately 95% of the electrical current is accounted for by the current pattern pictured in Figure 8.32. What about the ionosphere? Roughly 5% of the magnetosphere/magnetopause current is diverted, via Birkeland currents, to the ionosphere. The total ionospheric current is of the order of $I_{\mathrm{iono}} \approx 5\% \times 10\,\mathrm{MA} \approx 1\,\mathrm{MA}$. How this current is distributed depends on how the ionosphere maps along magnetic field lines to different regions of the magnetosphere. The magnetospheric (i.e., solar wind) electric field is mapped down to

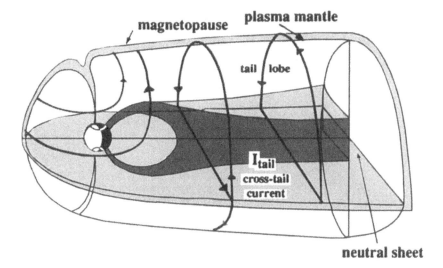

Figure 8.32. Schematic of current flow on the magnetopause and in the plasmasheet region of the terrestrial magnetotail.

the ionosphere in the open model and results in ionospheric plasma convection. This convection is resisted by ion–neutral drag so that the resulting stress balance requires some $\mathbf{J} \times \mathbf{B}$ force, which necessitates electrical current coming from the magnetosphere.

The current system involving the "direct" connection along open magnetic field lines of the solar wind and magnetopause region with the polar cap ionosphere is called the *Region 1 current system.* The current system connecting the plasmasheet to the ionosphere (near the auroral oval) along closed magnetic field lines is called the *Region 2 current system.* The ionospheric currents responsible for closing the electric circuit with the magnetosphere and solar wind are primarily Pederson currents and flow in the E-region, where the Pederson conductivity is largest, although Hall currents make some contribution (see Chapter 7). The greatest ionospheric current densities for both the Region 1 and 2 current systems are found in or near the auroral oval, partly because the electrical conductivity is especially high there due to excess ionization resulting from auroral particle precipitation, but also because the most highly stressed (i.e., large $\mathbf{J} \times \mathbf{B}$) magnetospheric regions connect to the auroral oval. Electrical current systems also exist at lower latitudes in the E-region, but these are associated with neutral wind dynamos rather than with the solar wind.

8.5.2 Region 1 current system

The Region 1 current system can be understood using the simple plasma slab scenario discussed in Section 8.3, with the magnetopause/solar wind being the

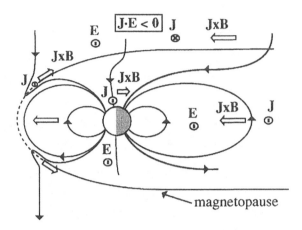

Figure 8.33. Cross-sectional view of the magnetosphere in the noon–midnight meridian showing the current density, the magnetic stress, and the electric field direction.

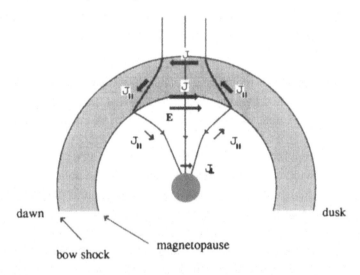

Figure 8.34. Cross-sectional view of the dayside magnetosphere from the Sun showing both perpendicular and field-aligned Region 1 currents.

upper slab and the ionosphere being the lower slab, although our slabs are now quite different in size unlike in the simple scenario. Figure 8.33 shows a cross-sectional view of the magnetosphere in the noon–midnight meridian. Near the dayside magnetopause just downstream of the reconnection region, the magnetic tension force (i.e., $\mathbf{J} \times \mathbf{B}$) is such as to accelerate the plasma, thus converting magnetic energy into kinetic/mechanical energy. The current density in this region must point from dawn to dusk (out of the page). Figure 8.34 shows a view of this region from the front of the magnetosphere. The current in the ionosphere must also be from dawn to dusk so that the $\mathbf{J} \times \mathbf{B}$ force can "drive" the plasma through the neutral atmosphere counteracting ion–neutral friction. Field-aligned currents

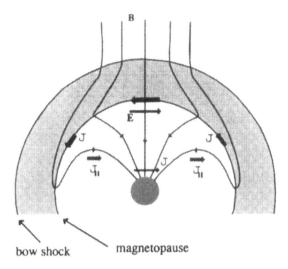

Figure 8.35. Cross section of the near-Earth tail region of the magnetosphere as viewed from the Sun showing both perpendicular and field-aligned Region 1 currents. Current flows along the magnetopause from dusk to dawn (right to left here), opposite the direction of the electric field.

connecting the magnetopause and ionosphere are also pictured schematically in Figure 8.34. The plasmasheet region and Region 2 currents will be discussed later.

Of importance is the extensive magnetopause boundary region located tailward of the polar region (also shown in Figure 8.32). The field line curvature is such that the current density $\mathbf{J}$ is into the page and the $\mathbf{J} \times \mathbf{B}$ force acts to oppose the solar wind flow. Figure 8.35 shows a front view of this region. The current in the ionosphere is in the dawn-to-dusk direction and again is such that $\mathbf{J} \times \mathbf{B}$ is antisunward and counteracts the ion–neutral friction force. The field-aligned current needed to close the circuit is upward on the duskside of the polar cap and downward on the dawnside. This is indeed very much like the slab scenario, except that quantitatively only a few percent of the magnetopause region current is diverted into the ionosphere; the rest circulates on the magnetopause or closes into the neutral sheet (and eventually part of it joins the Region 2 current system). What keeps the solar wind moving in the face of the opposing magnetic force? The answer is dynamic pressure and thermal pressure gradients.

Figure 8.36 is a highly idealized, and somewhat misleading as we will see later, schematic of the Region 1 currents in the polar cap ionosphere. The current flows down into the ionosphere on the dawnside of the polar cap, flows across the polar cap E-region as Pederson currents, and then flows up to the magnetosphere on the duskside of the polar cap. Note that the mapped magnetospheric convection electric field and the Pederson current are in the same direction. The $\mathbf{E} \times \mathbf{B}$ drift is antisunward and the $\mathbf{J} \times \mathbf{B}$ force is also antisunward. The $\mathbf{J} \times \mathbf{B}$ force is such as to maintain the ionospheric plasma convection caused by the mapped down

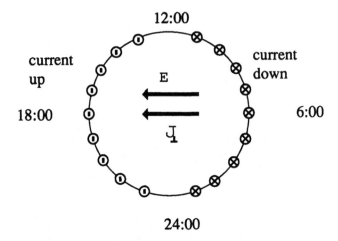

Figure 8.36. Schematic of Region 1 field-aligned currents in the polar cap ionosphere, with noon at the top and midnight at the bottom. This schematic is highly oversimplified.

electric field. Why is this picture somewhat misleading? Because the ionospheric conductivity is so much greater in the auroral oval than in the middle of the polar cap that most of the current closure takes place within the auroral oval region itself rather than crossing the polar cap. Furthermore, the Region 2 system as well as Hall currents were omitted from this picture.

Now let us consider the energetics of the Region 1 system using the slab scenario as a guide. The chapter in *Solar–Terrestrial Physics: Principles and Theoretical Foundation* by T. Hill provides much insight on this issue; I recommend it to the student who wishes to go into this topic in more detail. The flow of electromagnetic energy is described by Equation (A.45) (also see Equation (4.119)); this equation can be easily derived from Maxwell's equations (Problem 8.16). This equation states that the rate of change of the electromagnetic (EM) energy density (i.e., $\partial/\partial t[\varepsilon_0 E^2/2 + B^2/2\mu_0]$) equals the local net production rate, or source, of EM energy (i.e., $-\mathbf{E} \cdot \mathbf{J}$ with units of W/m^3) minus the divergence of the Poynting vector $\mathbf{S} = \mathbf{E} \times \mathbf{H} = (\mathbf{E} \times \mathbf{B})/\mu_0$. The Poynting vector is the directional flux of EM energy and has units W/m^2.

It is apparent from Figures 8.33–8.36 that in the portion of the magnetopause located tailward of the polar cap region, the current density and electric field point in opposite directions. Hence, $-\mathbf{E} \cdot \mathbf{J}$ is positive and this region is a source of EM energy; that is, this region acts as a *MHD generator* in a manner analogous to the upper slab discussed in Section 8.3. The *electromotive force (emf)*, or electrical potential, generated from dawn to dusk across the magnetopause is just Ψ_m (≈ 50–$100\,\text{kV}$) as discussed earlier, and the power generated can be estimated using an equation like Equation (8.67). In this equation we interpret $u_0 B_0 L_x$ as Ψ_m and we identify H as the thickness of the magnetopause current layer, in which most of the current flows. We also have to integrate the power given by Equation (8.67) over the "length" of the magnetopause (call this L_y):

$$power \approx \Psi_m H J_{\perp 0} L_y = \Psi_m I_{mp}, \qquad (8.91)$$

where the total current flowing through a length L_y of the magnetopause is just the integral of $J_{\perp 0}$ over the cross-sectional area of the magnetopause current layer (approximately, $I_{mp} \approx H L_y J_{\perp 0}$). If we are mainly concerned with the near-Earth tail (so that $L_y \approx 1\, r_{mp}$), then $I_{mp} \approx 1\,\text{MA}$ as discussed earlier and the *power* $\approx 10^{11}$ W. We can identify this power as the P_{sw-m} discussed in Section 8.2.3, and because we used a value of Ψ_m for rather active conditions, we can also identify this as approximately being $P_{sw-m,0}$. For a much larger section of the tailward magnetopause, with $L_y \approx 10\, r_{mp}$, the total current is about 10 MA. This is just I_{tail}; thus, the great majority of this current flows through the neutral sheet rather than through the ionosphere. In this case, the MHD generator produces *power* $\approx 10^{12}$ W. This power can be identified as the P_{tail} given by Equation (8.26).

Where is the power generated by the MHD generator dissipated? In the simple slab model all this power is dissipated in the lower slab (the resistor in Figure 8.21). However, most of the magnetopause current closes either on the magnetopause itself or through the neutral sheet rather than through the ionosphere. Using the circuit analogy, at least for a steady-state scenario, the circuit must contain other branches, and the ionospheric branch only includes about 5% of the total current. That is, $I_{iono} \approx 0.05 I_{tail} \approx 1\,\text{MA}$. The rest of the energy must be dissipated in the magnetotail itself.

Note that $-\mathbf{E} \cdot \mathbf{J}$ is negative in *both* the plasmasheet and in the ionosphere (see Figure 8.33) so that electromagnetic energy is *dissipated* in both of these regions, generating heat or mechanical energy (i.e., convection motion). The Poynting vector $\mathbf{S}$ must point from the region of EM energy production to the region of EM energy dissipation, and an examination of $\mathbf{E} \times \mathbf{B}$ does show that EM energy is transported from the magnetopause region toward the tail lobes and neutral sheet. In a steady-state scenario, steady magnetic reconnection in the tail prevents the buildup of magnetic flux in the tail lobes that would result from convection (or equivalently, the Poynting vector) and acts to create mechanical energy in the tail from EM energy. In a non–steady state scenario (a more realistic situation) the magnetic field in the lobes builds up as the MHD generator plus the convection (or EM energy transport) pumps in magnetic flux. The magnetic flux in the tail builds up and is not immediately dissipated.

Why does the ionosphere dissipate only about 5% of the power produced by the solar wind–magnetosphere interaction? To interpret this let us use an equation like Equation (8.71) for the slab. An accurate calculation would take into account Region 2 currents, but for now let us artificially restrict ourselves to the Region 1 currents. The voltage drop across the ionospheric polar cap (Figure 8.36) is $\varepsilon_{MHD} = \Psi_m \approx 60{,}000$ V. The current is just this voltage divided by the "slab" resistance R_{slab}, given by Equation (8.70). For our simple Region 1 model, both L_x and

L_y are about equal to the diameter of the polar cap, and they cancel. The height-integrated Pederson conductivity, $\Sigma_\perp$, can be estimated as the ionospheric scale height multiplied by the peak value of the Pederson conductivity $\sigma_\perp$ shown in Figure 7.13; that is, $\Sigma_\perp \approx \sigma_\perp H \approx (10^{-3}\,\text{S/m})(3 \times 10^4\,\text{m}) \approx 30\,\text{S}$. However, note that the Figure 7.13 conductivities apply to mid-latitude daytime conditions. Typically, in the polar cap, especially during the night, the E-region electron density (and hence the electrical conductivity) is much less than it is in the daytime and at mid-latitudes regions. A value of $\Sigma_\perp \approx 3\,\text{S}$ is more reasonable. But in the auroral oval, at the edge of the polar cap, the conductivity is quite variable and can be as large as 30 S! Using a compromise value of 15 S, we find that the polar cap ionospheric resistance is $R_{\text{iono}} \approx 1/\Sigma_\perp \approx .06\,\text{ohms}$, and the Region 1 ionospheric current is roughly

$$I_{\text{iono}} = \Psi_{\text{m}}/R_{\text{iono}} \approx 6 \times 10^4\,\text{V}/6 \times 10^{-2}\,\text{ohms} \approx 1\,\text{MA}. \tag{8.92}$$

The power dissipated is then $P_{\text{iono}} = I_{\text{iono}}\Psi_{\text{m}} \approx 10^{11}\,\text{W}$.

Now we estimate the Region 1 field-aligned current density at the ionosphere for this simple scenario. The electrical current flows along magnetic field lines from the magnetopause/solar wind region to the portion of the ionosphere located near the edge of the polar cap. This Region 1 current flows into a region that is roughly 100–200 km wide (the inner edge of the auroral oval); dividing the total Region 1 ionospheric current by a cross-sectional area of about $1\,R_{\text{E}}$ by 200 km (area $\approx 10^6\,\text{km}^2$) gives a current density for the *Region 1 Birkeland current* of

$$J_\| \approx 10^6\,\text{A}/10^{12}\,\text{m}^2 \approx 1\,\mu\text{A/m}^2. \tag{8.93}$$

8.5.3 Region 2 current system and simple magnetotail dynamics

The plasma flow is sunwardly directed in the plasmasheet region of the magnetosphere and is driven by a dawn-to-dusk electric field $\mathbf{E}_{\text{m}}$. For the simple steady-state scenario, the electric field "gains access" to the closed field lines by means of magnetic reconnection in the tail neutral sheet. Alternatively, from a circuit point of view, most of the cross-tail current I_{tail} comes from the magnetopause region. The plasma flow in the tail should be consistent with the MHD force balance as represented by the momentum Equation (8.28). The dynamic pressure and friction terms in the momentum balance can be neglected for the relatively slowly moving plasma in the equatorial plane within $20\,R_{\text{E}}$ or so of the Earth (Problem 8.17). Equation (8.28) simplifies to one in which the magnetic stress is balanced by the pressure gradient force:

$$\mathbf{J} \times \mathbf{B} = \nabla p. \tag{8.94}$$

The $\mathbf{J} \times \mathbf{B}$ force in the neutral sheet and plasmasheet is in the same direction as the plasma convection – sunward. Hence, the pressure p must increase with decreasing distance from the Earth.

Note that momentum balance must also exist in the direction perpendicular to the plasmasheet. In the mid and distant magnetotail where the magnetic tension force is not too important, Equation (8.94) (or equivalently, Equation (4.100)) tells us that the thermal pressure in the plasmasheet where the total magnetic field is small (and hence magnetic pressure is small) must balance the magnetic pressure in the tail lobes. That is, we must have p (plasmasheet) $\approx B^2/2\mu_0$ (lobes).

We can represent the pressure variation in the plasmasheet using the simple polytropic energy relation, (4.83), which can be written in terms of the convective derivative as

$$\frac{D}{Dt}\left(\frac{p}{\rho^\gamma}\right) = 0, \tag{8.95}$$

where γ is the polytropic index ($\gamma = 5/3$ for an ideal gas) and ρ is the mass density. For steady-state conditions in the region near the equatorial plane, D/Dt is just $-u_x(\partial/\partial x)$, where the x axis is directed from the Earth tailward. Equation (8.95) then becomes $p/\rho^\gamma = constant$ as plasma convects toward the Earth.

The analysis of Voigt and Wolf (1988) can be used to explain the slow convective plasma flow in the near-Earth plasmasheet. We shall present a highly simplified version of this analysis. Instead of considering an infinitesimal volume, let us work with a whole magnetic flux tube identified with either the x coordinate or with L-shell. The field lines far from Earth are not dipolar, but the flux tubes for closed field lines still attach to the Earth in the polar regions. A flux tube with a cross-sectional area in the equatorial plane, A_{tube}, defined such that the magnetic flux is unity ($BA_{\text{tube}} = 1$ Wb, where a weber, denoted Wb, is the SI unit of magnetic flux), has a volume designated V_{tube} given by

$$V_{\text{tube}} = \int_{\substack{\text{field} \\ \text{line}}} \frac{ds}{B(s)}, \tag{8.96}$$

where s is the distance along the field line and $B(s)$ is the magnetic field strength as a function of s.

Within a factor of two or three the flux tube volume is given by its length ($l_{\text{tube}} \approx 2LR_E$) multiplied by (1 Wb)/$B$, which can be approximated with its equatorial value; that is, $V_{\text{tube}} \approx l_{\text{tube}}/B \approx 2LR_E/B$. As a flux tube in the plasmasheet region convects sunward, B generally increases and L decreases; hence, the volume (V_{tube}) decreases. Now we can rewrite the polytropic energy relation (8.95) by recognizing that the average mass density in the flux tube, $\langle\rho\rangle$, is equal to the total mass in the tube divided by the volume: $\langle\rho\rangle = mass/V_{\text{tube}} \approx CB/(2L)$, where the constant C is the total mass (assumed to be conserved) divided by R_E. Equation (8.95) becomes

$$\frac{d}{dx}(pV_{\text{tube}}^\gamma) = 0 \quad \text{or} \quad pV_{\text{tube}}^\gamma = constant. \tag{8.97}$$

The pressure in the flux tube varies inversely as the γ power of the volume. As the plasma convects toward the Earth, V_{tube} decreases and the pressure increases. Thus, the pressure gradient force is directed away from the Earth, and to balance this, according to the MHD momentum Equation (8.28), we need a $\mathbf{J} \times \mathbf{B}$ force directed toward the Earth.

How does the magnetic field strength vary with L-shell? It depends on magnetic activity level, which partially controls the intensity of both the cross-tail current and the ring current. Within a distance of about $10\ R_{\text{E}}$, the magnetic field is dipole-like, but further down the tail, B varies only slowly with radial distance due to the magnetic field produced by the cross-tail current. Voigt and Wolf (1988) indicate that the field strength in the neutral sheet (that is, B_z) actually has a "local" minimum value for an L-shell located somewhere between 10 and 15. In the following example, we make some very simple assumptions concerning the magnetic field and see what the implications are for the pressure variation.

Example 8.4 (Simple pressure variation in the inner plasmasheet) Let us make the simple assumption that the magnetic field strength in the equatorial plane is constant. Due to the cross-tail current this is not such a bad assumption in the L-shell range of 10–30. V_{tube} is then proportional to L and the pressure varies with L as

$$p \approx const.\ [V_{\text{tube}}]^{-\gamma} = const.\ L^{-\gamma}. \tag{8.98}$$

From Table 8.1 we can estimate that a typical pressure in the plasmasheet (near $L \approx 20$ say) is roughly 0.2 nPa. Equation (8.98) then suggests that very roughly $p \approx 0.2\,\text{nPa}\,(20/L)^{5/3}$ for $\gamma = 5/3$. $\gamma = 5/3$ for an ideal gas and $\gamma = 2$ if the perpendicular (gyromotion) pressure is dominant.

However, if we make the very different assumption that the field is dipolar (usually not a bad assumption inside $L \approx 10$ or so), then $B \approx const./L^3$ for a dipole field, so that $V_{\text{tube}} \approx const.\ L^4$, and the pressure varies with L as

$$p \approx const.\ [V_{\text{tube}}]^{-\gamma} = const.\ L^{-4\gamma}. \tag{8.99}$$

This equation implies that $p \approx 1\,\text{nPa}\,(10/L)^{20/3}$ for $L < 10$ or so, using for the normalization a pressure appropriate for the inner edge of the plasmasheet from Lui and Hamilton (1992). Actually, this predicted pressure variation is too extreme, except perhaps very close to Earth in the ring current region, where other processes are important in any case.

A similar result (i.e., increase of pressure with decreasing L) can also be obtained by simply using the conservation of the first adiabatic invariant for single particle motion (Chapter 3). The average perpendicular pressure for the flux tube can be written in terms of an equation of state: $p_\perp = \langle \rho \rangle \tilde{R} \langle T \rangle$, where $\langle T \rangle$ is an average temperature for the perpendicular motion. $\langle \rho \rangle$ varies as $[V_{\text{tube}}]^{-1}$ and $\langle T \rangle$ is proportional to the average perpendicular kinetic energy of the particles, $\langle mv_\perp^2/2 \rangle$. However, as a particle convects inward and B changes, then the conservation of the adiabatic invariant indicates that $mv_\perp^2/(2B) = constant$. Hence $\langle mv_\perp^2/2 \rangle$, and therefore $\langle T \rangle$, is proportional to B and increases as B increases. This is just betatron acceleration as discussed in Chapter 3. If the plasma is able to convect to small enough L-shells, it experiences a large increase in B and the resulting betatron acceleration can create high-energy ring current particles. Because $\langle T \rangle$ is proportional to B (and therefore to $1/A_{tube}$), p is proportional to $1/(A_{tube}V_{tube})$, or proportional to $1/L^7$ for a dipole field.

Let us now reconsider the dynamics. The momentum equation, (8.94), can be rewritten to give a *diamagnetic current density*,

$$\mathbf{J} = \frac{-\nabla p \times \mathbf{B}}{B^2}. \tag{8.100}$$

Given that p increases with decreasing L value as discussed in Example 8.4, this equation indeed predicts that a current across the tail must be present and must flow in the dawn-to-dusk direction. However, a somewhat more complicated expression, which can be derived from single particle motion (i.e., recall Equation (3.88)), applies if the pressure is not isotropic:

$$\mathbf{J} = \frac{\hat{\mathbf{b}} \times \nabla p_\perp}{B} + (p_\parallel - p_\perp)\frac{\hat{\mathbf{b}} \times \nabla B}{B^2}, \tag{8.101}$$

where $p_\parallel$ and $p_\perp$ are the parallel and perpendicular pressures discussed in Chapter 3. This current density was discussed with respect to the ring current in Chapter 3. Note that for isotropic pressure, only the first term contributes to the current density and Equations (8.100) and (8.101) are identical.

We can use the simple pressure variations derived in Example 8.4 to explore how the current density varies with distance from Earth. The current density expression can be simplified to the following expression if the pressure is assumed to be isotropic (Lui and Hamilton, 1992):

$$J_\perp \approx -\frac{1}{BR_E}\frac{\partial p_\perp}{\partial L}. \tag{8.102}$$

Taking $\gamma = 5/3$, we find that for the constant magnetic field case of Example 8.4, the current density is proportional to $1/L^{8/3}$. That is, in going from $L = 20$ to $L = 10$ we expect the current density to increase by about a factor of 6. In contrast, if we assume that the magnetic field is dipolar, then the current density is proportional to $L^3 \partial p_\perp/\partial L \approx const./L^{(4\gamma-2)} = const./L^{14/3}$. Thus, in going from $L = 10$ to $L = 5$, we expect that the current density increases by about a factor of 30, if this simple model is to be believed. However, the actual pressure variation is much less than this, at least during quiet times.

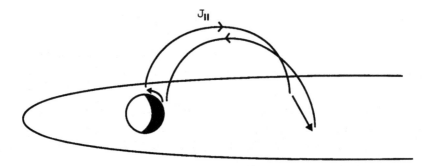

Figure 8.37. Schematic Region 2 current system. Shown are the cross-tail current and the Region 2 field-aligned currents closing in the ionosphere. Most of this cross-tail current actually closes on the magnetopause (not shown here – see Figure 8.32).

Using the simple pressure variation just discussed in Example 8.4 for the constant magnetic field case, the current density Equation (8.102) indicates that $J_\perp \approx$ .8 nA/m^3 near $L = 20$. The central plasmasheet is a couple of R_E thick, so that the total cross-tail current per unit length is $K_\perp \approx$ .01 A/m, giving a total current of roughly $I \approx r_{mp}K_\perp \approx$ .5 MA over a 10 R_E piece of the tail near $L \approx 20$.

The dynamics we have just been discussing necessitates an electrical current in the plasmasheet that is directed from dawn to dusk. As illustrated schematically in Figure 8.32, most of the cross-tail current closes via the magnetopause. However, some fraction of the current is diverted to the ionosphere via field-aligned currents, as shown in Figure 8.37. These Birkeland currents are, during relative quiet times, downward on the duskside of the polar cap and upward on the dawnside. This current system is known as the *Region 2 current system*. Near the inner edge of the plasmasheet ($L \approx 10$) the cross-tail current has more of a tendency to close in the ionosphere, and the current also begins to curve around the Earth due to the gradient and curvature drift of the plasma – this is known as the *partial ring current*. This results in Region 2 Birkeland currents existing around on the dayside. Even closer to Earth ($L < 6$), the current makes a complete circuit around the Earth – this is the *ring current*, discussed in Chapter 3.

In the polar ionosphere the Region 2 Birkeland currents are located equatorward of the Region 1 currents. Figure 8.38 is the well-known (in space physics) field-aligned current pattern from Iijima and Potemra (1978). This electrical current pattern was derived from an extensive set of magnetic field measurements. Notice that both the Region 1 and 2 currents are concentrated near the edge of the polar cap (i.e., basically coincident with the auroral oval).

What do spacecraft observations tell us about the plasma pressure and electrical currents in the inner magnetosphere and in the plasmasheet? Figures 8.39 and 8.40 show plasmasheet pressures measured by spacecraft. The pressure is approximately equal to 0.1 nPa in the $L \approx 20$–40 region and varies roughly as $p \sim L^{-1}$, whereas

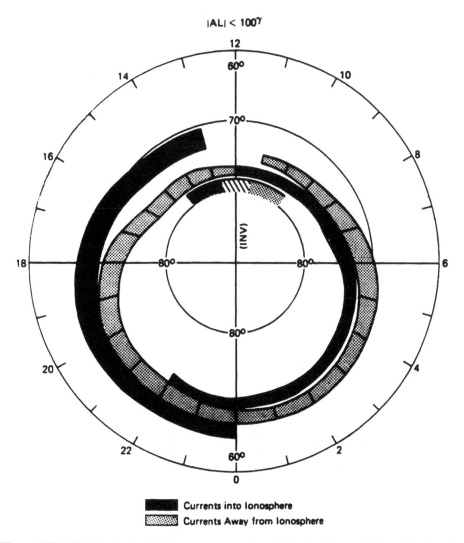

Figure 8.38. Birkeland electrical current pattern for relatively quiet conditions. The Region 1 currents are located on the polarward edge and the Region 2 currents are on the equatorward edge of the polar cap. (From Iijima and Potemra, 1978.)

Equation (8.98) (see Example 8.4) suggested that $p \approx 0.2 \,\text{nPa} \, (20/L)^{5/3}$ for $\gamma = 5/3$ if the neutral sheet magnetic field were independent of distance. Near $L \approx 10$ the data suggest a pressure variation more like L^{-2}, which is much less extreme than the L-dependence expected from the simple adiabatic pressure relation discussed in the example.

Lui and Hamilton (1992) used measured plasma data from the *AMPTE CCE* spacecraft to obtain plasma pressures and current densities for L-values between about 3 and 9 on both the day and nightsides. These data are from magnetically quiet time periods and are relevant to the quiet-time ring current, the partial ring current,

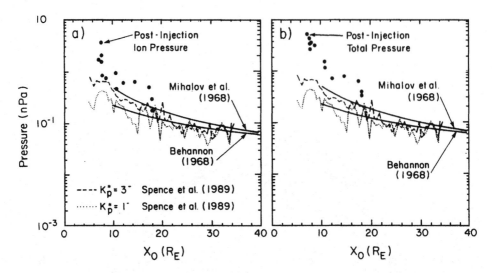

Figure 8.39. Plasma pressure in the plasmasheet plotted versus distance from the Earth, derived from plasma parameters measured by instruments aboard several spacecraft. (a) Ion pressure. (b) Total pressure. (From Kistler et al. (1992); details about this figure can be found in this paper.) Notice that the pressure gradient is larger closer to the Earth and that the pressure inside of about 15 R_E is greater for magnetically active times (i.e., the $K_p = 3^-$ (dashed) curve and also the "post-injection ion pressure" points) than it is during quiet times (i.e., $K_p = 1^-$, the dotted curve). (From Kistler et al., 1992)

and the inner part of the plasmasheet. The *inner edge of the plasmasheet* is roughly located in the tail at $L \approx 7$–10. Depending on the magnetic activity level (i.e., on the electric field $\mathbf{E}_m$), the plasma outside the inner edge of the plasmasheet is able to convect sunward, but inside this boundary the plasma is energetic enough that gradient and curvature drift motions (as discussed in Chapter 3) carry the plasma around the Earth rather than sunward and the particles do not readily move closer to the Earth.

The average proton density for $L \approx 6$–9 is about $1\,\mathrm{cm}^{-3}$, the average proton energy increases from about 7 keV at $L = 9$ to 20 keV at $L = 6$, and the plasma beta varies roughly as $\beta \approx (L/7)^4$ (Lui and Hamilton, 1992). Near $L \approx 7$, the drop in β below unity indicates departure from the plasmasheet. Figure 8.40, from Lui and Hamilton, shows plasma pressures for midnight local time. The perpendicular proton pressure (the parallel pressure is about 50–90% of this) was shown to vary as $L^{-3.2}$, which is a much slower variation than the $L^{-20/3}$ variation expected from the simple theory for a dipole field. The measured pressure at $L \approx 9$ is about 1 nPa; the current density associated with the measured pressure gradient was shown by Lui and Hamilton to be about $J \approx 1$–$2\,\mathrm{nA/m^2}$ in the $L \approx 4$–9 region (including both ring current and partial ring currents). What explains the discrepancy between the measured pressure variation and the variation deduced from the purely adiabatic theory for a shrinking flux tube?

The average magnetic moment (average perpendicular energy divided by mag-

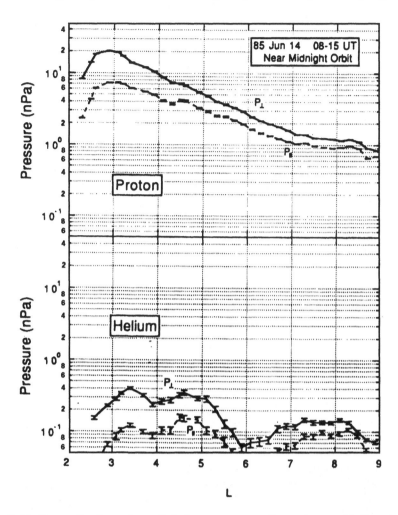

Figure 8.40. Perpendicular and parallel plasma pressures for protons and helium ions near the inner edge of the plasmasheet and in the ring current region of the magnetosphere. (From Lui and Hamilton, 1992.)

netic field strength) was shown by Lui and Hamilton to decrease by about a factor of two from $L = 9$ to $L = 7.5$, but it does remain roughly constant from $L = 7.5$ to $L = 6$. That is, the magnetic moment is *not* exactly conserved. Actually, one does not expect the first adiabatic invariant to be exactly conserved, at least in the neutral sheet itself where the field strength is fairly small and the proton gyroradii are fairly large. For example, the gyroradius of a 7 keV proton in a 5 nT field is about 1,000 km, which is a fair fraction of $1 R_E$, the rough width of the neutral sheet. Furthermore, the plasma is not able to convect Earthward inside of the inner plasmasheet edge, but instead the particles also start to gradient and curvature drift around the Earth. Hence, the simple interpretation of the convective derivative in Equations (8.95) and (8.97) is somewhat misleading.

The partial *shielding of the magnetospheric electric field* mentioned in Section 8.4

can be qualitatively explained in terms of the partial ring current and particle drifts. With decreasing L-value, the plasma no longer merely $\mathbf{E} \times \mathbf{B}$ drifts sunward but starts to be diverted via gradient and curvature drifts, as in the ring current. The protons and other positive particles drift westward and the electrons eastward. Within a few R_E of the Earth these drifts completely encircle the Earth, but near the inner edge of the plasmasheet the drifting particles "run into" the boundary layer of the magnetopause and electrical charge builds up in spite of current diversion to both the magnetopause and ionosphere (Region 2 currents). In fact, the Region 2 Birkeland currents reduce the effectiveness of this shielding. The electrical charge that results from the partial ring current is positive on the duskside and negative on the dawnside of the magnetosphere, resulting in a dusk-to-dawn polarization electric field that partially counteracts the dawn-to-dusk magnetospheric field $\mathbf{E}_m$. However, if $\mathbf{E}_m$ is greatly and rapidly increased, as during a magnetospheric substorm, then the $\mathbf{E} \times \mathbf{B}$ drift is more important than the gradient and curvature drifts even at smaller L-shells. This results in less shielding of $\mathbf{E}_m$ and in the energization of the ring current due to the betatron acceleration of plasmasheet particles that get "*injected*" into the inner magnetosphere.

8.5.4 Ionospheric currents

The electrical currents diverted from the magnetosphere to the polar ionosphere via Birkeland currents close via perpendicular currents in the E-region. However, the current closure is more complicated than the simple picture given by Figure 8.36. Instead of looking at the ionospheric currents as just the beginning of the "resistive" portion of the magnetospheric current systems, we could determine what field-aligned currents *to* the magnetosphere/solar wind are necessitated by the ionospheric current flow. Volland (1984) takes this approach, and a short simple summary is provided here.

Equations (8.34) and (8.36) are the starting point for this ionospheric approach to determining the Birkeland currents. The field-aligned current density out of the top of the ionosphere ($J_{\parallel \text{top}}$) is determined from the divergence of the height-integrated horizontal current density ($\mathbf{K}_{\text{hor}}$) in the ionosphere (mainly in the E-region). The current density $\mathbf{K}_{\text{hor}}$ can be determined using the Ohm's law, Equation (8.36), plus a knowledge of the ionospheric electric field and the height-integrated Pederson and Hall conductivities ($\Sigma_\perp$ and Σ_H, respectively). As discussed earlier, the Pederson conductivity is roughly $\Sigma_\perp \approx \sigma_\perp H \approx 15\,\text{S}$, and the Hall conductivity is comparable, $\Sigma_H \approx 15\,\text{S}$. The ionospheric electric field ($\mathbf{E}'$ or just $\mathbf{E}$ if we neglect the neutral wind) in the polar cap region is just the magnetospheric electric field mapped down into the polar ionosphere, which can be found by evaluating the gradient of the electric potential. The electric potential pattern during moderately active time periods looks like that shown in Figure 8.23, and, as we saw earlier, the cross-polar cap potential Ψ_m is just the cross-tail magnetospheric potential arising from the solar wind interaction.

If the electrical conductivities were spatially uniform, then the divergence of the Hall current would be zero and only the Pederson current would contribute to the Birkeland current. In this case, we could relatively simply estimate $J_{\parallel \text{top}}$ using Equations (8.34) and (8.36) in the manner presented by Volland. One obtains a Birkeland current pattern very much like the observed pattern, although it is somewhat more spread out in latitude. The peak values of $J_{\parallel \text{top}}$ obtained with this approach are about $1 \ \mu \text{A/m}^2$ near the edge of the polar cap.

However, in reality, the electrical conductivity in the polar cap is nonuniform and results in both Hall and Pederson contributions to $J_{\parallel \text{top}}$. In fact, the electrical conductivity is generally considerably larger within the auroral oval than outside it. Recall from Chapter 7 that the ionospheric conductivities are proportional to the ionospheric electron density n_e, and n_e depends on the balance between ion production (i.e., the ionization rate) and ion loss. Energetic auroral electrons deposit most of their energy in the E-region and most of this energy goes into ionization of neutral atoms and molecules. Hence, as the level of magnetic activity goes up, the flux of auroral electrons into the auroral oval region increases, resulting in an increase in the ionization rate and hence in the E-region electron density. The dependence of the high latitude Pederson conductivity on the auroral energy flux (or power density) was discussed more quantitatively by Lyons and Williams (1984). Equation (4.18) in Lyons and Williams is

$$\Sigma_\perp = 0.5 + 160[\Phi_{\text{Energy}}(\text{W/m}^2)]^{1/2} \ \text{S}, \qquad (8.103)$$

where Φ_{Energy} is the power density associated with auroral particle precipitation. The power density during active times is as high as $\Phi_{\text{Energy}} \approx 10^{-2} \ \text{W/m}^2$, which results in a height-integrated conductivity in the auroral oval of $\Sigma_\perp \approx 17 \ \text{S}$. The conductivity is much smaller than this outside the auroral oval and/or during magnetically quiet time periods. The Hall conductivity is somewhat greater than the Pederson conductivity during intense auroral activity.

Both $\Sigma_\perp$ and Σ_H are spatially nonuniform. This means that the Hall currents make some contribution to the current closure in the Region 1 and 2 current systems. The Hall currents tend to flow *along* the oval. In fact, it can be shown (Problem 8.19) that the Hall current $\mathbf{K}_H$ flows along equipotential lines (Figure 8.23) in a direction opposite to the plasma convection, but these Hall currents are much more intense within the auroral oval than outside it due to the larger values of Σ_H. Figure 8.41 illustrates the polar Hall currents schematically for quiet times (Figure 8.41a) and during a substorm (Figure 8.41b). Noon is at the top of this figure. These currents are called the *auroral electrojet*, and they are the electrical currents that produce most of the magnetic perturbations observed by ground-based magnetometers at high latitudes. The Pederson currents, although comparable to the Hall currents and very important dynamically, tend to have canceling magnetic perturbations at the Earth's surface. The Pederson currents tend to close within the oval, taking a "shorter" route than the Hall current, which flows along the

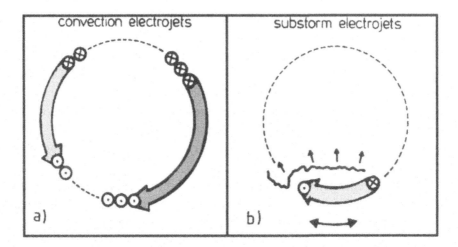

Figure 8.41. Schematic of the Hall current in the polar electrojet and the field-aligned currents during (a) relatively quiet times and (b) transient substorm events. (From Baumjohann (1982); also shown in Volland (1984).) The xs and dots denote field-aligned currents into and out of the ionosphere, respectively. (From Baumjohann, 1982.) Reprinted from *Advances in Space Research*, vol. 2, no. 10, W. Baumjohann, p. 55, copyright 1982, with kind permission from Elsevier Science Ltd., The Boulevard, Langford Lane, Kidlington OX5 1GB, UK.

oval. (See the discussion in Volland's book if you are interested in this topic.) The auroral electrojet associated with substorms is transient and is also much more intense than the steady-state "convection electrojets." A typical total current for the steady-state convection electrojet is roughly 100 kA. For comparison, the total ionospheric current (including Pederson currents) is several times greater than this. The substorm field-aligned currents pictured in Figure 8.41b are associated with intense electron precipitation and will be discussed more in the next section.

8.6 Substorms and aurorae

A simple steady-state picture of the magnetosphere current system was presented in Section 8.5; however, the magnetospheric plasma convection and currents are really time dependent. In this section we discuss this time dependence and, in particular, we discuss magnetospheric substorms. More specifically, the return flow convection, discussed in Section 8.4, and the associated Region 2 currents, discussed in Section 8.5, are not steady state but are observed to occur sporadically with a characteristic time separation between events of about three hours. These events of enhanced convection are called magnetospheric substorms. Substorm phenomena occur in both the ionosphere and the magnetosphere. A number of good references for this topic can be found in the AGU monograph *Magnetospheric Substorms* (1991) or in Elphinstone et al. (1996).

8.6.1 Substorm morphology

Observed substorm phenomena have been organized chronologically into three principal stages or phases: (1) the *growth phase*, (2) the *expansion phase*, and (3) the *recovery phase*. The growth and expansion phase each typically take about 30 minutes, and the recovery phase occupies about 2 hours. The article by Lui in the *Magnetospheric Substorms* (1991) volume provides a good summary of the phenomena that occur during each of these phases, which I now summarize.

8.6.1.1 Growth phase

The magnetic flux contained in the magnetotail markedly increases during this phase. We can see why this might happen by examining Figures 8.4 or 8.33. The plasma convection associated with the solar wind electric field mapped down into the open magnetosphere naturally moves magnetic flux into the tail lobes. In a steady-state scenario magnetic reconnection in the neutral sheet prevents the buildup of magnetic flux, but if for some reason this reconnection does not take place over some time period, or is not efficient enough, then the magnetic flux in the magnetotail naturally increases. The rate of increase of this tail flux is naturally greatest during time periods when the solar wind dynamic pressure is high and when the IMF is southward, because in the latter case the dayside reconnection process is more efficient and the magnetosphere is more "open." An increase in the cross-sectional area of the tail and a thinning of the plasmasheet are also associated with this phase. The magnetic field lines in the tail become more stretched and taillike (and less dipolelike) during this phase. The diameter of the polar cap, as delineated by the auroral oval and the field-aligned currents, becomes larger during the growth phase. The auroral electrojet is enhanced.

8.6.1.2 Expansion phase

The magnetic flux previously built up during the growth phase starts to "relax" during the expansion stage of the substorm. During this phase, the tail cross section in the near-Earth tail, within 15 R_E or so, decreases, and the magnetic field becomes more dipolar; however, in the tail beyond 15 R_E the plasmasheet continues to thin out. Increased plasma convection is present, implying an increase of the cross-tail electric field $\mathbf{E}_m$, at least in the near-Earth magnetotail. Plasma is "injected" into the inner magnetosphere during this phase, resulting in the energization of the ring current. And due to the increased magnetospheric field $\mathbf{E}_m$ and the associated plasma convection, the plasmapause moves closer to the Earth.

In the polar ionosphere the *onset of the expansion phase* of a substorm is marked by the sudden brightening of auroral arcs on the nightside. The Region 2 currents intensify as shown in Figure 8.41b. Much of the enhanced field-aligned electrical current in the midnight local time region is carried by energetic electrons precipitating into the ionosphere. This region of enhanced auroral activity and luminosity is called the *auroral bulge* (or *surge*), and during the expansion phase this portion of the auroral oval "expands" both poleward and equatorward, as well as westward. The auroral bulge is part of what is called the *substorm current wedge*. An enhanced local auroral electrojet is also part of this substorm current wedge. The Region 2 currents flowing in this current wedge system are thought to connect to the near-Earth portion of the plasmasheet. In the polar cap and parts of the auroral oval, enhanced outflows of ionospheric ions into the magnetosphere take place during this phase of a substorm.

8.6.1.3 Recovery phase

Although none of the substorm phases are well understood, the recovery phase is particularly poorly understood. The near-Earth magnetosphere and the ionosphere appear to "recover" their presubstorm configurations during this phase. That is, the polar cap/auroral oval contracts and the auroral luminosity weakens. In the mid-tail between about 15 and 80 R_E, the plasmasheet gradually thickens during this phase. In the far tail (beyond 80 R_E), a perturbation in the z component of the magnetic field is often observed, with first a northward B_z perturbation being observed and then a southward perturbation. This signature plus some measured plasma perturbations have been used to deduce the existence of plasmoids in the distant tail, which we will discuss later. These plasmoids appear to propagate away from the Earth.

8.6.2 Substorm models

Magnetic flux must be moved through the magnetosphere in the open model through the dayside magnetic reconnection process. In practice this does not happen in a steady-state manner; instead, the reconnection process in the magnetotail operates only sporadically so that magnetic flux builds up in the tail and is released at roughly 2–3 hour intervals. This buildup and release process constitutes a substorm and has certain natural consequences for the large-scale electric field as well as for the convection and current patterns. What "triggers" the substorm and allows the "release" (or "unloading") of the built-up magnetic flux is not currently known and is a quite controversial topic in the space physics community. Here, I merely list some of the suggested substorm mechanisms, and in Section 8.6.3 we will go over a simpler picture that conveniently does not dwell on "details" such as the triggering mechanism.

The article by Lui (1991) lists a number of the substorm mechanisms that have been proposed; Lui also presented a "synthesis model" of his own. The following substorm models are currently popular:

(1) **The near-earth neutral line (NENL) model** In this model (or in one of its many variants – this is a particularly popular model) a near-Earth (10 to 20 R_E downstream) neutral line (or x-line) forms during substorm expansion phase onset. The plasmasheet

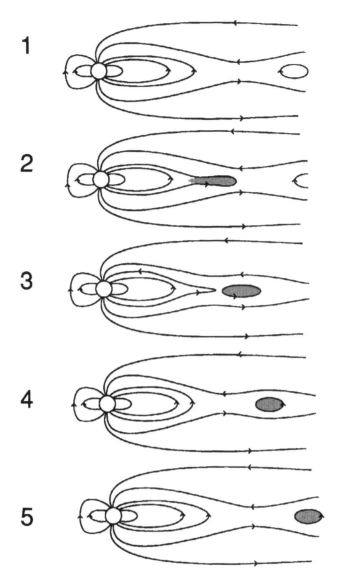

Figure 8.42. Substorm evolution in the near-Earth neutral line model. A plasmoid forms and flows tailward in this model. Also notice the thinning of the plasmasheet that takes place in the course of the substorm. Loosely adapted from Hones (1984).

downstream of this neutral line splits off and a detached chunk of plasma and magnetic field, called a plasmoid, flows tailward. Figure 8.42 depicts with cartoons how a plasmoid evolves.

(2) **Boundary layer model** Here reconnection in the distant magnetotail generates plasma flow Earthward in the low latitude boundary layer, which then triggers the Kelvin–Helmholtz instability in the near-Earth tail, which in turn results in the disruption of the cross-tail current, thus initiating the expansion phase.

(3) **Thermal catastrophe model** In this model turbulent Alfvén waves are generated in the plasmasheet boundary layer during the growth phase. When the wave power reaches

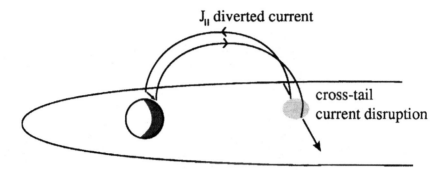

Figure 8.43. Schematic of substorm current disruption and diversion in the magnetotail. This electrical circuit is called the substorm current wedge.

some critical stage the waves are absorbed and heat the plasmasheet, initiating the expansion phase.

(4) **Magnetosphere–ionosphere (M–I) coupling model** In the M–I model intense Birkeland currents originate in the ionosphere from the divergence of both Hall and Pederson currents, and these set up the substorm current wedge in the near-Earth magnetotail.

(5) **Current disruption model** The observational evidence for a near-Earth neutral line is weak, yet current disruption is clearly present during the substorm; this model postulates current disruption without a NENL and then suggests that other observed substorm phenomena follow from this. For example, the plasmasheet thinning is due to a rarefaction wave generated by the sudden reduction in the near-Earth cross-tail current. Figure 8.43 shows a schematic of the disruption of the cross-tail (Region 2) electrical current and its diversion into the ionosphere via Birkeland currents. Note that this field-aligned current pattern is consistent with the substorm ionospheric current pattern (and auroral bulge on the nightside oval) depicted in Figure 8.41.

As discussed by Lui, some combination of the above models might actually be able to account for the observed evolution of a substorm in both the magnetosphere and ionosphere, but even this is controversial. In any case, any model that eventually appears will need to include something very much like the disruption of the Region 2 currents in the near-Earth current sheet and the associated auroral bulge in the ionosphere. This model will also have to account for the removal of magnetic flux and plasma in the tail with something that functions like a plasmoid, even if the details differ from those in the currently popular models.

8.6.3 Simple Faraday loop model

Let us not worry about what triggers a substorm but just assume that it somehow happens and consider the consequences for plasma convection of the buildup and release of magnetic flux in the tail. We will accomplish this by using a simple *Faraday loop model* (see the Pudovkin paper in *Magnetospheric Substorms*).

In the steady-state scenario, an electric field $\mathbf{E}_m$ is mapped into the magnetosphere

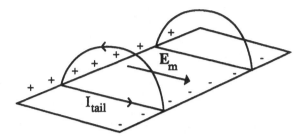

North Tail Lobe and Neutral Sheet

Figure 8.44. Schematic of the magnetotail showing the magnetospheric (cross-tail) convection electric field and the associated charge distribution. The plus side is the dawnside. A typical electrical current path on the magnetopause and in the neutral sheet is shown. The tail lobe magnetic field (not shown) is directed towards the planet (towards the left in the diagram) which is consistent with solenoid-like electrical current pattern shown.

from the solar wind and can be associated with charge separation, as pictured in Figure 8.44 for the magnetotail. As discussed earlier, power is generated near the magnetopause where $-\mathbf{E} \cdot \mathbf{J}$ is positive (current and electric field are oppositely directed), and power is dissipated in the plasmasheet where $-\mathbf{E} \cdot \mathbf{J}$ is negative (i.e., the current I_{tail} and $\mathbf{E}_m$ are in the same direction).

During the growth phase of a substorm, the magnetic flux in the tail lobes increases both because the cross-sectional area of the tail increases and because the field in the tail, B_{tail}, increases. The total magnetic field increase in the tail is limited, however, by the pressure balance perpendicular to the tail axis. That is, the thermal pressure in the plasmasheet must equal the magnetic pressure in the tail lobes, which must in turn equal the thermal plus magnetic pressure in the magnetosheath and solar wind. The magnetic flux increase in the tail occurs during the substorm growth phase because magnetic reconnection on the dayside magnetopause and the subsequent tailward plasma convection over the poles moves magnetic flux into the tail lobes, which is not removed fast enough for some reason, perhaps because magnetic reconnection in the tail is not effective. However, during the substorm expansion phase, some unknown mechanism disrupts the current (and/or allows magnetic reconnection to take place in the neutral sheets, and this results in removal of magnetic flux faster than it can be replaced in the lobes.

We know from *Faraday's law* that an *emf* (or *induced electric field*) is associated with a temporal change of magnetic flux. The integral form of this law states that for a closed path C, the "induced" emf around the path must equal the time rate of change of the magnetic flux through the area S, delineated by this path:

$$emf = -d\Phi_m/dt, \qquad (8.104)$$

where Φ_m is the magnetic flux through the area. The *emf* is just the path integral of the induced electric field $\mathbf{E}_{ind}$ around the path C:

$$emf = \oint_C \mathbf{E} \cdot d\mathbf{l} = \oint_C \mathbf{E}_{ind} \cdot d\mathbf{l}. \tag{8.105}$$

According to *Lenz's law*, the direction of the emf (and $\mathbf{E}_{ind}$) is such that the rate of increase of the magnetic flux is opposed by the magnetic field that would be created by a current in the same direction as the induced electric field. The total electric field is then the sum of the conservative field due to charge separation ($\mathbf{E}_m$) and the induced field:

$$\mathbf{E}_{total} = \mathbf{E}_m + \mathbf{E}_{ind}. \tag{8.106}$$

Figures 8.44 and 8.45 illustrate the application of Faraday's law to a closed path surrounding the north magnetotail lobe. The path runs along the magnetopause and through the neutral sheet. The magnetic flux is just the cross-sectional area of the tail, A_{tail}, multiplied by the tail lobe field B_{tail}; that is, $\Phi_m = A_{tail} B_{tail}$. During the growth phase of a magnetic substorm, Φ_m increases and an induced electric field is generated as shown in Figure 8.45. The induced electric field acts to reduce the total electric field in the equatorial region of the magnetosphere and to enhance the electric field near the magnetopause. This has several consequences. First, the sunward magnetospheric convection in the plasmasheet is weaker since $u_\perp = E_{total}/B$. Related to this reduced level of plasma convection is a reduction in the electromagnetic power dissipation in the neutral sheet ($-\mathbf{E}_{total} \cdot \mathbf{J}$) relative to a steady-state scenario. But the power generated near the magnetopause is actually larger. Because the electromagnetic energy generated exceeds that which is lost,

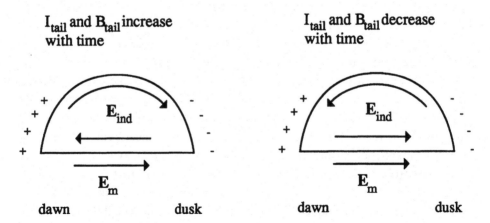

Figure 8.45. Faraday loop model of substorms. A cross-sectional view from the Sun of the north part of the magnetotail is shown. The standard convection electric field, $\mathbf{E}_m$, and an induced electric field, $\mathbf{E}_{ind}$, are shown for two cases: (left panel) increasing tail magnetic flux (growth substorm phase); (right panel) decreasing tail magnetic flux (expansion substorm phase).

there is an increase, or growth, in the amount of electromagnetic energy stored in the magnetotail.

In contrast, during the expansion phase of a substorm, the magnetic flux built up during the growth phase is "released" and Φ_m decreases. The right-hand panel of Figure 8.45 then illustrates that the induced electric field in this case is in the same direction as $\mathbf{E}_m$ in the plasmasheet and the total electric field increases. As a consequence, the sunward plasma convection in the equatorial plane is enhanced during the expansion stage of a magnetospheric substorm. Other phenomena associated with this phase of a magnetic substorm, such as enhanced magnetospheric convection, follow from the increased $\mathbf{E}_{total}$. For example, the increased convection is responsible for plasma "injection" into the inner magnetosphere and energization of ring current plasma. This increased convection also produces a smaller plasmasphere, as discussed earlier in this chapter. Larger Region 2 field-aligned current strengths are also present during this substorm phase.

Let us apply the Faraday loop model a little more quantitatively now and consider a reasonable magnitude for the induced electric field E_{ind} (or, equivalently, for the potential Ψ_{ind}). The cross-tail potential associated with the steady-state field is just $\Psi_m = E_m L_x$ ($\approx 75\,\text{kV}$ for moderately active conditions), where we earlier defined the width of the tail as $L_x \approx 3\,r_{mp}$. Similarly, the portion of the induced *emf* across the plasmasheet should roughly be half of the total *emf* around path C; that is, $\Psi_{ind} = E_{ind}L_x \approx 0.5\,emf$. The total cross-tail potential drop across the neutral sheet is $\Psi_{total} = \Psi_m + \Psi_{ind}$.

Now we estimate the induced emf in a couple of ways. A typical magnetic flux in the tail is $\Phi_m = A_{tail}B_{tail} \approx 5 \times 10^8$ Wb using (for a half-circle) $A_{tail} \approx (\pi/2)(1.5\,r_{mp})^2 \approx 300\,R_E^2 \approx 10^{16}\,\text{m}^2$ and using $B_{tail} \approx 20\,\text{nT}$. A typical substorm growth stage lasts for a couple of thousand seconds (30 minutes), and assuming that about half the typical tail lobe magnetic flux is added during this time period we find that

$$emf \approx \frac{\Delta \Phi_m}{\Delta t} \approx \frac{\frac{1}{2}\Phi_m}{\Delta t} \approx \frac{2.5 \times 10^8\,\text{Wb}}{2000\,\text{s}} = 120\,\text{kV}. \qquad (8.107)$$

The portion of this *emf* across the plasmasheet is roughly half this value – that is, 50 kV or so.

Alternatively, we can treat the tail as an inductor with inductance L_{induct}, in which case we can write the magnetic flux as $\Phi_m = L_{induct}I_{tail}$. Approximating the tail lobe as a solenoid, this inductance is approximately

$$L_{induct} \approx \mu_0 A_{tail}/L_y \qquad (8.108)$$

for a length L_y of the tail. Estimating the length of the tail that is Earthward of any x-line in the plasmasheet as $L_y \approx 10\,r_{mp} \approx 60\,R_E$, we find that the inductance is $L_{induct} \approx 20\,\text{H}$. The *emf* is then given by

$$emf = -L_{induct}(dI_{tail}/dt). \qquad (8.109)$$

We earlier estimated that total cross-tail current for this length of tail is roughly 10 MA. Assuming the change in the current is about 50% and that the time scale is again $\Delta t \approx 2{,}000$ s, we find from Equation (8.109) that the *emf* is $\approx (20\,\mathrm{H})(5\,\mathrm{MA}/2000\,\mathrm{s}) \approx 50\,\mathrm{kV}$, which is roughly the same as, within a factor of two, the earlier estimate. During the growth phase, the electrical current increases but this is opposed by the *emf* of the inductor. The magnetic field in the inductor increases during the growth phase. During the expansion phase, the inductor supplies *emf* to the magnetosphere, which is linked with the ionosphere via the Region 2 current system. The ionosphere acts as a load.

Let us simply assume $\Psi_{\mathrm{ind}} \approx 50\,\mathrm{kV}$, in which case we estimate the total potential difference across the tail of

$$\Psi_{\mathrm{total}} = \begin{cases} \Psi_{\mathrm{m}} + \Psi_{\mathrm{ind}} \approx 75\,\mathrm{kV} - 50\,\mathrm{kV} = 25\,\mathrm{kV} & \text{(for the growth phase)} \\ \Psi_{\mathrm{m}} + \Psi_{\mathrm{ind}} \approx 75\,\mathrm{kV} + 50\,\mathrm{kV} = 125\,\mathrm{kV} & \text{(for the expansion phase).} \end{cases}$$
$$(8.110)$$

Clearly, the total potential, and thus the total electric field, in the plasmasheet region is much larger during the expansion phase than it is during the growth phase.

8.6.4 *The aurora*

The *aurora borealis*, also known as the *northern lights*, are upper atmospheric displays of light observed in the northern polar region. The aurora appears as moving sheets and filaments of light of many different colors, although green and red displays are especially common. Aurorae are more intense and are observed at lower latitudes during magnetically active time periods (i.e., during magnetic substorms) than during magnetically quiet time periods. In the southern hemisphere the *aurora australis* is observed. Aurorae have also been observed from space since the start of the space age. A picture of the auroral oval taken by the auroral imager on board the *Dynamics Explorer 1 (DE-1)* is shown in Figure 8.46. The auroral oval is quite apparent in this picture and the latitudinal extent of the emissions is clearly greatest in the midnight region. The bright area on the left side of the picture is the atomic oxygen dayglow and is not associated with the aurora but with excitation due to solar EUV photons or due to photoelectrons.

It has been known for many decades that aurorae are associated with the precipitation of energetic particles into the upper atmosphere. In fact, particle detectors on satellites circling the Earth have measured the charged particle fluxes responsible for different types of aurorae. Auroral precipitation can be organized into the following categories:

(1) **Cusp precipitation** The magnetic field lines in the noon region of the auroral oval open almost directly into the magnetosheath, thus allowing some particles to move from the magnetosheath down to the ionosphere (Smith and Lockwood, 1996). The magnetosheath electrons that "precipitate" are those with pitch angles within the atmospheric

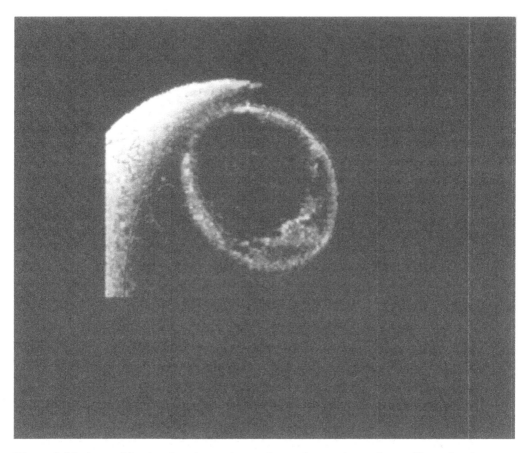

Figure 8.46. Auroral luminosity observed over the northern polar region on November 8, 1981, by the auroral imager on *DE-1*. The image was taken in the ultraviolet part of the spectrum and consists of atomic oxygen 130.3 nm wavelength emission and some band emission from molecular nitrogen. See Frank and Craven (1988) or Frank et al. (1985). (Courtesy of L. A. Frank, The University of Iowa.)

loss cone (see Chapter 3 on single particle motion). The energy spectrum of cusp auroral electrons is obviously about the same as the energy spectrum of magnetosheath electrons – energies of about 100 eV dominate and the distribution function is more or less isotropic. Solar wind protons tend not to precipitate in the cusp region since they convect downstream more rapidly, due to the convection electric field, than they can move down the magnetic field.

(2) **Polar cap aurora – Polar rain and proton precipitation** A rather weak and low-energy electron precipitation exists over the polar cap and is similar to cusp precipitation. The precipitation of protons with energies of several keV sometimes occurs over the polar cap. As well as *downward* particle precipitation, upward ion flows are common over both the polar cap and cusp regions, as discussed earlier with respect to sources of magnetospheric plasma.

(3) **Diffuse aurora** Relatively isotropic fluxes of electrons have been found to precipitate around most of the auroral oval. The energy distributions tend to be Maxwellian with

temperatures of about 1 keV. The spatial distribution of the diffuse aurora is rather smooth, or diffuse, and it tends to be rather steady in time. The electrons responsible for this type of aurora are thought to be relatively unaltered plasmasheet electrons that are in the loss cone or are scattered into the loss cone.

(4) **Discrete auroral arcs (or inverted-V events)** The most dramatic auroral displays take place during magnetic substorms and tend to occur in the nightside auroral oval in localized regions. These discrete auroral arcs have complicated spatial and temporal structures. Spacecraft observations have demonstrated that the electrons responsible for discrete arcs have energies of several keV or more and are highly field aligned. The electrons originate in the plasmasheet, but somewhere between the plasmasheet and the top of the ionosphere these electrons are accelerated from plasmasheet energies (≈ 1 keV) to the observed energies (2–20 keV). These auroral precipitation events are sometimes called inverted-V events because of their appearance on the energy-time plots of particle fluxes made from data measured by satellite-borne detectors. Inverted Vs have latitudinal extents of a couple of degrees, but inbedded within them are smaller structures, where the acceleration actually takes place, that map down to quite narrow (≈ 1 km or less) auroral arcs in the upper atmosphere.

What causes discrete auroral arcs? Simply put, the answer is field-aligned, or Birkeland, currents ($J_\parallel$). However, there are many "details" that need to be considered. Review articles by Lyons (1992) and Burke and Heinemann (1983) provide a nice explanation of these details, some highlights of which we will go over now.

We have seen in Sections 8.5 and 8.6 that the large-scale magnetospheric dynamics plus the coupling of the magnetosphere and ionosphere require the existence of field-aligned currents. That is, field-aligned currents are a necessary consequence of the MHD force balance in the magnetosphere and ionosphere. The largest values of $J_\parallel$ are found in the auroral oval. As we saw earlier, values of $J_\parallel$ at the top of the ionosphere are typically about 1 μA/m^2, which corresponds to about 1 MA spread out over the entire auroral oval. However, during magnetic substorms and in localized regions, the dynamics requires values of $J_\parallel$ in excess of 10 μA/m^2!

How are these large electrical currents carried by the plasma? Clearly, these currents consist of moving positive and negative charged particles: $J_\parallel = \Sigma n_s q_s u_{\parallel s}$, where the sum is over all charged particle species. But it is also clear that n_s and $u_{\parallel s}$ depend on the nature of the particle distribution functions in each region of the magnetosphere/ionosphere system. In fact, it can be demonstrated that if one makes the usual assumption (and a quite good one *almost* everywhere in space plasmas) that the field lines are equipotentials (i.e., parallel electric field $E_\parallel = 0$) then an upper limit of

$$J_{\max} = n_s q_s u_{\text{th}, s}, \qquad (8.111)$$

exists on the current that can be carried by a species s (Lyons, 1992), where for this purpose the thermal speed for a Maxwellian distribution is defined as $u_{\text{th}, s} = [k_B T_s / 2\pi m_s]^{1/2}$.

Electrons are more important as charge carriers than ions because electron thermal speeds greatly exceed ion thermal speeds. Downward field-aligned currents are not a problem because these can be carried by upward flowing ionospheric electrons and there is a large reservoir of such electrons. However, supporting upward $J_{\parallel}$ can be more of a problem because these require upward ion fluxes from the ionosphere and/or downward flow of ambient magnetospheric electrons.

A fully isotropic plasmasheet electron distribution, in the absence of collisions or small-scale fields, remains isotropic as it is mapped down the magnetic field to the ionosphere. The density and temperature remain the same along the magnetic field, and hence the maximum current, as given by Equation (8.111), also remains the same. Applying Equation (8.111) (Problem 8.20) to the plasmasheet electron population (from Table 8.1, $n_e \approx 1$ cm^{-3} and $T_e \approx 1$ keV), we can find that $J_{max} \approx 1$ μA/m^2. We can also apply this equation to upward flowing ionospheric O$^+$ ions originating in the topside ionosphere, in which case we find that (Problem 8.20) $J_{max} \approx 0.5$ μA/m^2. These maximum $J_{\parallel}$ values suggest that typical geomagnetically quiet levels of upward or downward $J_{\parallel}$ can indeed be supported by ambient particle populations. However, substorm values of upward $J_{\parallel}$ are a factor of 10 greater than the values of J_{max} we just estimated! In order to meet the demands that the dynamics place on $J_{\parallel}$ under these conditions, the magnetosphere must find another way to supply the electrical current.

So far in our treatment of the magnetosphere, magnetic field lines have been treated as equipotentials lines. In other words, the parallel electric field has been assumed to be zero ($E_{\parallel} = 0$) and the parallel electrical conductivity has been assumed to be infinite. Obviously, this cannot be exactly true, but overall it is an excellent assumption. However, an important exception to this ideal scenario exists for those magnetic field lines linking the plasmasheet and the auroral ionosphere. As we have just seen, during geomagnetically active time periods the large-scale dynamics of the magnetosphere demands, for some regions and field lines, parallel electrical currents that exceed the current-carrying capacity of the ambient charged particle populations. It is thought that charge separation takes place somewhere along such magnetic field lines in the form of electrical potential structures called *double-layers*. Parallel electric fields are present in these structures, and evidence has accumulated that these double-layer structures exist on some auroral field lines (as "demanded" by the large-scale dynamics) at distances of 1–2 R_E above the ionosphere. Several kilovolt field-aligned potential drops are present in these structures, thus strongly violating (locally) the earlier assumption of equipotential field lines. These big potential drops are located in small structures embedded within the larger inverted-V structure.

Now we follow the treatment given in the review of Lyons and show how enhanced field-aligned currents are produced by parallel electric fields. Recall that enhanced $J_{\parallel}$ values, in excess of J_{max}, are needed in the nightside Region 2 current system during substorms. Upward electric currents are needed, which requires a downward flow of electrons. If we start with a typical electron from the electron population in the region above a double-layer, then an upward parallel electric field increases the parallel velocity component of the electron (i.e., $v_\parallel$ increases). This parallel acceleration cannot by itself significantly increase the parallel current density beyond the value of J_{max}, because the current density is proportional to the electron flux $n_e v_\parallel$, and for a one-dimensional geometry an examination of the continuity relation shows that this flux remains constant along the field line even if $v_\parallel$ increases. That is, as $v_\parallel$ increases, the electron density decreases (Problem 8.21). But we must also consider the loss cone. Those electrons that actually carry the parallel electrical current all the way down to the ionosphere from the plasmasheet must reside within the atmospheric loss cone. Recall from Chapter 3 that all particles that start out with pitch angles lying outside the loss cone magnetically mirror and are reflected. Only electrons with small pitch angles therefore contribute to $J_\parallel$. A parallel electric field increases the flux of electrons inside this loss cone, and hence increases $J_\parallel$, because the electrons have their $v_\parallel$ boosted and their pitch angle α decreased.

The following algebraic expression quantitatively describes the relation between the field-aligned current density and the field-aligned potential drop $\Phi_\parallel$ associated with a double-layer (see the Lyons article):

$$\frac{J_\parallel}{J_{max}} = R\left[1 - \left(1 - \frac{1}{R}\right)\exp\left(-\frac{e\Phi_\parallel}{k_B T_e}\frac{1}{R-1}\right)\right], \qquad (8.112)$$

where $R = B_{ionosphere}/B_{top}$ is the mirror ratio of the magnetic field in the ionosphere to the field at the top of the potential structure. Conveniently, this relation is independent of the details of how the electric potential drop is distributed along the field line. A reasonable value of R is 30 for a double-layer located a couple of R_E above the ionosphere; Figure 8.47 shows $J_\parallel$ versus the potential drop (relative to the electron thermal energy $k_B T_e$). Clearly, the current density increases as $e\Phi_\parallel$ increases. A value of $J_\parallel \approx 10 J_{max}$ was suggested earlier to satisfy the Region 2 requirements for active times. Figure 8.47 indicates that to achieve this we need to have $e\Phi_\parallel/k_B T_e \approx 10$. Plasmasheet electrons typical have $k_B T_e \approx 1$ keV, so that the field-aligned potential is $\Phi_\parallel \approx 10$ kV. By the time the electrons reach the top of the ionosphere their energy has reached about 10 keV.

A measured electron spectra associated with a discrete auroral arc (or inverted-V event) is displayed in Figure 8.48. Note that the peak electron fluxes occur near an energy of 6 keV and at very small pitch angles. The lower energy electrons are mainly secondary electrons produced in the upper atmosphere by the electron impact ionization associated with the primary 6 keV beam. These lower energy electrons move up the magnetic field but are electrostatically reflected back into the atmosphere by the same electric potential structure that created the auroral beam in the first place.

When the energetic auroral electrons we have been discussing precipitate into the atmosphere, they lose their energy primarily by inelastic collisions with neu-

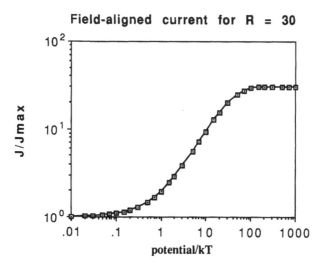

Figure 8.47. Field-aligned current density relative to the maximum possible current density versus voltage-drop along a magnetic field line in the magnetosphere for a mirror ratio $R = 30$.

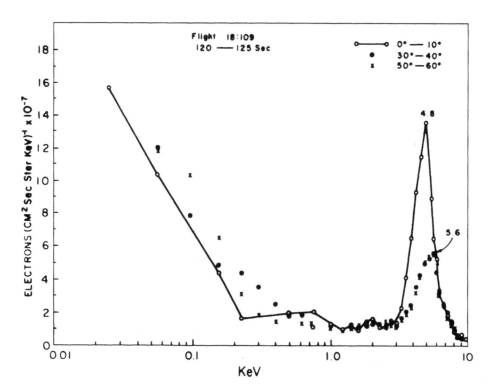

Figure 8.48. Rocket measurements of electron flux versus energy for three different pitch angle ranges for a discrete auroral arc. (From Arnoldy et al., 1974.)

tral atoms and molecules. The relevant collisional processes include ionization, electronic excitation, and rotational and vibrational excitation of molecules. Some energy loss to the ionospheric thermal electrons by Coulomb collisions also takes place. An electron loses energy in each collision. For example, each time an auroral electron collides with an O atom and excites the ^3S state of atomic oxygen from its ^3P ground electronic state the electron loses 9 eV. This process can be represented by the reaction

$$e^* + O(^3P) \rightarrow O(^3S) + e^*, \tag{8.113}$$

where e^* denotes an auroral electron. The $O(^3S)$ state quickly deexcites by emitting a 130.4 nm photon in the far UV part of the spectrum:

$$O(^3S) \rightarrow O(^3P) + h\nu \,(130.4\,\text{nm}). \tag{8.114}$$

These photons are not visible from the ground since the atmosphere absorbs radiation at this wavelength (fortunately for us), but rocket and satellite experiments have observed this and other auroral radiation (Figure 8.46). However, many other transitions of O, N_2, or O_2 in the visible portion of the spectrum can be excited by auroral electrons and are observed from the ground as the *aurora borealis* or *aurora australis*. For example, the $O(^1D)$ metastable state produces 630 nm light (the oxygen "red line") and the $O(^1S)$ state produces 557.7 nm light (the "green line") via the $O(^1S) \rightarrow O(^1D)$ transition.

The most important energy loss process for auroral electrons involves ionization of atmospheric neutrals. Ionization of atomic oxygen and molecular nitrogen by auroral electrons (denoted e^*) can be represented by the reactions

$$e^* + N_2 \rightarrow N_2^+ + e + e^* \tag{8.115}$$

and

$$e^* + O \rightarrow O^+ + e + e^*, \tag{8.116}$$

where the other electron produced from this process is called a *secondary electron* and typically has an energy of a few electron volts. An auroral electron typically loses about 25 eV in an ionizing collision – about 13 eV (the ionization potential) for the actual ionization event and about 10 eV for the secondary electron produced. This auroral ionization process is the main source of ions in the auroral E-region during geomagnetically active times, outstripping ion production due to photoionization by solar EUV photons. The auroral E-region ionospheric electron density is usually significantly enhanced relative to adjacent nonauroral regions. As we discussed earlier (Equation 8.103), the enhanced auroral electron densities lead to higher electrical conductivities in the auroral oval; this in turn affects the magnetosphere–ionosphere coupling and the required levels of $J_\parallel$, which itself generated the auroral precipitation – we are obviously dealing with a nonlinear "feedback loop."

For energetic electrons ($E > 100\,\text{eV}$) an average energy loss of $35\,\text{eV}$ takes place for each ion pair produced (i.e., for each ionization event). This *mean energy loss per ion pair* statistically takes into account the nonionizing collisions as well as the ionizing collisions. The depth of penetration into the atmosphere of an energetic electron depends on its intial energy as well as on the electron impact cross sections for all relevant transitions. The electron flux as a function of altitude can be determined by solving a transport equation for energetic electrons that takes into account all loss processes. This equation is analogous to the radiative transfer equation (5.30). We do not discuss this here, but you can learn more by reading the book by Rees (1989). The bottom line is that high-energy electrons penetrate deeper into the atmosphere than do lower-energy electrons. The altitude of peak energy deposition for the diffuse aurora (electron energies $E \approx 1\,\text{keV}$) is about $140\,\text{km}$, whereas the peak altitude for discrete auroral arcs ($E \approx 10\,\text{keV}$) is about $105\,\text{km}$. In Problem 8.23 you will explore some of the ionospheric consequences of auroral precipitation for the discrete auroral arc event whose incident electron flux was shown in Figure 8.48.

8.6.5 Numerical global MHD models of the magnetosphere

The advent of high-speed computing, including the availability of supercomputers, has allowed the development of increasingly sophisticated numerical models of space plasmas. Both kinetic and fluid numerical models have been developed for space plasma environments such as the solar corona, the solar wind, and the magnetosphere. In particular, several "global," three-dimensional, numerical models of the terrestrial magnetosphere and its interaction with the solar wind have been constructed, as discussed in the recent review paper "The Magnetosphere in the Machine: Large-Scale Theoretical Models of the Magnetosphere" by Walker and Ashour-Abdalla (1995).

The MHD equations as given in Chapter 4 can be solved numerically on a three-dimensional spatial grid. The fluid variables, including density, pressure, velocity components, and magnetic field components are "discretized," meaning that values are calculated only on the grid. Time is also discretized in that the solutions are advanced one time step each iteration. The numerical methods used to accomplish this vary widely in their sophistication. These methods are the subject of a whole discipline called *computational fluid dynamics*. Most of this field is devoted to solving the ordinary hydrodynamic equations (that is, no magnetic field), but many of the methods have also been developed for MHD. The solution of the equations for a specific problem requires boundary conditions. For MHD models of the magnetosphere, one boundary must be located in the solar wind upstream of the bow shock and the fluid variables must be maintained at solar wind values. Particularly troublesome is the "inner" boundary near the Earth where the intrinsic magnetic field dominates over the field generated by magnetospheric currents. Most models set this boundary at a radial distance from the Earth of a few Earth radii. In some models, this boundary is maintained as an equipotential, thus "short-circuiting" the ionospheric current systems, whereas in other models the ionospheric currents are mapped back out into the magnetosphere.

Global MHD models can provide snapshots or even time series of the three-dimensional magnetosphere. This global view cannot be obtained from satellite measurements since only a few satellites capable of making the necessary plasma and field measurements are operating simultaneously, even with the recent launches of NASA spacecraft such as *Polar* and *Wind*. Of course, not all the "physics," especially small-scale kinetic processes, can be included in MHD models, but they seem to be successful in describing the large-scale dynamics of the magnetosphere. Global pictures of diagnostic variables, such as the Poynting vector, can be produced by the models, and these can give the researcher insight into the dynamics. We will briefly consider here some results from just one global MHD model: that of Walker et al. (1993). This model focuses on explaining the substorm process. The abstract of the Walker et al. (1993) paper reads.

We have used a new high-resolution global magnetohydrodynamic simulation model to investigate the onset of reconnection in the magnetotail during intervals with southward interplanetary magnetic field (IMF). After the southward IMF reaches the dayside magnetopause reconnection begins and magnetic flux is convected into the tail lobes. After about 35 min., reconnection begins within the plasmasheet near midnight at $x = -14 R_E$. Later the x line moves toward dawn and dusk. The reconnection occurs just tailward of the region where the tail attaches onto the dipole-dominated inner magnetosphere. The simulation shows that prior to the onset of reconnection, the Poynting flux is concentrated in this region. The time required for the start of reconnection depends on the component of the magnetic field normal to the equator (B_z). Reconnection occurs only after the B_z component has been reduced sufficiently for the tearing mode to grow. Later, when all the plasmasheet field lines have reconnected, a plasmoid moves down the tail.

To illustrate the process described in this quotation, Figure 5 from the Walker et al. paper is reproduced in Figure 8.49 here. In the top panel, the region of reconnection is evident where the tail field lines dip toward the neutral sheet. Some closed field lines deeper in the tail are also apparent and are associated with a newly formed plasmoid. Tail lobe field lines are visible in the bottom panel, and field lines newly reconnected at the dayside magnetopause are identified by the sharp kinks in them. These results clearly show how useful global MHD models can be for "visualizing" macroscopic dynamical processes in the magnetosphere.

8.6.6 Space weather

The terrestrial space environment is often geomagnetically "disturbed." That is, as we have seen in this chapter, the magnetosphere and polar ionosphere often experience episodes of enhanced convection and electrical current flow associated

$$B_z = -5 \text{ nT}, B_z (t = 0) = 0 \text{ nT}, \; t = 47 \text{ m}$$

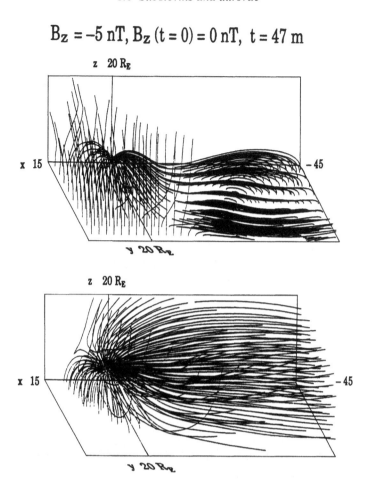

Figure 8.49. Magnetic field lines from the numerical global MHD model of the terrestrial magnetosphere by Walker et al. (1993). Figure 5 from that paper is reproduced here. The information at the top indicates that these results are for a time 47 minutes after the start of the simulation at which time ($t = 0$) the IMF was changed from a 0 nT value to southward with $B_z = -5$ nT. At a time of 47 min reconnection is just starting in the plasmasheet about 14 R_E from the Earth down the tail and reconnection has been taking place at the dayside magnetopause for about 30 min already. Only magnetosheath field lines not attached to Earth and closed magnetic field lines are displayed in the top panel. Only open field lines attached to both the Earth and the IMF are shown in the bottom panel. (From Walker et al., 1993.)

with favorable solar wind conditions such as southward-directed IMF and coronal mass ejections (CMEs). Episodes of especially high geomagnetic activity are called *magnetic storms*. These episodes of *space weather* are evident in perturbations of the geomagnetic field that can be characterized by a variety of *geomagnetic indices* including the D_{st}, K_p, and A_p indices. The D_{st} index is constructed from magnetometer measurements made at lower latitudes and is sensitive to magnetic perturbtions caused by the ring current. This was discussed in Chapter 3. The other two indices are constructed from averages of the fields measured by various higher latitude magnetometer networks or chains and are sensitive to polar region current systems including the auroral electrojet.

With the increasing importance of satellites and spacecraft for commerical and military purposes has come an increasing interest in *forecasting* space weather. The interested reader is referred to the recent review, "Geomagnetic Activity Forecasting: The State of the Art," by J. Joselyn (1995). Joselyn writes:

> The level of disturbance of the space environment is of more than academic interest. Such private and public endeavors as communication and navigation systems, electric power networks, geophysical exploration, spacecraft control, and science research campaigns are affected by geomagnetic fluctuations.

In order to make space weather forecasts, solar conditions, such as flares, sunspot groups, and especially CMEs, are closely monitored as is the solar wind. The Space Environment Laboratory (SEL) of the National Oceanic and Atmospheric Administration (NOAA) in Boulder, Colorado, issues daily reports and forecasts of geomagnetic activity.

8.7 Magnetospheres of the outer planets

8.7.1 A brief overview

Our knowledge of the outer planets has largely been derived from data generated by NASA's *Pioneer* and *Voyager 1* and *2* missions. Jupiter was visited by the *Pioneer 10* and *11* spacecraft and also the *Voyager 1* and *2* spacecraft. *Pioneer 11* and *Voyagers 1* and *2* flew by Saturn, and both Uranus and Neptune were encountered by *Voyager 2*. More recently, the *Ulysses* spacecraft, launched in October 1990, encountered Jupiter in February 1992 on its way out of the ecliptic plane, and in December 1995 NASA's *Galileo* spacecraft went into orbit around Jupiter and the probe plunged into the Jovian atmosphere. All these spacecraft carry thermal and energetic particle detectors, magnetometers, plasma wave detectors, and radio receivers. Reviews of our understanding of outer planet magnetospheres are listed at the end of this chapter (Bagenal, 1985, 1992; and articles in the book *Physics of the Jovian Magnetosphere* edited by Dessler).

The outer planets (Jupiter, Saturn, Uranus, and Neptune) all possess large intrinsic magnetic fields that play central roles in the interactions of these planets with the solar wind. (See Table 7.1 for planetary characteristics including magnetic dipole moments and rotation periods.) The planetary fields are not purely dipole fields, but include higher-order moments, and the magnetic dipoles are not in general aligned with the rotation axes. Figure 8.50, which was adapted from a figure in the article by Bagenal (1992), summarizes schematically the differences between the outer planets in this regard. The *obliquity* is the angle between the rotation axis and the vector normal to the ecliptic plane. For Earth, the obliquity is 23.5 degrees. The tilt angle of the magnetic dipole with respect to the rotation axis is also indicated

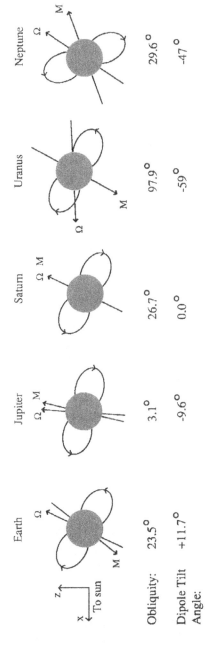

Figure 8.50. Schematic of magnetic field and rotation of the outer planets. **M** is the dipole moment; Ω is the angular frequency. The dipole tilt angle is the angle between **M** and Ω. Adapted from Bagenal (1992).

in Figure 8.50. Clearly, the Jovian and Saturnian magnetic fields are qualitatively similar to Earth's, whereas the magnetic fields of Uranus and Neptune are rather irregular. The magnetic field patterns of Uranus and Neptune change drastically over the course of a rotation period, unlike Earth, Jupiter, and Saturn. Uranus and Neptune are said (cf. Bagenal, 1992) to have *asymmetric magnetospheres*. The topologies of these two magnetospheres are quite complex.

The magnetospheres of the outer planets are all quite large. Putting aside details of the structure of each magnetosphere, we estimated the distance to each outer planet magnetopause using Chapman–Ferraro theory in Chapter 7. The radial distance to the Jovian magnetopause is observed to be $r_{mp} \approx 50\text{--}100\ R_J$. The Jovian magnetosphere is indeed huge considering that the Jovian radius R_J is 12 R_E. Simple Chapman–Ferraro theory, which balances internal magnetic pressure against solar wind dynamic pressure, gave (in Chapter 7) a Jovian magnetopause distance of only about 30 R_J. Internal plasma pressure associated with the plasmasheet (with $\beta \approx 5$) was invoked (Equation (7.14)) to supply extra internal pressure and push the magnetopause further out. In contrast, the distance to the terrestrial subsolar magnetopause ($r_{mp} \approx 10\ R_E$) was quite well described by the simple theory. It happens that the magnetopause distances of Saturn, Uranus, and Neptune are all reasonably accurately determined by the simple theory (you were asked to do this back in Problem 7.3 using Equation (7.10)). The observed magnetopause distances are approximately: 20 R_S for Saturn, 18 R_U for Uranus, and 25 R_N for Neptune. The planetary radii of Saturn, Uranus, and Neptune are denoted R_S, R_U, and R_N, respectively.

The outer planets all rotate rapidly, and the magnetospheres are largely dominated by rotational plasma flow. This can be seen by comparing the *corotational electric field* $\mathbf{E}_{cor}$ with the *convection electric field* $\mathbf{E}_m$, as was done for the terrestrial magnetosphere in Section 8.4.3. The magnitude of $\mathbf{E}_{cor}$ for the terrestrial magnetosphere was given by Equation (8.85). Generalizing this formula to planet "x" (with angular frequency Ω_x, surface field B_x, and radius R_x) we obtain for the equatorial plane ($\theta = 90°$)

$$E_{cor} = \Omega_x B_x R_x^3 / r^2, \tag{8.117}$$

where r is the radial distance from the center of the planet. As discussed in Section 8.4.3, E_{cor} decreases as r^{-2} whereas E_m is roughly a constant; this means that at a small enough radial distance corotation must dominate over magnetospheric plasma convection. The convection electric field can be estimated using Equation (8.3) and is $E_m = \zeta u_{sw} B_0$, where B_0 is the IMF strength at the heliocentric distance of planet x, u_{sw} is the solar wind speed, and ζ is the efficiency of the dayside magnetic reconnection process for the magnetosphere.

Recall that the region of the terrestrial magnetosphere where the plasma corotates with the Earth is called the plasmasphere. The plasma density in this region remains high relative to the portion of the magnetosphere outside where plasma is lost by

Table 8.2. *Theoretical plasmapause L-shell values*

	Jupiter	Saturn	Uranus	Neptune
L_{pp}	240	70	50	50

being convected to the dayside magnetopause. The radial distance to the boundary of the plasmapause was given by Equation (8.88), which was found by equating E_{cor} and E_m and then solving for the plasmapause L-shell value, L_{pp}. Generalizing this equation using Equation (8.117) and recognizing that $\Psi_m \approx 3\,r_{mp}E_m$, we obtain the following relation for L_{pp}:

$$L_{pp} \cong \sqrt{\frac{\Omega_x B_x R_x}{E_m}} = \sqrt{\frac{\Omega_x B_x R_x}{\zeta u_{sw} B_0}}. \tag{8.118}$$

This formula can be used to estimate the sizes of the corotating region (i.e., the plasmasphere) for the outer planets by using the parameters listed in Table 7.1. (See Problem 8.25.) The surface fields of Jupiter, Saturn, Uranus, and Neptune are approximately 40, 2, 2.5, and 1.4×10^{-5} T, respectively. Note that the IMF field strength B_0 decreases with heliospheric distance as discussed in Chapter 6. A question arises as to what values of the reconnection efficiency ζ are appropriate for the outer planets, but simply using a reasonable terrestrial value ($\zeta = 0.2$) for moderately active times we find the values of L_{pp} for the outer planets as shown in Table 8.2.

For Earth, Equation (8.118) gives $L_{pp} = 5.5$. Now we compare these plasmapause L-values with the size of the magnetosphere (i.e., with r_{mp} or r_{mp}/R_x). For Earth, $r_{mp} = 10\,R_E$, but the distance to the plasmapause is about half of this or 5.5 R_E. A large volume of the magnetosphere lies outside the plasmapause, even on the dayside, and for most of the magnetosphere the plasma dynamics is controlled by solar wind–driven convection. However, from the above comparisons for the outer planets we see that the radial size of the corotating region is about 2–4 times greater than the radial distance to the magnetopause, which is clearly impossible. What this means is that corotation dominates the plasma dynamics of these planets all the way out to the magnetopause! There is "no space" left for solar wind–driven plasma convection to take place, at least in the equatorial plane. This is an over-simplification, and solar wind–driven convection must still play some role at very high magnetic latitudes, deep in the magnetotail, and/or in a narrow boundary layer just inside the magnetopause. Clearly, the magnetospheric dynamics of the outer planets is quite different from the dynamics at the Earth or Mercury.

Bagenal (1985, 1992) has summarized the properties of the magnetospheric plasma of the outer planets. The thermal plasma in the magnetospheres of the outer

planets, or at Earth for that matter, is approximately Maxwellian with temperatures in the several eV to several keV range, depending on location. However, the
typical densities in these magnetospheres are quite different. The maximum magnetospheric plasma density is $\approx 1,000$–$5,000\,cm^{-3}$ at both Jupiter and the Earth (i.e.,
in the plasmasphere), whereas for Saturn the maximum density is $n_e \approx 100\,cm^{-3}$.
The maximum thermal density in the magnetospheres of both Uranus and Neptune
is quite small – only about $3\,cm^{-3}$.

Where does this plasma come from? Recall that the source of plasma for the
terrestrial magnetosphere is both the solar wind and the ionosphere. For the outer
planets these two sources still contribute plasma, but an even more important source
exists: the satellites of these planets, most of whose orbits lie within these very large
magnetospheres. For example, for Jupiter the most important source of both thermal,
and ultimately the energetic plasma, is the innermost Galilean moon Io, although
the icy satellites Ganymede, Europa, and Callisto are also thought to act as plasma
sources. The *Voyager* spacecraft measured large abundances of sulfur and oxygen
ions both in the thermal plasma and in the energetic radiation belt ion population;
this is a clear indicator of an Io plasma source since we know that Io has a thin
SO_2 atmosphere that appears to be replenished by volcanic activity. The important
plasma sources at Saturn include the satellite Titan, which has a dense nitrogen
atmosphere and contributes nitrogen ions to the outer magnetosphere; the icy satellites; and the rings, which contribute both oxygen ions and protons to the inner
magnetosphere. Similarly, Neptune's satellite Triton acts as a source of nitrogen
ions for that magnetosphere, although the thinner, colder atmosphere of Triton, in
comparison with Titan, somewhat limits this source. The satellite source of plasma
for Uranus is rather weak, consisting of rather small satellites and thin ring arcs.

The outer planets all have radiation belts containing energetic ions and electrons.
The phase space densities of MeV energy range ions (heavy ions as well as protons)
in the radiation belts of Earth and Saturn are comparable, but the radiation intensities
at Jupiter are more than a factor of 10 greater than they are at Earth. The radiation
belts of Uranus and Neptune are rather puny with intensities 10 times less than at
Earth.

8.7.2 The Jovian magnetosphere and the Io plasma torus

The following ion species have been observed by spacecraft in the thermal plasma
of the Jovian magnetosphere (e.g., Geiss et al., 1992): H^+, He^{++}, and O^{6+} of
solar wind origin; H^+, H_2^+, and H_3^+, originating from the Jovian ionosphere; and
S^{2+}, S^{3+}, S^{4+}, S^{5+}, O^+, O^{2+}, and O^{3+}, originating from Jupiter's satellite Io. The
S and O ion species are especially abundant, particularly in the inner ($L < 10$) and
middle ($L < 40$) regions of the magnetosphere. The densest part of the inner magnetosphere is called the *Io plasma torus*. The thermal electron gas in this torus has
a temperature of about 200 eV. The composition of the energetic plasma population

(i.e., the plasma whose particles have energies in excess of about 10 keV/amu) is also dominated by S and O species (cf. Lanzerotti et al., 1992). Energetic electrons with energies in excess of 100 keV are also present in the Jovian magnetosphere, and radiation belt particles with energies in the 1–100 MeV range have also been observed.

Io is a very important source of plasma for the Jovian magnetosphere. The orbital radius of Io is 5.9 R_J. Sulfur dioxide (SO_2) is released into the tenuous atmosphere of that body by volcanoes; some of these volcanoes were actually observed erupting by the *Voyager* television camera. Exospheric SO_2 molecules are able to escape the gravitational field of Io and move into the magnetosphere of Jupiter where they are ionized and dissociated by magnetospheric electrons, resulting in the creation of sulfur and oxygen ions. Charge transfer and electron removal collisions largely determine the distribution of ion charge states for S and O in the Io plasma torus. In some respects this is analogous to the solar corona, where collisions act to set up a distribution of charge states for the various ion species. The ultraviolet spectrometers on the *Voyager* spacecraft observed ultraviolet emissions from the Io plasma torus associated with the various S and O charge states.

Neutral particles in the Jovian magnetosphere do not "see" Jupiter's magnetic field or the corotation electric field. However, once the neutrals are ionized, the ions that are produced experience the Lorentz force and are "picked up" by the corotating Jovian magnetic field. The pickup process is essentially the same as we discussed in Chapter 7 with respect to cometary ions picked up by the solar wind. The newly born ions $\mathbf{E} \times \mathbf{B}$ drift at a speed slightly less then the corotation speed. In the case of comets, this speed was the solar wind speed rather than the corotation speed. The newly born ions are "hot" because the amplitude of their gyration is approximately equal to the corotation speed (73 km/s) minus the orbital speed of Io (16 km/s). The orbital speed is the average speed the neutrals have before they are ionized. New oxygen and sulfur ions acquire energies of 260 eV and 540 eV, respectively. The Io plasma torus is not collisionless, however, and this pickup gyration energy is eventually shared with the rest of the plasma via Coulomb collisions. Thus, the plasma as a whole is heated via the pickup process. The total energy input into the Io plasma torus from this process is just the total ion addition rate multiplied by the average pickup energy.

The total Io source of heavy ions is $S_{Io} \approx 10^{28}$–10^{29} ions/s, and the resulting heating rate is ≈ 400 eV $\times S_{Io} \approx (0.5$–$5) \times 10^{31}$ eV/s $\approx (0.8$–$8) \times 10^{12}$ W. The total energy loss associated with the observed radiation for S and O species from the torus is roughly 10^{13} W, which is comparable to the estimated energy input rate. The energetics of the Io plasma torus are not fully understood, but as with all such problems it is necessary that energy inputs, energy losses (such as radiation), and energy transport all be balanced.

Figure 8.51 shows *Voyager 1* electron density data (same as total ion density) and ion temperature (T_i) data for the Io plasma torus. The torus can be divided into

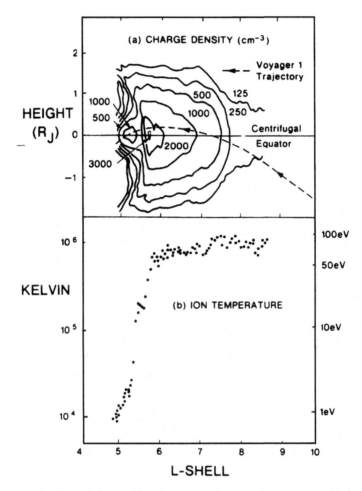

Figure 8.51. (Top panel) Contours of electron density versus *L*-shell and height from the equatorial plane of the magnetosphere, as determined from plasma measurements made by *Voyager 1* in the Io plasma torus. (Lower panel) Ion temperature in the equator. (From Bagenal, 1985.) Copyright 1985 Kluwer Academic Publishers. With kind permission from Kluwer Academic Publishers.

two parts: (1) an *inner cold torus* where the ion temperature $T_i \approx 1$ eV for $L < 5.7$ and (2) an *outer warm torus* for $L > 5.7$ where $T_i \approx 100$ eV. The ion temperature (or thermal energy) in the warm torus is somewhat less than the pickup energy, which is not surprising given that energy is lost by processes such as radiation. The electron density profile exhibits a maximum on the "rotational" equator of the magnetosphere, and the density decreases with height above or below this equatorial plane. The vertical scale height in the cold torus is quite small, but in the outer torus the scale height is larger with $H \approx 1 R_J$. The maximum electron density in the torus is about 2,000 cm^{-3} and occurs at $L \approx 6$. The density decreases to about 250 cm^{-3} by an *L*-shell of $L \approx 8$.

8.7.3 *Dynamics of the Jovian magnetosphere*

In this section, we briefly consider the dynamics of the Jovian magnetosphere using the same methods that were used earlier to explain the terrestrial magnetospheric dynamics. We will consider the force balance or dynamics of the inner and middle Jovian magnetosphere and will not discuss the outer magnetosphere ($L > 40$) or the magnetotail. The plasma essentially corotates, but we need to determine how the corotation is maintained and what electrical currents flow in the magnetosphere in response to the momentum balance. Again, the slab example presented in Section 8.3 can help us to understand this problem.

Before proceeding, we should ask ourselves if the currents can be found using simple MHD theory as discussed in Section 8.3. In the Earth's plasmasheet this theory with isotropic pressure worked reasonably well, although the pressure is not entirely isotropic (Figure 8.40); however, to understand the ring current in the inner magnetosphere we needed to include gradient and curvature drifts, which required (as discussed in Chapter 3) a separate consideration of perpendicular and parallel pressures for the energetic ring current plasma. Furthermore, in the inner magnetosphere of the Earth most of the total particle density (as opposed to the pressure) is associated with the colder (≈ 1 eV) thermal plasma within the plasmasphere for which convection (i.e., $\mathbf{E} \times \mathbf{B}$ drift motion) is more important than gradient and curvature drifts. We now derive a critical energy (or speed) for ions at which their $\mathbf{E} \times \mathbf{B}$ drift speed equals their gradient and curvature drift speed. Ions with speeds less than this critical speed mainly undergo $\mathbf{E} \times \mathbf{B}$ drift, and the dynamics can be described with simple single-fluid MHD theory with isotropic pressure, whereas ions with speeds greater than this critical speed also undergo gradient and curvature drift.

We use the corotation speed for the $\mathbf{E} \times \mathbf{B}$ drift speed ($v_E = v_{corot}$), in which case the critical speed can be determined from the condition $v_B = v_{corot}$. Recall from Chapter 3 that the gradient and curvature drift speed, v_B, depends on the total particle energy (Equation (3.34) or (3.76)). You will find from doing Problem 8.24 that the critical particle speed is roughly given by $v_{crit}^2 \approx \Omega_{gyro}\Omega_{rotation}R_x^2L^2$, where we have distinguished in our notation between the gyrofrequency Ω_{gyro} and planetary rotational frequency $\Omega_{rotation}$. R_x is the relevant planetary radius for planet "x" and L is the L-shell. For both protons and oxygen ions at $L \approx 6$ in the terrestrial magnetosphere the "critical" energy is roughly $E_{crit} \approx 10$ keV. Above this "critical" energy, the proton motion, and any associated electrical currents, is dominated by single particle gradient and curvature drifts; below this energy, the proton motion is determined mainly by $\mathbf{E} \times \mathbf{B}$ convection. The L-shell value $L \approx 6$ corresponds to the outer part of the terrestrial ring current region or to the inner edge of the plasmasheet. At Earth the total plasma pressure near $L \approx 6$ is dominated by the ring current population, and the electrical current (and "dynamics") for this component is determined by gradient and curvature drift motion since the average ring current energy $E \gg E_{crit}$. However, at larger L-values where $E < E_{crit}$, the pressure in the plasmasheet of Earth is carried by ions with energies of a few keV, and the simpler MHD theory with isotropic pressure works reasonably well, albeit not perfectly.

For Jupiter, do we need to worry about the electrical currents associated with single particle gradient and curvature drifts, or will the simpler MHD theory work reasonably well? Consider an L-shell of 10, which is roughly the boundary between the outer Io plasma torus and the middle magnetosphere. The critical energy for an oxygen ion at $L \approx 10$ is about 15 MeV! "Cold" thermal plasma particles near $L \approx 10$ have energies of about $T_i \approx 100\,\text{eV}$, and even the more energetic particles that carry most of the pressure near $L \approx 10$ (the equivalent of terrestrial ring current plasma) have typical energies of only 100 keV/amu; thus, $E \ll E_{\text{crit}}$. In other words, convection due to E_{corot} is more important for this Jovian "ring current" energetic plasma than are gradient and curvature drifts, unlike the Earth's ring current plasma. Only radiation belt particles in the Jovian magnetosphere, with energies in excess of about 15 MeV, are controlled by gradient and curvature drifts for $L \approx 10$, but these particles do *not* contribute a significant fraction of the total pressure.

Now we continue with our discussion of the Jovian magnetospheric dynamics, restricting ourselves to simple MHD theory. The momentum equation (8.28) is a good starting point. This equation includes a mass-loading term that is proportional to the ion production rate P_i multiplied by $(\mathbf{u} - \mathbf{u}_n)$. For the case of the Io plasma torus, $\mathbf{u}_n$ should be identified as the orbital speed of Io (the speed at which the neutrals are injected into the magnetosphere), and $\mathbf{u}$ is the plasma flow velocity; at the orbit of Io we have $|\mathbf{u} - \mathbf{u}_n| = 57$ km/s. Recall from the section on comets in Chapter 7 that mass-loading slows down a plasma; hence, the addition of sulfur and oxygen ions to the Io plasma torus should result in plasma flow that is at least a little slower than the corotation speed. Actually, the plasma flow is very close to corotation, at least for $L < 40$. That is $\mathbf{u} = \mathbf{\Omega} \times \mathbf{r}$ is a good description of the plasma motion, where $\mathbf{\Omega}$ is the same as $\mathbf{\Omega}_{\text{rotation}}$, and the momentum balance in the corotation frame of reference is an almost static balance. The plasma density and particle distribution functions are largely controlled, not by bulk radial plasma flow, but by slow radial diffusion. The right-hand side of Equation (8.28) for steady-state conditions and in a *corotating coordinate system* becomes

$$\rho \mathbf{\Omega} \times (\mathbf{\Omega} \times \mathbf{r}) \cong -\nabla p + \mathbf{J} \times \mathbf{B} - P_i m_i (\mathbf{u} - \mathbf{u}_n). \qquad (8.119)$$

The left-hand side of Equation (8.119) is just the centripetal force per unit volume of the plasma. This MHD force balance equation can be separated into its radial, azimuthal, and vertical components. Vertical is with respect to the rotational equatorial plane. We now consider each of these components individually.

8.7.3.1 Azimuthal force balance

If we assume that the Jovian magnetosphere (and plasma pressure) is symmetric about the rotation axis, then $\partial p / \partial \phi = 0$ and since $-\mathbf{\Omega} \times (\mathbf{\Omega} \times \mathbf{r})$ is directed radially outward and does not contribute to the azimuthal force balance, the azimuthal

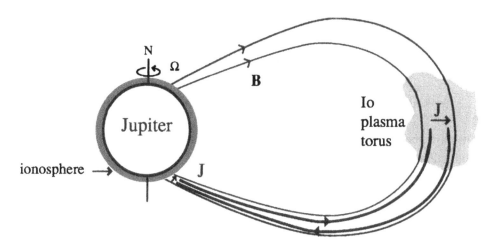

Figure 8.52. Current system associated with azimuthal MHD force balance in the Io plasma torus. Mass-loading from ion production requires a radial electrical current, which then flows as Birkeland currents along the magnetic field and closes in the Jovian ionosphere as Pederson currents.

component of Equation (8.119) simply becomes

$$J_r B \cong P_i m_i (u - u_n) \cong P_i m_i \Omega r, \qquad (8.120)$$

where J_r is the radial current density and where we have also made the approximation that $u \gg u_n$. Here Ω is Ω_{rot}. Clearly, the radial current density depends on both the rotational frequency of Jupiter and on the ion production rate. Equation (8.120) just states that a radially directed electrical current is needed so that the $\mathbf{J} \times \mathbf{B}$ force can balance the momentum addition due to ion production. Figure 8.52 illustrates the current system associated with mass-loading in the Io plasma torus. Charge conservation (i.e., Kirchoff's current law) requires that the electrical current come from somewhere and go somewhere. Parallel electrical currents, or Birkeland currents $(J_\parallel)$, are thus required, just as was discussed earlier in this chapter for the simple example with two slabs connected by a magnetic field. The electrical "circuit" closes in the ionosphere as Pederson currents. The resulting $\mathbf{J} \times \mathbf{B}$ force in the Jovian ionosphere acts to slow the plasma flow down relative to the corotating neutral atmosphere. This actually reduces the corotation electric field a little, and thus the plasma flow in the Io plasma torus is also a little slower than (that is, it lags) the corotation speed. In the ionosphere, the $\mathbf{J} \times \mathbf{B}$ force is balanced by ion–neutral drag. Some quantitative details on mass-loading effects in the Io torus are provided in the next two examples.

Example 8.5 (Total current flowing in "mass-loading electrical circuit" pictured in Figure 8.52) The total radial electrical current can be found from Equation (8.120) by integrating J_r over the cross-sectional area of the plasma torus, which can roughly be described as a circle around Jupiter with circumference $2\pi L R_J$, where $L \approx 6$, and with a vertical extent $\Delta z \approx 2R_J$. Thus,

$$I \approx 2\pi L R_J \Delta z J_r \approx 4\pi L^2 R_J^3 \Omega P_i m_i / B. \qquad (8.121)$$

Approximately, we can represent the magnetic field strength with a dipole value, $B \approx B_J / L^3$. The rotational frequency is denoted $\Omega = \Omega_J$. The total ion production rate throughout the entire Io plasma torus is just the volume integral of the local production rate P_i, or $S_{Io} \approx 2\pi L R_J^2 \Delta z \Delta L P_i$, where ΔL is the radial extent of the region with significant mass addition. Equation (8.121) now becomes

$$I \approx \frac{L^4}{\Delta L} \frac{m_i S_{Io} \Omega}{B_J}. \qquad (8.122)$$

Substituting reasonable values for the parameters ($L = 6$, $\Delta L = 2$, $S_{Io} = 3 \times 10^{28}$ s^{-1}, $\langle m_i \rangle \approx 25$ amu, $B \approx 5 \times 10^{-4}$ T), Equation (8.122) predicts that the total current in this system is $I \approx 1$ MA.

Example 8.6 (Departure of the plasma convection speed from the corotation speed in the Io plasma torus) In the ionospheric part of the "electric circuit" shown in Figure 8.52 the current density can be denoted $J_{\perp iono}$ and flows in the part of ionosphere connected via magnetic field lines to the Io plasma torus (i.e., at about 65° latitude). The force balance equation in the ionosphere can be written as $J_{\perp iono} B_J = \rho_{iono} \nu_{iono} (u - u_n)$, where the neutrals can be assumed to be, in the absence of any upper atmospheric winds, corotating with Jupiter: $u_n \approx \Omega R_J$. The ion–neutral collision frequency is denoted ν_{iono}. The departure from corotation is $\Delta u = u - u_n \approx J_{\perp iono} B_J / (\rho_{iono} \nu_{iono})$. The current density can be found by using the total current estimated in the previous example and by applying an appropriate ionospheric geometry. If there were no mass-loading in the magnetosphere, then there would be no current and $J_{\perp iono} = 0$. In this case, $u = u_n$; no departure from corotation would be present. On the other hand, the presence of a current results in a finite value of the corotation lag Δu.

A better way (but equivalent) of determining Δu is to calculate the height-integrated Pederson current density $K_\perp \approx J_{\perp iono} H_{iono}$ using the height-integrated conductivity $\Sigma_{\perp iono}$ for the Jovian ionosphere and using the electric field in the corotation frame of reference $E' = \Delta u B_J$:

$$K_\perp = \Sigma_{\perp iono} E' \quad \text{or} \quad \Delta u = K_\perp / (\Sigma_{\perp iono} B_J). \qquad (8.123)$$

Note that the E-region ionospheric scale height is approximately $H_{iono} \approx 200$ km. Reviewing Section 7.3.8 on ionospheric conductivity we find that the height-integrated conductivity can be approximated as $\Sigma_{\perp iono} \approx H_{iono} \sigma_{\perp iono(maximum)}$. The maximum Pederson conductivity is roughly $\sigma_{\perp iono(maximum)} \approx n_e e / B_J \approx 3 \times 10^{-16} n_e$ in SI units, where n_e is the electron density at the altitude where the ion gyrofrequency equals the collision frequency ($\nu_i \approx \Omega_i$), which for Jupiter is where the neutral density is about 10^{14} cm^{-3}. This is in the 300–400 km region, which is in the lower ionosphere. The electron density in the Jovian E-region is very uncertain, especially at these altitudes and at high latitudes that are subject to auroral precipitation. We adopt $n_e \approx 10^4$ cm^{-3}. The major ion species probably is H$_3^+$ or hydrocarbon ions. A quick estimate gives $\sigma_{\perp\text{iono(maximum)}} \approx 3 \times 10^{-6}$ S/m, and using $H_{\text{iono}} \approx 200$ km, we find that $\Sigma_{\perp\text{iono}} \approx 0.6$ S. The height-integrated current density determined from the total current is $K_\perp \approx I/\pi R_J \approx 0.005$ A/m, where I is the total current found in Example 8.5. Putting this all together, Equation (8.123) predicts that $\Delta u \approx 200$ m/s so that $\Delta u/u = \Delta u/u_n \approx 1\%$. The departure from corotation in the Io plasma torus and in the magnetically connected ionosphere is quite small. Note that this is really a departure from the motion of the neutral gas. If zonal winds exist then this also gives a departure from strict corotation even if $K_\perp = 0$.

8.7.3.2 Vertical force balance

Now we consider the component of Equation (8.119) orthogonal to the equatorial plane of the magnetosphere. We assume that the magnetic and rotational equators are identical, which is not quite true since the dipole tilt angle is 9.6°. We need to analyze the force balance along the magnetic field direction, which is *almost* the same as being in the normal/vertical direction. In the direction along the magnetic field $\mathbf{J} \times \mathbf{B} \approx \mathbf{0}$, so that we can write the force balance as

$$\rho \mathbf{\Omega} \times (\mathbf{\Omega} \times \mathbf{r}) \cdot \hat{\mathbf{b}} \cong -\frac{\partial p}{\partial z}. \tag{8.124}$$

We see from Figure 8.51 that a plasma density gradient exists in the z direction. For an isothermal gas the plasma pressure gradient is also in the z direction. The pressure gradient force is directed *outward* (up and down) from the equatorial plane. This is balanced by the component of the centrifugal force parallel to the magnetic field, which is primarily directed inward toward the equatorial plane.

Equation (8.124) can be used to derive (Bagenal, 1992) an expression for the electron density as a function of height above the equatorial plane (z). We first assume that the plasma pressure (electron plus ion) is given by $p \approx 2n_e k_B T_i$, where $T_e = T_i$ is assumed. Then, the component of the centrifugal force along the magnetic field direction can be determined by employing expressions for the components of a dipole magnetic field component given in Chapter 3. In Problem 8.27 you will show that near the equatorial plane, $\hat{\mathbf{n}} \cdot \hat{\mathbf{b}} \cong 3z/r$, where the $\hat{\mathbf{n}}$ vector is parallel to the centrifugal force, which is directed outward from the rotation axis. You can also then show that for an isothermal gas (and working with positive z only), Equation (8.124) becomes $n_e m_i \Omega^2 3z = -2k_B T_i (\partial n_e/\partial z)$. Integrating, we find that the electron density as a function of distance z above the plane varies exponentially with scale height H_z as

$$n_e(z) = n_{e0}\exp(-z^2/H_z^2) \quad \text{with } H_z = \sqrt{\frac{4k_B T_i}{3m_i \Omega^2}}. \tag{8.125}$$

If the electron pressure were omitted in this derivation, then a factor of 2 would appear in the numerator of H_z in Equation (8.125) rather than the factor of 4 that appears there now.

The value of H_z depends on temperature and is obviously greater in the outer plasma torus than it is in the inner torus because T_i is higher in the former than in the latter. Equation (8.125) predicts that $H_z \approx 1 \, R_J$ in the outer torus and that H_z is only about $0.1 \, R_J$ in the cold inner torus. These scale height values basically agree with the electron density data displayed in Figure 8.51.

8.7.3.3 Radial momentum balance

Now let us consider the radial component of the momentum Equation (8.119): $[\mathbf{J} \times \mathbf{B}]_r = \partial p/\partial r + [\rho\mathbf{\Omega} \times (\mathbf{\Omega} \times \mathbf{r})]_r$. We restrict ourselves to the equatorial plane of the magnetosphere for simplicity. The mass-loading term is assumed to be zero because we made the assumption that the plasma flow is entirely azimuthal. The relevant component of $\mathbf{J}$ is the azimuthal component J_ϕ. In terms of J_ϕ the momentum equation can now be written as

$$J_\phi = \frac{1}{B}\left[-\frac{\partial p}{\partial r} + \rho\Omega^2 r\right], \tag{8.126}$$

where we can also write $\partial p/\partial r = (1/R_J)\,\partial p/\partial L$.

The force balance represented by Equation (8.126) basically states that an outward pressure gradient force (assuming that p decreases with increasing radial distance) plus an outward centrifugal force on the plasma are countered by an inward $\mathbf{J} \times \mathbf{B}$ force associated with an azimuthal electrical current. This current is a *ring current*, in that it encircles Jupiter in the plasmasheet. Figure 8.53 illustrates this force balance schematically.

We now consider Equation (8.126) in more detail. In Example 8.7 we consider the Io plasma torus region ($L \approx 6$–8) and use the data shown in Figure 8.51 (assuming $T_e = T_i$) to find the pressure p of the thermal population and the density ρ. We find that the total "ring" current in the Io plasma torus is roughly 20 MA.

Example 8.7 (Azimuthal electrical current estimate at $L \approx 7$ in the Io plasma torus) Equation (8.126) and the data from Figure 8.51 can be used to estimate a typical current density in the Io plasma torus and the total azimuthal current I_ϕ between about $L = 6$ and $L = 8$.

The pressure in the $L = 6$–8 region is roughly 5×10^{-8} N/m^2 taking into account both T_e and T_i, and the radial pressure gradient is roughly $(1/R_J)\partial p/\partial L \approx \Delta p/(\Delta L R_J) \approx (5 \times 10^{-8}$ N/m$^2)/(2 \times 7 \times 10^7m) \approx 4 \times 10^{-16}$ N/m^3. The mass

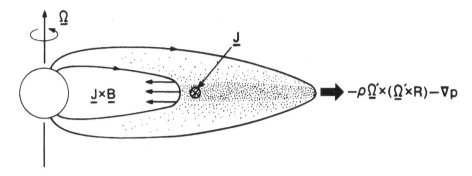

Figure 8.53. Schematic showing radial force balance in the Jovian plasmasheet. An inward **J** × **B** force balances an outward pressure gradient force plus an outward centrifugal force on the corotating plasma. (From McNutt, 1983.) Reprinted from *Advances in Space Research*, vol. 3, R. L. McNutt, p. 55, copyright 1983, with kind permission from Elsevier Science Ltd., The Boulevard, Langford Lane, Kidlington OX5 1GB, UK.

density near $L = 7$ is roughly $\rho = m_i n_e \approx 25\,\text{amu} \times 1.66 \times 10^{-27}\,\text{kg/amu} \times 10^3\,\text{cm}^{-3} \times 10^6\,\text{cm}^3\text{m}^{-3} \approx 4 \times 10^{-17}\,\text{kg/m}^3$, and the centrifugal force per unit volume can be estimated as $\rho\Omega^2 r \approx \rho(1.77 \times 10^{-4}\,\text{s}^{-1})^2\,7R_J \approx 6 \times 10^{-16}\,\text{N/m}^3$. Our average mass ($\langle m_i \rangle \approx 25\,\text{amu}$) assumed roughly comparable abundances of S and O ions. The centrifugal force term is comparable to the pressure gradient term, although the contribution of energetic particles to the pressure has *not* been taken into account. Equation (8.126) can now be used to estimate J_ϕ in the Io plasma torus: $J_\phi = [4 \times 10^{-16}\,\text{N/m}^3 + 6 \times 10^{-16}\,\text{N/m}^3]/B = 10^{-15} \times L^3/B_J \approx (3.5 \times 10^{-13}\,\text{N/m}^3)/(4 \times 10^{-4}\,\text{T}) \approx 10^{-9}\text{Am}^{-2}$. The cross-sectional area of the Io torus for $L = 6-8$ is $area \approx \Delta z \Delta L R_J \approx 4R_J^2 \approx 2 \times 10^{16}\,\text{m}^2$. The total "ring" current for this region is then

$$I_\phi \approx area \times J_\phi \approx 20\,\text{MA}. \tag{8.127}$$

This Jovian ring current encircles Jupiter just as the terrestrial ring current encircles the Earth, but in the opposite direction, although the dynamical "drivers" are different for these currents.

Energetic ion populations are present in the Jovian magnetosphere as well as the relatively cold but dense thermal ion population, especially beyond $L = 10$, and we know that for the terrestrial ring current the energetic ion population makes the dominant contribution to the pressure and to the electrical current. Information on the Jovian energetic ion population is minimal, which makes it difficult to determine the associated electrical current accurately. But as discussed by Vasyliunas (1983), instead of using Equation (8.126), the procedure can be reversed and Equation (8.126) can be used to estimate the pressure if the current density can be estimated independently. Comparison of magnetic field vectors measured along the *Pioneer* and *Voyager* spacecraft trajectories with the magnetic field expected for a simple tilted dipole provides an empirical determination of the current density **J**, from which the variation of the pressure can be estimated. This method was used by several researchers to estimate the pressure in the plasmasheet (and current sheet), as discussed in the book *Physics of the Jovian Magnetosphere* (Dessler, 1983). For the range $L \approx 8$–80 this empirical pressure variation can crudely be represented by the following function of L-shell:

$$p \approx 150 \, \text{keV/cm}^3 \quad [10/L]^4. \tag{8.128}$$

The actual deduced pressure variation (e.g., Vasyliunas, 1983) depends on the empirical magnetic field model that is used and has a somewhat more complex variation than is exhibited by Equation (8.128). In SI units, Equation (8.128) is $p = 2.5 \times 10^{-8} [10/L]^4 \, \text{N/m}^2$.

At $L \approx 7$ the simple formula (8.128) predicts that $p \approx 10^{-7} \, \text{N/m}^2$, which is about a factor of two greater than the value we estimated from the measured thermal plasma parameters in this region; however, the more energetic plasma also makes some contribution to the pressure, even at $L \approx 7$. The energetic plasma becomes especially important for L values greater than about 10. Equation (8.128) includes pressure contributions from *all* particle populations. The pressure contribution from ions (e.g., H^+ and O^+) in the 28 keV to 4 MeV energy range has been evaluated for the nightside portion of the Jovian plasmasheet using ion fluxes measured by instruments on board *Voyagers 1* and *2* (see Paranicas et al., 1991). This pressure roughly agrees with Equation (8.128), confirming the importance of the contribution of the more energetic ions.

Equation (8.128) can be used to estimate the plasma beta as a function of L with the assumption of a dipole magnetic field:

$$\beta = p/(B^2/2\mu_0) \approx 2 \times 10^{-3} L^2. \tag{8.129}$$

At $L \approx 7$, Equation (8.129) predicts that $\beta \approx 0.1$. Equation (8.129) indicates that β exceeds unity only for $L > 22$. Hence, beyond about $L \approx 20$ the magnetic field is strongly affected by the electrical current flowing in the plasmasheet, and the dipole field formula used to derive Equation (8.129) is no longer valid. Nonetheless, if we use it anyway to estimate β near the magnetopause ($L \approx 60$), we find that $\beta \approx 6$, which is roughly correct. This is also the value of β we used in Chapter 7 to estimate the magnetopause location for Jupiter.

The simple pressure formula (8.128) can also be used together with Equation (8.126) to estimate the azimuthal electrical current in the plasmasheet of the "middle magnetosphere" ($L \approx 10$–20). This is an oversimplification, which we will overlook for now. To use Equation (8.128) we need to know how the density ρ varies with L in order to evaluate the centrifugal force contribution to the current.

Using *Voyager* data as reviewed by Bagenal (1985), the following formula can be devised and provides a crude estimate of the L-shell variation of the electron density in the plasmasheet for $L \approx 10$–20:

$$n_e \approx 50\,\mathrm{cm}^{-3}\,[10/L]^3. \tag{8.130}$$

This formula underestimates the electron density for $L < 10$ (by about a factor of four at $L = 7$). Using this equation plus the pressure formula, as well as adopting a dipole magnetic field, we can derive an approximation for the current density of

$$J_\phi \approx 3 \times 10^{-8}L^{-2} + 10^{-11}L \quad [\mathrm{Am}^{-2}], \tag{8.131}$$

where the first term is from the pressure gradient force and the second term is from the centrifugal force. At $L \approx 10$ we find that $J_\phi \approx 4 \times 10^{-10}$ A/m^2 with the pressure term being a factor of three greater than the centrifugal term. By $L \approx 20$, Equation (8.131) indicates that $J_\phi \approx 3 \times 10^{-10}$ A/m^2 with the centrifugal term being three times greater than the pressure term. The two terms cross over in importance at about $L \approx 15$.

The total ring current carried in the region between $L = 10$ and 20 can be found by integrating Equation (8.131) from $L = 10$ to 20 over a suitable cross-sectional area (e.g., 2 R_J). This procedure gives a total current of about 30 MA.

The above picture is oversimplified. For example, the magnetic field is not really dipolelike due to the large electrical currents flowing in the plasmasheet, and the pressure is not really isotropic. And there is evidence that, at least on the nightside, the relative importance of the centrifugal force term is less than is indicated by Equation (8.131) (Paranicas et al., 1991; McNutt, 1983). The momentum equations (2.59) and (2.61) contain the divergence of a *pressure tensor* for species s, $\nabla \cdot \tilde{\mathbf{P}}_s$, but with one exception (Section 3.10.3) we have adopted isotropic pressure ($p_\parallel = p_\perp$) throughout this book. Recall that $p_\parallel$ and $p_\perp$ are the parallel and perpendicular pressures, respectively. Paranicas et al. (1991) demonstrated that the radial component of the pressure force $-\nabla \cdot \tilde{\mathbf{P}}_s$ can be expressed as

$$F_r = -\frac{\partial p_\perp}{\partial r} + \frac{p_\perp}{R_c}\left(\frac{p_\parallel}{p_\perp} - 1\right), \tag{8.132}$$

where R_c is the radius of curvature of the magnetic field lines (also see Equation (3.42)). The first term of Equation (8.132) was already included in Equation (8.126) but the second term is new, and according to Paranicas et al. (1991), it is several times more important than the first term for the nightside Jovian neutral sheet. These authors estimate that $p_\parallel/p_\perp \approx 1.2$ and that $R_c \approx 0.5\,R_J$ near $L \approx 20$. Hence the second term in Equation (8.132) is $\approx 0.4(p_\perp/R_J)$, whereas the first term is $\approx 4p_\perp/(LR_J) \approx 0.2(p_\perp/R_J)$.

8.7.4 Radial plasma transport in the Jovian magnetosphere

In considering the momentum balance in the way we did, the plasma was assumed
to be purely corotating, in which case the plasma density is solely determined
by local production and loss processes. Local production and loss processes are
indeed very important in the Io plasma torus, but radial plasma transport is also
important (cf. Bagenal, 1992). However, radial transport is thought to take place
primarily as a diffusion process rather than as "organized convection." An important
exception to this is thought to exist in the tail for $L > 50$, where the centrifugal force
overwhelms the capacity of the inward $\mathbf{J} \times \mathbf{B}$ force to counteract it such that bulk
plasma transport down the Jovian magnetotail takes place.

The radial transport Equation (3.105) for cross-L-shell diffusion is applicable to
the Jovian magnetospheric plasma, although source and sink terms must be added. A
steady-state version of Equation (3.105) can be written for the pitch-angle-averaged
distribution function in the equatorial plane:

$$L^2 \frac{\partial}{\partial L} \left(\frac{D_{\text{LL}}}{L^2} \frac{\partial f}{\partial L} \right) - \frac{f}{\tau_{\text{net}}} = 0, \tag{8.133}$$

where D_{LL} is the cross-L-shell diffusion coefficient. The distribution function,
$f(E, L)$, for any given particle species (mainly S or O ions in this case) can be
expressed as a function of particle energy and L-shell. At lower energies, the distribution function is that of the thermal plasma, but at higher energies, in the "tail" of
the distribution, the ring current and radiation belt populations are being represented.

The "net" local production time constant, or lifetime, τ_{net}, takes into account all
local production and loss processes as a function of particle energy. For the thermal
particles, collisional production and loss are the most important processes, and
these are mainly associated with the Io plasma torus. However, for the energetic
particles, the loss mechanisms include encounters with the satellites of Jupiter and
losses due to pitch-angle scattering into the loss cone. The latter process results
in the precipitation of energetic particles into the upper atmosphere of Jupiter. In
fact, auroral emissions from the polar regions of Jupiter have been observed by
the ultraviolet spectrometers on the *Voyager 1* and *2* spacecraft as well as by the
Earth-orbiting *International Ultraviolet Explorer (IUE)* satellite and the *Hubble
Space Telescope* (c.f. Nagy et al., 1995). The total power of the Jovian aurora
estimated from these observations is about 10^{13}–10^{14} W, which is a factor of 10
to 100 greater than the total power of the terrestrial aurora. Radio emissions in
several frequency bands have also been observed from the Jovian magnetosphere,
and some of these emissions are associated with the Io torus and with the aurora
(cf. Bagenal, 1992).

It is not known how the energetic particles in the Jovian magnetosphere, some
of which are responsible for the aurora, are accelerated, although several different
theories have been put forth. Particles can be accelerated in a variety of ways. In
particular, two processes can alter the energy of a charged particle undergoing radial
transport:

(1) *Betatron acceleration* can occur – a particle transported to a lower L-shell has its perpendicular kinetic energy increased as the magnetic field strength increases if the first adiabatic invariant ($\mu_{\mathrm{m}} = \frac{1}{2}mv_{\perp}^2/B$) is conserved during the transport process. If wave–particle scattering makes the distribution function isotropic so that parallel and perpendicular energy are equipartitioned, then the adiabatic relation (Example 8.4) applies and predicts plasma energization.

(2) The *corotation electric field* can also energize plasma. We saw that ions born in the Io plasma torus gain a few hundred eV from the corotating Jovian magnetic field and the associated corotation electric field. The corotation energy, $E_{\mathrm{corot}} = 1/2m\Omega^2 r^2$, increases as r^2 (or L^2). Outward diffusion should lead to energy gain from this process. Of course in a corotating reference frame this is *not* an energy gain. Furthermore, adiabatic cooling from process (1) can counteract this energy gain.

Radial diffusion transport times estimated for the Jovian magnetosphere are in the range of about 10–100 days for the thermal part of the plasma for $L > 5.7$ and are a few years for $L < 5.7$. The cross-L diffusion coefficients used by researchers have been mainly determined empirically, but speculation concerning the responsible physical processes has centered on the *magnetic flux tube interchange instability*, which is a macroscopic MHD instability that can be driven by the outwardly directed centrifugal force. This topic will not be considered here, but given a diffusion coefficient you recall that the transport time can be estimated as $\tau_{\mathrm{trans}} \approx \Delta L^2/D_{\mathrm{LL}}$, where ΔL is the interval over which the diffusive transport takes place. An average radial speed for the plasma can be estimated as $\langle u_r \rangle \approx R_{\mathrm{J}}\Delta L/\tau_{\mathrm{trans}}$. Using reasonable values of $\tau_{\mathrm{trans}} = 30$ days and $\Delta L \approx 5$, we find that $\langle u_r \rangle \approx 0.1$ km/s. Our earlier assumption that the corotational motion is dominant is indeed valid given that $\langle u_r \rangle/u_{\mathrm{corot}} = \langle u_r \rangle/(\Omega R_{\mathrm{J}}L) \approx 10^{-3}$ near the Io plasma torus. Nonetheless, although $\langle u_r \rangle \ll u_{\mathrm{corot}}$, the radial motion plays an important role in determining the particle distribution function in the Jovian magnetosphere.

The processes operating in the Saturnian magnetosphere are very similar to what has just been discussed for Jupiter, except that instead of an Io plasma source Saturn has plasma sources associated with its rings and with Titan. NASA's *Galileo* spacecraft went into orbit around Jupiter in December 1995 and is now making measurements of the particles and fields in the Jovian magnetosphere. And it is expected that the *Cassini* spacecraft will go into orbit around Saturn in the year 2004. Our knowledge of the magnetospheres of Jupiter and Saturn will certainly greatly increase because of the data returned by these two missions.

Problems

8.1   Derive Equation (8.5) for $d\rho/\rho$ from information found in Section 6.2.1. For strongly subsonic flow ($M^2 \ll 1$), show that this is equivalent to $dM/M = -dA/A$.

8.2   Show that in steady state and for cylindrical geometry, $\nabla \cdot (B\mathbf{u}) = 0$, for infinite electrical conductivity. The magnetic field is assumed to be aligned with the axis of the cylinder. For incompressible flow, demonstrate that this implies that $B \approx constant$.

8.3 **Determination of Alfvénic Mach number for the plasma flow into the reconnection region** Combine Equations (8.14) and (8.17) to obtain an algebraic equation for $M_{A0\,max}$:

$$M_{A0\,max} = \frac{\pi}{8} \frac{1}{\ln\left(\frac{r_{mp}}{r_L}\right) - \ln(M_{A0\,max})}.$$

Solve this equation iteratively and show that $M_{A0\,max} \approx 0.04$ for the values of the parameters discussed in the text. What happens to $M_{A0\,max}$ if larger values of either r_{mp} or the width of the diffusion region a are employed instead of the values tried in the text?

8.4 Estimate values for the parameters b, d, and L that appear in Figure 8.7.

8.5 Demonstrate that the *Poynting vector* (see the appendix for a review of electromagnetic theory) can be used to derive Equation (8.22). Also derive Equation (8.25).

8.6 Review Chapter 3 and demonstrate that the total amount of energy in the ring current is about 10^{15} J for moderately active conditions.

8.7 Use the reconnection formalism (e.g., u_0, etc.) to demonstrate that the power conversion of magnetic to kinetic energy at the dayside magnetopause via reconnection is, indeed, for southward IMF, about equal to $P_{sw-m} \approx 10^{11}$ W.

8.8 **Derivation of the electron inertial length** Derive Equation (8.27) for a sharp interface between a magnetized region that is a vacuum and an unmagnetized plasma region. Assume equal electron and ion temperatures. Consider what electric potential is needed to repulse the incident protons and use a simple generalized Ohm's law, $\mathbf{E} = -\mathbf{u}_e \times \mathbf{B}$, where $\mathbf{u}_e$ is the bulk flow velocity of the electrons. Sketch the current layer showing $\mathbf{E}$, $\mathbf{u}_e$, and the current density $\mathbf{J}$.

8.9 Demonstrate that for an ionosphere with uniform electrical conductivity the Hall current makes a negligible contribution to the field-aligned current into (or out of) the top of the ionosphere.

8.10 Repeat the analysis of Section 8.3 for a scenario in which the top plasma slab has a thickness of $3H$ rather than H. Assume that all other parameters remain the same. Assume that for the top slab $\partial u_0 / \partial t = 0$ but $\partial p / \partial y \neq 0$. Determine both u_0 and u_1 (i.e., the flow speeds in the top and bottom slabs, respectively) as well as the electric field strength and the current densities.

8.11 Derive an ordinary differential equation for the velocity function $u_0(t)$ in Section 8.3.3 and show that Equation (8.50) is the solution. Also derive Equation (8.66).

8.12   (a) Demonstrate that Equations (8.69)–(8.71) are consistent with Equation (8.60).

        (b) Derive Equations (8.73) and (8.74) given the simplifying assumptions made in Section 8.4.

8.13   Derive Equations (8.86) and (8.88). Then find the plasmapause L-shell, L_{pp}, for *all* local times rather than just 18:00 LT as was done in the text.

8.14   **Plasmaspheric refilling times** The ionospheric source (i.e., upward flux) of H^+ ions produced by the reaction of O^+ ions with neutral atomic hydrogen can be found by integrating over altitude the production rate, $P = k[O^+][H]$, where the reaction rate coefficient $k = 6 \times 10^{-10}\,cm^3 s^{-1}$, where the hydrogen density is approximately $[H] \approx 10^5\,cm^{-3}$ (almost constant in F-region altitude range), and where the peak O^+ density is $[O^+] \approx 10^6\,cm^{-3}$. Show that the H^+ flux upward into the plasmasphere is approximately given by: $flux \approx PH_p \approx 10^9\,cm^{-2}\,s^{-1}$. Estimate plasmaspheric refilling times for $L = 2, 3, 4$, and 5.

8.15   Estimate the trajectories of H^+, He^+, and O^+ ions that start out in the polar ionosphere with an average energy (i.e., temperature) of 30 eV and then move out into the magnetosphere. Sketch the trajectories (in the noon–midnight plane) you have found. Assume a cross-tail magnetospheric potential of 100 kV and use dipole magnetic field lines.

8.16   Derive the electromagnetic energy conservation equation (A.45) from Maxwell's equations.

8.17   Show that the dynamical pressure term, the frictional force term, and the gravitational term can be neglected in the momentum balance for the terrestrial plasmasheet at a radial distance of about 20 R_E. Take information as needed from the tables and figures of Chapter 8. Assume a neutral hydrogen density of about $1\,cm^{-3}$ at this location.

8.18   Estimate the gyroradius of a typical plasmasheet proton near the neutral sheet (a) for a constant field $|B_z| = 3\,nT$ and (b) as a function of L from $L = 6$ to $L = 20$ for a dipole field. Show that the adiabatic invariants (Chapter 3) might not be conserved right in the neutral sheet where the length scale for significant field changes is roughly 1 R_E. The resulting proton trajectories are "chaotic" and could play an important role in the plasmasheet dynamics during magnetic substorms when the plasmasheet is especially thin.

8.19   Estimate the height-integrated Hall current density in the auroral oval, K_H, and also the total Hall current near the auroral oval. Assume a magnetospheric potential of $\Psi_m = 100\,kV$ and a width of the oval of about $5°$ latitude. Making the approximation that all this current runs in a thin filament (like a wire) along the auroral oval at an altitude of 100 km, estimate the magnetic perturbation at the ground due to this *auroral electrojet* current. Demonstrate that the Hall currents flow along equipotential lines in a direction opposite to the polar cap plasma convection.

8.20 Apply Equation (8.111) to the plasmasheet electron and proton populations and find the maximum current density that can be carried by these particle populations (e.g., for electrons from Table 8.1, $n_e \approx 1\,\text{cm}^{-3}$ and $T_e \approx 1\,\text{keV}$). Also apply this equation to upward flowing ionospheric O^+ ions originating in the topside ionosphere and show that $J_{\text{max}} \approx 0.5\,\mu\text{A/m}^2$.

8.21 Review the treatment of kinetic theory in Chapter 2 and then demonstrate that charged particles accelerated by a one-dimensional electrostatic field (either $B = 0$ or only motion parallel to B is considered) conserve their flux in the direction of this field. That is, the flux of particles of species s is given by $n_s u_{sx} = constant$ (in the x direction) if the electrostatic field is in the x direction.

8.22 Consider two locations on a magnetic field line with radial distances $r_2 > r_1$. The magnetic field strength at r_2 is only 20% of the field strength at r_1; that is, $B(r_2) = 0.2B(r_1)$. Suppose we start with an isotropic flux of monoenergetic 1 keV electrons at r_2. The electron density at r_2 is $1\,\text{cm}^{-3}$. Calculate the loss cone angle for the distribution at r_2 assuming that all electrons are somehow physically absorbed at r_1. Also calculate the electron flux parallel to the field at both r_1 and r_2. Now assume that the electrons are accelerated by an electrostatic field in the direction parallel to the magnetic field. Assume that the field-aligned potential drop occurs in a narrow region ($\Delta r \ll r_2$) located just below r_2. Estimate the electron flux parallel to the magnetic field both at r_1 and r_2.

8.23 Estimate the energy flux into the atmosphere associated with the differential electron flux shown in Figure 8.48. Near what altitude is most of the energy deposited? Use Equation (8.103) to estimate the electrical conductivity of the auroral ionosphere associated with this particular auroral precipitation event. About how many ion pairs are produced in a vertical column with cross-sectional area $1\,\text{cm}^2$? Dividing this column ion production rate by a typical scale height in the lower thermosphere (15 km), you can also get a crude estimate of the total ion production rate, and then by assuming that dissociative recombination ultimately removes ionospheric plasma in the E-region you can also photochemically estimate the peak ionospheric electron density. How does this compare with nonauroral E-region electron densities?

8.24 Derive for planet "x" the following formula for the critical speed at which the corotation speed at an L-shell matches the magnetic curvature and gradient drift speed v_B:

$$v_{\text{crit}}^2 \approx \Omega_{\text{gyro}}\Omega_{\text{rotation}}R_x^2 L^2,$$

where R_x is the radius of the planet, Ω_{rotation} is the rotational frequency, and Ω_{gyro} is the gyrofrequency at that L value. Evaluate the critical energy ($(1/2)mv_{\text{crit}}^2$) for protons at $L = 6$ in the terrestrial plasmasphere and for oxygen ions at $L = 10$ in the Jovian magnetosphere.

8.25 Determine the surface magnetic field strength at the magnetic equators of all the outer planets given the data in Table 7.1 and assuming that the field is given by a simple magnetic dipole.

8.26 Calculate the gyroperiod, bounce period, and drift period of 50 MeV protons at $L = 10$ in the Jovian radiation belts. Assume a pitch angle of $45°$. Compare your results with the terrestrial values in Table 3.1.

8.27 Show that $\hat{n} \cdot \hat{b} \cong 3z/r$ near the equatorial plane of a magnetosphere of a planet with a dipole magnetic field, where the $\hat{n}$ vector is parallel to the centrifugal force, which is directed outward from the rotation axis, r is the radial distance, and z is the height above the rotational (and magnetic) equatorial plane.

8.28 Derive Equation (8.131) and then use Equation (8.132) to find a new version of Equation (8.131). Estimate and then plot the current density versus L value for the Jovian magnetosphere.

8.29 Calculate the corotation speed (and energy) for Jovian oxygen ions and Saturnian nitrogen ions and plot these speeds versus L value. Do this for $L = 1$ to 60 for Jupiter and $L = 1$ to 25 for Saturn.

8.30 Consider a hypothetical Saturn whose rotational frequency is only 0.1% that of the real Saturn; all other parameters are the same. Estimate the distance to the plasmapause (L_{pp}) for this planet. Estimate the magnetospheric electric field due to the solar wind assuming an open magnetosphere model and a reasonable dayside magnetopause magnetic reconnection efficiency. What is the direction of this electric field? What is the cross-tail magnetospheric potential? Sketch the magnetospheric plasma convection pattern in the equatorial plane.

8.31 Let us suppose that Titan supplies 10^{25} N atoms per second to a torus centered R_s at the orbital distance of Titan (20.3 R_S) and with an extent (or minor radius) of ± 1 R_S. These ions are all eventually ionized and thus contribute to the mass-loading of the magnetosphere near $L \approx 20$. Estimate the total electrical current associated with this mass-loading. Assume that Saturn's ionospheric properties are the same as was assumed for Jupiter in the text.

Bibliography

Arnoldy, R. L., P. B. Lewis, and P. O. Isaacson, Field-aligned auroral electron fluxes, *J. Geophys. Res.*, **79**, 4208, 1974.

Axford, W. I. and C. O. Hines, A unifying theory of high-latitude geophysical phenomena and geomagnetic storms, *Can. J. Phys.*, **39**, 1433, 1961.

Bagenal, F., Planetary magnetospheres, p. 224 in *Solar System Magnetic Fields*, ed. E. R. Priest, D. Reidel Publ. Co., Dordrecht, The Netherlands, 1985.

Bagenal, F., Giant planet magnetospheres, *Annu. Rev. Earth Planet. Sci.*, **20**, 289, 1992.

Baumjohann, W., Ionospheric and field-aligned current systems in the auroral zone: A concise review, *Adv. Space Res.*, Vol. 2, No. 10, 55, 1982.

Burke, W. J. and M. Heinemann, Origins and consequences of parallel electric fields, p. 393 and p. 609 in *Solar-Terrestrial Physics: Principles and Theoretical Foundations*, ed. R. L. Carovillano and J. M. Forbes, D. Reidel Publ. Co., Dordrecht, The Netherlands, 1983.

Carpenter, D. L. and R. R. Anderson, An ISEE/Whistler model of equatorial electron density in the magnetosphere, *J. Geophys. Res.*, **97**, 1097, 1992.

Chappell, C. R., The terrestrial plasma source: A new perspective in solar-terrestrial processes from Dynamics Explorer, *Rev. Geophys.*, **26**, 229, 1988.

Chappell, C. R., K. K. Harris, and G. W. Sharp, The morphology of the bulge region of the plasmasphere, *J. Geophys. Res.*, **75**, 3848, 1970.

Cowley, S. W. H., "Magnetic reconnection", p. 121 in *Solar System Magnetic Fields*, ed. E. R. Priest, D. Reidel Publ. Co., Dordrecht, The Netherlands, 1985.

Crooker, N. U., Dayside merging and cusp geometry, *J. Geophys. Res.*, **84**, 951, 1979.

Dessler, A. J., ed., *Physics of the Jovian Magnetosphere*, Cambridge Univ. Press, Cambridge, UK, 1983.

Dungey, J. W., Interplanetary magnetic field and the auroral zones, *Phys. Rev. Lett.*, **6**, 47, 1961.

Elphinstone, R. E., J. S. Murphee, and L. L. Cogger, What is a global auroral substrorm?, *Rev., Geophys.*, **34**, 169, 1996.

Frank, L. A. and J. D. Craven, Imaging results from Dynamics Explorer 1, *Rev. Geophys.*, **26**, 249, 1988.

Frank, L. A., J. D. Craven, and R. L. Rairden, Images of the Earth's aurora and geocorona from the Dynamics Explorer mission, *Adv. Space Phys.*, **5**, 53, 1985.

Geiss, J., G. Gloeckler, H. Balsiger, L. A. Fisk, A. B. Galvin, F. Gliem, D. C. Hamilton, F. M. Ipavich, S. Livi, U. Mall, K. W. Ogilvie, R. von Steiger, and B. Wilken, Plasma composition in Jupiter's magnetosphere: Initial results from the solar wind ion composition experiment, *Science*, **257**, 1535, 1992.

Haerendal, G. and G. Paschmann, Interaction of the solar wind with the dayside magnetosphere, chap. 2 in *Magnetospheric Plasma Physics*, ed. A. Nishida, Center for Academic Publ. Japan, D. Reidel Publ. Co., Dordrecht, The Netherlands, 1982.

Hill, T. W., Solar-wind magnetosphere coupling, p. 261 in *Solar-Terrestrial Physics: Principles and Theoretical Foundations*, ed. R. L. Carovillano and J. M. Forbes, D. Reidel Publ. Co., Dordrecht, The Netherlands, 1983.

Hones, E. W., Jr., Plasma sheet behaviour during substorms, p. 178 in *Magnetic Reconnection in Space and Laboratory Plasmas*, ed. E. W. Hones Jr., Geophysical Monograph **30**, American Geophys. Union, Washington DC, 1984.

Iijima, T. and T. A. Potemra, Large-scale characteristics of field-aligned currents associated with substorms, *J. Geophys. Res.*, **83**, 599, 1978.

Joselyn, J. A., Geomagnetic activity forecasting: The state of the art, *Rev. Geophys.*, **33**, 383, 1995.

Kan, J. R., Synthesizing a global model of substorms, p. 73 in *Magnetospheric Substorms*, Geophysical Monograph **64**, American Geophysical Union, Washington, DC, 1991.

Kistler, L. M., E. Mobius, W. Baumjohann, and G. Paschmann, Pressure changes in the plasmasheet during substorm injections, *J. Geophys. Res.*, **97**, 2973, 1992.

Kivelson, M. G. and C. J. Russell, eds., *Introduction to Space Physics*, Cambridge Univ. Press, Cambridge, UK, 1995.

Lanzerotti, L. J., T. P. Armstrong, R. E. Gold, K. A. Anderson, S. M. Krimigis, R. P. Lin, M. Pick, E. C. Roelof, E. T. Sarris, G. M. Simnett, C. G. Maclennan, H. T. Choo, and S. J. Tappin, The hot plasma environment at Jupiter: Ulysses Results, *Science*, **257**, 1518, 1992.

Lui, A. T. Y., A synthesis model for magnetospheric substorms, p. 43 in *Magnetospheric Substorms*, Geophysical Monograph **64**, American Geophysical Union, Washington, DC, 1991.

Lui, A. T. Y. and D. C. Hamilton, Radial profiles of quiet time magnetospheric parameters, *J. Geophys. Res.*, **97**, 19325, 1992.

Lyons, L. R., Formation of auroral arcs via magnetosphere-ionosphere coupling, *Rev. Geophys.*, **30**, 93, 1992.

Lyons, L. R. and D. J. Williams, *Quantitative Aspects of Magnetospheric Physics*, D. Reidel, Publ. Co., Dordrecht, The Netherlands, 1984.

Magnetospheric Substorms, Geophysical Monograph **64**, American Geophysical Union, Washington, DC, 1991.

McNutt, R. L., Force balance in the magnetospheres of Jupiter and Saturn, *Adv. Space Res.*, **3**, 55, 1983.

Moore, T. E., Origins of magnetospheric plasma, *Rev. Geophys. Suppl.*, U.S. National Report to IUGG 1987–1990, p. 1039, 1991.

Nagy, A. F., T. E. Cravens, and H. J. Waite Jr., All ionospheres are not alike: Reports from other planets, *Rev. Geophys. Suppl.* p. 525, U.S. National Report to International Union of Geodesy and Geophysics 1991–1994, 1995.

Paranicas, C. P., B. H. Mauk, and S. M. Krimigis, Pressure anisotropy and radial stress balance in the Jovian neutral sheet, *J. Geophys. Res.*, **96**, 21135, 1991.

Parks, G. K., *Physics of Space Plasma – An Introduction*, Addison-Wesley Publ. Co., Redwood City, CA, 1991.

Paschmann, G., Plasma structure of the magnetopause and boundary layer, p. 25 in *Magnetospheric Boundary Layers*, ed. B. Battrick, ESA SP-148, Paris, 1979.

Paschmann, G., B. U. Ö Sonnerup, I. Papamastorakis, N. Jekopke, G. Haerendal, S. J. Bame, J. R. Asbridge, J. T. Gosling, C. T. Russell, and R. C. Elphic, Plasma acceleration at the earth's magnetosphere: Evidence for reconnection, *Nature*, **282**, 243, 1979.

Perrault, P. and S.-I. Akasofu, A study of geomagnetic storms, *Geophys. J. R. Astron. Soc.*, **54**, 547, 1978.

Pudovkin, M. I., Physics of magnetospheric substorms: a review, p. 17 in *Magnetospheric Substorms*, Geophysical Monograph **64**, American Geophysical Union, Washington, DC, 1991.

Ratcliffe, J. A., *An Introduction to the Ionosphere and Magnetosphere*, Cambridge Univ. Press, Cambridge, UK, 1972.

Rees, M. H., *Physics and Chemistry of the Upper Atmosphere*, Cambridge Atmospheric and Space Science Series, Cambridge Univ. Press, Cambridge, UK, 1989.

Rostoker, G., Observational constraints for substorm models, p. 61 in *Magnetospheric Substorms*, Geophysical Monograph **64**, American Geophysical Union, Washington, DC, 1991.

Russell, C. T., The magnetopause of the Earth and planets, *Adv. Space Res.*, **1**, 67, 1981.

Smith, M. F. and M. Lockwood, Earth's magnetospheric cusps, *Rev. Geophys.*, **34**, 233, 1996.

Southwood, D. J., An introduction to magnetospheric MHD, chap. 2 in *Solar System Magnetic Fields*, ed. E. R. Priest, D. Reidel Publ. Co., Dordrecht, The Netherlands, 1985.

Spreiter, J. R., A. L. Summers, and A. Y. Alksne, Hydromagnetic flow around the magnetosphere, *Planet. Space Sci.*, **14**, 223, 1966.

Vasyliunas, V. M., Comparative magnetospheres, p. 479 in *Solar-Terrestrial Physics: Principles and Theoretical Foundations*, ed. R. L. Carovillano and J. M. Forbes, D. Reidel Publ. Co., Dordrecht, The Netherlands, 1983.

Vasyliunas, V. M., Plasma distribution and flow, chap. 11 in *Physics of the Jovian Magnetosphere*, ed. A. J. Dessler, Cambridge Univ. Press, Cambridge, UK, 1983.

Voigt, G.-H. and R. A. Wolf, Quasi-static magnetospheric MHD processes and the "ground state" of the magnetosphere, *Rev. Geophys.*, **26**, 823, 1988.

Volland, H., *Atmospheric Electrodynamics*, Physics and Chemistry in Space Vol. 11,

Springer-Verlag, Berlin, 1984.

Walker, R. J. and M. Ashour-Abdalla, The magnetosphere in the machine: Large scale theoretical models of the magnetosphere, *Rev. Geophys. Suppl.* p. 639, U.S. National Report to International Union of Geodesy and Geophysics 1991–1994, 1995.

Walker, R. J., T. Ogino, J. Raeder, and M. Ashour-Abdalla, A global magnetohydrodynamic simulation of the magnetosphere when the interplanetary magnetic field is southward: The onset of magnetotail reconnection, *J. Geophys. Res.*, 98, 17235, 1993.

Appendix: Review of electromagnetic theory

Space plasmas all contain electrons and ions. Electrons and ions, being charged particles, have electric and magnetic fields associated with them. Electric and magnetic fields in classical physics are described by Maxwell's equations. This appendix provides a brief review of electromagnetic theory, emphasizing Maxwell's equations. For a more complete treatment of this subject, the reader is referred to any standard text on electromagnetic theory.

A.1 Units

Unfortunately, a mixture of units has been used in the field of space/solar system physics, with cgs-Gaussian units being used most often. However, usage of the International System of Units (SI) is becoming increasingly common. The SI system is basically the rationalized mks (meters-kilograms-seconds) system of units. SI units will be used throughout this book, with a couple of exceptions. Number densities and lengths are often given in cgs units (i.e., cm^{-3} and cm, respectively) in order to agree with common usage. And a few important formulae are written in cgs-Gaussian units as well as in SI units. The unit of magnetic field intensity in the cgs-Gaussian system is the gauss.

There are four fundamental units in the SI system:

length	meters	[m]
mass	kilograms	[kg]
time	seconds	[s]
current	amperes	[A].

All other SI units can be constructed from these four units. The values of a few basic physical constants will now be given both in the SI and cgs-Gaussian systems. The speed of light in vacuum is

$$c = 2.998 \times 10^8 \text{ m/s} \, (= 2.998 \times 10^{10} \text{ cm/s in cgs}).$$

The mass of an electron is $m_e = 9.1095 \times 10^{-31}$ kg. Another important quantity is electrical charge, which has units of coulombs (C): 1 C = 1 ampere-second (1 A-s). The charge of a single electron is

$$e^- = 1.6022 \times 10^{-19} \text{ C} (= 4.8 \times 10^{-10} \text{ statcoulombs in cgs})$$

The SI unit for energy is the joule (J), and for power it is the watt (W). Another useful unit of energy is the electron volt (eV):

$$1 \text{ eV} = 1.602 \times 10^{19} \text{ J}.$$

The unit of temperature in either system of units is the kelvin (K), and Boltzmann's constant has the following value in SI:

$$k_B = 1.38 \times 10^{-23} \text{ J/K}.$$

The electromagnetic field quantities have the following names, symbols, and units:

electric field intensity	**E**	volts/meter	[V/m]
electric flux density	**D**	coulomb/meter^2	[C/m^2]
magnetic field intensity	**H**	ampere/meter	[A/m]
magnetic flux density	**B**	weber/meter^2	[Wb/m^2]
		(or tesla)	(T).

Note that the field quantities are vectors and are printed in bold-faced type. The relationships between the field quantities are described in the next section.

A.2 Field quantity relationships – Permittivity and permeability

The field quantities **D** and **E** basically refer to the same physical quantity in a vacuum, differing only by a multiplicative constant called the *permittivity of vacuum*, ε_0:

$$\mathbf{D} = \varepsilon_0 \mathbf{E}, \tag{A.1}$$

where $\varepsilon_0 = 8.854 \times 10^{-12}$ F/m. $\varepsilon_0 = 1$ in cgs-Gaussian units.

The relationship between **D** and **E** in a medium other than a vacuum is, in general, somewhat more complicated than (A.1). The electric field in a material such as a dielectric includes contributions from both the "free" charges in the system and the "bound" charges contained in the atoms and molecules of the material. The bound charge contribution is usually represented by the *polarization* vector **P** (C/m^2). **P** represents the integrated effect of the polarization of the bound charge in each of the atoms or molecules of the material. An expression for the electric flux density in terms of **P** and **E** is

$$\mathbf{D} = \varepsilon_0 \mathbf{E} + \mathbf{P} \quad \text{(in cgs-Gaussian: } \mathbf{D} = \mathbf{E} + 4\pi \mathbf{P}). \tag{A.2}$$

D is determined solely from the free charges, as will be seen from Gauss's law, which will be described in the next section. For most materials, **P** is proportional to the electric field applied to the medium ($\mathbf{P} = \chi_e \varepsilon_0 \mathbf{E}$), but its behavior in plasmas is more complex.

The constant χ_e is called the *electric susceptibility* of the medium. χ_e can be written in terms of the *relative permittivity*, ε_R, as $\chi_e = (\varepsilon_R - 1)\varepsilon_0$. ε_R is also called the *dielectric constant.*. The *permittivity* of a material is $\varepsilon = \varepsilon_R \varepsilon_0$.

The following simple equation now relates **D** and **E**:

$$\mathbf{D} = \varepsilon \mathbf{E}. \tag{A.3}$$

The relative permittivity of air is $\varepsilon_R = 1.0006$.

The magnetic field quantities, **B** and **H**, differ only by a multiplicative constant in free-space:

$$\mathbf{B} = \mu_0 \mathbf{H}, \tag{A.4}$$

where $\mu_0 = 4\pi \times 10^{-7}$ H/m is the *permeability of vacuum*. $\mu_0 = 1$ in cgs-Gaussian units. In a magnetic material, each atom or molecule has a magnetic moment (i.e., a current "loop") due to unbalanced electron orbital motions and/or due to the intrinsic electron spin. These electrical currents can be called "bound" currents. The cumulative sum of these individual magnetic moments gives rise to an overall *magnetization* **M** (A/m) for the material. The magnetic field intensity, **H**, can be described solely in terms of the "free" currents in a system, as will be seen in the next section, but the magnetic flux density, **B**, must also include the effects of the bound currents as represented by **M**:

$$\mathbf{B} = \mu_0(\mathbf{H} + \mathbf{M}). \tag{A.5}$$

M itself depends on the applied magnetic field, and this dependence is linear for linear, isotropic media. In this case, $\mathbf{M} = \chi_m \mathbf{H}$, where χ_m is the *magnetic susceptibility*. Equation (A.5) becomes

$$\mathbf{B} = \mu \mathbf{H}. \tag{A.6}$$

The *permeability* can be written as $\mu = \mu_R \mu_0$, where the *relative permeability* μ_R is equal to $1 + \chi_m$.

Plasmas are obviously not a vacuum, and they could, in principle, be described using the permittivity ε and the permeability μ. However, the electrical charges in a plasma do not behave as simply as the bound charges and currents in normal materials. It is easier to treat all charges and currents in a plasma as "free." The field quantities then obey the vacuum relations (A.1) and (A.4). Nonetheless, some of the concepts just reviewed, such as the dielectric constant, often appear in plasma physics.

A.3 Lorentz force

The force, $\mathbf{F}$, on a moving charge, q, is the sum of the force due to the electric and magnetic fields at the position of the charge. This force is given by the *Lorentz force* equation:

$$\mathbf{F} = q(\mathbf{E} + \mathbf{v} \times \mathbf{B}) \quad [\text{N}], \tag{A.7}$$

where $\mathbf{v}$ is the velocity of the charge. Obviously, the force on a stationary charge ($\mathbf{v} = \mathbf{0}$) is due to the electric field alone. Equation (A.7) can be generalized to find the net force on a collection of charges such as one finds in a fluid containing both electrons and ions. For those readers more familiar with cgs-Gaussian units, the Lorentz force equation in those units is

$$\mathbf{F} = q\left(\mathbf{E} + \frac{1}{c}\mathbf{v} \times \mathbf{B}\right) \quad [\text{dynes}]. \tag{A.8}$$

The Lorentz force equation combined with Newton's law ($\mathbf{F} = m\mathbf{a}$, where m is the particle mass and $\mathbf{a}$ is the acceleration) determines how a particle with mass m and charge q moves in specified electric and magnetic fields. The electric and magnetic field intensities are determined by Maxwell's equations; these are reviewed in the next section.

A.4 Maxwell's equations

The four field quantities ($\mathbf{D}, \mathbf{E}, \mathbf{B}, \mathbf{H}$) are determined by the four Maxwell's equations plus the subsidiary relations between $\mathbf{D}$ and $\mathbf{E}$ and between $\mathbf{B}$ and $\mathbf{H}$ given in Section A.2. James Clerk Maxwell derived these equations late in the nineteenth century, and they represent the culmination of over a hundred years of experimental and theoretical work by scientists such as Charles Coulomb, Joseph Henry, Michael Faraday, and André Marie Ampère. The two equations describing the electric field are

$$\nabla \cdot \mathbf{D} = \rho_c \quad \text{(Gauss's law)} \tag{A.9}$$

and

$$\nabla \times \mathbf{E} = -\frac{\partial \mathbf{B}}{\partial t} \quad \text{(Faraday's law).} \tag{A.10}$$

ρ_c is the *electric charge density* (C/m^3). The two equations describing the magnetic field are

$$\nabla \times \mathbf{H} = \mathbf{J} + \frac{\partial \mathbf{D}}{\partial t} \quad \text{(Ampère's law)} \tag{A.11}$$

and

$$\nabla \cdot \mathbf{B} = 0. \tag{A.12}$$

$\mathbf{J}$ is the *current density* (units of A/m^2) and $\partial \mathbf{D}/\partial t$ is the *displacement current*.

Gauss's law, Equation (A.9), states that the divergence of the electric flux density is equal to the electric charge density. Equation (A.9) together with Equation (A.1) provides the starting point for the field of *electrostatics*. In a vacuum, (A.9) becomes

$$\nabla \cdot \mathbf{E} = \rho_c / \varepsilon_0. \tag{A.13}$$

The cgs-Gaussian version of (A.15) is

$$\nabla \cdot \mathbf{E} = 4\pi \rho_c. \tag{A.14}$$

The integral form of Gauss's law is obtained by applying Gauss's integral theorem to (A.9):

$$\oint_S \mathbf{D} \cdot d\mathbf{S} = Q_S \quad [\text{C}], \tag{A.15}$$

where S is a closed surface and Q_S is the total electric charge enclosed by S:

$$Q_S = \int_v \rho_c \, dV. \tag{A.16}$$

V is the volume enclosed by the surface S. dV is a differential volume element and is sometimes written as d^3x. $d\mathbf{S}$ is a differential surface element on surface S, which has a direction normal to the surface in the outward direction. Equations (A.15) and (A.16) state that the total electric flux (C) crossing a closed surface must equal the total electric charge enclosed by that surface.

A time-varying magnetic field will produce an electric field – even if the charge density ρ_c is zero. The Maxwell's equation that describes this time-dependent behavior of the electric field is Equation (A.10) – Faraday's law, which states that the curl of the electric field intensity is proportional to the time rate of change of the magnetic flux density. The cgs-Gaussian version of Faraday's law is

$$\nabla \times \mathbf{E} = -\frac{1}{c} \frac{\partial \mathbf{B}}{\partial t}. \tag{A.17}$$

The integral form of Faraday's law can be found by applying Stokes's theorem to Equation (A.17), which gives

$$\oint_C \mathbf{E} \cdot d\mathbf{l} = -\frac{d\Phi_S}{dt}, \tag{A.18}$$

where Φ_S is the magnetic flux through that surface S enclosed by the closed contour C:

$$\Phi_S = \int_S \mathbf{B} \cdot d\mathbf{S}. \tag{A.19}$$

The line integral of the electric field intensity around the closed contour C is the electric potential, or emf (units of volts), induced by the time-varying magnetic flux enclosed by C. The electric potential is reviewed in Section A.5.

Equation (A.18) was derived from Equation (A.17) by assuming a contour C that does not move; the surface S was assumed to be constant in time. But it can be shown that Equation (A.18) also applies to a contour that is moving or parts of which are moving. That is, it is possible for an emf to exist around a contour even for a constant magnetic field, due to the time variation of the magnetic flux associated with a time variation of the surface area S. This contribution to the emf is a consequence of the *motional electric field*:

$$\mathbf{E} = -\mathbf{v} \times \mathbf{B}, \tag{A.20}$$

where $\mathbf{v}$ is the velocity of a particular location on C and $\mathbf{B}$ is the magnetic flux density. This expression can be found from the Lorentz force equation (A.8) by defining the motional electric field as that field required to make the net force on a moving test charge in a magnetic field equal to zero.

The next Maxwell's equation is Ampère's law (Equation (A.11)), which relates the curl of $\mathbf{H}$ to the sum of the current density, $\mathbf{J}$, and the displacement current, $\partial \mathbf{D}/\partial t$. Neglecting the displacement current (*magnetostatics*) Ampère's law can be written

$$\nabla \times \mathbf{B} = \mu_0 \mathbf{J}. \tag{A.21}$$

The integral form of Ampère's law is derived from Equation (A.21) by applying Stokes theorem:

$$\oint_C \mathbf{H} \cdot d\mathbf{l} = I_S, \tag{A.22}$$

where I_S [A] is the total current flowing through the surface S that is enclosed by C:

$$I_S = \int_S \mathbf{J} \cdot d\mathbf{S}. \tag{A.23}$$

Example A.1 (The magnetic field associated with an infinite sheet of current)
Relatively extensive thin sheets of current are often present in space plasmas, such as at the terrestrial magnetopause. This configuration can be approximated, for locations close to the current sheet, as an infinite sheet of current. Suppose that an infinitesimally thin current sheet is present in the x–y plane with current density $\mathbf{K} = K\hat{\mathbf{x}}$, where K is a constant with units of A/m. It is easy to show that the magnetic field associated with this current sheet is

$$\mathbf{B} = \begin{cases} -\dfrac{1}{2}\mu_0 K \hat{\mathbf{y}} & z > 0 \\[2mm] -\dfrac{1}{2}\mu_0 K \hat{\mathbf{y}} & z < 0. \end{cases} \tag{A.24}$$

The magnetic field above and below the x–y plane is uniform and in opposite directions.

The final Maxwell's equation is (A.12) ($\nabla \cdot \mathbf{B} = 0$). The integral form of this equation, which follows from applying Gauss's integral theorem, states that for a closed surface S, $\Phi_B = 0$; that is, the net magnetic flux through a closed surface must be zero. In other words, magnetic field lines do not begin or end (unlike electric field lines, which begin and end on electrical charges) and thus must always exist as closed loops. This equation reflects the empirical finding that magnetic monopoles have never (at least up to the present) been observed.

A.5 Scalar and vector potentials

Instead of working directly with the field quantities ($\mathbf{E}$, $\mathbf{D}$, $\mathbf{B}$, $\mathbf{H}$) themselves, it is often more convenient, although not necessary, to work with potential functions that are defined from the field quantities. The vector potential $\mathbf{A}$ is defined such that

$$\mathbf{B} = \nabla \times \mathbf{A}. \tag{A.25}$$

Note that the relation $\nabla \cdot \mathbf{B} = 0$ is automatically satisfied with this definition of vector potential because the divergence of a curl is identically zero. From Faraday's law and Equation (A.25), we can show that $\nabla \times (\mathbf{E} + \partial \mathbf{A}/\partial t) = 0$. Another vector identity states that the curl of a gradient of a scalar function is zero; hence, $\mathbf{E} + \partial \mathbf{A}/\partial t$ can be set equal to $-\nabla U$, where U is a scalar potential function (denoted V in some places in the text). We can now express the electric field intensity in terms of both the scalar and vector potential functions:

$$\mathbf{E} = -\nabla U - \frac{\partial \mathbf{A}}{\partial t}. \tag{A.26}$$

The scalar potential U can include an arbitrary additive constant. The vector potential $\mathbf{A}$ is not fully defined by these equations, and another condition (the "Gauge" condition) on $\mathbf{A}$ is needed (e.g., $\nabla \cdot \mathbf{A} = 0$). $\mathbf{A}$ and U can be used in place of $\mathbf{E}$ and $\mathbf{B}$, whenever it is convenient to do so.

For static conditions, (A.26) becomes

$$\mathbf{E} = -\nabla U. \tag{A.27}$$

Taking the curl of both sides of (A.27), we find that $\nabla \times \mathbf{E} = \mathbf{0}$ because the curl of a gradient is identically zero. This can also be found directly from Faraday's law for static conditions. In this case, the integral form of Faraday's law (A.18) indicates that the path integral of $\mathbf{E}$ around a closed path is always equal to zero, and the electric field is said to be *conservative*.

Gauss's law (A.9) can be rewritten using the definition of the potential. For a homogeneous and isotropic electrical medium we have

$$\nabla^2 U = -\rho_c/\varepsilon. \tag{A.28}$$

This is *Poisson's equation*, which relates the *Laplacian* ($\nabla^2 = \nabla \cdot \nabla$) of the potential to the electric charge density. In free space, where $\rho_c = 0$, Poisson's equation becomes *Laplace's equation*:

$$\nabla^2 U = 0. \tag{A.29}$$

Both Laplace's and Poisson's equation are elliptic second-order partial differential equations, and to obtain a solution of these equations throughout some volume V, boundary conditions must be applied on the closed surface S, enclosing the volume. These conditions can take the form of conditions either on U or on the derivative of U in a direction normal to S at each point.

A.6 Source-free wave equations

Maxwell's equations not only encompass the areas of electrostatics and magnetostatics, they also predict the existence of electromagnetic radiation. Consider a situation in free space (or vacuum) where there are no sources (i.e., $\rho_c = 0$ and $\mathbf{J} = \mathbf{0}$). Faraday's law and Ampère's law become

$$\nabla \times \mathbf{H} = \varepsilon_0 \frac{\partial \mathbf{E}}{\partial t} \tag{A.30}$$

and

$$\nabla \times \mathbf{E} = -\mu_0 \frac{\partial \mathbf{H}}{\partial t}. \tag{A.31}$$

We obtain an equation for $\mathbf{E}$ alone by taking the curl of both sides of Equation (A.31) and replacing $\nabla \times \mathbf{H}$, which appears on the right-hand side, with $\varepsilon_0 \partial \mathbf{E}/\partial t$ (as a consequence of (A.30)):

$$\nabla \times \nabla \times \mathbf{E} = -\mu_0 \varepsilon_0 \frac{\partial^2 \mathbf{E}}{\partial t^2}$$

$$= -\frac{1}{c^2} \frac{\partial^2 \mathbf{E}}{\partial t^2}, \tag{A.32}$$

where $c^2 = 1/\mu_0 \varepsilon_0$ and $c = 2.998 \times 10^8$ m/s is the speed of light in vacuum. This equation is the source-free wave equation for $\mathbf{E}$, and we can put it into a more recognizable form by using the vector identity $\nabla \times \nabla \times \mathbf{E} = \nabla(\nabla \cdot \mathbf{E}) - \nabla^2 \mathbf{E}$. For a source-free region ($\rho_c = 0$), $\nabla \cdot \mathbf{E} = 0$, and Equation (A.32) becomes

$$\nabla^2 \mathbf{E} - \frac{1}{c^2} \frac{\partial^2 \mathbf{E}}{\partial t^2} = 0. \tag{A.33}$$

The same wave equation can be derived for $\mathbf{H}$. To avoid errors, the Laplacian of a vector, which appears in Equation (A.33), should be evaluated for each Cartesian component of the vector. Solutions of the wave equation represent electromagnetic waves propagating at the speed of light. For a medium with uniform permittivity and permeability, the wave equation is the same as (A.33) but with c^2 replaced with $V_{\text{ph}}^2 = 1/\mu\varepsilon$. V_{ph} is the phase speed of the wave in the medium and can be written as $V_{\text{ph}} = c/N$, where $N = \sqrt{\mu_R\varepsilon_R}$ is the *index of refraction* of the medium, which can be expressed in terms of the relative permeability and relative permittivity.

The plane wave solution to the wave equation is

$$\mathbf{E}(\mathbf{x}, t) = \mathbf{E}_0 e^{+i[\mathbf{k}\cdot\mathbf{x}-\omega t]}. \tag{A.34}$$

The position vector is denoted $\mathbf{x}$. By substitution we can show that (A.34) is a solution to (A.33) only if the following condition is met:

$$k^2 = \omega^2/c^2. \tag{A.35}$$

$\mathbf{k}$ is the wavenumber vector; it points in the direction of wave propagation and its magnitude is related to the wavelength of the wave (λ) by $k = 2\pi/\lambda$. The relation (A.35) can also be written as $\omega^2 = k^2 c^2$. In general, for wave motion, ω is some function of k given by the *dispersion relation*, $\omega = \omega(k)$.

The amplitude of the plane-wave solution for the electric field intensity is $\mathbf{E}_0$ (V/m), which is complex in general. For a source-free region, the electric field must obey Gauss's law, $\nabla \cdot \mathbf{E}(\mathbf{x}, t) = 0$, or just $\nabla \cdot \mathbf{E}(\mathbf{x}) = 0$. Using Equation (A.34) and performing the divergence operation, one obtains $\nabla \cdot \mathbf{E}(\mathbf{x}) = i\mathbf{k} \cdot \mathbf{E}(\mathbf{x}) = 0$. This is true only if the wave amplitude satisfies the following condition:

$$\mathbf{k} \cdot \mathbf{E}_0 = 0. \tag{A.36}$$

Equation (A.36) states that the electric field vector is orthogonal to the direction of wave propagation $\mathbf{k}$; that is, electromagnetic waves are *transverse* waves.

Information on the polarization and phase of the plane wave is contained in the complex amplitude $\mathbf{E}_0$, which can be written as $\mathbf{E}_{0A}e^{i\delta}$, where $\mathbf{E}_{0A}$ is a vector and δ is the wave phase. We can now express the electric field intensity as

$$\mathbf{E}(\mathbf{x}, t) = \mathbf{E}_{0A} \exp\{i[\mathbf{k} \cdot \mathbf{x} - \omega t + \delta]\}. \tag{A.37}$$

We must also take the real part of this to find the "actual" electric field intensity. For example, for waves linearly polarized in the x direction we get

$$\mathbf{E}(\mathbf{x}, t) = E_{0A}\hat{\mathbf{x}}\cos[\mathbf{k} \cdot \mathbf{x} - \omega t + \delta]\}. \tag{A.38}$$

And for a circularly polarized plane wave traveling in the $+z$ direction (i.e., $\mathbf{k} = k\hat{\mathbf{z}}$), the amplitude is

$$\mathbf{E}_{0A} = E_{0A}[\hat{\mathbf{x}} \pm i\hat{\mathbf{y}}]. \tag{A.39}$$

E_{0A} is a real amplitude. The $+$ sign corresponds to right-hand circularly polarized waves, with respect to $\mathbf{k}$, and the $-$ sign to left-hand circularly polarized waves. $\mathbf{E}_{0A}$ is a real vector for linearly polarized waves.

An examination of the argument of the exponential in (A.38) shows that the phase of the plane wave (that is, everything in the brackets) remains constant on a wave front moving in the **k** direction at the following *phase speed*:

$$V_{\text{ph}} = \frac{\omega}{k}. \tag{A.40}$$

Equation (A.40) is general and applies to any type of wave. However, for an electromagnetic wave in free space, the dispersion relation (A.35) indicates that the phase speed is just the speed of light, $V_{\text{ph}} = 1/(\mu_0 \varepsilon_0)^{1/2} = c$. A wave packet does not move at the phase speed but at the *group speed*, given by

$$V_{\text{g}} = \frac{\partial \omega}{\partial k}. \tag{A.41}$$

For electromagnetic waves propagating in a vacuum $V_{\text{g}} = V_{\text{ph}} = c$, but in general (and specifically for a plasma) $V_{\text{g}} \neq V_{\text{ph}} \neq c$.

We now describe the magnetic field associated with electromagnetic radiation. The magnetic field intensity is expressed as the product of the spatially dependent part and a time-harmonic part: $\mathbf{H}(\mathbf{x}, t) = \mathbf{H}(\mathbf{x}) \exp(-i\omega t)$. Note that we can also write $\mathbf{E}(\mathbf{x}, t) = \mathbf{E}(\mathbf{x}) \exp(-i\omega t)$. $\mathbf{H}(\mathbf{x}, t)$ and $\mathbf{E}(\mathbf{x}, t)$ in a source-free region are related by Faraday's law:

$$\nabla \times \mathbf{E}(\mathbf{x}) = i\omega\mu_0 \mathbf{H}(\mathbf{x}). \tag{A.42}$$

From (A.38) we see that $\nabla \times \mathbf{E}(\mathbf{x}) = i\mathbf{k} \times \mathbf{E}(\mathbf{x})$ for a plane wave. We can use this to show that the magnetic field amplitude is

$$\mathbf{H}_0 = \frac{\mathbf{k} \times \mathbf{E}_0}{\omega\mu_0} = \frac{1}{V_{\text{ph}}\mu_0} \hat{\mathbf{k}} \times \mathbf{E}_0. \tag{A.43}$$

The combination $Z = V_{\text{ph}}\mu$ is the *intrinsic impedance* of a medium and has the SI units of ohms [Ω]. In free space where $\mu = \mu_0$ and $V_{\text{ph}} = c$, we have $Z = \sqrt{\mu_0/\varepsilon_0}$. Equation (A.43) tells us that the direction of the wave magnetic field is orthogonal to *both* the direction of the electric field and the wave propagation direction.

The wave equation for electromagnetic waves propagating in a plasma without a background magnetic field is different from Equation (A.33) due to the presence of electrical currents (i.e., sources). The dispersion relation is $\omega^2 = k^2 c^2 + \omega_{\text{p}}^2$, where the *plasma frequency* ω_{p} is defined by

$$\omega_{\text{p}}^2 = \frac{n_e e^2}{m_e \varepsilon_0}, \tag{A.44}$$

where n_e is the electron density and m_e is the electron mass.

A.7 Energy conservation and the Poynting vector

Electric and magnetic fields contain energy, and electromagnetic radiation carries power. An energy conservation relation for electromagnetic waves can be derived from Maxwell's equations. This derivation will be left to the student as an exercise. We now relax some of the assumptions of the last two sections and permit the existence of sources. The following expression can be derived from Maxwell's equations:

$$\frac{\partial}{\partial t}\left[\frac{1}{2}\varepsilon E^2 + \frac{1}{2}\mu H^2\right] + \nabla \cdot [\mathbf{E} \times \mathbf{H}] = -\mathbf{E} \cdot \mathbf{J}. \tag{A.45}$$

The first term on the left-hand side of (A.45) represents the time rate of change of the energy density [W/m^3] of the electric and magnetic fields. The *electric and magnetic field energy densities* are

$$w_E = \frac{1}{2}\varepsilon E^2 = \frac{D^2}{2\varepsilon} \quad [\text{J/m}^3]$$

and

$$w_B = \frac{1}{2}\mu H^2 = \frac{B^2}{2\mu} \quad [\text{J/m}^3]. \tag{A.46}$$

The $(-\mathbf{E} \cdot \mathbf{J})$ term in (A.45) represents the dissipation of electromagnetic energy (i.e., a resistor) if $\mathbf{E}$ and $\mathbf{J}$ are in the same direction and a source of electromagnetic energy (i.e., a generator) if they are in opposing directions. For example, for simple "ohmic" current flow, $\mathbf{J}$ is proportional to $\mathbf{E}$ and energy dissipation rate is the product of a constant, called the conductivity, and E^2.

The change of electromagnetic energy due to energy transport is described by the divergence of $\mathbf{E} \times \mathbf{H}$, which must therefore be an energy flux. The energy flux, or power density, is just the *Poynting Vector*, given by

$$S = \mathbf{E} \times \mathbf{H} \quad [\text{W/m}^2]. \tag{A.47}$$

$\mathbf{S}$ has the units of power per unit area and represents the energy flux associated with electromagnetic waves. The integral form of the energy conservation relation can be found by integrating over a volume V and using Gauss's integral theorem:

$$\frac{\partial}{\partial t}[W_E + W_B] = \oint_S S \cdot d\mathbf{S} - \int_V \mathbf{E} \cdot \mathbf{J}\, dV,$$

$$W_E = \int_V w_E dV \quad [\text{J}],$$

and

$$W_B = \int_V w_B dV \quad [\text{J}]. \tag{A.48}$$

The sum $W_E + W_B$ is the total energy associated with the E and B fields in the volume V, and the change in time of this energy is equal to the net inflow (or outflow) of energy via waves plus the internal creation or loss of electromagnetic energy due to currents.

S is the *instantaneous* power density, which is not really measurable for a high-frequency wave. For time-harmonic waves, S is also time harmonic, and a useful quantity is the time-averaged (over one period) Poynting vector, which can be written in terms of the spatial parts of the fields:

$$S_{AV} = \frac{1}{2}\text{Re}\{E \times H^*\} \quad [W/m^2].$$

(A.49)

H^* is the complex conjugate of $H(x)$. S_{AV} is a real and measurable power density.

Index

Printed in the United States
By Bookmasters